J Gill

Text-Book on Navigation and Nautical Astronomy

J Gill

Text-Book on Navigation and Nautical Astronomy

ISBN/EAN: 9783337396176

Printed in Europe, USA, Canada, Australia, Japan

Cover: Foto ©berggeist007 / pixelio.de

More available books at **www.hansebooks.com**

TEXT-BOOK ON NAVIGATION

AND

NAUTICAL ASTRONOMY

BY

J. GILL, F.R.A.S.

HEAD-MASTER OF THE LIVERPOOL CORPORATION NAUTICAL COLLEGE

LONGMANS, GREEN, AND CO.
39 PATERNOSTER ROW, LONDON
NEW YORK AND BOMBAY
1898

PREFACE

THIS work is designed to meet the special requirements of officers of the Merchant Service, and to be a complete Text-Book on Navigation, etc., for the general student.

Candidates for Board of Trade Certificates will find in it all that is needed to cope with the new regulations for the Examinations of Mates and Masters, from the lowest to the highest grades, as regards both practice and principles.

The usual cumbrous rules are entirely dispensed with, long experience having convinced the author that a mastery of the problems is best acquired from worked-out examples and illustrations. Every variety of problem is so treated and followed up by numerous exercises, whilst the large number of Examination Papers at the end of the book will provide a sufficient test of the accuracy of the student's knowledge, and his expertness in computation.

Special attention is given to "Star Work," which is coming into more general use in Practical Navigation.

So much depends on the taking out and reduction of the "elements" of celestial bodies from the Nautical Almanac, that a special chapter is devoted to these exercises.

The subject of Magnetism and Deviation is treated in its relation to merchant ships, and includes the Board of Trade Syllabus of Examination for Masters "Ordinary" and "Extra."

Part II. contains an introduction to Plane and Spherical Trigonometry, sufficient to enable the student to understand the principles from which the rules for working the various problems are derived. Concise mathematical investigations of these rules are also given.

The author gratefully acknowledges the valuable assistance given by his colleagues, Captain E. W. Owens and Mr. W. V. Merrifield, B.A., in the preparation and working out of the exercises and examination papers.

J. GILL.

CONTENTS

PART I.

CHAPTER		PAGE
I.	DEFINITIONS	3
II.	LOGARITHMS	21
III.	THE SAILINGS	28
IV.	TIDES	41
V.	REDUCTION OF SOUNDINGS	43
VI.	TIME	47
VII.	ELEMENTS FROM THE NAUTICAL ALMANAC	54
VIII.	SIDEREAL AND SOLAR TIME	67
IX.	LATITUDE BY MERIDIAN ALTITUDE	80
X.	LATITUDE BY EX-MERIDIAN ALTITUDE	90
XI.	AMPLITUDES	98
XII.	AZIMUTHS	102
XIII.	CHRONOMETERS AND LONGITUDE	108
XIV.	SUMNER'S METHOD BY PROJECTION	118
XV.	THE CHART	122
XVI.	USE OF NAPIER'S DIAGRAM	128
XVII.	GREAT CIRCLE SAILING	134
XVIII.	LONGITUDE BY LUNAR DISTANCES	146
XIX.	LATITUDE BY DOUBLE ALTITUDES	155
XX.	FINDING ERROR OF CHRONOMETER	165
XXI.	CONSTRUCTION OF CHARTS	173

PART II.

CHAPTER		PAGE
XXII.	SOLUTION OF PLANE TRIANGLES	177
XXIII.	SOLUTION OF SPHERICAL TRIANGLES	191
XXIV.	APPLICATION OF FORMULÆ TO PROOF OF RULES	200
XXV.	INSTRUMENTS	224
XXVI.	PROJECTIONS	226

PART III.

XXVII.	LAWS OF STORMS	229
XXVIII.	MAGNETISM AND DEVIATION OF THE COMPASS	235
XXIX.	SYLLABUS	256
	APPENDIX	291

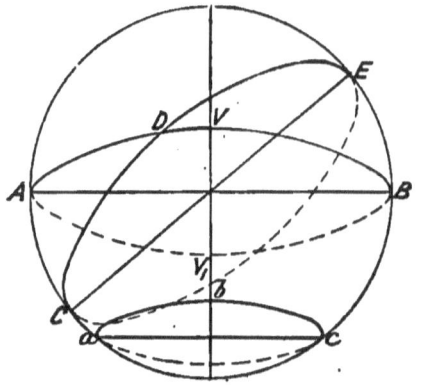

Fig. 1.—Great and Small Circles.
AVB and CDE, Great circles.
abc, Small circle.
V and V_1, Vertexes of AVB.

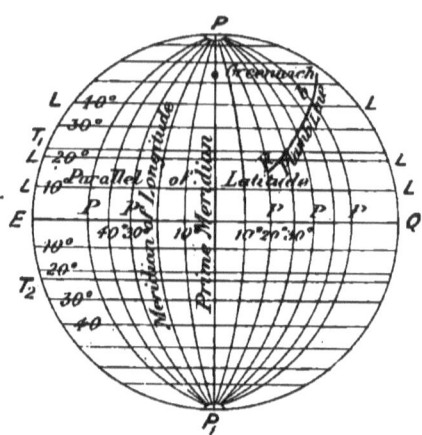

Fig. 2.—Terrestrial Globe.
P and P_1, Poles.
EQ, Equator.
T_1 and T_2, Tropics (Cancer and Capricorn).
LL, Parallels of latitude.
PP_1, Prime meridian.
PpP_1, Meridians of longitude.

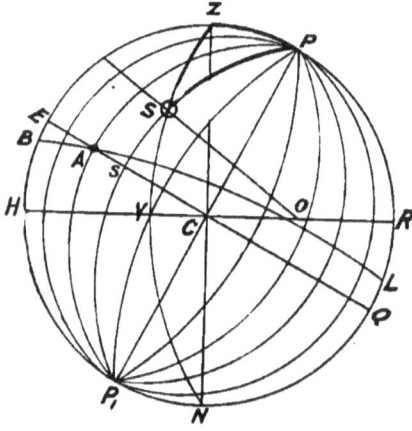

Fig. 3.

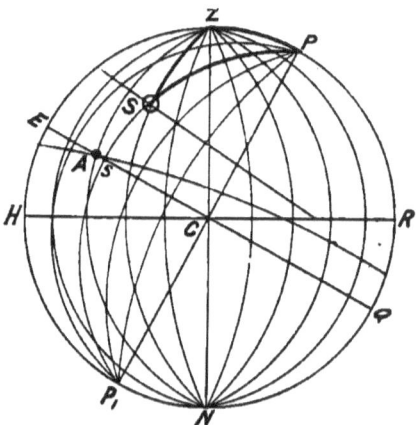

Fig. 4.—Celestial Sphere on Plane of Meridian.
P and P_1, Celestial poles.
EMQ, Equinoctial.
BAL, Ecliptic.
A, First Point of Aries.
ZCN, Prime Vertical.
ZPS, Hour angle.
PZS, Azimuth.
Co, Amplitude.
ZS, Zenith distance.
PS, Polar distance.
sS, Declination.
VS, True altitude.
As, Right ascension.
II., IV., etc., Hour Circles.
ZCN, ZSN, ZAN, etc., Vertical circles.
PCP_1, PVP_1, PSP_1, etc., Declination circles.
N.B.—The letters apply to all projections.

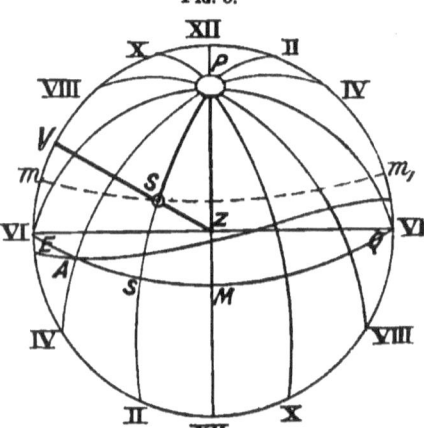

Fig. 5.—Celestial Sphere on Plane of Horizon.

PART I.

CHAPTER I.

DEFINITIONS.

1. **Great Circles.**—Circles whose planes pass through the centre of a sphere or globe (Fig. 1).
2. **Vertex of a Great Circle.**—That point in it which is farthest from the Equator.
3. **Small Circles.**—Circles whose planes do not pass through the centre of the sphere (Fig. 1, *abc*).
4. **Right Angle.**—An angle measuring 90°, or one quarter of a circle (Fig. 6).
5. **Oblique Angle.**—An angle not a right angle (Fig. 7).
6. **Obtuse Angle.**—An angle greater than a right angle (Fig. 7).
7. **Acute Angle.**—An angle less than a right angle (Fig. 7).
8. **Spherical Angle.**—An angle included between two great circles of a sphere (Fig. 8, *a*, *b*, and *c*).
9. **Arc.**—A part of the circumference of a circle (Fig. 9).
10. **Complement of an Arc or Angle.**—Its difference from 90° (Fig. 9).
11. **Supplement of an Arc or Angle.**—Its difference from 180° (Fig. 9).
12. **The Poles.**—The points where the earth's axis meets the surface (Fig. 2).
13. **Equator.**—A great circle of the earth 90° distant from the poles (Fig. 2).
14. **A Meridian.**—Half a great circle extending from pole to pole of the earth (Fig. 2).
15. **Tropics.**—Two circles parallel to the Equator and distant from it 23° 28' (Fig. 2): Cancer on the north side, and Capricorn on the south.
16. **Parallels of Latitude.**—Circles parallel to the Equator (Fig. 2).
17. **Difference of Latitude.**—The arc of a meridian intercepted between two parallels of latitude (Fig. 2, LL).
18. **Meridional Parts.**—The distance from the Equator in geographical miles of a given latitude on Mercator's Chart.

They may be found by adding the secants of every minute of latitude up to the given latitude.

19. **Prime Meridian.**—The meridian from which longitudes are measured; in this country it is the meridian of Greenwich (Fig. 2).

20. **Longitude.**—Distance in arc east or west of the meridian of Greenwich. It is measured on the Equator, or on a graduated parallel (Fig. 2).

21. **Difference of Longitude.**—An arc of the Equator intercepted between two meridians of longitude (Fig. 2).

22. **Departure.**—The distance in geographical miles between two places in the same latitude. In the traverse table, it is the number of miles of *easting* or *westing* made on a given course and distance (Figs. 11, 12, 13).

23. **Nautical Mile.**—The one-sixtieth part of a degree at the Equator, or 6080 feet nearly.

24. **Rhumb Line.**—A line which makes equal angles with all the meridians it crosses (Fig. 2).

25. **Zenith.**—The point in the heavens vertically over the observer's head (Figs. 3 and 4).

26. **Nadir.**—The point in the heavens diametrically opposite to the Zenith (Figs. 3 and 4).

27. **Equinoctial.**—A great circle in the heavens traced by the plane of the earth's Equator produced (Figs. 3 and 4).

28. **Ecliptic.**—A great circle in the heavens representing the sun's annual apparent path (Figs. 3, 4, 5).

29. **Vertical Circles.**—Circles passing through the Zenith and Nadir, and therefore perpendicular to the Horizon. They are also called "altitude circles" and "azimuth circles" (Figs. 3 and 4).

30. **Prime Vertical.**—The vertical circle which cuts the Horizon in the true East and West points (Figs. 3, 4, 5).

31. **Celestial Meridian of Observer.**—The vertical circle which cuts the Horizon in the true North and South points (Figs. 3, 4, 5).

32. **First Point of Aries.**—The point where the sun crosses the Equinoctial about the 21st of March (Figs. 3, 4, 5).

33. **Right Ascension.**—Distance eastward from the first point of Aries measured on the Equinoctial and expressed in Time (Figs. 3, 4, 5).

34. **Declination.**—Distance in arc north or south from the Equinoctial (Figs. 3, 4, 5).

35. **Polar Distance.**—The distance in arc from the elevated pole, and therefore the complement of the Declination (Figs. 3, 4, 5).

36. **Civil Time.**—The time used in civil life. The day begins at midnight, and is divided into two parts of twelve hours each, a.m. from midnight to noon, and p.m. from noon to midnight.

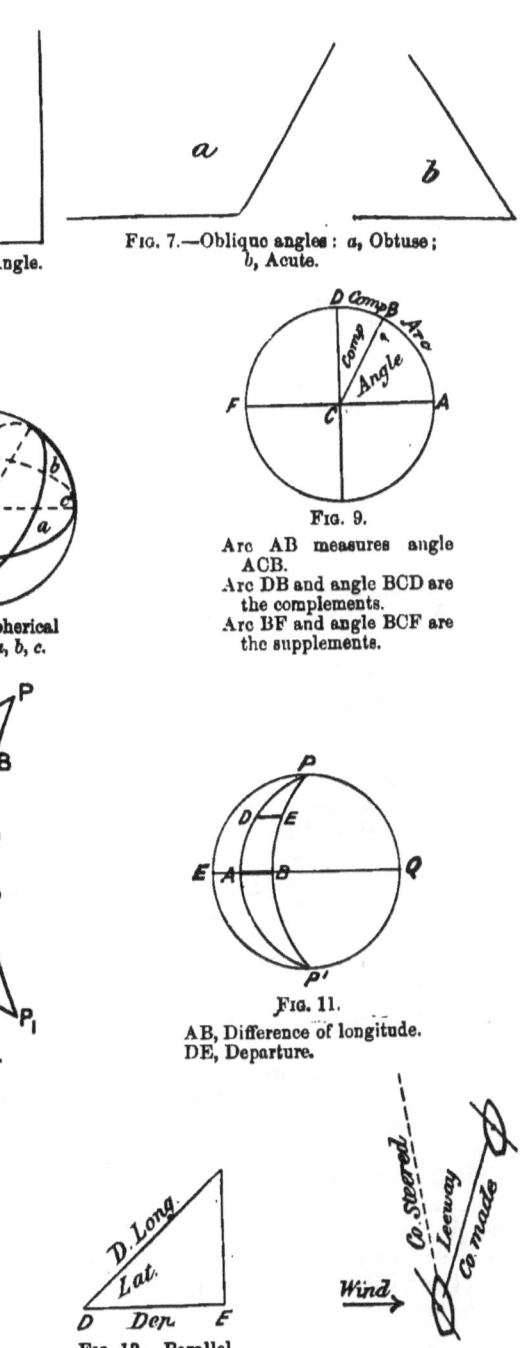

FIG. 6.—Right Angle.

FIG. 7.—Oblique angles: *a*, Obtuse; *b*, Acute.

FIG. 8.—Spherical angles; *a, b, c*.

FIG. 9.
Arc AB measures angle ACB.
Arc DB and angle BCD are the complements.
Arc BF and angle BCF are the supplements.

FIG. 10.

FIG. 11.
AB, Difference of longitude.
DE, Departure.

FIG. 12.—Plane sailing.

FIG. 13.—Parallel sailing.

FIG. 14.

37. **Astronomical Time.**—Time used for astronomical calculations. The day begins at noon, and the hours count from 0 to 24.

38. **Hour Circles.**—Halves of great circles which pass through the celestial poles, one hour being equal to 15° (Figs. 3, 4, 5).

39. **Hour Angle.**—The angle at the pole between the observer's Meridian and an Hour Circle (Figs. 3, 4, 5).

40. **Sidereal Time.**—The Hour Angle westward of the first point of Aries (Figs. 3, 4, 5).

41. **Mean Time.**—The time shown by a correctly going clock or watch, and regulated by the sun's mean motion in Right Ascension. A Chronometer shows mean time at Greenwich.

42. **Apparent Time.**—The time shown by a sun-dial, and corresponding to the apparent motion of the sun.

43. **Equation of Time.**—The difference between Apparent and Mean times.

44. **Azimuth.**—The bearing of an elevated object reckoned from the N. or S. point (Figs. 3, 4, 5).

45. **Amplitude.**—The bearing of an object when rising or setting, reckoned from the E. or W. point (Figs. 3 and 5).

46. **Visible Horizon.**—The line where the sea and sky appear to meet (Fig. 18).

47. **Sensible Horizon.**—A plane touching the Earth at the observer's position and extended to the Heavens (Fig. 18).

48. **Rational Horizon.**—A plane passing through the Earth's Centre parallel to the Sensible Horizon and extended to the Heavens (Fig. 18).

49. **Artificial Horizon.**—A horizontal reflecting surface used for observing altitudes in the absence of a good sea-horizon.

50. **Dip.**—The angular depression of the Visible horizon below the Sensible, due to the observer's elevated position (Fig. 18).

51. **Refraction.**—The error of altitude caused by the bending of rays of light in passing through the atmosphere (Fig. 16).

52. **Parallax.**—The correction for reducing an observation to the earth's centre. It is the angle subtended by the earth's radius at the centre of the object (Fig. 17).

53. **Semidiameter.**—The angle subtended at the earth's centre by the radius of the sun, moon, or planet (Fig. 19).

54. **Augmentation of Moon's Semidiameter.**—An apparent increase due to the fact that the moon's distance from the observer decreases as the altitude increases (Fig. 19).

55. **Reduction of Horizontal Parallax.**—A correction necessary on account of the shortening of the earth's radii towards the poles (Fig. 20).

56. **Observed Altitude.**—The angular height of an object above the Visible horizon as measured by a Sextant.

57. **Apparent Altitude.**—The angular height of an object

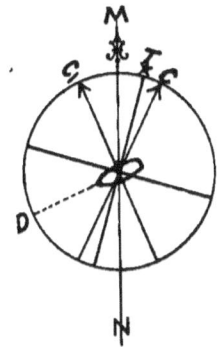

T, True north.
MN, Magnetic meridian.
C and C_1, Different directions of compass (north).
TM, Variation (west).
MC, Deviation (east).
MC_1, " (west).
TC, Error of compass (east).
TC_1, " " (west).
CD or C_1D, Compass course from north.
TD, True course from north.
MD, Magnetic course from north.

FIG. 15.

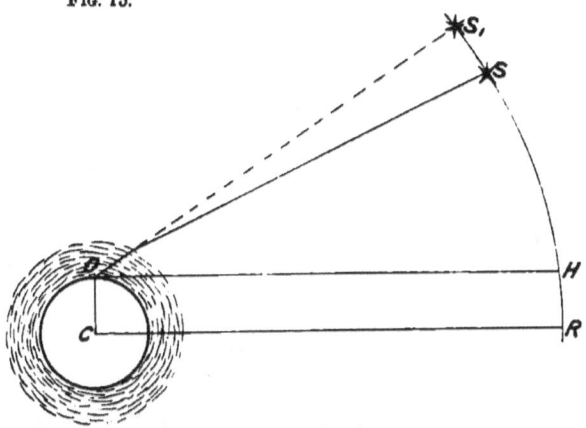

FIG. 16.—Refraction.

O, Observer's position.
S, True position of star.
S_1, Apparent position of star.
SS_1, Refraction.

OH, Horizon.
HS_1, Apparent altitude of a star.
RS, True altitude of a star.

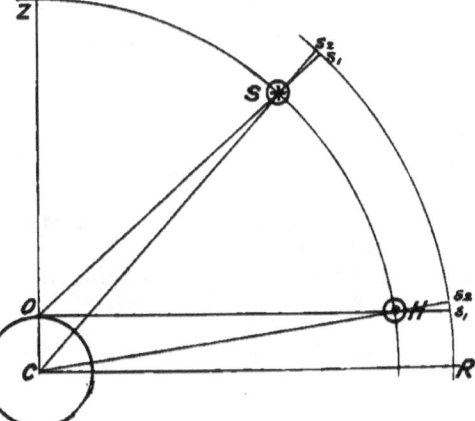

O, Observer's position.
⊙H, Sun's position on horizon.
⊙S, Sun's position in altitude.
∠ OHC, Horizontal parallax.
∠ OSC, Parallax in altitude.
Z, The zenith of observer.
Rs_2, True altitude of sun.

FIG. 17.—Parallax.

above the Sensible horizon, found by correcting the observed altitude for Index Error and Dip (Fig. 16).

58. **True Altitude.**—The angular height of an object above the Rational Horizon, found by correcting the Apparent Altitude for Refraction and Parallax (Fig. 17).

59. **Zenith Distance.**—The angular distance from the Zenith, or the complement of the true altitude (Figs. 3, 4, 5).

60. **Magnetic Meridian.**—A line passing through all places having the same Variation.

61. **Variation.**—The angular difference between the True and Magnetic meridians, measured from True North (Fig. 15).

62. **Deviation.**—The deflection of the needle from Magnetic North caused by the influence of iron (Fig. 15).

63. **Error of Compass.**—The combined result of Variation and Deviation; that is, the angular difference between the Compass North and True North (Fig. 15).

64. **Leeway.**—The angle between the ship's keel and her " wake " (Fig. 14).

NOTE.—*The foregoing comprise the definitions required by the Board of Trade Regulations for the examination of mates and masters.*

65. **Latitude.**—Distance in arc north or south from the Equator measured on a Meridian (Fig. 2).

66. **Arctic Circle.**—The parallel of 66° 32′ N. lat.

67. **Antarctic Circle.**—The parallel of 66° 32′ S. lat.

68. **Obliquity of Ecliptic.**—The angle between the planes of the Ecliptic and Equinoctial (Fig. 4).

69. **Equinoxes.**—The points where the Ecliptic intersects the Equinoctial. The Vernal Equinox is where the sun crosses the Equinoctial in March, and the Autumnal where it crosses in September (Figs. 3, 4, 5).

70. **Precession.**—A slow motion of the Equinoxes along the Equinoctial westwards, amounting to about 50″ annually.

71. **Solstices.**—The points on the Ecliptic farthest from the Equinoctial.

N.B.—The terms " Equinoxes " (Spring and Autumn) and " Solstices " (Summer and Winter) are also applied to the times of the year when the sun is on the Equinoctial or at the Solstices.

72. **Celestial Latitude.**—Distance in arc north or south from the Ecliptic.

73. **Celestial Longitude.**—Distance in arc from the "first point of Aries " measured on the Ecliptic.

74. **Circumpolar Stars.**—Stars whose polar distance is less than the latitude of the place, and consequently never set.

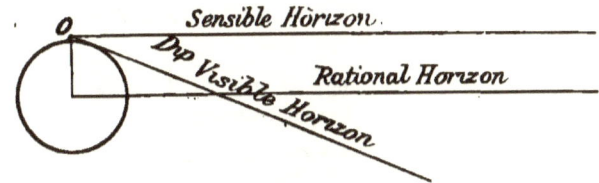

Fig. 18.—Horizon.

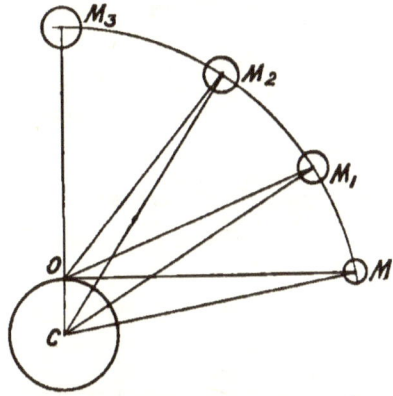

Fig. 19.—Augmentation of Moon's Diameter.

O, Observer's position.
C, Centre of the earth.
M, The moon.
OM, The distance of moon when on the horizon.
OM_1, OM_2, etc., The decreasing distance from observer as altitude increases. Hence the apparently augmented diameter.

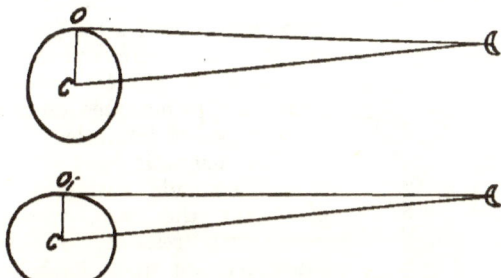

Fig. 20.—Reduction of Moon's Horizontal Parallax.

O, Observer's position on longest diameter of earth.
O_1, Position on a shorter diameter.
O ☾ C, Equatorial horizontal parallax.
O_1 ☾ C, Horizontal parallax for latitude of observer, which is less because subtended by a shorter radius.

QUESTIONS ON VARIATION AND DEVIATION, FROM THE
BOARD OF TRADE REGULATIONS.

1. *Q.*—Does the variation change with time?
A.—Yes; it slowly changes in amount. At London it now decreases about 9′ annually.
2. *Q.*—Is the variation the same all over the world?
A.—No; all amounts from 0° to 180° may be found.
3. *Q.*—Where do you find the variation?
A.—On a variation chart.
4. *Q.*—Does the deviation change? if so, when?
A.—It changes with a change of course; also, for the same course with lapse of time, because the magnetic condition of the ship is liable to change, and from change of cargo.
5. *Q.*—Where is the north magnetic pole situated?
A.—In about 70° N. lat. and 92° W. long.
6. *Q.*—Where is the south magnetic pole situated?
A.—In about lat. 75° S., and long. 150° E.
7. *Q.*—Should the compass-needle point to the magnetic or the true pole of the earth?
A.—To the magnetic pole.
8. *Q.*—When is the altitude of an object most seriously affected by refraction?
A.—When nearest to the horizon.
9. *Q.*—Where is the pole star situated?
A.—At the distance of 1° 18′ from the true celestial pole.
10. *Q.*—Which is the most favourable time for determining the hour angle of a celestial body, and thence the longitude, and state the reason why?
A.—When it is on the prime vertical, because the hour angle is then least affected by errors in altitude or latitude.

INSTRUMENTS.

Mariner's Compass.—The Mariner's Compass consists of one or more magnetized steel needles or bars, attached to a circular card, which is graduated to quarter-points or to degrees, and is mounted on gimbals in a suitable box or binnacle.

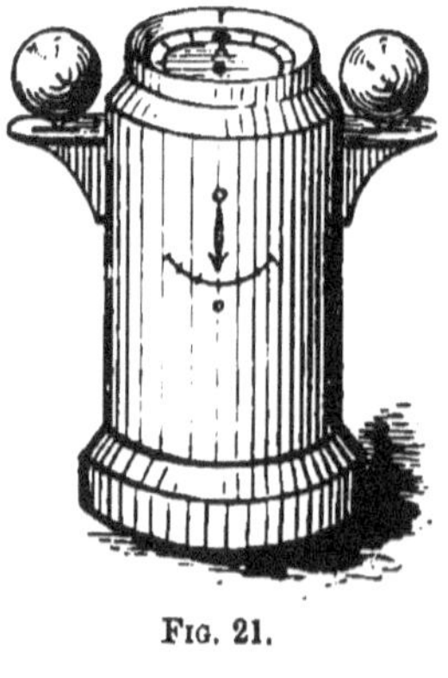

The magnetic axis of the needles should coincide with the N. and S. points of the card, whose jewelled centre cap rests on a sharp-pointed support rising from the bottom of the compass bowl, which should be of copper.

On the inside of the bowl is marked the "lubber" line, which indicates on the card the Course of the ship. Two brackets at the sides support cast-iron correctors, and provision is made inside the binnacle for

FIG. 21.

DEFINITIONS.

compensating magnets and a Flinders Bar. A Clinometer shows the angle of "heeling."

Azimuth Compass.—The Azimuth Compass is like an ordinary Steering-compass in the mounting of the needles and card. There is a moveable ring on the bowl, provided with sight-vanes at opposite ends of a diameter. One of these should have a reflecting prism for reading the graduations on the card, and the other a mirror for reflecting elevated objects.

It is mounted in a box or on a tripod, and in using it the observer looks through the narrow slit, and moves the ring

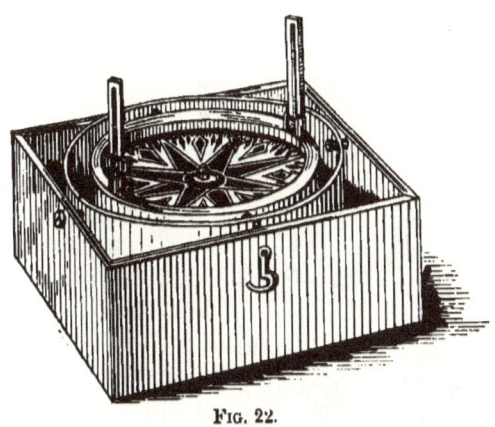

FIG. 22.

round until the object is bisected by the wire in the other sight-vane, and then reads off the bearing by means of the prism.

Sometimes a vertical style is fixed on the centre of the glass cover of an ordinary compass, for showing the sun's compass bearing by means of the shadow.

Pelorus.—The "Pelorus" resembles an Azimuth Compass, but has no magnets nor bowl. The "card," usually of metal, is not suspended, but is free to move round its centre, and is mounted on gimbals and weighted to keep it horizontal. A bar carrying the two sight-vanes is pivoted at the centre of the card, and may be clamped in any position. Four "lubber-points," 90° distant, are usually marked outside the edge of the "card."

The uses of the "Pelorus" are—

(*a*) To find the Deviation on any course. The sight-vane being clamped to the known *correct magnetic* bearing of a distant object, the card is turned round until, on looking through the slit, the object is seen bisected by the wire. The *correct magnetic* course is shown at the lubber-point, and the difference between this course and the course by compass is the Deviation.

(*b*) To place the ship's head in any desired direction for the purpose of compensating the compasses.

The sight-vanes being clamped to a known *correct magnetic* bearing, and the card turned round until the desired direction of ship's head is at the lubber-point, then the ship is swung until the object is seen bisected by the wire. Her head is now on the desired point, and the magnets or correctors may be placed

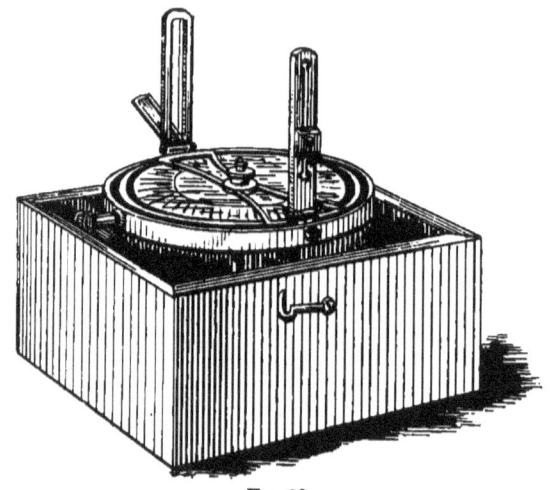

Fig. 23.

in position so that the Compass shall show the same course as the Pelorus.

(*c*) For taking Bearings when the Standard Compass is not available.

Care must be taken that the ship is upright, and that a line through the centre of the card and the lubber-point shall be parallel to the line of the keel.

Barometers.—The atmosphere surrounds the earth to a height of two hundred miles or more, and presses on the surface with an average weight of nearly 15 lbs. on a square inch. The pressure varies with changes of temperature and moisture and from other causes, and decreases from the sea-level upwards. A Barometer is an instrument for measuring this pressure.

Barometers are generally of two kinds, viz. "Aneroid" and "Mercurial."

DEFINITIONS.

The **Aneroid Barometer** has a shallow metal cylinder with a flexible cover, and partially exhausted of air. The tendency to collapse in consequence of the partial vacuum is balanced by a strong spring attached to the cover and fastened to the bottom of the containing box. As the pressure of the atmosphere varies, the movements of the flexible cover inwards and outwards are, by a mechanical contrivance, conveyed to a pointer which shows on a dial the pressure of the atmosphere.

The spaces on the dial, corresponding to inches of mercury, are determined by comparison with a mercurial barometer.

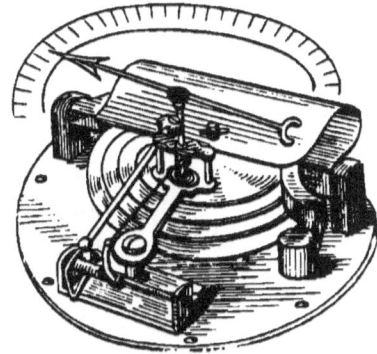

Fig. 24.

The **Mercurial Barometer** consists of a glass tube about 33 in. long (whose open end dips into a quantity of mercury contained in a trough or "cistern"), enclosing a column of mercury of varying height. The weight of this column measured from the surface of the mercury in the cistern at any moment is the measure of the pressure of the atmosphere on a surface equal to the section of the tube. This pressure is indicated in inches of mercury on a scale near the top.

The portable barometers (Fig. 25) used at sea have the glass tube protected by a casing of wood or iron, and the mercury in the cistern is prevented from escaping by a cover or bag of washleather, which, however, does not interfere with the air-pressure. In this form of barometer the "inches" on the scale are altered, so as to compensate for the changing level of the mercury in the cistern as the column rises or falls.

For scientific purposes, corrections of the barometer-readings are made for temperature and height above sea-level.

Thermometer.—A Thermometer is an instrument for measuring temperatures. The common Thermometer is a glass tube of exceedingly small bore, blown out at one end

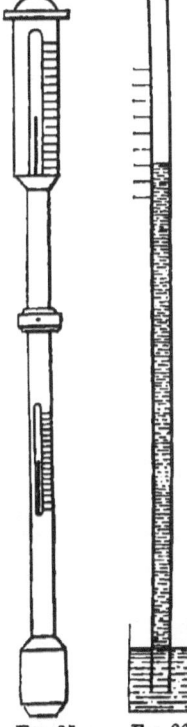

Fig. 25.   Fig. 26.

into a bulb, which is filled with mercury, the other end being hermetically sealed after the air is driven out. The mercury in the bulb expands with heat and contracts with cold, which causes the column in the tube to advance or retire. The degree of the scale indicated by the end of the column is the temperature.

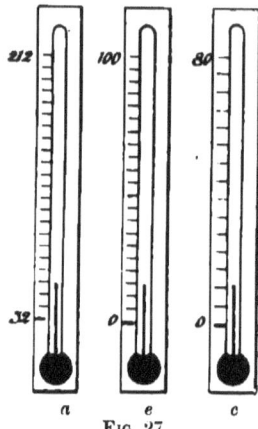

Fig. 27.

Three different scales of temperature are in use.

(a) The **Fahrenheit** thermometer (Fig. 27, a), which is the one generally used in this country, marks the freezing-point of water 32° and the boiling-point 212°, a difference of 180°. Zero (0) on this scale is the temperature of a mixture of salt and snow.

(b) The **Centigrade** or **Celsius** thermometer (Fig. 27, e) marks the freezing-point zero (0) and the boiling-point 100°.

(c) The **Reaumur** thermometer (Fig. 27, c) marks freezing-point zero (0) and boiling-point 80°.

These three scales are therefore in the ratio of 180, 100, and 80, or 9, 5, 4, from which it is easy to compare the readings of any one of these instruments with each of the other two.

Examples.—1. What are the readings of the Centigrade and the Réaumur scales corresponding to 72° Fahrenheit.

```
        72°                40              40
Freezing 32°                5               4
        ___              _____           _____
Above freezing 40°       9)200           9)160
                         _____           _____
                         22·2 Cent.      17°·7 Reaumur.
```

2. Convert 32° Cent. and 20° Reaumur into the corresponding Fahrenheit temperatures.

```
   32° Cent.              20° Reau.
       9                      9
     ___                    ___
   5)288                  4)180
     ___                    ___
   57·6 above freezing.   45 above freezing.
   + 32                   + 32
   _____                  _____
   89·6 Fahr.             77 Fahr.
```

Chronometer.—A Chronometer is an accurately made and

DEFINITIONS.

compensated timepiece, mounted on gimbals in a box, and carefully protected from dust, damp, and vibration. It is used

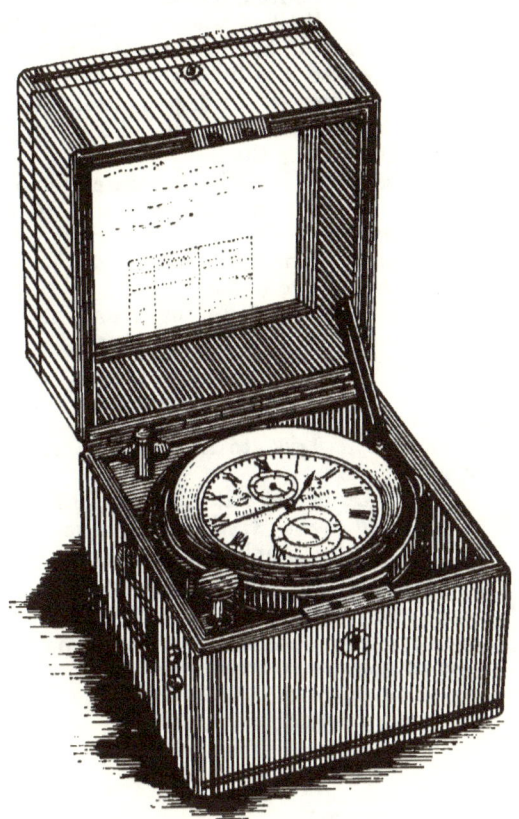

Fig. 28.

on board ship for the purpose of obtaining correct mean time at Greenwich (the error and rate being known), and thence the longitude.

Hydrometer.—The Hydrometer is an instrument for measuring Density or Specific Gravity.

Specific gravity of a solid or liquid is its weight compared with the weight of an equal bulk or volume of distilled water at the temperature of 62° Fahr., or as some scientists prefer, 39° 2 Fahr., which is the point of maximum density.

The Hydrometer used on board ship for obtaining the density of sea-water, and the water in Docks or Rivers, is a glass or metal stem with two bulbs, the lower of which is weighted with mercury or small shot to make the instrument float upright. The scale on the stem reads from 0 downwards.

When in distilled water (temp. 62°), the level of the water is at 0, but in sea-water at about 26. In brackish water it will be somewhere between those marks, according to the degree of saltness.

One thousand must be added to the figures on the scale when stating the S.G.

Hence the S.G. of distilled water (temp. 62° Fahr.) is 1000, and the S.G. of sea-water 1025 or 1026, whilst the S.G. of the water in a river or estuary or dock where fresh water enters might be anything between these two limits.

These facts have an important bearing on the *loading of ships*, since the *law* fixes the *load-line* for seagoing vessels.

It is obvious that a ship may be loaded deeper than her "load-mark" in fresh or brackish water, for she will "rise to her marks" when she goes to sea.

The S.G. of the water at the loading-place may be obtained by a Hydrometer, and, her *draught* being known, a simple proportion will give the "sea draught" nearly. Also when the "draught" at sea is known, the "draught" in dock or river can be found.

FIG. 29.

Examples.—1. The S.G. of the water in dock is 1015, and the ship's draught 23 ft. 6 in.: required her draught at sea. Taking S.G. of sea-water as 1025—

 ft.   ft.
1025 : 1015 :: 23·5 : 23·3 nearly
                 23·5
                 ————
                 5075
                 3045
                 2030
                 ————
1025)23852·5(23·27 ft. nearly
     2050
     ————
      3352
      3075
      ————
       2775
       2050
       ————
        7250
        7175
        ————
          75

Answer, 23·3 ft. nearly.

DEFINITIONS.

2. At sea a ship draws 26 ft. 9 in. : what will her draught be in dock, where the S.G. is 1010?

```
              ft.  in.  ft.  in.
1010 : 1025 :: 26   9 :  27   2
              12
              ───
              321 in.
             1025
             ────
             1605
              642
              321
             ─────
      1010)329025(326 inches nearly
           3030        = 27 ft. 2 in. Answer.
           ────
            2602
            2020
           ─────
            5825
```

The Station Pointer.—The Station Pointer is a graduated circle having three *radial arms*, one of which is fixed, and the other two moveable, the latter being provided with verniers and clamp and tangent screws.

The use of the Station Pointer is to locate the ship's position on a chart from the angles subtended by three lighthouses or other fixed objects on shore.

With a sextant, the horizontal angles between the middle object and each of the other two are measured, and the moveable arms are set to these angles.

Then the instrument is laid on the chart with the fixed arm on the middle object, and moved about until the other two arms fall on the other two objects. The centre of the circle gives the ship's position.

The Sextant.—The "full" Sextant for sea use consists of a metal frame in the form of a Sector of about 60°, with an *arc* or limb graduated to degrees and 10′ spaces. An index bar, pivoted

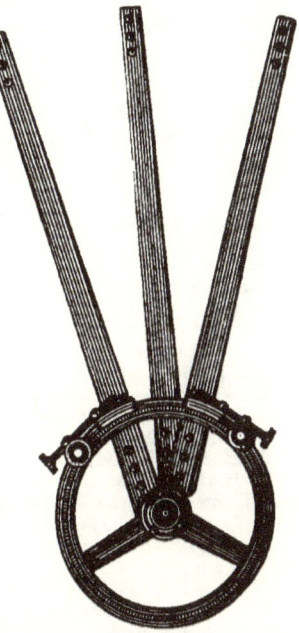

Fig. 30.

at the centre of the circle, carries the silvered *Index Glass*, and at the other end a Vernier or Nonius for subdividing to 10″.

The index bar is provided with clamp and tangent screws, and carries a reading microscope. The *Horizon Glass*—half of which is silvered—stands on the frame. Both glasses are provided with screens for taking off the glare of a strong light when the instrument is used. A handle at the back of the frame is convenient for holding, and a collar, which is moveable towards or from the frame by means of a milled-head screw, serves to hold a plain tube or a telescope.

FIG. 31.

When in correct adjustment, the index and horizon glasses are *perpendicular* to the plane of the instrument (the upper surface of the frame), and *parallel* to each other when the pointer on the vernier indicates 0 on the arc. The axis of the telescope when screwed into the collar is *parallel* to the plane of the instrument.

The first two adjustments are essential, but if the glasses are slightly out of parallelism, it is of little consequence, for the "index error" can be easily determined and applied. Screws are provided for making all the adjustments.

QUESTIONS ON THE ADJUSTMENTS OF THE SEXTANT IN BOARD OF TRADE EXAMINATIONS FOR MATES AND MASTERS.

The applicant will answer in writing, on a sheet of paper which will be given him by the examiner, all the following questions, numbering his answers with the numbers corresponding to the questions.
1. What is the first adjustment of the sextant?
2. How do you make that adjustment?
3. What is the second adjustment?
4. Describe how to make that adjustment?
5. What is the third adjustment?
6. How would you make the third adjustment?
7. In the absence of a screw, how would you proceed?
8. How would you find the index error by the horizon?
9. How is it to be applied?
10. Place the index at error of   minutes to be added, clamp it and leave it.

NOTE.—The examiner will see that it is correct.

11. The examiner will then place the zero of the vernier on the arc, not near any of the marked divisions, and the candidate will read it.

NOTE.—In all cases the applicant will name or otherwise point out the screws used in the various adjustments.

12. How do you find the index error by the sun?
13. How is the same applied?

DEFINITIONS.

Suppose the reading on the arc to be   , and the reading off the arc to be   , find the index error.

14. What proof have you that those measurements or angles have been taken with tolerable accuracy?

15. Describe the fourth adjustment.

Answers to the Questions on the Sextant.

1. To set the **Index Glass** *perpendicular* to the plane of the sextant.

2. The index should be clamped at about 60°, and the sextant held face upwards with the limb from the observer; then, on looking into the index glass towards the limb, if the reflected and real arcs do not coincide, turn the adjusting screws at the back of the glass until they do coincide.

3. To set the **Horizon Glass** *perpendicular* to the plane of the sextant.

4. The index being clamped at 0, the sextant should be held *horizontally*, face upwards—then, on looking through the telescope collar towards the horizon, if the reflected and real horizons do not coincide, turn the upper adjusting screw at the back of the glass until they do coincide.

5. To set the **Horizon Glass** *parallel* to the **Index Glass** when the index is at 0 on the arc.

6. The index being clamped at 0, the sextant should be held *vertically*; then, on looking through the tube or telescope at the horizon, if the reflected and real horizons do not coincide, turn the lower adjusting screw at the back of the glass until they do coincide.

7. Find the **Index Error**.

8. Placing the index at 0, and holding the sextant as for the third adjustment, make the reflected and real horizons coincide by means of the *Tangent Screw*. The reading "On" or "Off" is the Index Error.

9. It is *subtractive* if "on," and *additive* if "off."

10 and 11 are practical, and require no written answer.

12. Clamping the index at 40' "on" the arc and looking through the sight-tube or telescope at the sun, make the limbs of the reflected and real suns touch by means of the tangent screw, and note the reading; then clamp at 40' "off," and again make the contact and note the reading. Half the difference of the readings is the Index Error.

13. If the greater reading is "on," the error is *subtractive*; if "off," it is *additive*.

14. One-fourth the sum of the readings should be the Sun's Semi-diameter, as in the Nautical Almanac.

15. The Fourth Adjustment is to set the line of collimation or axis of the telescope parallel to the plane of the instrument. The adjustment is made as follows: Screw on the *Inverting Telescope*, and, getting it in correct focus, turn the eye-piece round until two parallel wires in the tube are parallel to the plane of the sextant. Then select two celestial objects not less than 90° apart, and holding the sextant with its face in the plane of the objects, move the index forward until the reflected image of one object comes into view in the centre of the field of the telescope with the direct image of the other. Bring both images in contact on the wire nearest the frame, and by slightly moving the instrument get them on the other wire: if they remain in contact, the adjustment is perfect; if they separate, the object end inclines towards the frame; but if they overlap, it declines from it. The adjustment is made by means of two screws in the collar which holds the telescope.

Star Maps and Planispheres.—A knowledge of the Star-groups or Constellations and the names of the principal bright Stars is essential to the Navigator who would make use of these objects to find the Latitude and Longitude at sea.

Many Star-maps and "Planispheres" are in use, but it is unnecessary to mention here more than the following:—
Proctor's Star Atlas.
Philip's Planisphere.
Poole's Celestial Planisphere.

On a Star-map or a Celestial Globe, the position of a star can be found from its R.A. and Dec. in the same manner as a place is located on a globe or chart by the Latitude and Longitude.

The *planispheres* above mentioned are designed to show the principal star-groups above the Horizon at a given hour. The hour of the night on the edge of the moveable disc is brought to coincide with the day of the month on the fixed disc. Then the part of the star-map within the elliptical figure shows the visible Heavens at the given hour.

A wire stretched across the middle of the ellipse represents the Meridian.

The Planisphere gives by inspection the approximate Time of Meridian Passage of a star as well as the times of Rising and Setting. It can also be used to find by inspection what bright stars are within a given time from the Meridian at any given hour of Greenwich or local Time.

Tables of Meridian Passage of Stars.—Table XLIV. of Norie gives the apparent times of Mer. Pass. of the principal stars at intervals of five days, and as a star crosses the Meridian about 4^m *earlier* on each successive day, it is easy to find the time for intermediate days.

Table XXVII. of Raper gives the time of Mer. Pass. of fifty-five bright stars on the first day of each month, and by subtracting 4^m for each day elapsed, as before, the required time is approximately found for any day.

The Meridian Altitude of a selected star being computed from its Dec. and the *Approximate* Latitude, the sextant can be set to this angle; and as the approximate time of Mer. Pass. is known, the observer can be in readiness to measure the altitude at the moment the star is on the Meridian, and so obtain the *Correct* Latitude.

CHAPTER II.

LOGARITHMS.

ART. 1.—By the aid of Logarithms, calculations which would otherwise be very difficult are rendered simple and easy. Multiplication is converted into addition, division into subtraction, involution into simple multiplication; and evolution into simple division.

A logarithm consists of two parts—the Index and the Decimal Part. For a whole number the index is one less than the number of figures, but for a decimal it is negative—that is, $-1, -2, -3$, etc., according to the place after the decimal point which the first figure (not 0) occupies.

ART. 2.—**Finding the Logarithms of Numbers.**

Examples—

Numbers.	Logarithms.
1. 6	0·778151
2. 34	1·531479
3. 420	2·623249
4. 7·856	0·895201
5. 38270	4·582858
6. ·5	$\bar{1}$·698970
7. ·065	$\bar{2}$·812913
8. ·0013	$\bar{3}$·119343
9. ·00000671	$\bar{6}$·826723

NOTE.—The negative index is generally changed into a positive for Nautical Computations by adding 10, so that $\bar{1}$ becomes 9, $\bar{2}$ becomes 8, etc.

When the number contains more than four figures, take out the log. of the first four; then find the correction for the remaining figures by multiplying the "Diff." by these figures, and cutting off as many to the right. The remaining quantity is the correction to be added.

Examples—

Number.	Logarithms.	Diff. 185
1. 2·340401	0·369216	401
	74	185
	———	740
	0·369290	74,185

TEXT-BOOK ON NAVIGATION.

```
       Numbers.              Logarithms.
    2. 55300·017        ...  4·742725           Diff.  78
                                   13                 0017
                             ─────────                 ───
                             4·742738                  546
                                                        78
                                                      ─────
                                                      13,26
```

3. 86430000   ...   7·936664

NOTE.—There is no correction for 0's at the end of a number.

EXERCISES.

1. Find the Logarithms of the following numbers :—

3	3·409
47	7·5
50	163·45
100	704·009
275	32636·04
8360	·5
57·29	·049
365·74	·0036
1·0009	·001204
75230	·006728
62·7014	·00028041
175·008	·0000753
23002·7	·0346009
164·3019	·075604
21550·007	·3670007
23460	

ART. 3.—**Finding the Numbers corresponding to Given Logarithms.**

Examples.—

```
       Given logs.            Numbers.
    1. 0·113943     ...   ...  1·3
    2. 1·791480     ...   ...  61·87
    3. 2·724931     ...   ...  530·8
```

When the Index exceeds three, and more than four figures are required, the log. *next less* in the tables is taken out, and the remainder divided (as in Division of Decimals) by the "Diff." gives the additional figures required.

Examples.—

```
       Given logs.              Numbers.
    1. 6·490379                3092993
       490239
       ──────
       Diff. 141)140·0(993
             1269
             ────
             1310
             1269
             ────
              410
              423
             ────
```

LOGARITHMS.

```
         Given logs.              Numbers.
    2.  5·820006       ...  ...  660703
        820004
        ──────
        Diff. 66)2·00(03
              198
              ───
               2
    3.  4·875003       ...  ...  74990
```

When the index is negative, and the number therefore a decimal, the position of the first figure (of value) is indicated by the index; thus $\bar{1}$ shows it will occupy the *first* place after the decimal point, $\bar{2}$ requires the first figure (of value) in the *second* place, etc.

Examples.—

```
         Logs.                Numbers.
    1̄·432488    ...   ...    ·2707
    2̄·715339    ...   ...    ·05192
    4̄·934550    ...   ...    ·00086
```

NOTE.—These index numbers when increased by 10 become 9, 8, 6.

EXERCISES.

2. Find the numbers of which the following are the Logarithms:—

```
    3·203577              5·175801
    2·145818              6·531481
    1·700271              7·000435
    0·715920              5·361917
    0·004382              4·041393
    4·681484              6·049606
    5·690196              0·000434
    6·148900              3·670154
    7·040203              7·276003
    4·806044              5·380032
```

ART. 4.—Multiplication by Logarithms.

Rule.—Add together the logarithms of the numbers, and find the number corresponding to the sum of logarithms.

Examples.—

1. Multiply 234 by 1·76 by common logarithms.

```
              234    log 2·369216
              1·76   log 0·245513

         Ans. 411·8  log 2·614729
```

2. Multiply 32640 by ·00876.

```
              32640  log 4·513750
              ·00876 log 7·942504

         Ans  285·9  log 2·456254
```

26   TEXT-BOOK ON NAVIGATION.

16. Divide ·000876 by ·0000438.
17.   ,,   ·012 by ·009847.
18.   ,,   ·76591 by ·13.
19. Find the value of $\dfrac{198\cdot79}{3\cdot46}$.
20.   ,,   ,, $\dfrac{2\cdot13 \times 856\cdot4}{\cdot087}$.
21.   ,,   ,, $\dfrac{\cdot00175 \times 3872\cdot4}{7\cdot69 \times \cdot0432}$.
22.   ,,   ,, $\dfrac{8\cdot32 \times \cdot096 \times 248\cdot109}{1\cdot74 \times 23\cdot42 \times 148\cdot6}$.
23.   ,,   ,, $\dfrac{100\cdot108 \times \cdot36749}{\cdot00984 \times 468\cdot06}$.
24.   ,,   ,, $\dfrac{\cdot0756 \times \cdot00998 \times \cdot247}{\cdot0001 \times \cdot0137 \times \cdot142}$.

ART. 6.—**Involution by Logarithms.**

Rule.—Multiply the log. of the number by the required Power, and then find the corresponding number.

Example.—Required the 7th power of 1·09.

1·09   log 0·037426
                    7
       ────────────
Ans. 1·828   log 0·261982

EXERCISES.

1. Required the square or 2nd power of 23·4.
2.   ,,   cube or 3rd power of 1·876.
3.   ,,   5th power of ·08967.
4.   ,,   9th power of ·1039.
5.   ,,   13th power of ·367.

ART. 7.—**Evolution by Logarithms.**

Rule.—Divide the log. of the given number by the required Root, and then find the corresponding number.

Examples.—

1. Required the square root of 10816.

10816   log 2)4·034066
         ──────────────
Ans. 104   log   2·017033

2. Required the cube root of 42875.

42875   log 3)4·632204
         ──────────────
Ans. 35   log   1·544068

3. Required the 5th root of 9·618.

9·618   log 5)0·983085
         ──────────────
Ans. 1·573   log   0·196617

4. Required the 7th root of 234·68.
   234·68   log 7)2·370476

   Ans. 2·181   log   0·338639

5. Required the cube (or 3rd) root of ·004913.
   ·004913   log 3)$\bar{3}$·691437

   Ans. ·17   log $\bar{1}$·230479

6. Required the cube root of 0·06174.
   ·06174   log $\bar{2}$·790567 or $-3 + 1\cdot790567$
   3) $-3 + 1\cdot790567$

   Ans. ·3952   log $\bar{1}$·   596856

7. Required the 5th root of ·007652.
   ·007652   log 5)$\bar{3}$·883775   ($\bar{3} = \bar{5} + 2$)

   Ans. ·3774   log $\bar{1}$·576755

EXERCISES.

1. Required the square root of 306·25.
2.   ,,   cube   ,,   46·656.
3.   ,,   5th   ,,   ·002435.
4.   ,,   cube   ,,   12167000.
5.   ,,   7th   ,,   ·00000002097152.

Remark.—With a Table of Logarithms containing only six places of figures, such as is used for Navigation, the answers in the preceding exercises cannot be expected to be correct to more than four or five places.

Of course, it must be understood that whenever logs are *added* in the calculations of problems in Navigation and Nautical Astronomy, it means that the terms are *multiplied* together, and when logs are *subtracted*, it means division. This is important to remember when one of the terms is zero (or 0).

ADDITIONAL EXERCISES.

Find by logs the values of the following :—
1. $\sqrt{20675}$; $\sqrt{769\cdot52}$; $\sqrt[3]{4697\cdot6482}$.
2. $(29\cdot865)^2$ and $(98\cdot7652)^3$.
3. $(91\cdot543)^2 \times \sqrt[3]{8794\cdot027}$; $\sqrt{\dfrac{4682}{99 \times 37987}}$
4. $\dfrac{\sqrt{8794\cdot08}}{\sqrt[5]{276509}}$.

CHAPTER III.

THE SAILINGS.

ART. 8. **The Day's Work and Traverse Sailing.**—The purpose of the Day's Work is to find the position of the ship at noon by "Dead Reckoning," that is, from the courses and distances sailed since the preceding noon. The direct course and distance made good are also determined.

The courses given are by compass as per ship's Log Book, but the *true* courses must be found for computing a " Traverse Table."

Directions for finding the True Courses.—(*a*) Allow the leeway to the Right (R.) when the wind blows on the Left or Port Side of the ship, and to the Left (L.) when on the Right or Starboard Side. (*b*) Allow E. deviation and variation (or total correction) to the Right (R.), and W. to the Left (L.).

NOTE.—The Leeway may be determined by towing the log-ship astern and observing the angle the line makes with the keel, but it is often merely estimated.

The Deviation is found by reference to the Deviation Card, and the Variation from a chart.

When the Departure is taken from a known point of land or light, the opposite course to the bearing of the point or light from the ship is called the "Departure Course," and the distance off is the distance belonging to that course.

The " Set " and " Drift " of the current are found by chart, and are treated as a course and distance made by the ship, because the effect is the same on the Day's Work.

THE SAILINGS.

Example 1.

Hour.	Course steered.	Knts.	10ths.	Wind.	Leeway.	Deviation.	Remarks.
1	N. 65° W.	5	6	N. by E.	20°	12° W.	
2		5	5				
3		5	8				A point of land in
4		6	1				lat. 46° 3′ S., long. 166° 15′
5	N. 77° E.	6	2	N. by E.	17°	3° W.	E., bore N.E. from ship, distant 12 miles, ship's head being W., and Deviation 4° W. Variation 15° E.
6		5	9				
7		5	7				
8		5	2				
9	N. 57° W.	5	2	N. by E. ¼ E.	16°	17° W.	
10		5	3				
11		5	0				
12		5	5				
1	E.	5	7	N.N.E.	19°	6° E.	
2		6	3				
3		6	8				
4		7	2				A current set N. magnetic 24 miles during the 24 hours from the time of departure to the end of the day.
5	N. 52° W.	7	2	N.N.E.	14°	15° W.	
6		7	3				
7		7	6				
8		7	9				
9	S. 80° E.	8	0	N.E. by N. ¼ N.	28°	14° E.	
10		8	1				
11		8	2				
12		8	7				

Correct the courses, and find the course and distance made good; also the Latitude and Longitude in by inspection.

Traverse Table.

True courses.	Distance.	D. Lat.		Dep.	
		N.	S.	E.	W.
S. 56° W.	12		6·7		9·9
N. 82° W.	23	3·2			22·8
S. 74° E.	23		6·3	22·1	
N. 75° W.	21	5·4			20·3
S. 50° E.	26		16·7	19·9	
N. 66° W.	30	12·2			27·4
S. 30° E.	33		28·6	16·5	
N. 15° E.	24	23·2		6·2	
		44·0	58·3	64·7	80·4
			44·0		64·7
		D. lat.	14·3	Dep.	15·7

The D. lat and dep. are found nearest together under the course 42

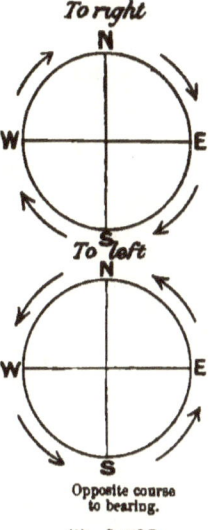

Illustrations and Minor Calculations.

To right

To left

Opposite course to bearing.

(1)   S. 45° R.
    Corr. 11 R.
      56 R.

(2)   N. 65° L.
    Leeway 20 L.
      N. 85 L.
    Corr. 3 R.
      N. 82 L.

(3)   N. 77° R.
      17 R.
      94 R.
      12 K.
      106 R.
      180
      S. 74 E.

(4)   N. 57° L.
      16 L.
      73 L.
      2 L.
      75 L.

(5)   S. 90° L.
      19 R.
      71 L.
      21 R.
      50 L.

(6)   N. 52° L.
      14 L.
      66
  Corr. 0

TEXT-BOOK ON NAVIGATION.

Hour.	Course steered.	Knts.	10ths.	Wind.	Lee-way.	Devia-tion.	Remarks.
1	N.E. by E. $\frac{1}{4}$ E.	13	3	S.E. by E.	$3\frac{1}{2}$	5° W.	
2		17	4				A current set cor-
3		10	5				rect magnetic
4		15	8				South 4 miles
5	S.W. $\frac{1}{4}$ W.	14	9	S. by E. $\frac{1}{2}$ E.	$\frac{1}{2}$	4° E.	from the time departure was
6		13	3				taken to the end
7		15	9				of the day.
8		17	9				
9	S.S.W.	11	4	West.	2	5° E.	
10		14	9				
11		13	8				
12		16	9				

Correct the courses for Deviation, Variation, and Leeway, and find the course and distance from the given point, and the Latitude and Longitude in.

2.

Hour.	Course steered.	Knts.	10ths.	Wind.	Lee-way.	Devia-tion.	Remarks.
1	N. by E. $\frac{3}{4}$ E.	9	1	S.E. by E.	$\frac{1}{4}$	18° E.	
2		9	3				
3		9	9				
4		8	7				A point of land in lat. 18° 59′ S., long.179°53′ W.,
5	S.E. by E.	6	4	S. by W.	$1\frac{1}{2}$	23° W.	bearing by compass E.S.E., distant 19 miles, with the ship's head at N. by E. $\frac{1}{4}$ E. Variation 9° E.
6		6	1				
7		6	3				
8		6	2				
9	E. by N. $\frac{1}{4}$ N.	6	0	N. $\frac{1}{2}$ E.	$1\frac{3}{4}$	3° W.	
10		5	9				
11		5	8				
12		5	3				
1	S.W. $\frac{3}{4}$ W.	5	3	S.S.E.	$2\frac{1}{4}$	0	
2		5	1				
3		5	2				
4		5	4				A current set correct magnetic S.S.E. $\frac{3}{4}$ E. 15 miles from the time departure was taken to the end of the day.
5	S.S.W.	7	9	W. by N.	1	11° W.	
6		8	2				
7		8	4				
8		9	5				
9	N.N.E.	9	6	N.W. $\frac{1}{4}$ W.	$1\frac{1}{4}$	16° E.	
10		10	0				
11		10	3				
12		10	1				

Correct the courses for Deviation, Variation, and Leeway, and find the course and distance from the given point, and the Latitude and Longitude in.

3.

Hour.	Course steered.	Knts.	10ths.	Wind.	Lee-way.	Devia-tion.	Remarks.
1	N. by E. ¼ E.	9	1	S.E. by E.	¼	18° E.	
2		9	3				A point of land in
3		9	9				lat. 18° 59' S.,
4		8	7				long. 179° 53'
5	S.E. by E.	6	4	S. by W.	1½	23° W.	W., bearing by compass E.S.E.,
6		6	1				distant 19 miles,
7		6	3				with the ship's
8		6	2				head at N. by
							E. ¼ E. Devia-
9	E. by N. ½ N.	6	0	N. ½ E.	1¾	3° W.	tion as per log.
10		5	9				Variation 9° E.
11		5	8				
12		5	3				
1	S.W. ¾ W.	5	3	S.S.E.	2¼	0	
2		5	1				
3		5	2				
4		5	4				A current set cor-
							rect magnetic
5	S.S.W.	7	9	W. by N.	1	11° W.	S.S.E. ¾ E., 15
6		8	2				miles from the
7		8	4				time departure
8		9	5				was taken to
							the end of the
9	N.N.E.	9	6	N.W. ¼ W.	1¼	16° E.	day.
10		10	0				
11		10	3				
12		10	1				

Correct the courses for Deviation, Variation, and Leeway, and find the course and distance from the given point, and the Latitude and Longitude in.

4.

Hour.	Course steered.	Knts.	10ths.	Wind.	Lee-way.	Devia-tion.	Remarks.
1	S. by W. ½ W.	16	3	W. ½ S.	1¼	7° E.	
2		17	1				A point of land in
3		10	2				lat. 58° 30' S.,
4		9	4				long. 64° 54' W.,
							bearing by com-
5	W. by S. ¼ S.	11	9	N.W. by N.	¾	9° E.	pass N.W. ¼ N.,
6		12	0				distant 60 miles,
7		13	0				with the ship's
8		14	1				head at S. by
							W. ½ W. Devi-
9	S.W. ¼ W.	16	9	S.S.E.	¼	8° W.	ation as per log.
10		14	8				Variation 21° E.
11		12	8				
12		13	5				

(*Continued on next page.*)

TEXT-BOOK ON NAVIGATION.

Hour.	Course steered.	Knts.	10ths.	Wind.	Lee-way.	Devia-tion.	Remarks.
1	N.E. by N.	10	9	W.S.W.	0	11° W.	
2		11	8				
3		12	8				
4		13	5				
5	N. by E.	14	8	N.W. by W.	½	12° W.	A current set correct magnetic W. ½ S. 39 miles, from the time departure was taken to the end of the day.
6		15	8				
7		16	8				
8		17	6				
9	W. by N.	9	7	S.W. by S.	¼	41° W.	
10		1	7				
11		2	7				
12		6	9				

Correct the courses for Deviation, Variation, and Leeway, and find the course and distance from the given point, and the Latitude and Longitude in.

5.

Hour.	Course steered.	Knts.	10ths.	Wind.	Lee-way.	Devia-tion.	Remarks.
1	N.N.W. ½ W.	4	1	N.E.	1¾	2° W.	
2		3	3				A point of land in lat. 29° 59' N., long. 132° 54' E., bearing by compass N.N.E. ½ E., distant 15 miles, with the ship's head at N.W. by W. Deviation 6° W. Variation 1° E.
3		3	1				
4		1	5				
5	E.S.E.	2	1	N.E.	2	7° E.	
6		3	1				
7		3	1				
8		4	7				
9	S. ¾ E.	6	3	E.S.E.	2¼	2° W.	
10		6	9				
11		5	4				
12		6	4				
1	N.E. ¼ N.	9	7	E.S.E.	1½	8° E.	
2		7	4				
3		0	2				
4		2	7				
5	W. ½ N.	3	5	S.S.W. ½ W.	1¼	9° W.	A current set correct magnetic N.E. 30 miles, from the time departure was taken to the end of the day.
6		4	3				
7		3	6				
8		3	6				
9	N. by E.	8	5	E. by N.	¼	6° E.	
10		1	1				
11		8	1				
12		9	3				

Correct the courses for Deviation, Variation, and Leeway, and find the course and distance from the given point, and the Latitude and Longitude in.

THE SAILINGS.

6.

Hour.	Course steered.	Knts.	10ths.	Wind.	Leeway.	Deviation.	Remarks.
1	S. by W.	12	1	W. by S.	$\frac{3}{4}$	9° W.	At noon a point of land in lat. 54° 22' N., long. 178° 26' E., bearing by compass N.W. $\frac{1}{4}$ W., distant 36 miles, with the ship's head at S. by W. Deviation as per log. Variation 12° E.
2		13	2				
3		11	3				
4		12	4				
5	N. $\frac{3}{4}$ E.	12	9	East	$\frac{1}{4}$	6° E.	
6		13	8				
7		14	7				
8		11	6				
9	East	11	2	S.S.E.	$\frac{1}{2}$	19° E.	
10		12	2				
11		12	2				
12		13	4				
1	S.E.	9	3	S.S.W.	$\frac{3}{4}$	11° E.	
2		11	4				
3		12	7				
4		10	6				A current set correct magnetic S. 44 miles from the time departure was taken to the end of the day.
5	S.S.W. $\frac{1}{4}$ W.	12	6	N.W.	0	12° W.	
6		11	2				
7		14	9				
8		15	3				
9	E. $\frac{3}{4}$ N.	13	9	North	$\frac{1}{4}$	18° E.	
10		12	9				
11		11	9				
12		13	3				

Correct the courses for Deviation, Variation, and Leeway, and find the course and distance from the given point, and the Latitude and Longitude in.

NOTE.—For additional exercises, see Examination Papers.

ART. 9. **Parallel Sailing.**—In parallel sailing, the ship's course is supposed to continue on the same parallel of latitude. There are three things concerned, viz. the Latitude, the Departure or Meridian Distance, and the Difference of Longitude.

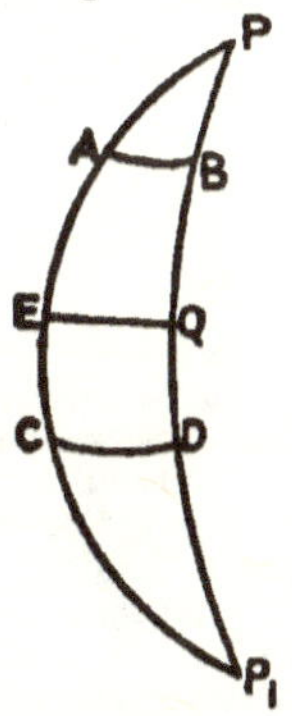

Definitions.—(1) *Latitude :* the distance of the ship north or south from the Equator.

(2) *Departure* (or Meridian Distance) : the distance in geographical miles between two places on the same parallel of latitude.

(3) *Difference of Longitude :* the arc of the Equator corresponding to the Meridian Distance.

Illustration.—AB and CD are Departures at different distances from the Equator; EQ is the D. long., or corresponding Arc of the Equator.

Remark.—It is seen that the Meridian Distance decreases as the Latitude increases, and vanishes altogether at the Poles.

FIG. 32.

The formulæ which express the relation between the three terms are—

1. D. long. = dep. × sec lat.
2. Dep. = diff. long. × cos lat.
3. Sec lat. = $\dfrac{\text{diff. long.}}{\text{dep.}}$

These are the three *cases* in Parallel Sailing, and show how the D. long. or Dep. or Lat. may be found when the other two terms are given.

NOTE.—For mathematical principles, see Appendix.

Examples.—

1. In lat. 46° 45' the Departure made good was 110·5 miles: required the Difference of Longitude.

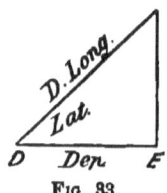
FIG. 33.

Dep. 110·5	...	log 2·043362
Lat. 46° 45' ...	...	sec 10·164193
D. long. = 161·3	...	log 2·207555

2. In lat. 53° 14' the D. long. being 3° 16', required the Meridian Distance or Departure.

D. long. 196'	...	log 2·292256
Lat. 53° 14' ...	...	cos 9·777106
Dep. = 117·3	...	log 2·069362

3. On what parallel of latitude would a Meridian Distance of 245 miles give a difference of longitude of 5° 54'?

D. long. 354	...	log 2·549063
Dep. 245	...	log 2·389166
Lat. 46° 12'	...	sec 10·159837

Remark.—In the first two cases 10 is rejected from the sum of indexes, and in the third case 10 is added to the upper index, because in the Table of Log Sines, etc., the proper index is always increased by 10.

EXERCISES.

1. In lat. 16° 30' N. the dep. made good was 167·6 miles: what will the difference of longitude be by parallel sailing?
2. The difference of longitude between two places in lat. 63° 19' 30" is 4° 20': required the distance as measured along the parallel.
3. In lat. 11° 56' N. the dep. made good was 365 miles: required the difference of longitude in minutes.
4. In what latitude is 200 miles of dep. equal to 600 minutes of longitude?
5. In lat. 35° 40' S. the dep. made good was 9·76: required the difference of longitude measured on the parallel.
6. What is the difference of latitude between two parallels of latitude on different sides of the Equator, when between the same two meridians the dep. on the northern parallel is 250 miles, on the southern 350 miles, and on the Equator 470 miles?

THE SAILINGS.

7. On the parallel of 48° 12' ran true west 1000·5 miles: if the longitude left was 9° 36' W., what was the longitude in?
8. From two ports in lat. 40° 20' N., distant 300 minutes of longitude, two ships sail N. for 300 miles: how many miles apart will they be, and in what latitude?
9. In lat. 33° 15' N. ran east 10·95 miles: how many minutes of longitude will that represent?
10. If two ships in lat. 51° S. should sail directly north until the departure changes from 400 to 500 minutes, what latitude will they be in?
11. In lat. 72° 59' the dep. was 370·5 miles: convert that into minutes of longitude.
12. In what latitude does 300 miles of dep. correspond to 600 minutes of longitude?
13. The latitude by Mer. alt. was 36° 35' N.; longitude by afternoon sights, 137° 38' E.; run since noon due west, 39 miles: find noon longitude.
14. A vessel sailed due east along the parallel of 45° 24' S. for 29 miles from long. 43° 40' E.: what is her longitude in?
15. What distance must a vessel sail on the parallel 48° 25' N. to change her longitude 2° 25'?
16. A vessel sailing due east finds her difference of longitude to be 1° 19', and distance made good 50': what is her latitude?
17. How far on the parallel of 38° S. must a vessel sail from long. 47° 29' E. to reach long. 49° 27' E.?
18. For every 1° diff. long. I have to sail 45 miles due west: along what parallel am I sailing?

ART. 10. **Mercator's Sailing.**—The problem in Mercator's Sailing commonly used, and the only one in the Board of Trade Examination for Mates and Masters, is to find the *Course* and *Distance* from place to place, the latitudes and longitudes of both places being given.

The data used in the calculation are—
The difference of latitude.
The difference of longitude.
The meridional difference of latitude.

Definitions.—(1) D. lat.: The sum or difference of the two latitudes (according as they are of different names or of the same name) reduced to miles.

(2) D. long.: The sum or difference of the two longitudes (according as they are of different names or of the same name) reduced to minutes of arc.

N.B.—When the sum of longitudes exceeds 180°, subtract from 360°.

(3) Mer. D. lat.: The sum or difference of the meridional parts for the two latitudes.

Formulæ.—(1) To find the Course—

$$\text{Tangent of course} = \frac{\text{D. long.}}{\text{Mer. D. lat.}}$$

(2) To find the Distance—

$$\text{Distance} = \text{secant of course} \times \text{D. lat.}$$

TEXT-BOOK ON NAVIGATION.

Examples.—

1. Find the Course and Distance from A to B on Mercator's principle.

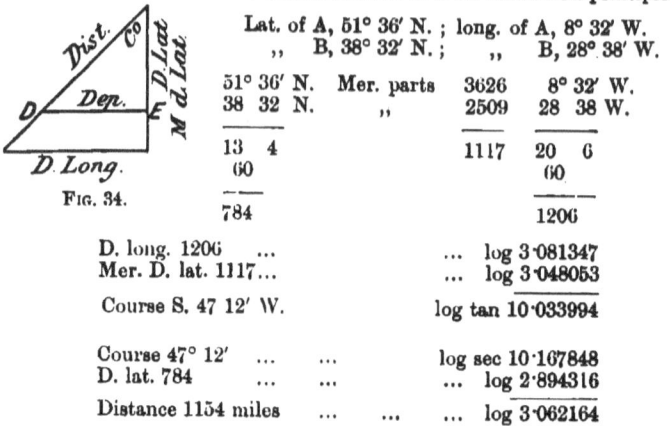

Fig. 34.

Lat. of A, 51° 36′ N. ; long. of A, 8° 32′ W.
,,   B, 38° 32′ N. ; ,,   B, 28° 38′ W.

51° 36′ N.	Mer. parts	3626	8° 32′ W.
38 32 N.	,,	2509	28 38 W.
13 4		1117	20 6
60			60
784			1206

D. long. 1206 ... ... log 3·081347
Mer. D. lat. 1117... ... log 3·048053

Course S. 47 12′ W.   log tan 10·033994

Course 47° 12′ ... ... log sec 10·167848
D. lat. 784 ... ... log 2·894316

Distance 1154 miles ... ... log 3·062164

NOTE.—The Course is reckoned from S because B is south of A, and towards W. because B is west of A.

2. Find the Course and Distance from A (Hawraki) to B (Hawaii).

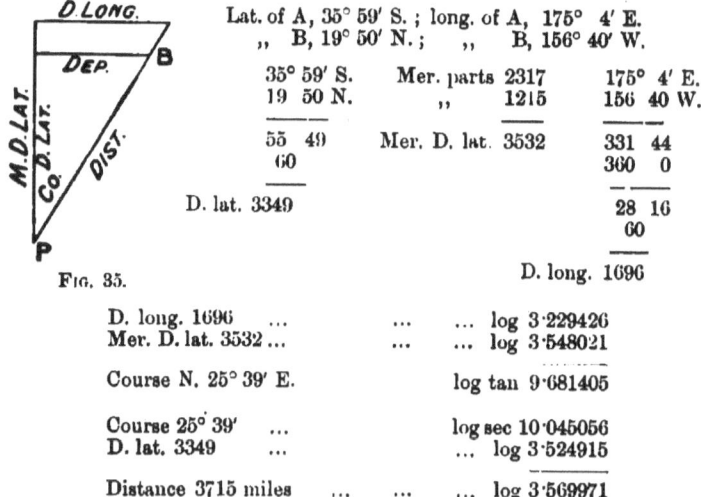

Fig. 35.

Lat. of A, 35° 59′ S. ; long. of A, 175° 4′ E.
,,   B, 19° 50′ N. ; ,,   B, 156° 40′ W.

35° 59′ S.	Mer. parts	2317	175° 4′ E.
19 50 N.	,,	1215	156 40 W.
55 49	Mer. D. lat. 3532		331 44
60			360 0
D. lat. 3349			28 16
			60
		D. long. 1696	

D. long. 1696 ... ... log 3·229426
Mer. D. lat. 3532 ... ... log 3·548021

Course N. 25° 39′ E.   log tan 9·681405

Course 25° 39′ ... ... log sec 10·045056
D. lat. 3349 ... ... log 3·524915

Distance 3715 miles ... ... log 3·569971

NOTE.—The Course is reckoned from N. because B is north of A, and E. because it crosses the meridian of 180°.

3. Required the Course and Distance by Mercator from A to B.

Lat. of A, 6° 15′ N. ; long. of A, 79° 30′ E.
,,   B, 36° 10′ S. ; ,,   B, 113° 45′ E.

THE SAILINGS.

Lat. 6° 15′ N.   Mer. parts 376   Long. 79° 30′ E.
„   36° 10′ S.   „   2330   „   113 45 E.

42 25   Mer. D. lat. 2706   34 15
60   60

D. lat. }2545   D. long. 2055

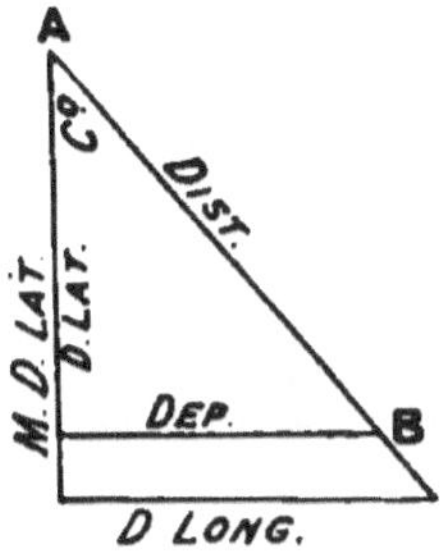

D. long. 2055   ...   ... log 3·312812
Mer. D. lat. 2706 ...   ... log 3·432328

Course S. 37° 13′ E.   log tan 9·880484

Course 37° 13′   ...   log sec 10·098894
D. lat. 2545   ...   ... log 3·405688

Distance 3196 miles   ... log 3·504582

FIG. 36.

4. Required the Course and Distance by Mercator's principle from A, in lat. 57° 42′ S.; long., 70° 3′ W., to B. in lat. 46° 58′ S.; long. 168° 20′ E.; also the Course to steer by compass, supposing the variation to be 22° E., and the deviation of the compass 7° 30′ W.

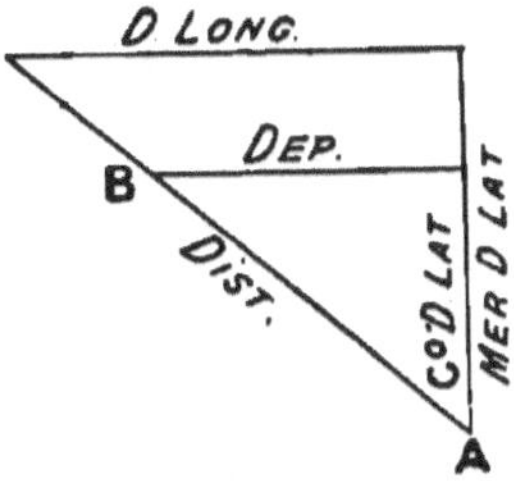

Lat. 57° 42′ S. Mer. parts 4260   Long. 70° 3′ W.
„   46° 58′ S.   „   3200   „   168° 20′ E.
    10 44   Mer. D. lat. 1060   238 23
    60   360   0

D. lat. }644   121 37
                                      60

                D. long. 7297   FIG. 37.

D. long. 7297   ...   ... log 3·863144
Mer. D. lat. 1060 ...   ... log 3·025306

N. 81° 44′ 5″ W. ...   log tan 10·837838
                                     764

                                   74 gives 5″

Course 81° 44′ 5″   log sec 10·842372
D. lat. 644   ...   ... log 2·808886

Distance 4480 miles   ...   . log 3·651258

NOTE.—When the Course is large, it may be advisable to work to seconds, as in this case, in order to ensure the accuracy of the distance within a mile.

To find the Course by compass :

True course   ...   . N. 81° 44′ W.
Variation   22   0′ E.

Magnetic course ...   N. 103 44 W.
Deviation   ...   7 30 W.

Compass course ...   . N. 96 14 W.
Or   ...   . S. 83 46 W.

40   TEXT-BOOK ON NAVIGATION.

Remark.—Mercator's method gives the Course and Distance on a rhumb line joining two places. It has the great advantage of giving a direct course, but it does not give the shortest distance, which would be on the Great Circle passing through the places.

EXERCISES.

Required the Compass Course and Distance from A to B by calculation on Mercator's principle.

1. Lat. A, 50° 49' N. ;   long. A, 0° 0'.
   ,, B, 48° 17' N. ;   ,, B, 54° 47' W.
   Variation 19° 45' W. ; deviation 20° W.

2. Lat. A, 11° 11' S. ;   long. A, 0° 37' W.
   ,, B, 0° 0' ;   ,, B, 0° 0'.
   Var. 23° 30' W. ; dev. 10° 17' E.

3. Lat. A, 52° 57' N. ;   long. A, 158° 33' W.
   ,, B, 39° 30' S. ;   ,, B, 178° 48' E.
   Var. 25° E. ; dev. 15° W.

4. Lat. A, 14° 9' S. ;   long. A, 180° 0'.
   ,, B, 0° 0' ;   ,, 174° 27' E.
   Var. 9° E. ; dev. 5° 30' E.

5. Lat. A, 35° 15' S. ;   long. A, 115° 15' E.
   ,, B, 51° 30' S. ;   ,, B, 178° 52 W.
   Var. 7½° W. ; dev. 15½° E.

6. Lat. A, 51° 24' N. ;   long. A, 9° 36' W.
   ,, B, 38° 36' S. ;   ,, B, 170° 24' E.
   Var. 25½° W. ; dev. 14½° E.

7. Lat. A, 0° 0' ;   long. A, 81° 27' W.
   ,, B, 61° 10' S. ;   ,, B, 170° 59' E.
   Var. 2° 20' E. ; dev. 20° 40' E.

8. Lat. A, 42° 56' S. ;   long. A, 100° 36' W.
   ,, B, 10° 36' N. ;   ,, B, 180° 0'.
   Var. 17° 10' W. ; dev. 13° 50' E.

9. Lat. A, 36° 19' S. ;   long. A, 30° 46' W.
   ,, B, 36° 12' S. ;   ,, B, 58° 23' W.
   Var. 31° 15' W. ; dev. 14° 26' W.

10. Lat. A, 0° 0' ;   long. A, 31° 17' W.
    ,, B, 63° 28' N. ;   ,, B, 28° 16' W.
    Var. 12° 20' W. ; dev. 15° 30' E.

11. Lat. A, 32° 22' N. ;   long. A, 33° 15' E.
    ,, B, 37° 14' N. ;   ,, B, 15° 40' E.
    Var. 5° 35' W. ; dev. 12° 29' W.

12. Lat. A, 0° 17' S. ;   long. A, 59° 47' W.
    ,, B, 1° 59' N. ;   ,, B, 120° 13' E.
    Var. 6° 40' E. ; dev. 13° 50' W.

NOTE.—For Mathematical Principles, see Appendix.

CHAPTER IV.

TIDES.

ART. 11. The Admiralty Tide Tables give the times of High Water and Heights above Mean Low Water of Spring Tides at twenty-four "standard" ports. An Index of these ports, with the page on which the Tide Table for each month may be found, is given at the beginning of the book. For other British and European ports a "Table of Tidal Constants" is given on pp. 101 to 105. These "time constants" show how much later or earlier high water occurs at the given port than at the standard port.

The times in the tables are stated to be *mean time at place*, and (*if required*) the mean time at Greenwich could be found by applying the longitude in time.

When the place is not in the Table of Constants, a constant may be found by turning to the Alphabetical List of Places at the end of the tables and taking the difference between the "full and change" times of H.W. for the given port and any standard port. It is evident this constant will be − (minus) when the full and change at given port is the lesser of the two, and + (plus) when it is the greater.

The heights can be found in a similar manner by using "height constants."

Examples.—

1. Required the time of H.W. at Valentia harbour on May 2, 1897.

Constant, −1^h 19^m Queenstown.

H.W. Queenstown, May 2 ... 5^h 5^m A.M. 5^h 22^m P.M.
−1 19   −1 19

H.W. Valentia harbour, May 2 ... 3 46 A.M.   4 3 P.M.

2. Required the time of H.W. at St. Malo on June 8, 1897.

Constant, +2^h 18^m Brest.

H.W. Brest, June 8 ... ... 9^h 1^m A.M.   9^h 31^m P.M.
+2 18   +2 18

H.W. St. Malo, June 8 ... ... 11 19 A.M.   11 49 P.M.

TEXT-BOOK ON NAVIGATION.

3. Required the time of H.W. at Penzance on January 13, 1897.

       Constant, $-1^h\ 13^m$ Devonport.
H.W. Devonport, Jan. 13 ... $0^h\ 21^m$ P.M. Jan. 14, $0^h\ 57^m$ A.M.
            $-1\ 13$ $-1\ 13$

H.W. Penzance, Jan. 13 ... 11 8 A.M. Jan. 13, 11 44 P.M.

4. Required the time of H.W. at West Cowes on February 4, 1897.

       Constant, $-1^h\ 26^m$ Portsmouth.
H.W. Portsmouth, February 4 ... $1^h\ 18^m$ P.M. No P.M. tide.
             $-1\ 26$

   H.W. West Cowes, February 4, 11 52 A.M.

NOTE.—The following tide is on February 5.

5. Required the time of H.W. at Antwerp on March 18, 1897.

       Constant, $+5^h\ 13^m$ Dover.
H.W. Dover, March 17 ... $10^h\ 16^m$ P.M. March 18, $10^h\ 37^m$ A.M.
           $+5\ 13$ $+5\ 13$

H.W. Antwerp, March 18 ... 3 29 A.M. March 18, 3 50 P.M.

6. Required the time of H.W. at Oban on December 3, 1897.

       Constant, $+5^h\ 20^m$ Greenock.
H.W. Greenock, February 3 ... $6^h\ 56^m$ A.M.
            $+5\ 20$

H.W. Oban, February 3 ... 0 16 P.M. No A.M. tide.
NOTE.—The previous tide belongs to February 2.

EXERCISES.

1897.

1. Find the time of H.W. at Bordeaux A.M. and P.M. on Jan. 11th and 28th.
2. ,, ,, H.W. at St. Nazaire A.M. and P.M. on Jan. 14th and 30th.
3. ,, ,, H.W. at Cadiz A.M. and P.M. on Jan. 15th and 30th.
4. ,, ,, H.W. at Ferrol A.M. and P.M. on Jan. 1st and 14th.
5. ,, ,, H.W. at Oporto A.M. and P.M. on Jan. 15th and 16th.
6. ,, ,, H.W. at Ecrehous A.M. and P.M. Jan. 11th and 28th.
7. ,, ,, H.W. at Poole A.M. and P.M. on Jan. 7th and 19th.
8. ,, , H.W. at Selsea Bill A.M. and P.M. on Jan. 4th and 19th.
9. ,, ,, H.W. at Dieppe A.M. and P.M. on Jan. 5th and 21st.
10. ,, ,, H.W. at Southampton A.M. and P.M. on Jan. 4th and 21st.
11. ,, ,, H.W. at Exmouth A.M. and P.M. on Jan. 12th and 26th.
12. ,, ,, H.W. at Penzance A.M. and P.M. on Jan. 14th and 27th.
13. ,, ,, H.W. at Brisbane A.M. and P.M. on Jan. 23rd and 29th.
14. ,, ,. H.W. at Astoria A.M. and P.M. on Jan. 2nd and 9th.

NOTE.—For the last two exercises, take Brest as the Standard Point.
   H.W. Full and change at Brest $3^h\ 47^m$
   H.W. ,, ,, Brisbane 11 0
   H.W. ,, ,, Astoria 0 42

CHAPTER V.

REDUCTION OF SOUNDINGS.

ART. 12.—Near to land in foggy weather, the greatest security for the navigator is the frequent use of the "lead," and feeling his way by "ground navigation." A line of soundings and specimens of the bottom (brought up by the "arming" of the lead) will generally enable him to fix his position on the chart, or give him due warning of danger.

In Board of Trade inquiries into cases of stranding, the first importance is properly attached to the taking of soundings, and woe to the captain who may be found to have neglected this precaution.

The soundings marked on charts are from the Mean Low-water Level of spring tides, but the actual depth at a place varies with Springs and Neaps and the interval from high water. The difference between the sea-level at the "time of cast" and the Mean Low water of Springs is the *Reduction of Soundings*.

The Heights of Tides in the Admiralty Tables are measured from the same zero or datum line as used for the chart soundings, but some local tables, *e.g.* the Liverpool or Holden's Tables, give the heights from a datum line five feet below the Mean Sea-level.

In the diagram, let O (zero) be the mean level of low water spring tides, H the mean high water springs, and M the mean level of the sea or half-tide level; also let h_1, h_2, h_3, h_4 be the high-water marks of various tides, and l_1, l_2, l_3, l_4 the corresponding low-water marks, and B the sea-bottom.

Now, if the lead were cast when the sea-level is at O, it would give the same sounding

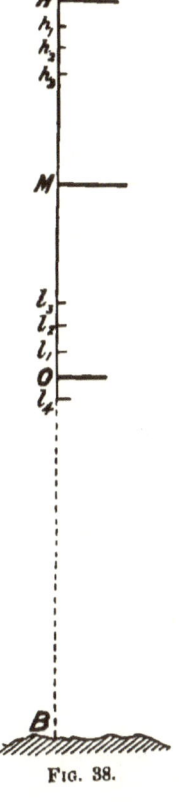

FIG. 38.

as the chart, but at other times there would be a difference depending on the height of the tide in the tables and the interval from high water. This difference is the Reduction of Soundings.

The Half Range of any tide is found by subtracting the half mean spring range (H.M.S.R.) from the height, *e.g.* $Mh_1 = Oh_1 - OM$.

When the half-range and the time from high water are known, Table B in the Admiralty Tables (p. 98) gives the difference of level between M and the actual surface of the sea, which, being added to or taken from the H.M.S.R. (according as the interval from high water is under or over three hours), gives the height above mean low water, or *zero*.

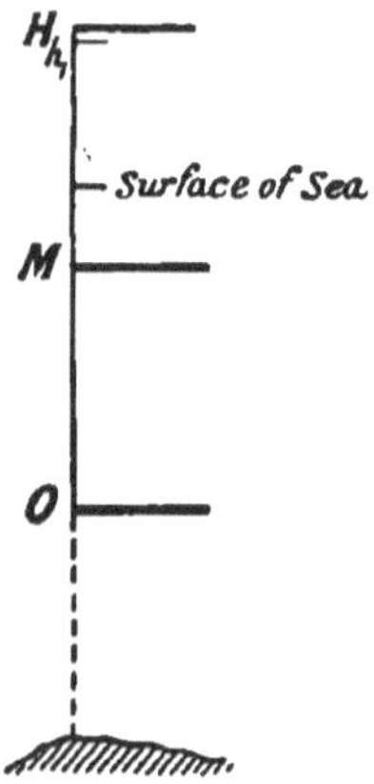

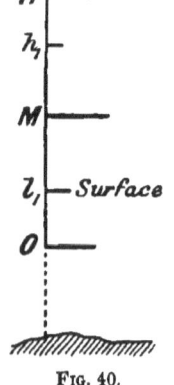

Fig. 39.

Fig. 40.

Examples.—

1. Required the Reduction of Soundings at Holyhead at 6 P.M. on May 15, 1897.

	h. m.		ft. in.
Time given ...	6 0 A.M.	Height	15 10
Time of H.W.	8 36 A.M.	H.M.S.R.	8 0
Time from H.W.	2 36	Half-range	7 10

Table B with $2\frac{1}{2}$ hrs. and 8 ft. gives +2 1
H.M.S.R. 8 0

Ans. 10 1 to be subtracted from cast.

2. 1897, June 21, 9^h 30^m P.M.: required the Correction to be applied to the soundings by lead-line before comparing with the chart at Liverpool.

	h. m.		ft. in.
Time of H.W.	3 56 P.M.	Height	21 11
Time of cast	9 30 P.M.	H.M.S.R.	13 9
Time from H.W.	5 34	Half-range	8 2

Table B with $5\frac{1}{2}$ hrs. and 8 ft. gives −7 9
H.M.S.R. ... 13 9

Ans. 6 0 to be subtracted from cast.

REDUCTION OF SOUNDINGS.

3. March 19, at 1ʰ 50ᵐ A.M., mean time at place: required the Correction of Soundings at Portishead.

	h. m.		ft. in.
Time of H.W.	7 18 A.M.	Height	45 0
Time of cast	1 50 A.M.	H.M.S.R.	21 0
Time from H.W.	5 28	Half-range	24 0

Table B with 5½ hrs. and 24 ft. gives   −23 2
                                  H.M.S.R.   21 0

                                       Ans. −2   2 to be added to cast.

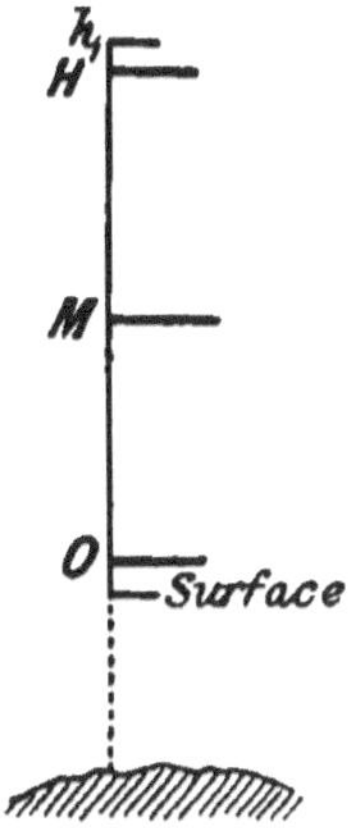

NOTES.—(a) In this case the level of the sea is below the mean low-water level.

(b) The above results may also be used to find the actual depth of water at a given time when the chart soundings are given.

When the place mentioned is not near any Standard port, the correction is first worked out for the Standard port, and then reduced by proportion to the given place.

FIG. 41.

Example.—Required the Reduction of Soundings at Barmouth at 5 A.M., on December 3, 1897.

                                                  h. m.
Time of H.W. at Holyhead   5 29 A.M.
        Constant for Barmouth   2 25

        Time of H.W. at Barmouth   3   4 A.M.
                   Time given   5   0 A.M.

    Time from H.W. at Barmouth   1 56

                                          ft. in.
                Height for Holyhead   13   2
                        H.M.S.R.   8   0

                            Half-range   5   2

Table B with 5 ft. and 2 hrs. gives   +2   6
                             H.M.S.R.   8   0

               Ans. for Holyhead   10   6

In the Alphabetical Table of Places at the end of the Admiralty Tide Tables the mean range of spring tides at Holyhead and Barmouth are found to be 16 and 14¼.

                              ft. in.
Then 16 : 14¼ : : 10 6 : Ans. for Barmouth.
      4   4   12
     ──   ──   ──────
     64   57   126 in.
                   57
             ──────
              882
              630
             ──────
              7182

46   TEXT-BOOK ON NAVIGATION.

$$64)7182(112 \text{ in.} = 9 \text{ ft. } 4 \text{ in.}$$
$$\underline{64}$$

$$78$$
$$\underline{64}$$

$$142$$
$$128$$

Therefore for Barmouth the required reduction is 9 ft. 4 in., to be subtracted from the cast.

Second method, by using a height constant.

         ft. in.
Height for Holyhead 13 2
Height constant   −2 0 p. 102, Ad. Tables.

Height for Barmouth 11 2
H.M.S.R. for Barmouth   7 0 p. 170, Ad. Tables.

Half-range   4 2
Table B with 2^h and 4 ft. 2 in. gives +2 1
H.M.S.R.   7 0

Height above zero at Barmouth   9 1

EXERCISES.

Find the Correction to be applied to the cast before comparing Soundings with the Chart at the following places and times :—

1. 1897, Jan. 14th, $5^h\ 37^m$ A.M. off Bembridge Point.
2.   ,,   22nd, $2^h\ 20^m$ A.M. off St. Helier, Jersey.
3.   ,,   2nd, $1^h\ 15^m$ P.M. off Needles Point.
4.   ,,   11th, noon, off Bordeaux.
5.   ,,   20th, $5^h\ 42^m$ P.M. off Ile d'Yeu.
6.   ,,   13th, midnight, off Ushant.
7.   ,,   21st, $1^h\ 38^m$ A.M. off St. Malo.
8.   ,,   5th, $8^h\ 50^m$ A.M. off Havre.
9.   ,,   8th, $2^h\ 38^m$ A.M. off Poole.
10. ,,   30th, $3^h\ 15^m$ P.M. off Fowey.
11. ,,   21st, $11^h\ 59^m$ A.M. off St. Mary, Scilly.
12. ,,   12th, $2^h\ 19^m$ P.M. off Selsea Bill.
13. ,,   13th, $11^h\ 25^m$ P.M. off Hurst Cambor.
14. ,,   17th, $10^h\ 49^m$ P.M. off Portland.
15. ,,   26th, $9^h\ 16^m$ A.M. off Gibraltar.
16. ,,   21st, $6^h\ 15^m$ P.M. off Fécamp.
17. ,,   22nd, $8^h\ 0^m$ A.M. off Torbay.
18. ,,   19th, $4^h\ 15^m$ P.M. off Falmouth.
19. ,,   31st, $5^h\ 9^m$ A.M. off West Cowes.
20. ,,   21st, $2^h\ 0^m$ P.M. off Granville.
21. ,,   1st, $3^h\ 5^m$ A.M. off Exmouth.
22. ,,   16th, $1^h\ 11^m$ A.M. off Santander.
23. ,,   18th, $9^h\ 12^m$ A.M. off Cadiz.
24. ,,   5th, $1^h\ 17^m$ P.M. off Bridport.

CHAPTER VI.

TIME.

Art. 13. Time and Arc.—It is often necessary to change Time (^{h} ^{m} ^{s}) into Arc (° ′ ″), or the contrary.

Twenty-four hours are the equivalent of 360°;

therefore 1^h is equivalent to 15°
1^m ,, ,, 15′
1^s ,, ,, 15″
also 4^m are ,, ,, 1°
4^s ,, ,, 1′

Two rules for converting Arc into Time are deduced from the above, viz.—
(1) Divide the ° ′ ″ by 15.
(2) Multiply ,, ,, by 4, and divide by 60.

NOTE.—The latter is generally preferred, because of the easier division.

Example.—Convert 34° 16′ 30″ into Time.

```
   3)34° 16' 30"              34° 16' 30"
                                       4
   5)11 25 30          or   ─────────────
   ─────────────             6,0)13,7 6 0
   2ʰ 17ᵐ 6ˢ                ─────────────
                              2ʰ 17ᵐ 6ˢ
```

EXERCISES.

Find the equivalent values in *Time* of the following *Arcs*:—
2° 59′ 30″ ; 179° 0′ 45″ ; 15° 0′ 15 ′ ; 67° 52′ 55″ ; 0° 17′ 20″ ; 10° 19′ 40″ ; 155° 11′ 36″ ; 93° 0′ 45″ ; 117° 38′ 30″ ; 144° 54′ ; 176° 39′ 35 ; 3° 23′ ; 0° 59′ 45″ ; 0° 0′ 15″ ; 0° 15′ ; 0° 48′ ; 13° ; 168° 9′ ; 61° 48′ 45″.

By reversing the above rules, Time may be changed into Arc, viz.—
(1) Multiply by 15.
(2) Multiply by 60 and divide by 4.

48 TEXT-BOOK ON NAVIGATION.

Example.—Change $7^h\ 32^m\ 15^s$ into arc.

```
  7ʰ 32ᵐ 15ˢ              7ʰ 32ᵐ 15ˢ
        3                       60
  ─────────       or      ─────────
  22 36 45                4)452 15 0
        5                 ─────────
  ─────────               113° 3′ 45″
  113° 3′ 45″
```

EXERCISES.

Change into Arc the following *Time intervals* :—
$1^h\ 12^m\ 30^s$; $10^h\ 19^m\ 55^s$; $8^h\ 0^m\ 10^s$; $6^h\ 59^m\ 18^s$; $11^h\ 59^m\ 58^s$; $3^h\ 0^m\ 19^s$;
$5^h\ 59^m\ 22^s$; $9^h\ 1^m\ 37^s$; $0^h\ 15^m\ 15^s$; $2^h\ 25^m$; $0^h\ 10^m\ 35^s$; $0^h\ 0^m\ 45^s$; $2^h\ 24^m\ 44^s$;
$11^h\ 0^m\ 19^s$; $11^h\ 13^m\ 30^s$; $5^h\ 0^m\ 5^s$; $8^h\ 28^m\ 52^s$; $10^h\ 0^m\ 12^s$; $0^h\ 0^m\ 12^s$;
$0^h\ 1^m\ 40^s$.

ART. 14. **Time and Longitude.**—The rules for Time and Arc may be employed to find—

(1) The difference between Greenwich Time and the Time at a place whose Longitude is given; and

(2) The Longitude corresponding to a given difference of Time.

But it must be borne in mind that, as the Earth rotates from west to east, places in E. longitude have sunlight earlier than Greenwich, and places in W. longitude later. Therefore the Time at a place in E. longitude is always *fast* of or *greater* than the Time at Greenwich, and at a place in W. longitude it is always *slow* of or *less* than Greenwich Time.

ART. 15.—**Difference of Time due to Longitude.**

Examples.—

1. What is the difference between the time at a place in 72° 50′ E. long. and the time at Greenwich ?

```
  3)72° 50′                  72° 50′
  ─────────                        4
  5)24 16 40      or         ─────────
  ─────────                  6,0)29,1 20
  4ʰ 51ᵐ 20ˢ                 ─────────
                             4ʰ 51ᵐ 20ˢ
```

Ans. The time at place is $4^h\ 51^m\ 20^s$ fast of Greenwich time.

2. What is the difference between Greenwich time and the time at a place in 124° 10′ 45″ W. long. ?

```
  3)124° 10′ 45″             124° 10′ 45″
  ─────────────                         4
  5)41 23 35      or         ─────────────
  ─────────────              6,0)49,6 43 0
  8ʰ 16ᵐ 43ˢ                 ─────────────
                             8ʰ 16ᵐ 43ˢ
```

Ans. The time at place is $8^h\ 16^m\ 43^s$ *behind*, or slow, of Greenwich time.

TIME.

EXERCISES.

Find the difference between time at Greenwich and corresponding time in the following longitudes, and say which is the *greater* :—
10° 20′ W. ; 17° 45′ E. ; 110° 17′ E. ; 172° 36′ W. ; 100° 16′ 30″ E. ; 90° 4′ 45″ W.; 0° 7′ 15″ E.; 179° 56′ 45″ W.; 1° 2′ 15″ W. ; 115° 0′ 30″ E. ; 0° 2′ 45″ E. ; 165° 0′ 15″ W.

ART. 16.—Longitude from Difference of Time.

Examples.—

1. The mean time at ship is January $5^d\ 2^h\ 14^m\ 40^s$, when mean time at Greenwich is January $5^d\ 9^h\ 45^m\ 20^s$: required the longitude.

```
M.T.S. January..  5ᵈ 2ʰ 14ᵐ 40ˢ
M.T.G.    ,,      5ᵈ 9  45  20
                  ─────────────
         Difference  7  30  40        or      7ʰ 30ᵐ 40ˢ
                            3                 60
                  ─────────────              ─────────────
                     22  32  0                4)450 40  0
                             5               ─────────────
                  ─────────────               112° 40′  0″
         Long. W. 112° 40′  0″
```

2. The mean time at ship is February 14, $17^h\ 16^m\ 25^s$ when mean time at Greenwich is February 14, $17^h\ 19^m\ 14^s$: required the longitude.

```
M.T.S. February 14ᵈ 17ʰ 16ᵐ 25ˢ
M.T.G.    ,,    14ᵈ 17  19  14
                ──────────────
       Difference   0  2  49       or      0ʰ  2ᵐ 49ˢ
                          3                60
                ──────────────             ─────────────
                    0  8  27                4)2 49  0
                           5               ─────────────
                ──────────────              0° 42′ 15″
       Long. W. 0° 42′ 15″
```

3. Given M.T.G. March $30^d\ 2^h\ 19^m\ 55^s$, and M.T.S. March $29^d\ 21^h\ 16^m\ 58^s$: required the longitude.

```
M.T.G. March 30ᵈ  2ʰ 19ᵐ 55ˢ
M.T.S.   ,,  29ᵈ 21  16  58
             ───────────────
   Difference     5  2  57
                  60
             ───────────────
                 4)302 57  0
             ───────────────
       Long. W.  75° 44′ 15″
```

4. Given M.T.G. December $31^d\ 19^h\ 11^m\ 25^s$, and M.T.S. January $1^d\ 3^h\ 17^m\ 50^s$: required the longitude.

TEXT-BOOK ON NAVIGATION.

$$\begin{array}{lrrrr}
\text{M.T.G. December} & 31^d & 19^h & 11^m & 25^s \\
\text{M.T.S. January} & 1^d & 3 & 17 & 50 \\
\hline
\text{Difference} & 8 & 6 & 25 \\
& & & 60 \\
\hline
& 4)486 & 25 & 0 \\
\hline
\text{Long. E.} & 121° & 36' & 15'' \\
\end{array}$$

5. Given M.T.S. July 2^d 5^h 15^m 45^s, and M.T.G. July 1^d 22^h 20^m 30^s: required the longitude.

$$\begin{array}{lrrrr}
\text{M.T.S. July} & 2^d & 5^h & 15^m & 45^s \\
\text{M.T.G. ,,} & 1^d & 22 & 20 & 30 \\
\hline
\text{Difference} & 6 & 55 & 15 \\
& & & 60 \\
\hline
& 4)415 & 15 & 0 \\
\hline
\text{Long. E.} & 103° & 48' & 45'' \\
\end{array}$$

NOTE.—The student may change Time into Longitude by any method he prefers.

EXERCISES.

Find the Longitudes from the following Simultaneous Times at ship and Greenwich :—

No.	M.T.S.					M.T.G.			
			h.	m.	s.			h.	m. s.
1	1899, January	10^d	7	40	25	1899, January	10^d	1	14 42
2	,, February	14^d	3	22	33	,, February	14^d	10	29 54
3	,, March	31^d	19	8	11	,, April	1^d	2	28 31
4	,, April	20^d	3	13	48	,, April	19^d	19	3 54
5	,, May	13^d	22	46	59	,, May	14^d	4	2 14
6	,, June	1^d	5	3	13	,, May	31^d	21	11 19
7	,, July	23^d	11	12	43	,, July	23^d	11	10 24
8	,, August	1^d	0	36	18	,, July	31^d	23	59 48
9	,, September	30^d	23	58	45	,, October	1^d	0	1 5
10	,, October	29^d	11	13	10	,, October	29^d	22	49 17
11	,, November	22^d	23	31	49	,, November	23^d	1	0 28
12	,, December	1^d	9	10	12	,, November	30^d	22	2 12
13	1898, December	31^d	18	50	40	,, January	1^d	0	11 50
14	,, December	31^d	13	20	25	,, January	1^d	1	16 33

ART. 17.—**To find the Time at Greenwich** when the Time at a place or at ship and the Longitude are known.

Examples.—

1. When it is apparent noon at ship on May 10th in long. 96° W., what is the time at Greenwich?

$$\begin{array}{lrrrr}
\text{A.T.S. May} & 10^d & 0^h & 0^m & 0^s & & 96° \\
& & 6 & 24 & 0 & & 4 \\
\hline
\text{A.T.G. May} & 10^d & 6 & 24 & 0 & & 6,0)38,4 \\
\hline
& & & & & & 6^h\ 24^m \\
\end{array}$$

TIME. 51

2. It is apparent noon at a place in 175° 40′ E. on July 30th : required the time at Greenwich.

A.T. at place, July 30ᵈ 0ʰ 0ᵐ 0ˢ   175° 40′
              11 42 40   4

A.T.G. July 29ᵈ 12 17 20   6,0)70,2 40

                                    11ʰ 42ᵐ 40ˢ

3. The sun is on the Meridian of ship's place in long. 100° 35′ E. on May 15th : required the time at Greenwich.

A.T.S. May 15ᵈ 0ʰ 0ᵐ 0ˢ   100° 35′
            6 42 20   4

A.T.G. May 14ᵈ 17 17 40   6,0)40,2 20

Or 15ᵗʰ 5 17 40 A.M.   6ʰ 42ᵐ 20ˢ

4. When it is 6ʰ 50ᵐ P.M. apparent time on February 10th at ship in long. 75° 50′ W, what is the time at Greenwich ?

A.T.S. February 10ᵈ 6ʰ 50ᵐ 0ˢ   75° 50′
                 6 3 20   4

A.T.G. February 10ᵈ 11 53 20   6,0)30,3 20

                                      5ʰ 3ᵐ 20ˢ

5. At a place in long. 115° 56′ E. it is 7ʰ 30ᵐ A.M. mean time on June 15th : required the Greenwich time.

M.T. at place, June 14ᵈ 19ʰ 30ʰ 0ˢ   115° 56′
                 7 43 44 E.   4

M.T.G. June 14ᵈ 11 46 16   6,0)46,3 44

                                    7ʰ 43ᵐ 44ˢ

NOTE.—A.M. Civil time is changed into Astronomical time by adding the interval of 12 hours between noon and midnight, which makes the Astronomical day one less than the Civil day ; but when it is P.M. the Astronomical day is the same as the Civil.

Astronomical time is always written thus : February 10ᵈ 6ʰ 50ᵐ 0ˢ, or June 14ᵈ 19ʰ 30ᵐ 0ˢ being reckoned from noon onwards to 24 hours.

Civil time is written thus : February 10ᵗʰ, 6ʰ 50ᵐ 0ˢ P.M., or June 15ᵗʰ, 7ʰ 30ᵐ 0ˢ A.M.

EXERCISES.

To find the Astronomical Time at Greenwich.

1889.	Civil time at ship. h. m. s.	Longitude of ship. ° ′ ″
1. May 16th	3 12 30 P.M.	59 16 30 W.
2. May 1st	5 8 10 A.M.	136 22 50 W.
3. June 13th	8 27 35 P.M.	174 14 45 E.
4. June 30th	11 2 49 A.M.	105 31 45 W.
5. June 1st	4 1 50 P.M.	96 17 15 E.
6. February 17th	7 43 28 P.M.	74 44 30 W.
7. February 28th	10 19 18 A.M.	112 14 45 W.
8. August 21st	6 16 40 A.M.	43 13 30 W.
9. August 1st	2 54 14 P.M.	165 55 0 E.
10. August 10th	1 19 55 P.M.	15 50 15 E.
11. October 22nd	9 38 12 A.M.	155 0 45 W.
12. October 29th	10 56 11 P.M.	82 17 15 W.

52   TEXT-BOOK ON NAVIGATION.

ART. 18.—**To find the Time at Ship** when the Time at Greenwich and the Longitude of Ship are known.

Rule.—Subtract the difference of time for West Longitude, and add for East, which is simply the reverse of the preceding rule. The reason has been previously explained (Art. 14).

Examples.—

1. Required the Time at ship in long. 65° E., when the A.T.G. is August 5th, 10ʰ 30ᵐ A.M.

```
A.T.G. August 4ᵈ 22ʰ 30ᵐ          65°
              4 20                 4
              ─────              ──────
A.T.S. August 5ᵈ  2 50 P.M.      6,0)26,0
                                 ──────
                                   4ʰ 20ᵐ
```

2. What is the A.T.S. in long. 120° 15′ W. when the A.T.G. is 2ʰ 35ᵐ P.M. on January 2nd?

```
A.T.G. January 2ᵈ 2ʰ 35ᵐ          120° 15′
               8  1                  4
               ────               ────────
A.T.S. January 1ᵈ 18 34           6,0)48,1·0
 Or January 2ᵈ  6 34 A.M.          ────────
                                     8 1
```

3. Required the Mean Time at ship in long. 64° W. when the M.T.G. is August 15ᵈ 5ʰ 17ᵐ 35ˢ.

```
M.T.G. August 15ᵈ 5ʰ 17ᵐ 35ˢ       Long. 64° W.
              4 16  0                    4
              ──────                 ────────
M.T.S. August 15ᵈ 1  1 35 P.M.       6,0)25,6
                                     ────────
                                       4ʰ 16ᵐ
```

4. Required the Apparent Time at ship in long. 172° 30′ E. when the M.T.G. is September 30ᵈ 7ʰ 31ᵐ 5ˢ.

	h. m. s.	Eq. of T.	H. var.	Long.
M.T.G. Sept. 30ᵈ	7 31 5			
	+10 6	+10ᵐ 0·07ˢ	·808	172° 30′
	────────	6·06	7·5	4
A.T.G. Sept. 30ᵈ	7 41 11			────────
	11 30 0 E.	10 6·13	4040	6,0)69,0 0
	────────		5656	────────
A.T.S. Sept. 30ᵈ	19 11 11		────	11ʰ 30ᵐ
			6 0600	
Or Oct. 1st	7 11 11 A.M.			

5. Required A.T.S. when M.T.G. is December 18ᵈ 14ʰ 17ᵐ 55ˢ, and the longitude of ship 122° 10′ W.

	h. m. s.		Eq. of T.	H. var.	Long.
M.T.G. Dec. 18ᵈ	14 17 55				
	+2 57	19th	2ᵐ 44·55ˢ	1·237	122° 10′
	────────		12·00	9·7	4
A.T.G. Dec. 18ᵈ	14 20 52			────	────────
	8 8 40		+2 56·55	8659	6,0)48,8 40
	────────			11133	────────
A.T.S. Dec. 18th	6 12 12 P.M.			────	8 8 40
				11·9989	

TIME.

EXERCISES.

1. What is the mean time at ship in long. 123° 27′ 30″ E. when the M.T.G. is October 14^d 9^h 15^m 24^s ?

2. Find the A.T.S. when M.T.G. is March 19^d 15^h 4^m 17^s in long. 63° 48′ 45″ W.

3. When the M.T.G. is June 1^d 23^h 19^m 14^s, what is the A.T.S. in long. 95° 46′ W. ?

4. Required the A.T.S. when the M.T.G. is April 30^d 17^h 49^m 54^s, the longitude of ship being 164° 48′ E.

5. What is the M.T.S. in long. 35° W. when the M.T.G. is May 6^d 2^h 3^m 46^s ?

6. Find the A.T.S. in long. 14° 15′ E. when the M.T.G. is January 12^d 12^h 15^m 37^s.

7. Required the A.T.S. in long. 97° 16′ 45″ E. when the M.T.G. is October 19^d 5^h 8^m 39^s.

8. What is the M.T.S. in long. 143° 27′ 30″ W. when the M.T.G. is August 31^d 16^h 37^m 23^s ?

9. When the M.T.G. is July 1 noon, what is the A.T.S. in long. 64° 30′ E. ?

CHAPTER VII.

ELEMENTS FROM THE NAUTICAL ALMANAC.

A GREAT part of the labour involved in the calculations of problems in Nautical Astronomy is the *Reduction of the Elements* from the Nautical Almanac; therefore it is most important that the student should master this part of the work at the outset. The advantage of a thorough acquaintance with the Nautical Almanac is incalculable.

ART. 19. **The Sun's Elements.**—These are given for *Noon at Greenwich*, therefore it is always necessary to know the *Greenwich Astronomical Time*.

When *Apparent Time* is given, the elements are taken from p. I. of the Month; but when Mean Time is given, they are taken from p. II.

The variations or changes in 1 hour are on p. I.

NOTE.—For ease of calculation the minutes and seconds should be reduced to decimals of an hour, but for Navigation purposes the *nearest tenth* is sufficient.

(a) **The Sun's Declination.**
Examples.—
1. Given Apparent Time at Greenwich (A.T.G.), 1899. January $10^d 14^h 24^m$: required the Sun's Declination.

		H. var.
Dec. at App. Noon, January 10th, p. I.,	21° 56' 54"·6 S.	22"·55
Cor. for $14^h 24^m$	−5 24·7	14·4
$14^h 24^m$		
9 36	Reduced dec. 21 51 29·9 S.	9020
		9020
		2255
		6,0)32,4·720
		5 24·7

Or it may be worked back from the next noon, thus—

Dec. at App. Noon, January 11^d	21° 47' 40"·6 S	23"·61
Cor. for $9^h 36^m$	+3 46 ·6	9·6
Reduced dec.	21 51 27·2	14166
		21249
		6,0)22,6·656
		3 46·6

ELEMENTS FROM THE NAUTICAL ALMANAC.

NOTE.—The slight discrepancy is due to the difference in the hourly variations, but it is preferable to work back when the time much exceeds twelve hours.

2. Given Mean Time at Greenwich (M.T.G.), 1899, February 27ᵈ 8ʰ 37ᵐ : required the Sun's declination.

```
                                              H. var.
Dec. at Mean Noon, February 27ᵈ  8° 18′ 31″·4 S.  56″·40
                    Cor. 8ʰ 37ᵐ      8  5·0        8·6
                                  ─────────────
Reduced dec.  8  10  26·4 S.                      33840
                                                  45120
                                                ─────────
                                              6,0)48,5·040
                                                ─────────
                                                  8′ 5″
```

3. Given App. Time at ship, 1899, March 15th, 5ʰ 40ᵐ P.M., in long. 40° 50′ W.: required the Sun's Declination.

```
A.T.S. March 15ᵈ  5ʰ 40ᵐ  0                     40° 50′
       Long.      2  43  20                         4
                ─────────────                  ─────────
A.T.G. March 15ᵈ  8  23  20                   6,0)16,3 20
                                                  2 43 20
                                       H. var.
Dec. at App. Noon  2° 6′ 3″·4 S.       59″·21
    Cor. for 8ʰ 23ᵐ    8 17·3           8·4
                   ──────────         ──────
Reduced dec.       1 57 46·1          23684
                                      47368
                                    ─────────
                                   6,0)49,7·364
                                    ─────────
                                     8′ 17″·3
```

4. Given Mean Time at ship, 1899, April 28th, at 7ʰ 48ᵐ A.M. in long. 120° 45′ W.: required the Sun's Declination.

```
                                       H. var.
M.T.S. April 27ᵈ 19ʰ 48ᵐ               47″·13        120° 45′
      Long.       8  3                  3·85             ·4
                ────────               ─────         ─────────
M.T.G. April 28ᵈ  3 51                 23565         60)483 0
                                       37704          8ʰ 3ᵐ
Dec. at Mean Noon 14° 10′ 18″·6        14139
    Cor. for 3ʰ 51ᵐ      3  1·4       ─────────
                     ──────────      6,0)18,1·4505
Reduced dec.      14 13 20·0          ─────────
                                        3′ 1″·4
```

5. Given A.T.G. March 20ᵈ 11ʰ 35ᵐ : required the Sun's Declination.

```
                                       H. var.
Dec. at App. Noon  0° 7′ 32″·3 S.      59″·24
    Cor. for 11ʰ 35ᵐ   11 27·2         11·6
                   ──────────          ─────
Reduced dec.       0  3 54·9 N.        35544
                                       65164
                                     ─────────
                                   6,0)68,7·184
                                     ─────────
                                     11′ 27″·2
```

56 TEXT-BOOK ON NAVIGATION.

6. Given M.T.G., 1899, September 22^d 21^h 54^m: required the Sun's Declination.

<pre>
24h 0m   Dec. at Mean Noon, Sept. 23d 0° 5' 14"·5 S.   H. var.
21 54   Cor. for 2h 6m   2 2·8   58"·46
 2   6   2·1
               Reduced dec.   0 3 11·7 S.   ─────
                                                                 5846
                                                                11692
                                                                ──────
                                                              6,0)12,2·766
                                                                ──────
                                                                 2' 2"·8
</pre>

EXERCISES.

1. Required the Sun's Declination for the following times at Greenwich :—

A.T.G. 1899, February 15^d 5^h 41^m ; March 19^d 22^h 54^m ; April 17^d 0^h 16^m ; May 25^d 18^h 10^m ; March 20^d 14^h 25^m.
M.T.G. 1899, June 13^d 8^h 37^m ; July 19^d 3^h 44^m ; September 22^d 19^h 48^m ; March 20^d 7^h 42^m ; December 21^d 10^h 13^m ; November 10^d 21^h 3^m.

(b) Equation of Time.

Examples.—
1. Given M.T.G., 1899, May 25^d 10^h 12^m 50^s: required the Equation of Time.

<pre>
                                                             H. var.
          Eq. of T. at Mean Noon 3m 17s·38   0s·238
                     Cor. for 10h 12m   2·43   10·2
                                        ──────   ─────
          Reduced Eq. of T.   3 14·95   476
                                                               238
                                                              ──────
                                                              2·4276
</pre>

To be added to Mean Time, or subtracted from App. Time.

2. At 5^h 15^m P.M. App. Time at ship on April 15th, 1899, in long. 56° 50' W., what is the Equation of Time?

<pre>
56° 50'   A.T.S. April 15d 5h 15m 0s
  4   Long.   3 47 20
──────   ──────────
60)22,7 20   A.T.G. April 15d 9 2·20
3h 47m 20s   H. var.
               Eq. of T. App. Noon 0m 3s·45   0s·615
60)2.3   Cor. for 9h 2m 20s   5·56   9·04
──────   ──────   ──────
 ·04   Reduced Eq. of T.   0 2·11   2460
                                                              5535
                                                             ──────
                                                             5·55960
</pre>

To be subtracted from App. Time, or added to Mean Time.

3. 1899, September 1st 3^h 47^m P.M., M.T.S. in long. 179° 30' E. : required the Equation of Time.

ELEMENTS FROM THE NAUTICAL ALMANAC. 57

```
                                           H. var.
M.T.S. September 1ᵈ  3ʰ 47ᵐ               0·782
        Long.    11 58                     8·2                179° 30'
                                          ─────                ─────────
M.T.G., August 31ᵈ  15 49                  1564               6,0)71,8   9
                                           6256               ─────────
     or    8 11 less Noon, September 1st  ─────                11ʰ 58ᵐ
                                           6·4124
Eq. of T. September 1ᵈ  0ᵐ 5·00ˢ
        Cor. for 8ʰ 11ᵐ     6·41
                          ─────
                          0 1·41
```
To be subtracted from Mean Time, or added to App. Time.

4. 1899, January 1st, 9ʰ 26ᵐ A.M., Mean Time at ship in long. 16° 30′ E. :
Required the Equation of Time.

```
                                           H. var.
M.T.S. December 31ᵈ  21ʰ 26ᵐ              1··178
        Long.     1   6                    3·7
                                          ─────                16° 30'
M.T.G. December 31ᵈ  20 20                 8246               ─────────
                                           3534               6,0)6,6  0
     or    3 40 less Noon, January 1.     ─────                ─────────
                                           4·3586              1ʰ 6ᵐ
Eq. of T. Mean Noon 1st  3ᵐ 47ˢ·24
        Cor. for 3ʰ 40ᵐ    - 4·36
                          ──────
Reduced Eq. of T.  3 42·88
```
To be subtracted from Mean Time, or added to App. Time.

EXERCISES.

1. Reduce the Sun's Declination and the Equation of Time to the following Mean Times at Greenwich, from the Nautical Almanac for 1899:—

(a) January 15ᵈ 6ʰ 24ᵐ. (d) April 15ᵈ 8ʰ 48ᵐ.
(b) February 26ᵈ 0ʰ 42ᵐ. (e) June 21ᵈ 19ʰ 36ᵐ.
(c) March 20ᵈ 10ʰ 18ᵐ. (f) September 22ᵈ 20ʰ 54ᵐ.

2. Correct the Sun's Declination and the Equation of Time for the following Apparent Times at Greenwich, and find the corresponding Mean Time at Greenwich :—

(a) 1899, March 20ᵈ 18ʰ 12ᵐ. (d) August 31ᵈ 19ʰ 2ᵐ.
(b)   ,,   April 15ᵈ 5ʰ 54ᵐ. (e) September 23ᵈ 4ʰ 20ᵐ.
(c)   ,,   June 30ᵈ 14ʰ 3ᵐ.   (f) December 24ᵈ 14ʰ 10ᵐ.

(c) The Sun's Right Ascension.
Explanation.—

(a) *Apparent Right Ascension* on p. I. of the Month means the Right Ascension (R.A.) of the Real Sun at *Apparent* Noon, Greenwich Time.

(b) *Apparent Right Ascension* on p. II. means the R.A. of the Real Sun at *Mean* Noon at Greenwich.

(c) *Sidereal Time* on p. II. is the same as the R.A. of the Mean Sun (or supposed Sun moving uniformly on the Equinoctial) at the moment of Mean Noon, Greenwich Time.

TEXT-BOOK ON NAVIGATION.

NOTE.—The variation in 1 hour for (a) and (b) is on p. I., but (c) is most conveniently corrected by the table on p. 586 of the Nautical Almanac (1899).

Examples.—

1. What is the R.A. of the Apparent Sun at A.T.G. 1899, June $24^d\ 17^h\ 54^m$?

		Var. 1 hour.
R.A. at Noon	$6^h\ 11^m\ 49^s\cdot54$	$10^s\cdot388$
Cor. for $17^h\ 54^m$	3 5·94	17·9
Reduced R.A.	6 14 55·48	93492
		72716
		10388
		6,0)18,5·9452
		$3^m\ 5^s\cdot94$

2. Required the R.A. of Apparent Sun for M.T.G. 1899, July $5^d\ 5^h\ 46^m\ 55^s$.

		H. Var.
R.A. at Noon	$6^h\ 57^m\ 21^s\cdot41$	$10^s\cdot295$
Cor. for $5\cdot8^h$	59·71	5·8
Reduced R.A.	6 58 21·12	82360
		51475
		59·7110

3. Find the Mean Sun's R.A. for 1899, October $22^d\ 9^h\ 40^m\ 45^s$ M.T.G.

Sid. Time or R.A. of Mean Sun at Noon	$14^h\ 2^m\ 47^s\cdot99$
P. 586. Acceleration for 9^h	1 28·71
,, 40^m	6·57
,, 45^s	·12
R.A. of Mean Sun	14 4 23·39

EXERCISES.

1. Find the R.A. of Apparent Sun and Mean Sun for January $15^d\ 6^h\ 24^m$ M.T.G.
2. What is the R.A. of App. Sun for May $17^d\ 18^h\ 25^m$ A.T.G. ?
3. Find the R.A.'s of App. Sun and Mean Sun for February $26^d\ 0^h\ 42^m$ M.T.G.
4. Required the R.A.'s of App. Sun and Mean Sun for March $20^d\ 10^h\ 18^m$ M.T.G.
5. Find the R.A. of App. Sun when the A.T.G. is July $27^d\ 4^h\ 39^m$.
6. Find the R.A.'s of App. and Mean Suns when the M.T.G. is September $22^d\ 20^h\ 54^m$.
7. What is the R.A. of App. Sun for A.T.G. October $19^d\ 21^h\ 45^m$?

ART. 20.—**The Moon's Elements.**—The elements used in the problems connected with Navigation are (a) Semi-diameter, (b) Horizontal Parallax, (c) Meridian Passage, (d) Right Ascension, (e) Declination. For their reduction the Mean Time at Greenwich is required.

(a) **Semi-diameter** and (b) **Hor. Parallax.**—These are found

ELEMENTS FROM THE NAUTICAL ALMANAC. 59

on p. III. of the Month, and as the corrections are similar they are usually worked out together. When the M.T.G. is less than 12 hours, the S.D. and H.P. are taken out for *Noon*, and corrected for the number of hours past Noon, but when the M.T.G. exceeds 12 hours, they are taken out for *Midnight* and corrected for the number of hours past Midnight.

The *Moon's Augmentation* is a correction depending on the *Altitude*, and the *Reduction* of H.P. depends on the *Latitude* (see definitions). Both are found in the Tables.

Examples.—

1. Find the Moon's S.D. and H.P. for M.T.G. August $15^d\ 9^h\ 36^m$, 1899, the Moon's app. altitude being 45° 40', and the latitude 56°.

S.D.

Noon 15' 48"·15		Noon 15' 48"·15	
Cor. for 9·6^h	6·04	Midnight 15 55·70	
	15 54·19	Diff. for 12 hrs.	7·55
Aug.	+12·00		9·6
	16 6·19		4530
			6795
			12)72·480
			6·04

H.P.

	57' 53"·80	Noon 57' 53"·80	
	22·13	Midnight 58 21·46	
	58 15·93	Diff. for 12 hrs.	27·66
Red.	−7·90		9·6
	58 8·03		16596
			24894
			12)265·536
			22·13

2. Given M.T.G. 1899, September $12^d\ 22^h\ 48^m$, Moon's app. altitude 64° lat. 48° : required the Moon's S.D. and H.P.

S.D.

Midnight 15' 51"·71		15' 51"·71	
Cor. for 10·8^h	5·77	15 58 ·12	
	15 57·48	Diff. for 12 hrs.	6·41
Aug.	+14·8		10·8
	16 12·28		5128
			641
			12)69·228
			5·769

TEXT-BOOK ON NAVIGATION.

H.P.

Midnight 58' 6"·85
Cor. for 10·8ʰ 21·13

58 27·98
Red.   − 6·4

58 21·58

Midnight 58' 6"·85
Noon 58 30 ·33

Diff. for 12 hrs. 23·48
                   10·8

18784
2348

12)253·584

21·132

EXERCISES.

With the following data, correct the Moon's Semi-diameter and Horizontal Parallax :—

No.	M.T.G.	App. altitude.	Latitude.
		°	°
1	January 17ᵈ 14 19 24	40	53 N.
2	February 5ᵈ 16 29 3	32	26 N.
3	March 25ᵈ 8 49 16	26	52 N.
4	April 7ᵈ 13 26 15	47	30 S.
5	May 4ᵈ 19 12 11	50	Equator
6	June 29ᵈ 6 47 16	60	19 S.
7	July 22ᵈ 3 14 0	36	57 S.
8	August 16ᵈ 20 36 48	5	60 N.
9	September 25ᵈ 15 48 30	52	80 N.

(c) **Meridian Passage.**—In the column headed "Meridian Passage," on p. IV. of the Month, is given the Mean Time at Greenwich (Astronomical), when the Moon crosses the Meridian of Greenwich; in other words, the *Interval* between the Sun's and the Moon's crossing. But as the Earth rotates from W. to E., and the Moon is constantly changing its angular distance eastward from the Sun (about 12° daily), it follows that this Interval is *less* when the Moon crosses the Meridian of a place in E. Longitude, and *greater* for a place in W. Longitude, than the time given in the Nautical Almanac. The change for 24 hours is found by comparing with the *following* day for W. *Longitude*, and the *preceding* day for *E. Longitude*. The proportion of this change for the "Longitude in Time" is the *correction* required.

The M.T.G. is found by applying the "Longitude in Time" in the ordinary way.

NOTE.—(a) It is usual to ask for the time of Mer. pass. on a given *Civil Day*; therefore, if the *Corrected Time* of Mer. pass. *exceeds* 12 *hours*, it must be taken out for the *preceding* day, but for the *same* day if *under* 12 hours.

(b) The "Lower" Mer. pass. is used only when the Moon is *below* the pole.

ELEMENTS FROM THE NAUTICAL ALMANAC.   61

Examples.—
1. 1899, February 14, Civil Time in long. 78° 15′ W. : required the time of Moon's Meridian passage ; also the M.T.G.

```
Mer. pass. Feb. 14ᵈ  3ʰ 40ᵐ·8   Increase to following day 50ᵐ·2        78° 15′
    Cor. for long.     +10·9                                5·2            4
                      ───────                              ─────         ─────
M.T. at place 14      3  51·7                              1004       6,0)31,3  0
    Long.             5  13·0                              2510          ──────
                      ───────                              ─────          5ʰ 13ᵐ
M.T.G.  14            9   4·7                            4)261·04
                                                          ──────
                                                         6)65·26
                                                          ──────
                                                          10·88
```

2. 1899, March 1, Civil Time at ship in long. 156° E. : required the ship's Mean Time of the Moon's Meridian passage ; also the M.T.G.

```
Mer. pass. Feb. 28ᵈ  14ʰ 33ᵐ·8  Decrease to preceding day 45ᵐ·0       156°
    Cor. for long.    −19·5                                10·4         4
                     ────────                             ─────       ─────
M.T.S. Feb. 28       14  14·3                             1800      6,0)62,4
    Long.            10  24·0                              450        ──────
                     ────────                             ─────        10ʰ 24ᵐ
M.T.G. Feb. 28        3  50·3                           4)468·00
                                                         ───────
                                                        6)117·00
                                                         ───────
                                                          19·5
```

NOTE.—When two asterisks (**) are seen in the upper Meridian passage column, it is near New Moon, and therefore no observation can be made.

EXERCISES.

From the following data find the M.T.S. of the Moon's Meridian passage, and the corresponding M.T.G. :—

1. 1899, April 18th, Civil Time in long. 172° 30′ W.
2.   ,,   May 5th,   ,,   ,,   96° 25′ W.
3.   ,,   June 28th,   ,,   ,,   125° 25′ E.
4.   ,,   June 13th,   ,,   ,,   89° 45′ E.
5.   ,,   July 1st,   ,,   ,,   142° 48′ E.
6.   ,,   August 21st,   ,,   ,,   162° 54′ W.
7.   ,,   September 19th, ,,   ,,   179° 20′ E.
8.   ,,   January 7th,   ,,   ,,   157° 32′ W.
9.   ,,   February 1st,   ,,   ,,   96° 42′ E.
10. ,,   March 12th,   ,,   ,,   100° 48′ E.
11. ,,   April 18th,   ,,   ,,   40° 54′ W.
12. ,,   July 28th,   ,,   ,,   33° 16′ W.
13. ,,   August 4th,   ,,   ,,   169° 40′ W.
14. ,,   October 18th,   ,,   ,,   24° 30′ E.
15. ,,   December 23rd, ,,   ,,   55° 35′ E.

(d and e) **The Moon's Right Ascension and Declination.**—The Right Ascension and Declination of the Moon are given on pp. V. to XII. of the Nautical Almanac for every hour of M.T.G. up to 23, and the "Variations" are given for 10^m.

In reducing these elements to M.T.G., it is convenient to find the var. in 1^m by removing the decimal point one place to the left, and then multiplying by the number of minutes and decimals of a minute.

Examples.—

1. Reduce the Moon's R.A. to M.T.G., 1899, May 3^d 16^h 35^m 42^s.

	h.	m.	s.	Var. in 1^m.
Moon's R.A. for 16^h	22	10	24·82	2·173
Cor. for $35·7^m$	1	17·58		35·7
Reduced R.A.	22	11	42·40	15211
				10865
				6519
				6,0)7,7·5761
				1 17·58

2. Find the Moon's R.A. for M.T.G., 1899, June 23^d 22^h 0^m 45^s.

	h.	m.	s.	Var. in 1^m.
Moon's R.A. for 22^h	19	0	11·54	2·624
Cor. for $0^m·75$			1·97	0·75
Reduced R.A.	19	0	13·51	13120
				18368
				19·6800

3. Required the Moon's Dec. for M.T.G., August 7^d 15^h 13^m 30^s.

	°	′	″	Var. in 1^m.
Moon's Dec. for 15^h	6°	3′	18″·8 N.	11″·615
Cor. for $13^m·5$		2	36·8	13·5
Reduced Dec.	6	0	42·0	58075
				34845
				11615
				6,0)15,6·8025
				2′· 36″·80

4. Required the Moon's Dec. for M.T.G., October 15^d 21^h 37^m 18^s.

	°	′	″	Var. in 1^m.
Moon's Dec. for 21^h	0°	4′	16″·1 S.	14·511
Cor. for $37^m·3$		9	1·3	37·3
Reduced Dec.	0	4	45·2 N.	43533
				101577
				43533
				6,0)54,1·2603
				9′ 1″·26

ELEMENTS FROM THE NAUTICAL ALMANAC. 63

EXERCISES.

Reduce the Moon's R.A. and Dec. to the following Mean Times at Greenwich:—

1. 1899 February 23^d 12^h 14^m 36^s.
2. ,, March 3^d 5^h 49^m 25^s.
3. ,, April 9^d 22^h 0^m 54^s.
4. ,, June 1^d 0^h 19^m 12^s.
5. ,, June 15^d 13^h 30^m 49^s.
6. ,, July 7^d 23^h 52^m 19^s.
7. ,, January 17^d 14^h 19^m 24^s.
8. ,, February 5^d 16^h 29^m 3^s.
9. ,, March 25^d 8^h 49^m 16^s.
10. ,, April 7^d 13^h 26^m 15^s.
11. ,, May 4^d 19^h 12^m 11^s.
12. ,, June 29^d 6^h 47^m 16^s.

ART. 21.—**Planets' Elements.**—The Planetary Elements used in Navigation are—(*a*) Right Ascension, (*b*) Declination, (*c*) Semi-diameter, (*d*) Horizontal Parallax, and (*e*) Meridian Passage. They are all given, on the same page of the Nautical Almanac, for Mean Noon at Greenwich.

For reducing the *R.A.* and *Dec.* to M.T.G., the "Var. in 1 hour" may be found by dividing the Change in 1 day by 24.

(*a*) **Right Ascension.**

Example.—Required the R.A. of the Planet Mars for M.T.G. 1899, April 10^d 14^h 40^m 50^s.

```
                   h.  m.   s.                           h.  m.   s.
R.A. April 10th    8   3   10·62         R.A. 10th       8   3   10·62
   Cor. for 14·7ʰ         56·01            ,,  11th     8   4   42·12
                   ─────────────                        ─────────────
Reduced R.A.       8   4    6·63         Change in 1 day    1  31·50
                                                                60
                                                           ─────────
                                                         4)91·50
                                                         ─────────
                                                         6)22·87
                                                         ─────────
                                         Var. in 1 hour    3·81
                                                          14·7
                                                         ─────────
                                                          2667
                                                          1524
                                                           381
                                                         ─────────
                                                         56 007
```

(*b*) **Declination.**

Example.—Required the Dec. of the Planet Mars for the same M.T.G. as above.

64   TEXT-BOOK ON NAVIGATION.

```
Dec. April 10th  22° 57' 25" N.        December 10th  22° 57' 25"
Cor. for 14ʰ·7   −3   37·6                  ,,    11th  22 51 29·8
                 ─────────                              ──────────
Reduced Dec.    22 53 47·4             Change in 1 day   5 55·2
                                                        60
                                                       ─────
                                                      4)355·2
                                                       ─────
                                                      6)88·8
                                                       ─────
                                       Var. in 1 hour   14·8
                                                        14·7
                                                       ─────
                                                       1036
                                                        592
                                                        148
                                                       ─────
                                                    6,0)21,7·56
                                                       ─────
                                                       3' 37"·56
```

EXERCISES.

Required the Right Ascensions and Declinations of Planets as under:—

Planet.	M.T.G.
1. Venus	June 1ᵈ 16ʰ 30ᵐ 30ˢ.
2. Mars	May 19ᵈ 8ʰ 22ᵐ 45ˢ.
3. Jupiter	January 6ᵈ 13ʰ 37ᵐ 40ˢ.
4. Saturn	July 4ᵈ 2ʰ 53ᵐ 50ˢ.
5. Uranus	May 9ᵈ 19ʰ 48ᵐ 10ˢ.
6. Mercury	September 28ᵈ 15ʰ 15ᵐ 0ˢ.
7. Jupiter	August 17ᵈ 12ʰ 35ᵐ 50ˢ.

(c) **Semidiameter**, (d) **Horizontal Parallax.**—These Elements do not require to be corrected for M.T.G., because the daily change is very slight.

The *Semidiameters* of Planets are often used in "Lunar Distance" problems.

The *Horizontal Parallax* is required for finding the *Parallax in Altitude*, either from the Table of *Par. in Alt. for Planets* or by Calculation.

Example.—Required the *Semidiameter* and *Horizontal Parallax* of the Planet Venus on January 20th; also the *Par. in Alt.* for 64° of Altitude.

Directions.—On p. 226 of Nautical Almanac (1899), for January 20th, the Semidiameter of Venus is 16'''·4, and the H.P. is 17'''·3. Then enter a Table of Par. in Alt. (XLVIII Norie) with alt. 64° and H.P. 17", and find Par. in Alt. 7".

EXERCISES.

With the following Greenwich dates and apparent altitudes, find the Semidiameter and Parallax in Altitude for each of the Planets named:—

ELEMENTS FROM THE NAUTICAL ALMANAC. 65

Planet.		G. date.		App. alt.
1. Venus	...	1899, February 25th	...	54°
2. Mars	...	,, January 7th	...	62°
3. Jupiter	...	,, April 30th	...	48°
4. Saturn	...	,, May 15th	...	59°
5. Mars	...	,, February 13th	...	65°
6. Venus	...	,, January 20th	...	45°

(e) "**Meridian Passage**" of **Planets.**—This column in the N.A. is consulted to obtain the time for observing the Mer. Alt. of a Planet, for the purpose of finding the Latitude. It also determines whether a planet is East or West of the Meridian at a given time.

The correction for Longitude (when necessary) is similar to that of the Moon's Mer. Pass.

Example.—
1. At what time will the Planet Mars pass the Meridian of a place in 60° E. long. on January 19th, 1899?

```
                        h.  m.                             m.
Mer. Pass. January 19th 12  9·0    Change to pre-  ⎫ +5·6
        Cor. for Long.      + ·9   ceding day     ⎭    4 (long.)
                        ─────────                     ─────
M.T. at place, 19ᵈ 12   9·9                           4)22·4
       Long. in time  4   0                           ─────
                        ─────────                     6) 5·6
       M.T.G. 19ᵈ   8   9·9                           ─────
                                                        ·9
```

2. State whether the Planet Jupiter is E. or W. of the Meridian of Ship in 126° W. at 5ʰ 30ᵐ A.M., M.T.G. on June 6th, 1899.

```
                d.  h.  m.                           m.
Mer. Pass. June 5   9   1·2    Change to  ⎫ -4·2         126°
                       -1·5    next day   ⎭  8·4           4
                    ─────────                             ───
     M.T.S. June 5  8  59·7                  168        6,0)50,4
           Long.    8  24·0                  336         ─────
                    ─────────                            8ʰ 24ᵐ
           M.T.G.  17  23·7                4)35·28
                    ─────────              ─────
        or 6th   5  23·7 A.M.              6)8·82
                                           ─────
                                             1·47
```

Therefore the planet is W. of meridian.

EXERCISES.

1. Required the time of transit of the Planet Saturn over the Meridian of the ship in long. 97° 40′ E. on May 26th, 1899.
2. At what time M.T.S. and M.T.G. would the Meridian Altitude of Jupiter be observed on April 12th, 1899, in long. 125° 45′ W.?
3. Determine whether the Planet Mars is E. or W. of the Meridian on March 6th at 3ʰ 30ᵐ A.M. mean time at ship in long. 165° 40′ E

F

4. At what time is the Planet Venus on the Meridian of 125° W. on December 30th, 1899 ?

5. What is the time for getting a Meridian Altitude of the Planet Jupiter M.T.S. and M.T.G. on March 14th, 1899, in long. 37° W. ?

6. On which side of the Meridian is the Planet Saturn at midnight of June 11th, 1899, in long. 155° E. ?

ART. 22.—**Stars' Elements.**—The only *Stars' Elements* used in Navigation are—(a) the *Right Ascension*, and (b) the *Declination*, and these do not require any corrections.

Directions.—Find the Star in the Table of "Mean Places of Stars," pp. 294–304, and note the *Right Ascension*. This R.A. may be used to find the Star in the Extended Table of *Apparent Places*. Then the R.A. and Dec. are taken out for the nearest date.

Example.—Required the R.A. and Dec. of Sirius (α Canis Majoris) on April 16th, 1899.

On p. 297 α Canis Majoris has R.A. $6^h\ 40^m$. Then, turning to the "Apparent Places" (commencing on p. 356), and following up the R.A.'s, we come to $6^h\ 40^m$ on p. 384. Opposite the date the R.A. is found to be $6^h\ 40^m\ 43^s$, and the Dec. 16° 34' 53" S.

EXERCISES.

Find the R.A. and Dec. of stars as under :—

Name of star.	Given date.
1. Aldebaran	December 8th, 1899
2. Procyon	February 20th, ,,
3. Achernar	September 17th, ,,
4. η Centauri	June 24th, ,,
5. α Scorpii	April 10th, ,,
6. γ Draconis	July 28th, ,,
7. α Pavonis	May 23rd, ,,
8. β Argûs	March 14th, ,,
9. Antares	May 13th, ,,
10. α Gruis	November 4th, ,,

CHAPTER VIII.

SIDEREAL AND SOLAR TIME.

ART. 23.—As a consequence of the Earth's rotation from W. to E., all Celestial bodies have an apparent motion from E. to W., but not all at the same rate. A star completes its apparent revolution in exactly the same time as the Earth takes to turn once on its axis, but the Sun, Moon, and Planets all change their positions with respect to the Stars. The Sun moves to the eastward about 1° daily, and the Moon about 13° daily, whilst a planet may have a motion to the E. or W., and sometimes stands still amongst the Stars.

The *Apparent Motions* of the *Stars* and the *Sun* are those used for measuring *Time*.

A Sidereal Day is the interval between two successive transits of a star over the Meridian of a place, and all Sidereal Days are equal.

A Solar Day is the interval between two successive transits of the Sun's centre, but Solar Days are *not* all equal, as they increase or decrease in length according to the time of the year. The Sun cannot therefore be a *uniform time-keeper*. Yet, as Day and Night depend on the Sun, it is the Natural Controller of time for ordinary purposes, and its position with respect to the Meridian marks *Apparent Time*, as indicated by a sun-dial.

To secure a *Uniform Solar Time* and *Equal Solar Days*, a *Mean Sun* is adopted, which gives *Equal Days* of 24 *Equal Hours* all the year round. Its position—sometimes to the E., sometimes to the W., and sometimes (four times in the year) coinciding in Right Ascension with the true Sun—marks *Mean Time*.

But on account of the progress Eastward (about 1° daily) of this *Mean Sun*, it is evident that it will take a longer time to come round to the Meridian again than would a fixed Star—nearly 4^m, in fact.

Hence the *Mean Solar Day* is nearly 4^m longer than a Sidereal Day, but as both kinds of days are divided into 24 hours, it follows that *Solar* hours, minutes, and seconds are longer than the corresponding divisions of *Sidereal* Time. The exact relative values are given in two Tables in the Nautical Almanac, on pp. 586 and 588.

Any star would do to measure a Sidereal Day, but astronomers have made the Sidereal Day begin at the moment when the point where the Ecliptic cuts the Equator—corresponding to the Sun's position on March 21st, and named the "First point of Aries"—is on the Meridian of a place.

Right Ascension is also measured from this point (see Definition).

N.B.—An hour of any kind, whether of Right Ascension, Sidereal Time, or Solar Time, corresponds to an arc of 15°, because 24 × 15° = 360°.

ART. 24.—**To find Sidereal Time.**—When the *Hour Angle* and the *Right Ascension* of any Celestial Object are known, the *Sidereal Time* may be found as follows:—

(a) **Hour Angle West.**—*Add* the Hour Angle to the Right Ascension (rejecting 24 if the sum exceeds 24 hours).

(b) **Hour Angle East.**—*Subtract* the H.A. from the R.A. (adding 24 hours if necessary).

Illustration.—PMP$_1$ is the Celestial Meridian of a place, and the other projected semicircles from P to P$_1$ are "Hour Circles" (15° apart). A is the "First point of Aries," and the Arc of the Equinoctial M.A. is the Sidereal Time (in this case, 4 hours). But the *same Arc* measured eastwards from A to M is the R.A. of the Meridian; therefore *Sidereal Time* and *Right Ascension of Meridian* are always *equal.*

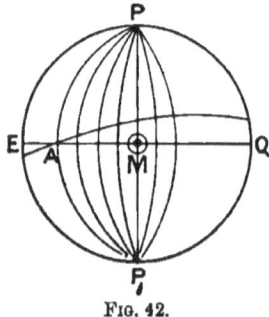

FIG. 42.

NOTE.—The student must regard the Meridian and Hour Circles as fixed, and the "First Point of Aries," as well as all celestial objects, in motion from E. to W. in the direction of the arrow.

Using the initials H.A. for Hour Angle, S.T. for Sidereal Time, R.A. for Right Ascension, and R.A.M. for Right Ascension of Meridian, the above rules may be stated concisely thus:

(a) H.A. West : S.T. or R.A.M. = R.A. + H.A.
(b) H.A. East : S.T. or R.A.M. = R.A. − H.A.

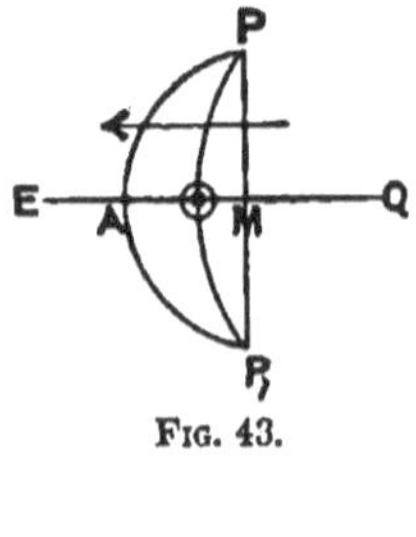

Examples.—

1. Given the Sun's H.A. 3^h 10^m 30^s W. of the Meridian, and Sun's R.A. 14^h 20^m 40^s: required the Sidereal Time.

Here S.T. = R.A + H.A.

Sun's R.A. 14^h 20^m 40^s
Sun's H.A.   3   10   30
————————————
S.T.   17   31   10

FIG. 43.

SIDEREAL AND SOLAR TIME.

2. Given the moon's H.A. 4ʰ 15ᵐ 50ˢ E. of Meridian, and moon's R.A. 10ʰ 19ᵐ 35ˢ : required the S.T.

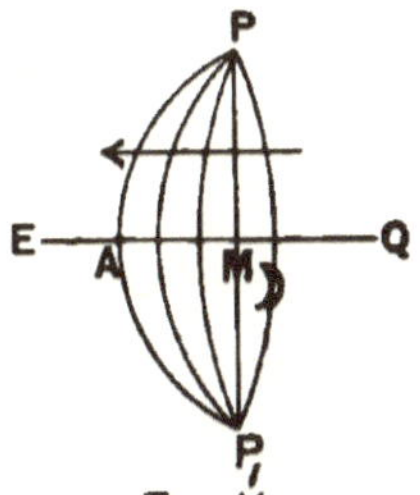

Here S.T. = R.A. − H.A.

Moon's R.A. 10ʰ 19ᵐ 35ˢ
Moon's H.A.   4 15 50
———————————
S.T.   6 3 45

3. Given a star's H.A. 6ʰ 50ᵐ 25ˢ E., and its R.A. 2ʰ 15ᵐ 45ˢ : required the S.T.

Star's R.A. 2ʰ 15ᵐ 45ˢ (add 24ʰ)
Star's H.A. 6 50 25
———————————
S.T. 19 25 20 (always counted westward)

4. Given a Planet's H.A. 4ʰ 12ᵐ 20ˢ W., and its R.A. 22ʰ 44ᵐ 14ˢ : required the S.T.

Planet's R.A. 22ʰ 44ᵐ 14ˢ
Planet's H.A. 4 12 20
———————————
S.T.   2 56 34

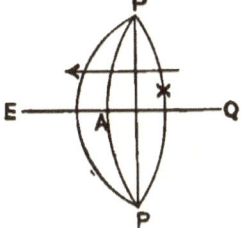

FIG. 44.

FIG. 45.

NOTE.—The above calculation occurs in the 'Lunar' and other problems, and should be clearly understood.

ART. 25.—**To find Sidereal Time from Solar Time.**—Since *Apparent Time* is the *Westerly Hour* Angle of the Sun, and Mean Time is the *Westerly* Hour Angle of the *Mean Sun*, the formula for finding Sidereal Time from the Solar Time will be S.T. = R.A. + H.A.

Examples.—

FIG. 46.

1. Given *Apparent* Time at ship in long. 78° W., 1899, May 25th, 8ʰ 20ᵐ A.M. : required the Sidereal Time.

A.T.S. May 24ᵈ 20ʰ 20ᵐ 0ˢ   78°
                5 12 0   4
              ——————————   ————————
A.T.G.   ,, 25ᵈ 1 32 0   6,0)31,2
                              ————
                              5ʰ 12ᵐ

                                          H. var.
Sun's R.A. at noon   4ʰ 8ᵐ 6ˢ·75   10·096
Cor. for 1ʰ·5   15·14   1·5
                    ————————————   ————————
Sun's R.A. (Reduced) 4 8 21·89   50480
Sun's H.A. (or A.T.S.) 20 20 0ˢ   10096
                    ————————————   ————————
           Sid. time 0 28 21·89   15·1440

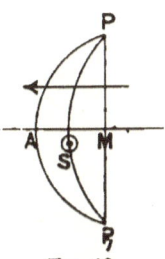

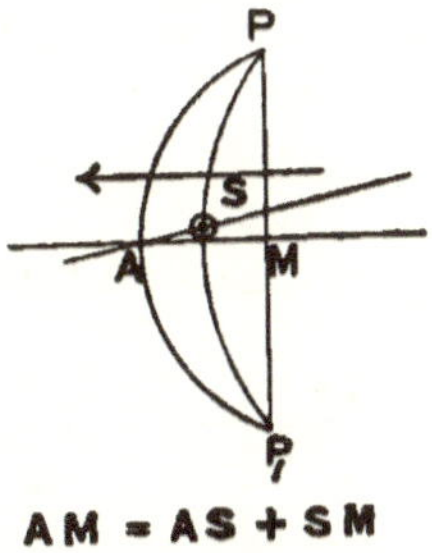

AM = AS + SM

FIG. 47.

TEXT-BOOK ON NAVIGATION.

2. Given Mean Time at ship in long. 84° 35' W., 1899, January 23rd, 4ʰ 36ᵐ A.M.: required the Sidereal Time.

$$\text{M.T.S. January } 22^d \ 16^h \ 36^m \ 0^s \qquad 84° \ 35'$$
$$\underline{ 5 \ 38 \ 20} \qquad \underline{ 4}$$
$$\text{M.T.G.} \quad ,, \quad 22^d \ 22 \ 14 \ 20 \qquad 6,0)33,8 \ 20$$
$$\phantom{\text{M.T.G.} \quad ,, \quad 22^d \ 22 \ 14 \ 20} \qquad \overline{5^h \ 38^m \ 20^s}$$

Mean Sun's R.A. at noon (or Sid. Time) 20ʰ 6ᵐ 28ˢ·51
Cor. for 22ʰ   3   36·84 ⎫
  ,,   14ᵐ   2·30 ⎬ p. 586 N.A.
  ,,   20ˢ   ·05 ⎭

Mean Sun's R.A. (reduced) 20 10   7·70
 ,, ,, H.A. (M.T.S.) 16 36   0

Sid. Time 12 46   7·70

3. Given M.T.G. 1899, February 25ᵈ 20ʰ 16ᵐ 50ˢ, long. 75° 20' E.: required the Sidereal Time.

$$\text{M.T.G. February } 25^d \ 20^h \ 16^m \ 50^s \qquad 75° \ 20'$$
$$\underline{ 5 \ 1 \ 20} \qquad \underline{ 4}$$
$$\text{M.T.S.} \quad ,, \quad 26^d \ 1 \ 18 \ 10 \qquad 6,0)30,1 \ 20$$
$$\phantom{\text{M.T.S.} \quad ,, \quad 26^d \ 1 \ 18 \ 10} \qquad \overline{5^h \ 1^m \ 20^s}$$

S.T. at noon 22ʰ 24ᵐ 27ˢ·94
Cor. for 20ʰ   3   17·13 ⎫
  ,,   16ᵐ   2·63 ⎬ p. 586 N.A.
  ,,   50ˢ   ·14 ⎭

R.A.M.S. 22 27   47·84
M.T.S.   1 18   10·

S.T. 23 45   57·84

NOTE.—The Mean Sun's R.A. at noon is taken from the Sidereal Time column on p. II. of Nautical Almanac because the Mean Sun being on the Meridian at Mean Noon, its R.A. at that moment is the same as the R.A.M. and the S.T. at Mean Noon.

EXERCISES.

Compute the S.T. from the following data :—

			h. m. s.	long. ° '
1. A.T.S.	1899,	February 4th ...	3 15 0 A.M.	162 30 W.
2. M.T.S.	,,	May 10th ...	7 10 0 P.M.	105 45 E.
3. M.T.G.	,,	June 19ᵈ ...	23 14 42	36 30 E.
4. M.T.G.	,,	August 1ᵈ ...	5 15 25	124 37 W.
5. M.T.S.	,,	September 14th ...	8 30 0 A.M.	42 25 W.
6. A.T.S.	,,	November 19th ...	5 40 0 P.M.	155 12 E.

SIDEREAL AND SOLAR TIME.

ART. 26.—**To find the Hour Angle of a Celestial Object.**—When the *Sidereal Time* and the *Right Ascension* of a Celestial object are known, its *Hour Angle* can be found by subtracting one from the other, so as to obtain the least difference between them (borrowing 24 hours when necessary). The object is *West* of the Meridian when R.A. is *less* than S.T., and *East* when *greater*.

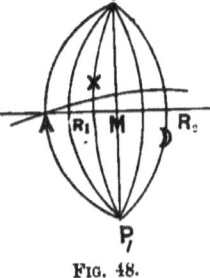

FIG. 48.

Illustration.—A is the first point of Aries. The arc MA is the S.T. of observation, AR_1 is the R.A. of a star west of the Meridian; its H.A. (the arc MR_1) is evidently $= MA - AR_1$, that is, S.T. − star's R.A.

Again, AR_2 is the R.A. of the Moon, and its H.A. (the arc MR_2) $= AR_2 - MA$, that is, Moon's R.A. − S.T., and it is East of the Meridian.

Examples.—

1. Given the S.T. $7^h\ 42^m\ 35^s$, and the R.A. of a star $6^h\ 40^m\ 13^s$: required its H.A.

$$S.T.\ \ 7^h\ 42^m\ 35^s$$
$$Star's\ R.A.\ \ 6\ \ 40\ \ 13$$
$$Star's\ H.A.\ \ 1\ \ \ 2\ \ 22\ W.\ of\ Mer.$$

2. Given S.T. $19^h\ 10^m\ 40^s$, and R.A. of an object $1^h\ 2^m\ 50^s$: required its H.A.

$$S.T.\ \ 19^h\ 10^m\ 40^s$$
$$R.A.\ \ \ 1\ \ \ 2\ \ 50$$
$$H.A.\ \ \ 5\ \ 52\ \ 10\ E.\ of\ Mer.$$

NOTE.—In this example the Sidereal Time is subtracted from the R.A. (which is therefore the greater), because it gives a difference less than 12 hours, and no object can be farther from the Meridian than 12 hours.

ART. 27.—All the Celestial bodies are so related to each other in position that when the H.A. of *one* of them is known, the H.A. of *any other* can be found by means of their known Right Ascensions; because it was shown in *Art.* 24 that S.T. is equal to the sum or difference of the R.A. and H.A. of a given object, and now it has been shown that the H.A. is the difference between the S.T. and the R.A. of any given object.

NOTE.—The H.A. can also be computed from the Altitude, as will be seen hereafter.

TEXT-BOOK ON NAVIGATION.

Examples.—

1. Given R.A. of Star Aldebaran 4ʰ 30ᵐ 7ˢ, and its H.A. 2ʰ 10ᵐ 25ˢ west of Meridian: what is the H.A. of the Star Regulus whose R.A. is 10ʰ 3ᵐ 0ˢ?

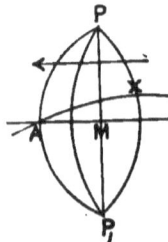

	R.A. Star Aldebaran	4ʰ 30ᵐ 7ˢ
	H.A. ,,	2 10 25 W.
	S.T. (or R.A.M.)	6 40 32
	R.A. Star Regulus	10 3 0
FIG. 49.	H.A. ,,	3 22 28 E. of Mer.

2. Given R.A. of Star Achernar 1ʰ 33ᵐ 57ˢ, and H.A. 4ʰ 50ᵐ 47ˢ east of Meridian: required the H.A. of the Moon when its R.A. is 18ʰ 10ᵐ 25ˢ.

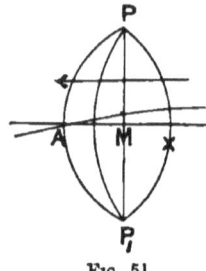

	R.A. Star Achernar	1ʰ 33ᵐ 57ˢ
	H.A. ,,	4 50 47 E.
	S.T. (or R.A.M.)	20 43 10
	R.A. of moon	18 10 25
FIG. 50.	H.A. ,,	2 32 45 W. of Mer.

3. Given M.T.S., 1899, June 13th, at 11ʰ 25ᵐ F.M. in long. 44° 30′ W.: required the Hour Angle of the Star Altair.

M.T.S. June 13ᵈ 11ʰ 25ᵐ	Long. 44° 30′
2 58	4
M.T.G. ,, 13ᵈ 14 23	6,0)17,8·0
	2ʰ 58ᵐ
S.T. at noon 5ʰ 26ᵐ 19ˢ·30	
2 17·99	
3·78	
R.A. Mean Sun 5 28 41·07	
H.A. ,, 11 25 0·00	
S.T. (or R.A.M.) 16 53 41·07	
R.A. Star Altair 19 45 51·28	
H.A. ,, 2 52 10·21 E. of Mer.	

FIG. 51.

4. Given M.T.G., 1899, January 17ᵈ 8ʰ 14ᵐ 20ˢ in long. 80° E.: what is the Meridian Distance or Hour Angle of the Planet Mars.

M.T.G. January 17ᵈ 8ʰ 14ᵐ 20ˢ	80°
5 20 0	4
M.T.S. or H.A.	
Mean Sun, January 17ᵈ 13 34 20	6,0)32·0
	5ʰ 20ᵐ

FIG. 52.

SIDEREAL AND SOLAR TIME.

	h. m. s.	R.A. Mars. h. m. s.	h. m. s.
S.T. at noon	19 46 45·72	8 9 51·92	8 9 51·92
Cor. for 8ʰ	1 18·85	35·01	8 8 9·36
,, 14ᵐ	2·30		
,, 20ˢ	·05	8 9 16·91	1 42·56
			60
R.A. Mean Sun	19 48 6·92		
H.A. ,,	13 34 20		4)102·56
S.T. (or R.A.M.)	9 22 26·92		6)25·64
R.A. Mars	8 19 16·91		For 1 hr. 4·27
			8·2
H.A. ,,	1 3 10·01 W. of Mer.		
			854
			3416
			35·014

ART. 28.—**To find the Solar Time corresponding to a given Sidereal Time.**

Illustration.—Let A be the position of the "First Point of Aries" west of the Meridian at Mean Noon (*i.e.* when the Mean Sun is *on* the Meridian), and A_1 its position at the *given Sidereal Time*. The difference AA_1 is the Sidereal Interval elapsed since Mean Noon.

The equivalent *Solar Interval* SS_1 or Mean Time, is found on p. 588 N.A., and is less than the Sidereal Interval, because the *Sun's apparent diurnal motion* is *slower* than that of the "First Point of Aries," on account of the Sun's R.A. increasing.

NOTE.—AS is the Mean Sun's R.A. at Mean Noon, and A_1S_1 at the given Sidereal Time.

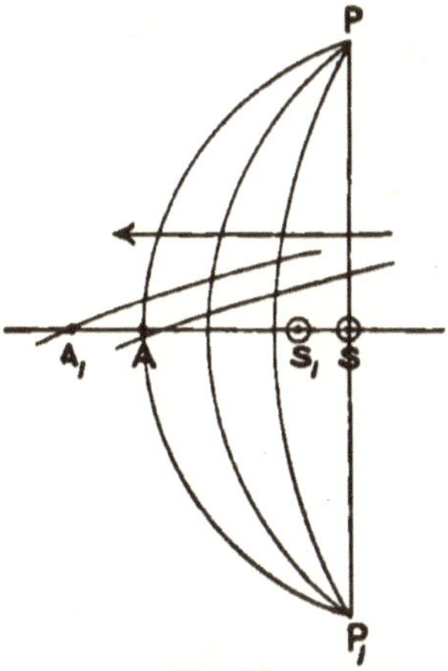

Examples.—

FIG. 53.

1. Required the Mean Time at Greenwich on May 24th, 1899, when the Sidereal Time is 13ʰ 32ᵐ 30ˢ at the same place.

Given S.T. 13ʰ 32ᵐ 30ˢ
R.A. Mean Sun or S.T. at Noon 4 7 28

Sid. Interval since Noon 9 25 2

8 58 31·53 } Equivalent Solar
24 55·90 } Time, p. 588 N.A.
1·99

Mean Solar Int. or M.T.G. 24ᵈ 9 23 29·42

74   TEXT-BOOK ON NAVIGATION.

2. Required the M.T.S. in long. 65° W. on April 15th, when the S.T. is 7ʰ 55ᵐ 20ˢ.

```
        Mean Noon  15ᵈ  0ʰ   0ᵐ   0ˢ
         Longitude  4   20   0
```

Corresponding M.T.G. April 15ᵈ 4 20   0

$$\begin{array}{lrrr} \text{S.T. at Noon} & 1 & 33 & 42\cdot50 \\ \text{Cor. for } 4^h & 0 & 39\cdot43 \\ \text{Cor. for } 20^m & & & 3\cdot28 \end{array} \Big\} \text{p. 586 N.A.}$$

```
    R.A. Mean Sun  1  34  25·21
         Given S.T. 7  55  20·00
```

Sid. Int. since Noon 6 20 54·79

$$\begin{array}{rrr} 5 & 59 & 1\cdot02 \\ & 19 & 56\cdot72 \\ & & 53\cdot85 \\ & & \cdot79 \end{array} \Big\} \begin{array}{l}\text{Equivalent Solar} \\ \text{Time, p. 588 N.A.}\end{array}$$

M.T.S. April 15th 6 19 52·38 P.M.

3. On October 11th civil time, in long. 82° E., the time by a Sidereal Clock being 6ʰ 22ᵐ 50ˢ, required the M.T.S.

```
Mean Noon at ship, October 10ᵈ  0ʰ  0ᵐ  0ˢ         82°
                       Long.    5  28   0            4
```

Corresponding M.T.G. October 9ᵈ 18 32   0   6,0)32,8

                                                    5ʰ 28ᵐ

$$\begin{array}{lrrr} \text{S.T. at Noon} & 13^h & 11^m & 32^s\cdot78 \\ & & 2 & 57\cdot42 \\ & & & 5\cdot26 \end{array}\Big\} \text{p. 586 N.A.}$$

```
    R.A. Mean Sun  13  14  35·46
          Given S.T.  6  22  50·00
```

Sid. Interval 17 8 14·54

$$\begin{array}{rrr} 16 & 57 & 12\cdot90 \\ 7 & 58\cdot69 \\ & 13\cdot96 \\ & \cdot54 \end{array} \Big\} \text{p. 588 N.A.}$$

M.T.S. October 10ᵈ 17 5 26·09

Or October 11th   5   5 26·09 A.M. Civ. T.

EXERCISES.

1. Required the M.T.S. on July 6th, 1899, in long. 124° 40′ W., when the Sidereal Time is 7ʰ 24ᵐ 48ˢ.

2. Required the M.T.G. on August 14th, 1899 (astronomical), when the S.T. is 22ʰ 24ᵐ 30ˢ.

SIDEREAL AND SOLAR TIME.

3. A Sidereal Clock showed $1^h\ 33^m\ 36^s$ at Greenwich on September 30th, 1899 : what is the M.T.G. ?
4. November 1st C.T. at ship in long. 115° 50′ W. : what is the M.T.S. corresponding to $11^h\ 33^m\ 20^s$ Sidereal Time ?
5. Required the A.T.S. on December 16th, 1899, in long. 32° 30′ E., when the Sidereal Time is $9^h\ 10^m\ 36^s$.
6. The Time by Sidereal Clock on January 18th, in long. 174° 30′ E., was $19^h\ 26^m\ 14^s$: required the M.T.S.

ART. 29.—**To compute the Time when a given Star is on the Observer's Meridian.**—The principle of this computation is the same as in *Art.* 28.

Illustration.—Let s be the position of a Star east of the Meridian and approaching it, when the Mean Sun is on the Meridian at M, and the "First Point of Aries" at A.

The Angular Distance SM between s and the Meridian is the Sidereal Interval to elapse before the Star comes on the Meridian at s_1, and this is evidently the difference between the Right Ascension of Meridian (AM) and the Right Ascension of the Star (AS), which, being turned into Solar Time (p. 588 N.A), will give the Mean Time at Ship when the Star is *on* the Meridian.

It is evident that, when the Star is on the Meridian at s_1 the "First Point of Aries" A will have advanced an equal angular distance to A_1 and the Mean Sun to M_1—a less angular distance, because of its slower apparent Diurnal Motion.

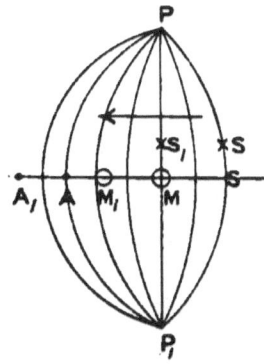

FIG. 54.

MM_1 represents the Hour Angle of the Mean Sun, *i.e.* the Mean Time at Ship.

Examples.—

1. Required the M.T.G. when the Star Procyon is on the Meridian of Greenwich on June 1st, 1899.

$$
\begin{array}{rr}
\text{R.A. of Star} & 7^h\ 34^m\ 2^s \\
\text{R.A. of Mean Sun} & 4\ 39\ 1 \\ \hline
\text{Sid. Interval} & 2\ 55\ 1 \\
\end{array}
$$

$$
\left.\begin{array}{r} 1\ 59\ 40\cdot34 \\ 54\ 50\cdot99 \\ \cdot997 \end{array}\right\} \text{p. 588 N.A.}
$$

M.T.G. June $1^d\ 2\ 54\ 32\cdot327$

2. Find the time when the Star Canopus is on the Meridian of the ship in long. 65° 50′ E. on February 20th, 1899.

76   TEXT-BOOK ON NAVIGATION.

Mean noon at ship February 20ᵈ 0ʰ 0ᵐ 0ˢ   65° 50'
                                4 23 20   4
                                -------   -------
Corresponding M.T.G. 19 19 36 40   6,0)26,3 20

                                       4ʰ 23ᵐ 20ˢ
S.T. at noon 21ʰ 56ᵐ 52ˢ·07
Cor. for 19ʰ   3   7·27⎫
    ,,   36ᵐ   5·91⎬ p. 586 N.A.
    ,,   40ˢ   ·11⎭

R.A. of Mean Sun 22 0   5·36
R.A. of Star   6 21 44·41

Sidereal interval 8 21 39·05

                  7 58 41·36⎫
                   20 56·56⎪ Equiv. solar time,
                       38·89⎬ p. 588 N.A.
                         ·05⎭

M.T.S.* 20ᵈ 8 20 16·86 P.M. civil time.

3. At what time will the Star Sirius be on the Observer's Meridian in long. 35° W. on November 15th, 1899, civil day?

Mean noon at ship, November 14ᵈ 0ʰ 0ᵐ 0ˢ   35°
                   Longitude   2 20 0   4
                              ----------   -----
Corresponding M.T.G. November 14 2 20 0   6,0)14,0

                                               2ʰ 20ᵐ
S.T. at noon 15ʰ 33ᵐ 28ˢ·74
                   19·71⎫
                    3·29⎬ p. 586 N.A.

R.A. Mean Sun 15 33 51·74
R.A. Star   6 40 46·02

Sidereal interval 15 6 54·28

              14 57 32·56⎫
               5 59·02⎪
                 53·85⎬ p. 588 N.A.
                  ·28⎭

M.T.S.* 14ᵈ 15 4 25·71

Or 15th 3 4 25·71 A.M. civil time.

4. At what time on December 30th, 1899, civil time, will the Star Regulus be on the Observer's Meridian, the longitude of the ship being 115° 48' E.?

Mean noon at ship, December 29ᵈ 0ʰ 0ᵐ 0ˢ   115° 48'
                                7 43 12   4
                                --------   -------
Corresponding M.T.G. December 28ᵈ 16 16 48   6,0)46,3 12
                                              ----
                                                7 43 12

SIDEREAL AND SOLAR TIME.

S.T. at noon, December 28 18ʰ 26ᵐ 57·26
                              2 37·70
                                 2·63 } p. 586 N.A.
                                  ·13

R.A. Mean Sun   18 29 37·72

R.A. Star Regulus 10   3   4·64

Sidereal interval 15 33 26·92

                   14 57 32·56
                       32 54·59
                           25·93 } p. 588 N.A.
                             ·92

M.T.S.* 29ᵈ 15 30 54·00

Or 30th 3 30 54 A.M. civil time.

M.T.S. 29ᵈ 15ʰ 30ᵐ 54ˢ
            7 43 12

M.T.G. 29ᵈ 7 47 42

EXERCISES.

1. Required the M.T.S. and M.T.G. when a given Star is on the Observer's Meridian from the following data:—

No.	Name of star.	Civil day at ship.	Longitude.
1	α Hydræ	1899, May 6th	41 48 W.
2	α Ursæ Majoris	„ June 24th	72 15 E.
3	Antares	„ March 31st	115 30 W.
4	Altair	„ April 15th	141 13 E.
5	Achernar	„ February 25th	89 30 W.
6	Aldebaran	„ January 26th	38 25 E.
7	Capella	„ October 10th	108 30 W.
8	β Aurigæ	„ November 9th	0 0
9	α Gruis	„ December 17th	0 0
10	Markab	„ March 1st	175 25 E.
11	α Centauri	„ April 25th	162 20 W.
12	β Orionis	„ July 31st	72 30 E.

ART. 30.—**Bright Stars and Planets.**—Stars which may be considered as "bright," and suitable for Navigation purposes, are those of the first and second magnitudes. In the Table of "Mean Places of Stars" in the Nautical Almanac, there are eighteen of the former and about twice as many of the latter, which gives the Navigator a choice of fifty-four objects suitable for observations, in addition to the Sun, Moon, and Planets.

It is not a work of much labour for the Student, by the help

* The M.T.S. must always be dated the same as the *assumed* Mean Noon, and it must fall on the civil day stated in the question.

of a Star Atlas or a "Planisphere," to make himself familiar with most of the bright Stars and the Constellations to which they belong.

The *Planets* Venus, Mars, Jupiter, and Saturn are sometimes very conspicuous objects, and well adapted for observations, but they vary very much in brightness, according to their distance from the Earth. They may easily be recognized by their wanderings amongst the Stars, or by the times of their Meridian passage as given in the Nautical Almanac.

With such an ample choice of objects, there is no excuse for the Navigator relying upon observations of the Sun only.

ART. 31.—**To find what Bright Stars are near the Meridian, or within a given Hour Angle at a Stated Time.**

Directions.—Find the S.T. corresponding to the given time, then look in the "Table of Mean Places" in the Nautical Almanac for the names of bright Stars whose Right Ascensions do not differ from the S.T. by more than the given Hour Angle.

N.B.—The star is E. of Mer. when its R.A. is *greater* than the S.T. (or R.A.M.), and W. when *less*.

It is N. of the observer when its Declination reads N. of the latitude, and S. when it reads S.

Remark.—As the object in view is to find what Stars would be suitable for finding the latitude by Meridian and Ex-Meridian Altitudes, or for finding the Time and Azimuth, it will be sufficient to take the Sun's R.A. at the *nearest* Noon at Greenwich, since the error would not amount to 2 minutes in any case.

Examples.—

1. What bright Stars are near the Meridian at midnight of December 2^d, 1899, in long. 30° W. ?

A.T.S. December 2^d 12^h 0^m Sun's R.A. 16^h 34^m
Long. 2 0   A.T.S. 12 0
                                           ─────────────
A.T.G. December 2^d 14 0   S.T. (and R.A.M.) 4 34

On p. 296 of Nautical Almanac are found the names of four bright stars, viz. α Aurigæ, β Orionis, γ Orionis, and β Tauri, whose Right Ascensions differ from the R.A.M. by from 22 to 33 minutes, all of them being E. of Mer., and therefore approaching it.

2. 1899, January 20th, 2^h 0^m A.M. A.T.S. in lat. 5° N., long. 48° 30' W.: required the names of bright Stars which are within 2 hours of the observer's Meridian, and state whether E. or W. of Meridian, and whether N. or S. of observer.

A.T.S. January 19^d 14^h 0^m Sun's R.A. 20^h 10^m
Long. 3 14   A.T.S. 14 0
                                           ─────────────
A.T.G. January 19 17 14   R A.M. 10 10

SIDEREAL AND SOLAR TIME. 79

Pages 297-8 (Nautical Almanac) give the following names of bright Stars whose Right Ascensions are within 2 hours of the R.A.M :—

Name of star.	R. A.	Dec.	From Mer.	From observer.
	h. m.	°	h. m.	
ε Argûs	8 20	59 S.	.1 50 W.	S.
β Argûs	9 12	69 S.	0 58 W.	S.
α Hydræ	9 22	8 S.	0 48 W.	S.
α Leonis	10 3	12 N.	0 7 W.	N.
α Ursæ Majoris ...	10 57	62 N.	0 47 E.	N.

3. 1899, May 29^d 3^h 35^m M.T.G. in lat. 23° 12′ S., long. 96° 15′ E. : required the names of bright Stars within 2½ hours of the observer's Meridian, and state whether E. or W. of Meridian, and whether N. or S. of observer.

M.T.G. May 29^d 3^h 35^m R.A. Mean Sun 4^h 27^m
    Long.   6 25   M.T.S. 10 0

M.T.S. May 29 10 00 P.M.   R.A.M. or S.T. 14 27

The Table of Mean Places of Stars (N.A.) gives the following names of bright Stars, whose R.A.'s are within 2½ hours of the R.A.M. as found above :—

Name of star.	R.A.	Dec.	From Mer.	From observer.
	h. m.	° ′	h. m.	
a^1 Crucis ...	12 21	62 32 S.	2 6 W.	S.
γ ,, ...	12 26	56 33 S.	2 1 W.	S.
β ,, ...	12 42	59 8 S.	1 43 W.	S.
α Virginis ...	13 20	10 38 S.	1 7 W.	N.
β Centauri ...	13 57	59 53 S.	0 30 W.	S.
θ ,, ...	14 1	35 52 S.	0 26 W.	S.
α Bootis ...	14 11	19 42 N.	0 16 W.	N.
a^2 Centauri ...	14 33	60 24 S.	0 6 E.	S.
Antares	16 23	26 12 S.	1 56 E.	S.

EXERCISES.

From the following data, give the names of bright Stars which are within 1 hour, 2 hours, and 3 hours respectively of the observer's Meridian, and state whether east or west of Meridian, and north or south of observer :—

No.	Time.	Longtitude.	Latitude.
1	1899, A.T.S. March 1st, 11 44 P.M.	155° W.	42° N.
2	,, M.T.G. April 20^d 19^h 40^m ...	52° W.	40° S.
3	,, M.T.S. June 12th, 2^h 50^m A.M.	168° E.	37° N.
4	,, M.T.G. October 23^d 1^h 25^m ...	145° E.	22° N.

CHAPTER IX.

LATITUDE BY MERIDIAN ALTITUDE.

ART. 32.—When the true *Meridian Zenith Distance* of a Celestial Object, together with its *Declination*, are known, then it can be proved that the *Latitude* is equal to the *Sum* of these, and of the *same* name as both when they are of the *same* name, and to the *difference* with the *name* of the *greater* when of *different* names (see Appendix).

The Meridian Zenith Distance is named opposite to the bearing of the object.

ART. 33.—**Meridian Altitude of the Sun.**

Examples.—

1. 1899, February 27th, in long. 55° 45′ W. the observed Meridian Altitude of the Sun's lower limb was 35° 10′ 30″ bearing south ; index cor., −4′ 50″ ; height of eye, 18 ft. Required the latitude.

	h. m. s.	Dec.	H. var.
A.T.S., February 27ᵈ	0 0 0	8° 18′ 19″ S.	56·40
	3 43 0	3 29	3·7
A.T.G., February 27ᵈ	3 43 0	8 14 50 S.	39480
			16920
			6,0)20,8·680
			3′ 29″

```
55° 45'
    4
6,0)22,3 0
  3h 43m
```

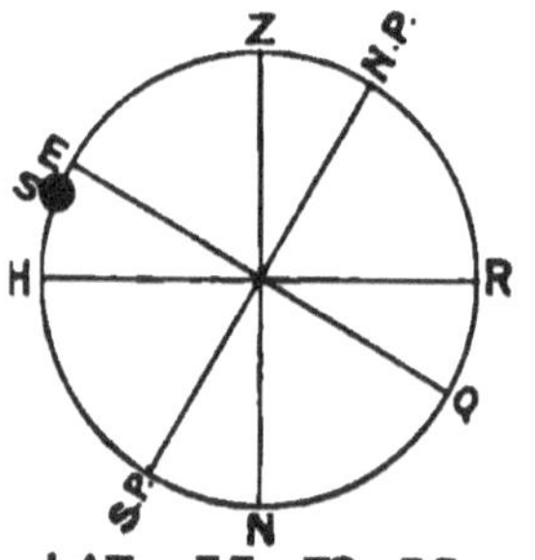

```
Obs. alt. Sun's lower limb   35° 10′ 30″
            Index Cor.        − 4  50
                             ───────────
                              35   5  40
            Dip.              − 4   9
                             ───────────
                              35   1  31
            Refr. − Par.      − 1  14
                             ───────────
                              35   0  17
            S.D.              16  11
                             ───────────
            True alt.         35  16  28
                              90   0   0
                             ───────────
            Mer. Zen Dist.    54  43  32 N.
            Dec.               8  14  50 S.
                             ───────────
            Lat.              46  28  42 N.
```

LAT. = EZ = ZS − ES

Fig. 55.

LATITUDE BY MERIDIAN ALTITUDE.

2. 1899, January 1st, in long. 87° W., the observed Altitude of the Sun's upper limb was 56° 30′ 30″ bearing north; height of eye, 16 ft.; index cor., + 3′ 20″. Required the Latitude.

```
                  h.  m.  s.                                         H. var.
A.T.S. January 1ᵈ  0   0   0    Dec. Jan. 1st 23° 0′ 14″ S.          12·51      87°
       Long. in Time 5  48   0   Cor. for 5ʰ 48ᵐ    −1 13             5·8        4
                                                                              ────
A.T.G. January 1ᵈ  5  48   0    Reduced Dec. 22 59  1 S.             10008   60)34,8
                                                                      6255    5ʰ 48ᵐ
Obs. Alt. Sun's L.L. 56° 30′ 30″                                     ─────
               I.C.   +3 20                                          0,0)7,2·558
                     ────────
                     56 33 50                                           1 13
                 Dip  −3 55
                     ────────
                     56 29 55

Sun's Cor. (Refr. − Par.) −0 33

Sun's Semidiam.           −16 18

True Alt. Sun's Centre 56 13  4
                       90  0  0

Mer. Zen. Dist. 33 46 56 S.
     Dec.       22 59 26 S.

Lat. 56 46 22 S.
```

LAT = EZ = ES + SZ.

FIG. 56.

3. 1899, January 1st, in long. 49° E., the observed Meridian Altitude of the Sun's upper limb was 76° 54′ 40″ bearing south; height of eye, 25 ft.; index cor., −5′ 10″. Required the latitude.

```
                  h.  m.  s.                                         H. var.
A.T.S. January 1ᵈ  0   0   0    Dec. Jan. 1st 23° 0′ 14″ S.          12·51      49°
       Long. in Time 3  16   0   Cor. for 3ʰ 16ᵐ    +41               3·3        4
                                                                              ────
A.T.G. December 31ᵈ 20 44  0    Reduced Dec. 23  0 55 S.             3753    6,0)19,6
                                                                     3753     3ʰ 16ᵐ
Obs. Alt. Sun's U.L. 76° 54′ 40″                                     ─────
               I.C.   −5 10                                          41·283
                     ────────
                     76 49 30
                 Dip  −4 54
                     ────────
                     76 44 36
Cor. (= Refr. − Par.) −0 12
                     ────────
                     76 44 24
         Semidiam.   −16 18
                     ────────
True Alt. Sun's Centre 76 28  6
                       90  0  0

Mer. Zen. Dist. 13 31 54 N.
     Dec.       23  0 55 S.

Lat.  9 29  1 S.
```

LAT = EZ = ES − SZ

FIG. 57.

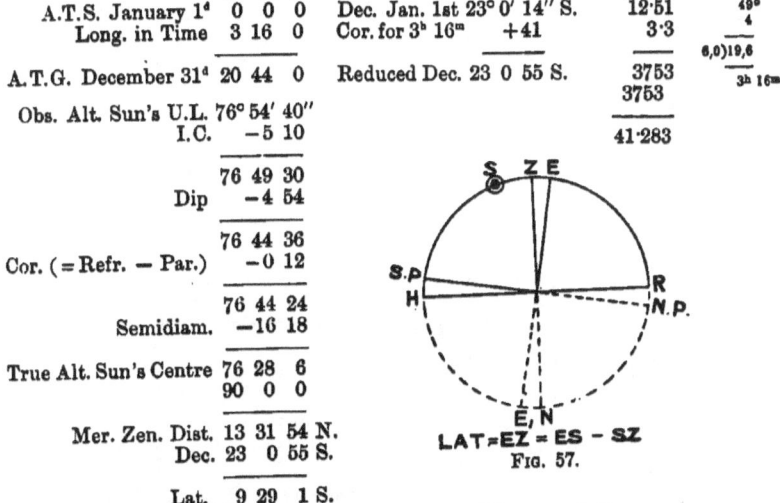

Note that the Dec. is taken out for the *nearest* Noon, and the cor. is added because the Declination is greater before Noon.

G

82   TEXT-BOOK ON NAVIGATION.

4. 1899, June 23rd in Long. 165° 45' W., the observed Meridian Altitude of the Sun's lower limb was 65° 14' 20'' bearing south; height of eye, 14 ft.; index cor., − 6' 25''. Required the latitude.

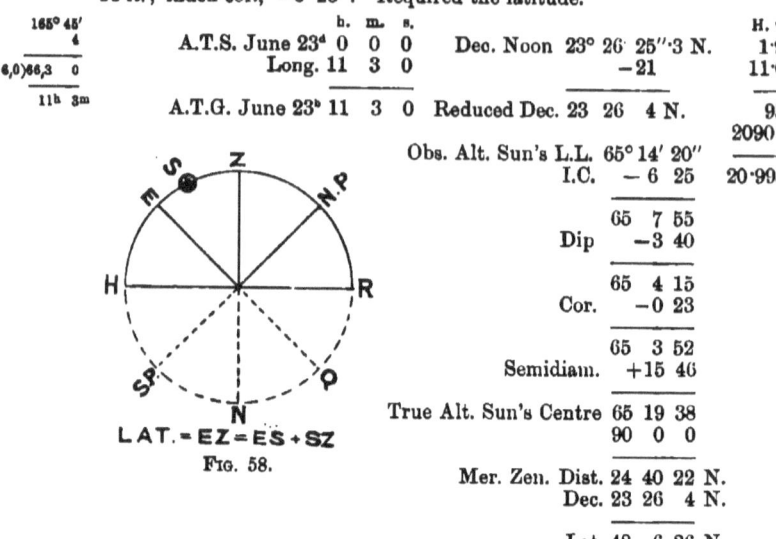

FIG. 58.

```
165° 45'                            h.  m.  s.                                    H. var.
    4            A.T.S. June 23ᵈ    0    0    0    Dec. Noon 23° 26 25''·3 N.      1·90
6,0)66,3  0                 Long. 11    3    0                  −21               11·05
  11ʰ 3ᵐ         A.T.G. June 23ʰ   11    3    0    Reduced Dec. 23 26   4 N.        950
                                                                                   2090
                                         Obs. Alt. Sun's L.L. 65° 14' 20''
                                                         I.C.  − 6  25           20·9950

                                                               65   7  55
                                                         Dip   −3  40

                                                               65   4  15
                                                         Cor.  −0  23

                                                               65   3  52
                                                         Semidiam. +15 46

                                         True Alt. Sun's Centre 65 19 38
       LAT.=EZ=ES+SZ                                            90  0  0

                                         Mer. Zen. Dist. 24 40 22 N.
                                         Dec.             23 26  4 N.

                                         Lat.             48  6 26 N.
```

5. 1899, August 2ᵈ, in long. 159° 30' E., the observed Altitude of the Sun's upper limb was 46° 14' 50'' bearing north; height of eye, 31 ft.; index cor. − 7' 15''. Required the latitude.

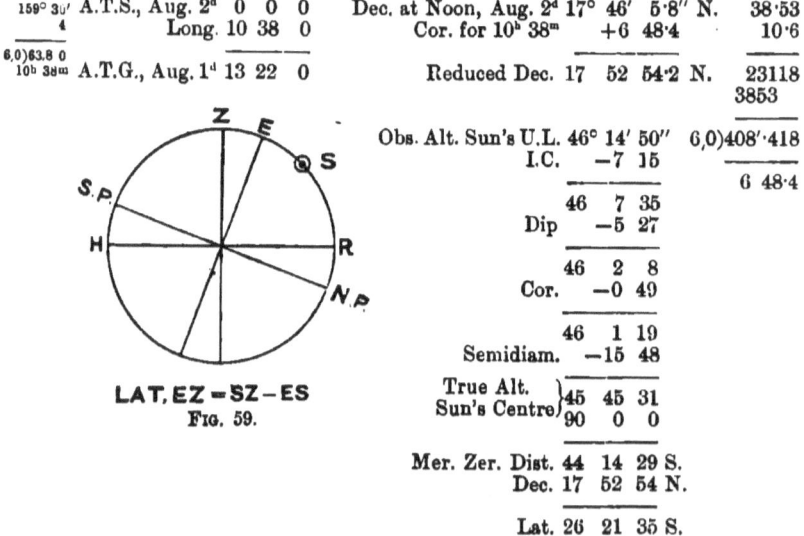

FIG. 59.

```
                                         h.  m.  s.                                      H. var.
159° 30'  A.T.S., Aug. 2ᵈ    0    0    0   Dec. at Noon, Aug. 2ᵈ 17° 46'  5·8'' N.        38·53
     4           Long. 10   38    0          Cor. for 10ʰ 38ᵐ     +6 48·4                 10·6
6,0)63.8 0
  10ʰ 38ᵐ A.T.G., Aug. 1ᵈ   13   22    0    Reduced Dec.       17 52 54·2 N.            23118
                                                                                          3853
                                         Obs. Alt. Sun's U.L. 46° 14' 50''    6,0)408'·418
                                                         I.C.  −7  15
                                                                                 6 48·4
                                                               46   7  35
                                                         Dip   −5  27

                                                               46   2   8
                                                         Cor.  −0  49

                                                               46   1  19
                                                         Semidiam. −15 48

      LAT, EZ =SZ−ES               True Alt.    }45  45  31
                                   Sun's Centre }90   0   0

                                         Mer. Zer. Dist. 44 14 29 S.
                                         Dec.            17 52 54 N.

                                         Lat.            26 21 35 S.
```

LATITUDE BY MERIDIAN ALTITUDE.

NOTE (a).—The student should draw a figure to illustrate each of the following exercises :—

NOTE (b).—When the altitude is observed by Artificial Horizon—apply the Index Cor. and divide by 2, which gives the altitude above the Sensible Horizon, and therefore the "Dip" is omitted.

EXERCISES.

1. 1889, March 20th, in long. 114° 31′ 30″ W., the observed Meridian Altitude of the Sun's lower limb, south of the observer, was 89° 43′ 54″; index error, +4′ 24″; height of eye, 20 ft. Required the latitude.
2. 1899, May 15th, in long. 100° 30′ W., the observed Mer. Alt. of the Sun's lower limb was 11° 14′ 16″, bearing north of the observer; index error, 0′ 55″ on the arc ; height of eye, 17 ft. Required the latitude.
3. 1899, June 21st, in long. 120° 30′ E., the observed Mer. Alt. of the Sun's lower limb north of the observer was 16° 5′ 10″; index error, −4′ 25″ ; height of eye, 38 ft. Required the latitude.
4. 1899, August 15th, in long. 30° 50′ W., the observed Mer. Alt. of the Sun's lower limb north of the observer was 75° 57′ 20 ; index correction, −7′ 35″ ; height of eye, 21 ft. Required the latitude.
5. 1899, September 23rd, in long. 150° 45′ E., the observed Mer. Alt. of the Sun's upper limb south of the observer was 37° 38′ 00″ ; index error, 6′ 25″ on the arc; height of eye, 16 ft. Required the latitude.
6. 1899, September 23rd, in long. 11° 45′ W., the observed Mer. Alt. of the Sun's lower limb south of the observer was 33° 44′ 40″ ; index error, −5′ 15″ ; height of eye, 23 ft. Required the latitude.
7. 1899, November 30th, in long. 179° 59′ W., the observed Mer. Alt. of the Sun's lower limb south of the observer was 68° 4′ 00″ ; index correction, −1′ 32″ ; height of eye, 11 ft. Required the latitude.
8. 1899, June 29th, in long. 178° 20′ E., the observed Mer. Alt. of the Sun's lower limb, 61° 40′ 25″ bearing south; index correction, +1′ 20″ ; height of eye, 36 ft. Find the latitude.
9. 1899, September 23rd, in long. 90° 15′ E., the observed Mer. Alt. of the Sun's lower limb, 52° 40′ 55″ bearing north ; index correction, +8′ 40″ ; height of eye, 9 ft. Required the latitude.
10. 1899, December 21st, in long. 105° 37′ W., the observed Mer. Alt. of the Sun's lower limb, 89° 39′ 16″ bearing south ; index correction, +8′ 36″ ; height of eye, 18 ft. Find the latitude.
11. 1899, July 1st, in long. 15° 45′ E., the observed Mer. Alt. of the Sun's upper limb, 38° 50′ 0″, taken in an Artificial Horizon, zenith south of the sun ; height of eye, 13 ft. ; index correction, +2′ 16″. Find the latitude.
12. 1899, September 22nd, in long. 179° 15′ W., the observed Mer. Alt. of the Sun's lower limb in an Artificial Horizon was 81° 17′ 50″ bearing south ; index error, −2′ 2″. Find the latitude.
13. 1899, October 1st, in long. 137° 30′ E., the observed Mer. Alt. of the Sun's lower limb in an Artificial Horizon was 62° 32′ 50″, zenith south of the Sun ; height of eye, 50 ft. ; index correction, +2′ 16″. Required the latitude.

ART. 34.—Meridian Altitude of a Star.

Example.—1899, June 5th, the observed Meridian Altitude of the Star Canopus (α Argûs) bearing south was 54° 17′ 30″ ; index correction, −6′ 10″ ; height of eye, 24 ft. Required the latitude.

TEXT-BOOK ON NAVIGATION.

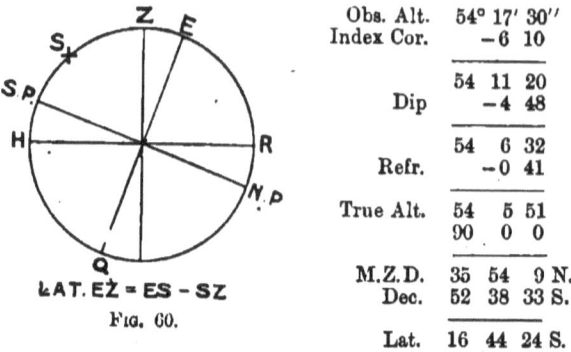

Obs. Alt.	54° 17' 30''
Index Cor.	−6 10
	54 11 20
Dip	−4 48
	54 6 32
Refr.	−0 41
True Alt.	54 5 51
	90 0 0
M.Z.D.	35 54 9 N.
Dec.	52 38 33 S.
Lat.	16 44 24 S.

LAT. EŻ = ES − SZ
FIG. 60.

1. 1899, July 30th, the observed Mer. Alt. of Capella was 54° 29' 15'' north of the observer; index error, −3' 10''; height of eye, 20 ft. Required the latitude.
2. 1899, August 10th, the observed Mer. Alt. of Rigel was 84° 54' 20' north of the observer; index error, +1' 10'; height of eye, 21 ft. Required the latitude.
3. October 19th, the observed Mer. Alt. of Sirius, zenith north of star, was 45° 30' 30''; index error, +1' 30''; height of eye, 23 ft. Required the latitude.
4. November 4th, the observed Mer. Alt. of Spica bearing south of observer was 36° 25' 20''; index error, −3' 10''; height of eye, 24 ft. Required the latitude.
5. 1899, December 29th, the observed Mer. Alt. of Fomalhaut bearing south of the observer was 10° 19' 30''; height of eye, 25 ft. Required the latitude.
6. 1899, January 12th, the Mer. Alt. of Aldebaran bearing south was 56° 23'; index correction, +1' 19''; height of eye, 28 ft. Required the latitude.
7. 1899, March 19th, the Mer. Alt. of Spica in an Artificial Horizon was 86° 47' 20''; index correction, −2' 18''; zenith north of the star. Find the latitude.
8. 1899, June 29th, the Mer. Alt. of Markab bearing north was 30° 59' 10''; index correction, +4' 2''; height of eye, 33 ft. Find the latitude.
9. 1899, July 18th, the Mer. Alt. of α Arietis bearing south was 75° 18' 40'; index correction, −4' 9''; height of eye, 19 ft. Required the latitude.
10. 1899, September 22nd, the Mer. Alt. of Algenib in an Artificial Horizon was 111° 57' 10''; index correction, +8' 6''; zenith south of the star. Find the latitude.
11. 1899, December 20th, the Mer. Alt. of Regulus in an Artificial Horizon was 114° 20' 30''; index correction, −5' 16''; zenith north of the star. Find the latitude.

ART. 35.—**Meridian Altitude of the Moon.**—It is first necessary to find the M.T.G. when the Moon is on the observer's Meridian (see p. 60), and then reduce the "Elements" (see pp. 58 and 61).

Example.—1899, July 20th, in long. 126° 5' W., the observed Mer. Alt. of the Moon's lower limb was 42° 26' 15'' bearing south; index error, − 3' 25''; height of eye, 16 ft. Required the latitude.

LATITUDE BY MERIDIAN ALTITUDE.   85

```
Mer. pass. 20ᵈ 10ʰ 6ᵐ·8        62ᵐ·3 (for 24ʰ)   Long. 126° 5'
Cor. for long.   +21·8          8·4                      4
                 ─────          ─────             ──────────
   M.T.S. 20ᵈ 10 28·6           2492              6,0)50,4 20
       Long.  8 24·3            4984              ──────────
              ─────                                8ʰ 24ᵐ 20ˢ
   M.T.G. 20ᵈ 18 52·9         4)523·32
                              ────────
                              6)130·83
                              ────────
                                21·8
```

```
         S.D.                        H.P.
      16' 28"·29      5"·46       60' 20"·87      20"·01      Alt. 42°
         +3·14         6·9         +11·51          6·9            90
         ─────        ────        ─────────        ─────     ─────────
Aug.  16  31·43      4914         60 32·38       18009      M.Z.D. 47 N.
         +11·8       3276   Red.    − 2          12006       Dec.  23 S.
         ─────       ────         ────────       ─────      ────────────
      16  43·2    12)37·674       60 30·4     12)138·069   Approx. lat. 24 N.
                     3·14                        11·51
                                                             Diff. for 1ᵐ
                                       Dec.                     4·743
Obs. Alt. lower limb 42° 26' 15"    22° 50' 18"·3 S.             52·9
              I.C.     − 3  25       − 4 10·9                ───────────
                     ─────────      ───────────                 42687
                       42 22 50      22 46  7·4                  9486
              Dip     − 3 55                                    23715
                     ─────────                               ───────────
                       42 18 55                             6,0)25,0·9047
              S.D.    +16 43                                 ──────────
                     ─────────                                  4 10·9
                       42 35 38
           Cor. for }     43  2
           Par. − refr. }     22
                              6
                     ─────────
            True Alt. 43 19  8
                      90  0  0
                     ─────────
             Mer. Z. D. 46 40 52 N.
               Dec.     22 46  7 S.
                       ───────────
               Lat.     23 54 45 N.
```

EXERCISES.

1. 1899, July 23rd A.M., in long. 15° 45' E., the observed Meridian Altitude of the Moon's upper limb bearing N. was 73° 21' 30"; index error, +3' 15"; height of eye, 10 ft. Required the latitude.

2. 1899, August 19th P.M., in long. 70° 42' W., the observed Mer. Alt. of the Moon's lower limb south of the observer was 33° 30' 15"; index error, −10"; height of eye, 12 ft. Required the latitude.

3. 1899, September 19th in long. 170° 47' E., the observed Mer. Alt. of Moon's lower limb was 39° 41' 20", zenith south of the Moon; index error, −1' 30"; height of eye, 14 ft. Required the latitude.

4. 1899, October 18th, in long. 126° 35' W., the observed Mer. Alt. of the Moon's lower limb was 67° 2' 15" south of the observer; index error, +2' 15"; height of eye, 16 ft. Required the latitude.

86   TEXT-BOOK ON NAVIGATION.

5. 1899, November 25th A.M. at ship in long. 0°., the observed Mer. Alt. of the Moon's lower limb was 45° 1′ 45″ south of the observer; index error, 0; height of eye, 18 ft. Required the latitude.

6. 1899, December 24th A.M. at ship in long. 128° 10′ E., the observed Mer. Alt. of the Moon's lower limb was 35° 31′ 42″, zenith south of Moon; index error, 2′ 15″ on the arc; height of eye, 20 ft. Required the latitude.

7. 1899, January 4th C.T. at ship in long. 156° 25′ W., the Mer. Alt. of the Moon's upper limb 61° 18′ 40″ bearing north; index correction, −3′ 10″; height of eye, 29 ft. Required the latitude.

8. 1899, March 19th C.T. at ship in long. 145° 42′ E., the Mer. Alt. of the Moon's lower limb in an artificial horizon bearing north 64° 56′ 20″; index correction, +3′ 20″. Required the latitude.

9. 1899, April 19th C.T. at ship in long. 34° 27′ W., the Mer. Alt. of the Moon's upper limb in an artificial horizon bearing south was 97° 25′ 40″; index correction, +2′ 18″. Required the latitude.

10. 1899, June 29th C.T. at ship in long. 175° 37′ W., the Mer. Alt. of the Moon's lower limb bearing north was 54° 17′ 20″; index correction, −2′ 25″; height of eye, 40 ft. Required the latitude.

11. 1899, October 3rd C.T. at ship in long. 179° 51′ E., the Mer. Alt. of the Moon's lower limb bearing south was 46° 14′ 10″; index correction, − 3′ 48″; height of eye, 16 ft. Required the latitude.

ART. 36.—**Meridian Altitude of a Planet.**

Example.—1899, April 2^d A.M., at Ship in long. 140° E., the observed Meridian Altitude of the Planet Jupiter's centre bearing south was 44° 15′ 40″; index correction, −1′ 30″; height of eye, 30 ft. Required the latitude.

	h. m.		
Mer. Pass. April 1^d	13 44·3	Change for 24^h +4″·3	Long. 140°
(P. 243 N.A.)	+ 1·7	9·3	4
M.T.S. 1^d	13 46·0	129	6,0)56,0
Long.	9 20 E.	387	
			9^h 20^m
M.T.G. April 1^d	4 26·0	4)39·99	
		6)9·99	
		1·66	

S.D.	H.P.	Dec.	Change in 24h.
20″·1	2″	12° 53′ 46·8 S.	2′ 3″·8
		− 22·7	60
		12 53 24·1	213·8
			4·4
			4952
			4952
			4)544·72
			6)136·18
			22·69

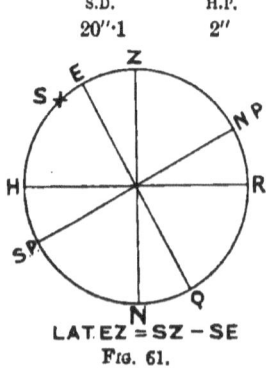

LAT EZ = SZ − SE
FIG. 61.

LATITUDE BY MERIDIAN ALTITUDE.

```
         Obs. Alt. 44° 15' 40"
         I.C.      -1 30
                  ─────────
                   44 14 10
         Dip       -5 22
                  ─────────
                   44  8 48
         Refr.     -0 59
                  ─────────
                   44  7 49
         Par. in Alt.  +1·5  (Tab. XLVIII. Norie)
                  ─────────
         True Alt. 44  7 50·5
                   90  0  0
                  ─────────
         M.Z.D. 45 52  9·5 N.
         Dec.   12 53 24   S.
                  ─────────
         Lat. 32 58 45·5 N.
```

Remark.—Of course the Parallax, being so small, might be neglected, but it is here introduced to show the method of dealing with the Parallax of Planets. The S.D. is not used, because in practice at sea the centre would be observed.

EXERCISES.

1. 1899, January 29th, in long. 150° E., the observed Mer. Alt. of the centre of Venus was 35° 59' 30" bearing south; index correction,—1' 8"; height of eye, 24 ft. Find the latitude.
2. 1899, February 18th, in long. 172° E., the observed Mer. Alt. of the centre of Mars was 33° 58' 30" bearing north; index correction,+3' 15"; height of eye, 40 ft. Required the latitude.
3. 1899, March 19th, in long. 148° 15' W., the observed Mer. Alt. of Jupiter's centre was 52° 38' 20" bearing north; index correction,+3' 52"; height of eye, 27 ft. Find the latitude.
4. 1899, January 15th, in long. 167° 27' E., the observed Mer. Alt. of the centre of Mars was 65° 55' 40"; zenith south of the object; index correction, —4' 16"; height of eye, 42 ft. Find the latitude.

ART. 37. Meridian Altitude of a Star below the Pole.—A Star whose Polar Distance is less than the elevation of the Pole above the Horizon will never set, and will therefore be seen on the Meridian *below* the Pole as well as *above* the Pole.

Now, the elevation of the Pole is equal to the latitude of the place, and this can be found by adding together the star's Meridian Altitude below the Pole and its Polar Distance;

i.e. Lat. = Alt. + P. Distance.

Example.—1899, October 10th, the observed Meridian Altitude below the Pole of the Star α Ursæ Majoris was 16° 33' 30"; index correction, +3' 30"; height of eye, 23 ft. Required the latitude.

TEXT-BOOK ON NAVIGATION.

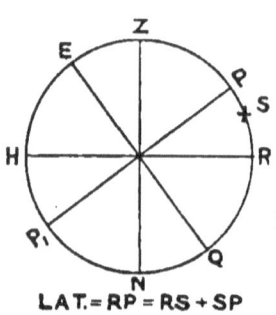

FIG. 62.

LAT.= RP = RS + SP

```
Obs. Alt. 16° 33' 30"    Dec. 62° 17' 19" N
I.C.      + 3  30              90
         ─────────
          16 37  0     P.  ⎫
Dip       − 4 42      Dist.⎬ 27  42  41
         ─────────         ⎭
          16 32 18
Refr.     − 3 10
         ─────────
True Alt. 16 29  8
P. Distance 27 42 41 N.
         ─────────
Lat.      44 11 49 N.
```

The Latitude is of the *same name* as the Declination, because, obviously, only stars having N. Declination can be seen below the Pole in N. Latitude, and stars having S. Declination in S. Latitude.

EXERCISE.

1. 1899, September 28th, the observed Meridian Altitude of Canopus below the Pole was 29° 14' 0"; index correction, −3' 30"; height of eye, 22 ft. Required the latitude.
2. 1899, July 22nd C.T. at ship in long. 46° 27' W., the Mer. Alt. of the Moon's lower limb in an artificial horizon below the Pole was 18° 53' 20"; index correction, −5' 48". Required the latitude.

ART. 38.—To find the Approximate Meridian Altitude of a Star, for setting the Sextant preparatory to observing the Meridian Altitude.

Referring to the Rule for finding the Latitude by a Meridian Altitude (Art. 32), it was shown that the Latitude is equal to the *sum* of the Meridian Zenith Distance and Declination when they are of the *same* name, and the *difference* when of *different* names.

Therefore it follows that the Meridian Zenith Distance is the *difference* of the Lat. and Dec. when they are of the *same* name, and the *sum* when of different names.

The Meridian Zenith distance subtracted from 90° gives the *Meridian Altitude*.

Examples.—

1. Given lat. 25° 30' N., and a Star's Dec. 52° 20' N., required the Mer. Alt. of the Star.

```
Lat. 25° 30' N. ⎫
Dec. 52  20  N. ⎬ Same name—subtract.
                ⎭
Mer. Zen. Dist. 26 50 S.
                90  0
               ─────────
Mer. Alt.      63 10   North of observer.
```

LATITUDE BY MERIDIAN ALTITUDE. 89

2. Suppose the latitude to be 24° 25' S., and the Star's Dec. 32° 10' N., required its Mer. Alt.

$$\left.\begin{array}{l}\text{Lat. } 24° 25' \text{ S.}\\ \text{Dec. } 32\ 10\ \text{ N.}\end{array}\right\} \text{Different names—add.}$$

Mer. Zen. Dist. 56 35 S.
                 90   0
                 ―――――
Mer. Alt.   33 25 North of observer.

3. Given latitude of Ship 10° 15' N., what is the Mer. Alt. of a Star whose Dec. is 20° 40' S., and how does it bear from observer?

$$\left.\begin{array}{l}\text{Lat. } 10° 15' \text{ N.}\\ \text{Dec. } 20\ 40\ \text{ S.}\end{array}\right\} \text{Different names—add.}$$

Mer. Zen. Dist. 30 55 N.
                 90   0
                 ―――――
Mer. Alt.   59   5 N. South of observer.

N.B.—A Star is north or south of observer according as its Declination reads north or south of the Latitude of observer.

EXERCISES.

Find the Meridian Altitudes of Stars from the following data, and state whether N. or S. of observer:—

No.	Latitude.	Name of star.	Declination.
1	14 40 N.	α Draconis	64 52 N.
2	28 12 S.	α Bootis	19 42 N.
3	44 10 N.	α Scorpii	26 12 S.
4	14 25 S.	β Gruis	47 25 S.
5	52 19 S.	α Piscis Aust.	30 9 S.
6	0 0	Achernar	57 45 S.
7	0 0	β Geminorum	28 16 N.
8	35 28 N.	Sirius	16 35 S.
9	49 50 N.	Polaris	88 46 N.
10	3 56 S.	α Ursæ Majoris	61 3 N.

CHAPTER X.

LATITUDE BY EX-MERIDIAN ALTITUDE.

ART. 39. By Reduction to the Meridian.—There are several methods for finding the Latitude from the Altitude of an object when *near* the Meridian, on the principle of finding its Altitude or its Zenith Distance when *on* the Meridian. Only one of these methods is here given, and that the one most commonly used.

Example.—1899, September 19th P.M. at ship, in latitude by account 54° 10′ N., long. 17° 30′ W., the observed Altitude of the Sun's lower limb bearing south, was 36° 34′ 10″; index correction, +2′ 20″; height of eye, 29 ft.; a chron. which had been found 1ʰ 4ᵐ fast of A.T.S., showed 19ᵈ 1ʰ 37ᵐ 29ˢ, and the difference of longitude made to the East since determination of error was 16′. Required the latitude.

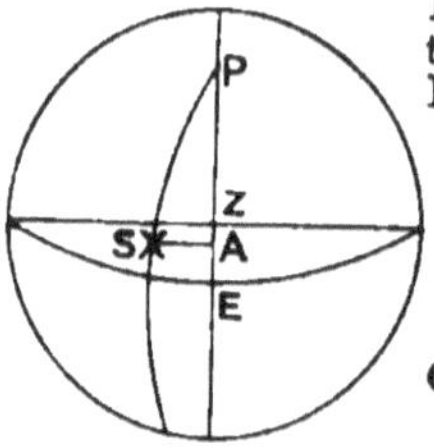

LAT EZ = EA + AZ
Fig. 63.

	h. m. s.			
Chron. 19ᵈ	1 37 29	Dec. 1° 28′ 8″ N.		58·21
	1 4 0	1 39		1·7
	0 33 29	1 26 29 N.		40747
Cor. for run	+1 4			5821
A.T.S. 19	0 34 33	(H.A.)		6,0)9,8·957
Long.	1 10 0			
A.T.G. 19	1 44 33			1 39

605184
294
―――
385)11000(29
770
―――
3300

			Constant 1·
Obs. Alt. 36° 34′ 10″	H.A. 34ᵐ 33ˢ	...	log rising 3·054760
I.C. +2 20	Dec. 1° 26′·29″	...	cosine 9·999863
36 36 30	Lat. 54 10	...	cosine 9·767475
Dip −5 16			
36 31 14	6639		log 3·822098
Ref. Par. −1 9	Nat. sin T.A. 598545		
36 30 5	Nat. cos M.Z.D. 605184		
S.D. +15 57	M.Z.D. = 52° 45′ 29″ N.		
	Dec. = 1 26 30 N.		
T.A. 36 46 2	Lat. = 54 11 59 N.		

ART. 40. Direct Method.—This method is independent of the Latitude by account, and is correct for *any* distance from the Meridian. It only involves the taking out of one more log

LATITUDE BY EX-MERIDIAN ALTITUDE.

than the "Reduction" method, and has the advantage that all the logs are found from the same table.

Examples of Computation.—

H.A. 34ᵐ 33ˢ   ... cos 9·995046
Dec. 1° 26½'   ... cot 11·599166 cosec 11·599304

Arc I. 1° 27¼'   ... cot 11·594212   sine 8·405687
T.A.   36° 46'   ...   .   sine 9·777106

Arc II. 52° 44½' N.   cos   9·782097
Arc I. 1° 47½' N.

Lat. 54 12 N.

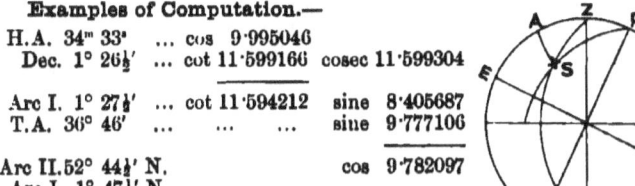

NOTE.—Arc I. is named the same as Declination, and Arc II. the same as the Zenith Distance; that is, opposite to the bearing.

LAT. EZ = EA + AZ
FIG. 64.

N.B.—When Dec. is 0, Arc I. disappears. The Latitude is then found as follows :—

	° ′ ″		
True Alt.	— — —	...	sine
H.A.	— — —	...	sec
Lat.	— — —	...	cos

ART. 41. **By Towson's Ex-Meridian Tables.**—This is a most simple and convenient method, based on the same formulæ as the preceding, and independent of the Latitude by account.

Directions.—Enter Table I. with the Hour Angle and Declination, and take out the first correction—to be added to the Declination, and the "Index Number" opposite. Next, enter Table II. with the Altitude and the "Index Number," and take out the second correction—to be added to the Altitude. Then proceed as in an ordinary Meridian Altitude Problem.

Example.—

H.A. 34ᵐ 33ˢ   Dec. 1° 26' 29"   Alt. 36° 46' 2"
                    Index No. 91 +1 2 (Tab. I.)   +29 27 (Tab. II.)

                    Augmented Dec. 1 27 31   Aug. Alt. 37 15 29
                                                       90 0 0

                                    Reduced Zen. Dist. 52 44 31 N.
                                    Augmented Dec.   1 27 31 N.

                                              Lat. 54 12 2 N.

EXERCISES.

1. 1899, March 20th A.M. at ship, in lat. by account 52° 35' S., long. 73° 2' W., the observed Alt. of the Sun's lower limb north of the observer was 36° 55' 12"; index error, −3' 30"; height of eye, 27 ft.; time by chron., 20ᵈ 5ʰ 26ᵐ 40ˢ, which was fast of M.T.G. 59ᵐ 45ˢ. Required the latitude.

2. 1899, May 27th A.M. at ship, in lat. by account 2° 37' N., long. 108° 15' E., the observed Alt. of the Sun's lower limb was 67° 41' 51" N.; index

error, 0; height of eye, 21 ft.; time by chron., 26^d 14^h 41^m 46^s, which was slow of M.T.G. 1^h 13^m 10^s. Required the latitude.

3. 1899, July 31st P.M. at ship, in lat. by account 0° 50' N., long. 179° 28' W., the observed Alt. of the Sun's upper limb was 71° 6' 7" N.; index error, +1' 20"; height of eye, 14 ft.; time by chron., 31^d 11^h 42^m 33^s, which was slow for M.T.G. 55^m 20^s. Required the latitude.

4. 1899, August 7th at ship, in lat. by account 33° 15' N., long. 10° 57' W., the observed Alt. of the Sun's upper limb was 72° 52' 54" S.; index error, 3' 0"; height of eye, 16 ft.; time by watch 6^d 22^h 57^m 58^s, which was slow of A.T.S. 44^m 10^s; the difference of long. made to the W. since the error was determined being 17¾'. Required the latitude by reduction to the Meridian; and, supposing the course and distance to be S. by W. 4 miles since observation, find the latitude at noon.

5. 1899, August 19th, in lat. by D.R. 17° 25' S., long. 177° 30' E., the observed Alt. of the Sun's lower limb north was 57° 15' 22"; sextant error, 1' 18" off the arc; height of eye, 18 ft.; time by chron., 18^d 12^h 51^m 39^s, which was slow of A.T.S. 11^h 55^m 0^s; the course and distance made since the error was determined, S. 54° E 23'. Required the latitude by reduction to the Meridian; and, supposing the same course to be continued 3 miles since noon, find the latitude at noon.

6. 1899, September 23rd, in lat. by D.R. 37° 22' N., long. 8° 50' E., the observed Alt. of the Sun's lower limb was 52° 15' 45" N.; index error, −0' 20"; height of eye, 19 ft.; time by watch, 23^d 0^h 0^m 0^s, which was slow of A.T.S. 9^m 32^s; the course and distance since the error was found was N. 74° E. 25'. Required the latitude by reduction to the Meridian; and, supposing the ship to steer N. 80° E. 4 miles since noon, required the latitude at noon.

7. 1899, October 9th A.M. at ship, in lat. by account 46° 44'. long. 140° 45' E., the observed Alt. of the Sun's lower limb north was 48° 34' 30"; index error, −7' 12"; height of eye, 17 ft.; time by chron., 8^d 13^h 48^m 25^s, which was slow of A.T.S. 9^h 25^m 10^s; the difference of long. made to the W. after the error was determined being 15¾'. Required the latitude by reduction to the Meridian; and, supposing the course and distance since observation to be west 4 miles, required the latitude at noon.

8. 1899, November 6th A.M. at ship, in lat. by account 15° 29' S., long. 0° 8' E., the observed Alt. of the Sun's lower limb south was 81° 14' 40"; index error, +6' 33"; height of eye, 18 ft.; time by watch, 5^d 23^h 45^m 4^s, which was fast of A.T.S. 21^m 16^s; the difference of longitude made to the east after the error was determined being 17¼'. Required the latitude.

9. 1899, December 31st at ship, in lat. by account 13° 40' N., long. 150° 15' W., the observed Alt. of the Sun's lower limb south was 54° 4' 10"; index error, −5' 8"; height of eye, 25 ft.; time by a watch, 31^d 12^h 18^m 5^s, which was fast of M.T.G. 2^h 2^m 37^s. Required the latitude by reduction to the Meridian; and, supposing the course and distance to be north 1½ mile since noon, required the latitude at noon.

10. 1899, July 16th A.M. at ship, in lat. by account 30° 35' S., long. 12° 30' W., the observed Alt. of the Sun's lower limb bearing north was 36° 52' 16"; index correction, −2' 15"; height of eye, 13 ft.; time by chron., 16^d 0^h 42^m 44^s, which had been found to be 50^m 59^s fast of A.T.S.; the difference of longitude made to the west after the error was determined being 30¾'. Required the latitude.

11. 1899, June 15th A.M. at ship, in lat. by account 58° 50' N., long. 173° 56' 45" E., the observed Alt. of the Sun's lower limb was 53° 7' 10" S. of the observer; index correction, +7' 26"; height of eye, 25 ft.; time by chron., 14^d 13^h 4^m 5^s, which was fast 50^m 35^s of G.M.T.; run since sights to noon, S.S.E. 3 miles. Required latitude at sights and noon.

12. 1899, November 13th at ship, in lat. by account 14° 20' N., long. 90° 35' E., the observed Alt. of the Sun's lower limb south of the observer

LATITUDE BY EX-MERIDIAN ALTITUDE.

was 56° 38' 10"; index correction, −0' 41"; height of eye, 33 ft.; time by a watch, $12^d\ 18^h\ 10^m\ 33^s$, which was slow $3^m\ 33^s$ of mean time at Greenwich; run from noon to sights, N.W. ¼ N. 10 miles. Required the latitude at sights and noon.

13. 1899, December 24th at ship, in lat. by account 11° 16' N., long. 90° 32' W., the observed Alt. of the Sun's lower limb south of the observer was 54° 11' 20"; index correction, +2' 21"; height of eye, 25 ft.; time by a chron., $24^d\ 4^h\ 28^m\ 57^s$, which was slow $1^h\ 0^m\ 30^s$ of mean time at Greenwich; run from sights to noon, W.S.W. 11 miles. Required the latitude at sights and noon.

14. 1899, October 21st P.M. at ship, in lat. by account 0°, long. 100° 30' W., the observed Alt. of the Sun's lower limb south of the observer was 77° 54' 10"; index correction, +2' 9"; height of eye, 22 ft.; time by a chron., $21^d\ 7^h\ 12^m\ 37^s$, which was fast $6^h\ 55^m\ 10^s$ of A.T.S.; the difference of longitude made to the east after the error was determined being 30'; run from noon to sights, N. by E. ¼ E. 6 miles. Required the latitude at sights and noon.

ART. 42. Ex-Meridian Altitude of a Star, Moon, or Planet.

1899, January 31st, in lat. by account 31° 40', long. 55° 35' W., the observed Alt. of the Star Sirius bearing south was 40° 44' 20"; index correction, +5' 24"; height of eye, 25 ft.; a chronometer which was $2^m\ 28^s$ slow of M.T.G., showed $31^d\ 12^h\ 56^m\ 27^s$. Required the latitude.

Chron. 31^d	h. m. s. 12 56 27	Dec. of Sirius. 16° 34' 40" S.	55° 35' 4
	+2 28	Obs. Alt. 40° 44' 20"	6.0)22.20
M.T.G. 31^d	12 58 55	I.C. +5 24	$3^h\ 42^m\ 20^s$
Long.	3 42 20	——	
	——	40 49 44	
M.T.S.	9 16 35	Dip −4 54	
S.T. at noon	20 41 57·51	——	
Acl. for 12^h	1 58·28	40 44 50	
58^m	9·43	Refr. −1 5	369 45
35^s	·10	——	
	——	T.A. 40 43 45	1845
R.A.M.	6 0 40		1476
R.A. Star	6 40 44	... ...	166.05
		Const. 1·000000	652319
H.A.	40 4	... ...	——
		log rising 3·183090	652485
	——	... ...	
Dec.	16° 34' 40"	log cosine 9·981562	
Lat. by act.	31 40 ...	log cosine 9·929989	

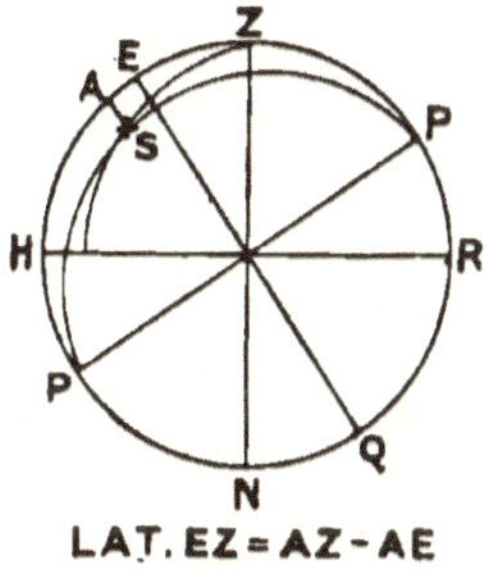

Nat. No. 12435   log 4·094641

T.A. 40° 43' 45"   ... Nat. sine 652485

M.Z.D. 48 19 26 N. ... Nat. cos 664920
Dec. 16 34 40 S.

Lat. 31 44 46 N.

LAT. EZ = AZ − AE

FIG. 65.

94   TEXT-BOOK ON NAVIGATION.

Direct Method.

6649
665013   H.A. 40ᵐ 4ˢ   ... cos 9·993329
───────   Dec. 16° 34' 40" ... cot 10·526235 ... cosec 10·544674
363)9300(2
726
───────   Arc. I. 16 49 12 ... cot 10·519564 ... sine 9·461364
2040   83
                  T.A. 40 43 45   ...   ...   sine 9·814570

                  Arc. II. 48 34 0 N.
                  Arc. I. 16 49 12 S.   ... cosine 9·820691

                  Lat. 31 44 48 N.

By Towson's Ex-Meridian Tables.

H.A. 40ᵐ 4ˢ   Dec. 16° 34' 40"   Alt. 40° 43' 45"
Index No. 100   Augⁿ.   14 58   Augn.   42 5

Augmented Dec. 16 49 38   Augd. Alt. 41 25 50
                                                     90 0 0

                                        Red. Z.D. 48 34 10 N.
                                        Augd. Dec. 16 49 38 S.

                                        Lat. 31 44 32 N.

Remark.—The Latitude may be computed in the same manner from an "Ex-Meridian" Altitude of a planet or the Moon, the Hour Angle, Declination, and True Altitude having been found.

EXERCISES.

1. 1899, February 24th, 6ʰ 3ᵐ 30ˢ A.M. A.T.S., in lat. by account 33° 55' N., long. 35° 30' E., the observed Alt. of the Star Antares was 29° 57' 50" south of the observer; index error, − 2' 10"; height of eye, 15 ft. Required the latitude.

2. 1899, May 24th, time by chron. 24ᵈ 10ʰ 13ᵐ 30ˢ, which was correct M.T.G.; lat. by D.R. 51° 22' N., long. 9° 35' W.; an Alt. of Star Spica was observed 27° 53' 30" south of the observer; index error, − 1' 30"; height of eye, 18 ft. Required the latitude.

3. 1899, July 4th, 8ʰ P.M. M.T.S., in lat. 0°, long. 180° 00' E., by D.R.; observed Alt. Star α Centauri S. was 29° 34' 30"; index error, + 2' 0"; height of eye, 23 ft. Required the latitude.

4. 1899, August 16th at ship, lat. 31° 30' S., long. 175° 30' 30" W.; time by chron. 16ᵈ 0ʰ 0ᵐ 10ˢ, which was fast of M.T.G. 10ˢ; observed Alt. of Star α Cygni bearing N. and W. of the observer, 11° 37' 50"; index error, + 2' 30"; height of eye, 26 ft. Required the latitude.

5. 1899, September 28th at ship, in lat. by D.R. 50° 44' 15" N., long. 12° 58' E.; time by watch, 28ᵈ 10ʰ 50ᵐ, which was slow M.T.S. 10ᵐ 52ˢ; the observed Alt. Star Fomalhaut, 8° 52' south of observer; index error, − 1' 39"; height of eye, 30 ft. Required the latitude.

6. 1899, April 27th A.M. at ship, in lat. by D.R. 0° 0', long. 25° W.; time by clock, 5ʰ 2ᵐ 23ˢ A.T.S.; observed Alt. of Star Vega bearing N. was 50° 0' 0"; error of sextant, 2' 41" on the arc; height of eye, 23 ft. Required the latitude.

7. 1899, December 14th A.M. at ship, in lat. by D.R. 53° 20' N., long.

LATITUDE BY EX-MERIDIAN ALTITUDE.

5° 30' W. A.T.G. 13ᵈ 19ʰ 37ᵐ; observed Alt. Star Spica bearing S. and E. was 25° 34' 34"; index error, +2' 10"; height of eye, 35 ft. Required the latitude.

8. 1899, March 31st P.M. at ship, in lat. by D.R. 42° 20' N., long. 70° 55' W.; the observed Alt. of Sirius south was 30° 54' 10" when the clock showed 6ʰ 30ᵐ 30ˢ M.T.S.; index error, −1' 10"; height of eye, 24 ft. Required the latitude.

9. 1899, November 18th at ship, in lat. by D.R. 34° 30' S., long. 18° 45' E.; time by chron. which was correct for M.T.G., 17ᵈ 16ʰ 30ᵐ; observed Alt. of Star Regulus was 42° 34' 31" north of observer; index error, +1' 30"; height of eye, 22 ft. Required the latitude.

10. 1899, October 8th A.M. at ship, in lat. by account 42° 30' N., long. 150° 14' E., the observed Alt. of Rigel south of the observer was 38° 57' 10"; index correction, −2' 5"; height of eye, 25 ft.; time by watch, 5ʰ 50ᵐ 37ˢ A.M., which had been found to be 1ʰ 14ᵐ 59ˢ fast of mean time at ship; run since the error was determined, N.W. by W. ½ W. 27·5 miles. Required the latitude.

11. 1899, November 19th A.M. at ship, in lat. by account 50° 20' N., long. 8° 26' W., the observed Alt. of Procyon near the Meridian south of observer was 45° 1' 40"; index correction, +1' 19"; height of eye, 31 ft.; time by chron., 18ᵈ 15ʰ 48ᵐ 54ˢ, which had been found to be 51ᵐ 13ˢ slow of mean time at Greenwich; since error was determined the difference of long. made to the westward was 45·5 miles. Required the latitude.

12. 1899, December 25th A.M. at ship, in lat. by account 10° 10' S., long. 125° 35' E., the observed Alt. of Pollux north of observer was 51° 0' 40"; index correction, −2' 11"; height of eye, 33 ft.; time by chron., 5ʰ 15ᵐ 31ˢ, which had been found to be 41ᵐ 38ˢ fast of apparent time at Greenwich; the difference of long. made to the eastward since determining the error, 22·5 miles. Required the latitude.

13. 1899, May 25th A.M. at ship, in lat. by account 27° 25' N., long. 135° E., the observed Alt. of Saturn's centre south of the observer was 40° 38' 20"; index correction, −2' 13"; height of eye, 26 ft.; time by chron., 2ʰ 40ᵐ 39ˢ, which had been found to be slow 1ʰ 55ᵐ 36ˢ of G.M.T.; the difference of long. made to the eastward was 36 miles, since error on G.M.T. was determined. Required the latitude.

14. 1899, October 13th P.M. at ship, in lat. by account 45° 12' S., long. 47° 15' E., the observed Alt. of the Moon's lower limb out of the Meridian was 57° 24' 10", being north; index correction, −3' 19"; height of eye, 29 ft.; time by chron., 5ʰ 42ᵐ 42ˢ, which had been found 59ᵐ 12ˢ fast of G.M.T.; run since error was determined, S.E. by E. 32 miles. Required the latitude.

15. 1899, June 17th P.M. at ship, in lat. by account 39° 14' S., long. 100° 30' W., the observed Alt. of Jupiter's centre was 60° 42' 30" being north; index correction, +3' 48"; height of eye, 32 ft.; time by chron., 10ʰ 49ᵐ 48ˢ, which had been found to be fast 2ʰ 12ᵐ 15ˢ of mean time at ship; run since error was determined, S.W. ¼ W. 33 miles. Required the latitude.

16. 1899, September 19th P.M. at ship, in lat. by account 34° 50' N., long. 135° 40' W., the observed Alt. of Altair south of observer was 63° 17' 10"; index correction, −2' 4"; height of eye, 29 ft.; time by watch, 19ᵈ 7ʰ 4ᵐ 54ˢ, which had been found to be 19ᵐ 48ˢ slow of mean time at ship; run since error was determined, E.S.E. 15·4 miles. Required the latitude.

ART. 43. **Latitude by an Altitude of the Pole Star out of the Meridian.**—If the Pole Star (Polaris or α Ursæ Minoris) were exactly in the Pole, then its true Altitude would be the latitude of the observer, but, as a matter of fact, it is about 1° 18' distant from the Pole, and appears to describe about it a small circle of this radius in the course of a diurnal revolution.

On pp. 583-85 of the Nautical Almanac are tables for deducing the latitude from the true Altitude of Polaris.

Example.—1899, August 3^d 1^h 10^m A.M. mean time at ship, in longitude 20° W., the observed Altitude of Polaris was 56° 46′ 30″; index error, +2′ 35″; height of eye, 20 ft. Required the latitude.

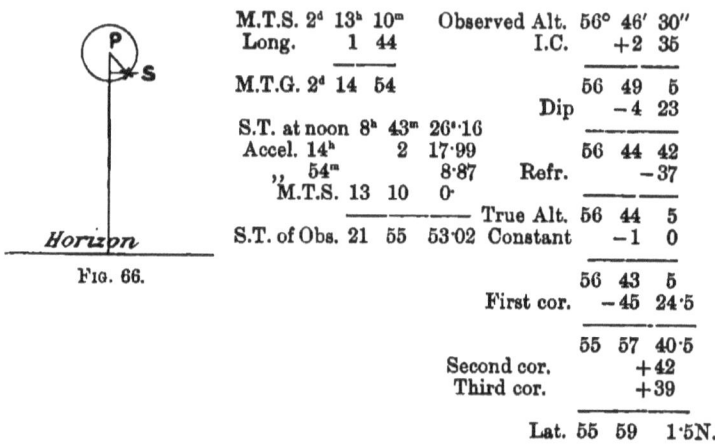

FIG. 66.

M.T.S. 2^d 13^h 10^m	Observed Alt.	56° 46′ 30″
Long. 1 44	I.C.	+2 35
M.T.G. 2^d 14 54		56 49 5
	Dip	−4 23
S.T. at noon 8^h 43^m 26^s·16		
Accel. 14^h 2 17·99		56 44 42
„ 54^m 8·87	Refr.	−37
M.T.S. 13 10 0·		
	True Alt.	56 44 5
S.T. of Obs. 21 55 53·02	Constant	−1 0
		56 43 5
	First cor.	−45 24·5
		55 57 40·5
	Second cor.	+42
	Third cor.	+39
	Lat.	55 59 1·5N.

EXERCISES.

1. 1899, July 24th, 3^h 15^m A.M. mean time at ship, in long. 72° 40′ W.; observed Alt. of Polaris, 12° 40′ 10″; index error, −10″; height of eye, 20 ft. Required the latitude.

2. 1899, August 10th, A.T.S. 10^h 35^m P.M., in long. 11° 20′ W.; observed Alt. of Polaris, 55° 14′ 30″; index error, +2′ 30″; height of eye, 21 ft. Required the latitude.

3. 1899, September 20^d, 19^h 20^m 15^s M.T.G. in long. 124° 37′ W.; observed Alt. of Polaris, 36° 39′ 20″; index error, 0; height of eye, 22 ft. Required the latitude.

4. 1899, October 29^d 23^h 24^m 10^s M.T.G. in long. 179° 14′ E.; observed Alt. of Polaris, 19° 20′ 30″; index error, +1′ 10″; height of eye, 23 ft. Required the latitude.

5. 1899, November 5th, A.T.S. 6^h 12^m A.M., long. 4° 15′ E.; observed Alt. of Polaris, 53° 32′ 15″; index error, +2′ 15″; height of eye, 24 ft. Required the latitude.

6. 1899, December 9th, M.T.S. 7^h 24^m 30^s P.M., long. 33° 30′ W.; observed Alt. of Polaris, 10° 20′ 30″; Height of eye, 26 ft. Required the latitude.

7. 1899, January 15th, at 9^h 56^m 15^s P.M. M.T., ship in long. 24° 19′ E., observed Alt. of Polaris out of the Meridian was 48° 57′ 10″; index correction, −2′ 35″; height of eye, 32 ft. Required the latitude.

8. 1899, March 26th, when the G.M.T. was 8^h 6^m 30^s in long. 30° 26′ E., the observed Alt. of Polaris out of the Meridian taken in an artificial horizon was 84° 52′ 35″; index correction, −3′ 15″. Required the latitude.

9. 1899, April 27th, when the G.M.T. was 27^d 14^h 13^m 26^s in long.

LATITUDE BY EX-MERIDIAN ALTITUDE.

42° 20' W., the observed Alt. of Polaris out of the Meridian in an artificial horizon was 78° 26' 50"; index correction, −4' 30". Required the latitude.

10. 1899, June 27th, at $10^h\ 54^m\ 16^s$ P.M. in long. 49° 26' W., the observed Alt. of Polaris out of the Meridian in an artificial horizon was 110° 37' 40"; height of eye, 56 ft.; index correction, +6' 30". Required the latitude.

11. 1899, July 7th, when the M.T.S. was $3^h\ 14^m\ 27^s$ A.M. in long. 125° 26' E., the observed Alt. of Polaris out of the Meridian was 39° 42' 50"; index correction, −4' 11"; height of eye, 33 ft. Required the latitude.

12. 1899, October 17th, at $4^h\ 47^m\ 15^s$ A.M. M.T. at ship in long. 43° 52' W., the observed Alt. of Polaris out of the Meridian was 47° 32' 40"; index correction, +4' 15"; height of eye, 26 ft. Required the latitude.

CHAPTER XI.

AMPLITUDES.

ART. 44.—When the Sun is on the Equator (*i.e.* when its Declination is 0), it rises in the true E. point and sets true W. At other times it rises and sets towards the N. or S., according as its Declination is N. or S.

The angle from the E. or W. point at which the Sun (or other Celestial object) rises or sets is called its *Amplitude*, and as this is very easy to calculate, it affords a convenient means for finding the error of the Compass, and thence the Deviation caused by the magnetic influence of iron in the ship.

Examples.—

1. 1899, March 20th, at $6^h\ 6^m$ P.M. apparent time at ship, in lat. 25° N., long. 23° W., the bearing of the Sun by compass at the moment of setting was found to be W. 25° 30' N. Required the error of the compass and the deviation, supposing the variation by chart to be $19\frac{1}{4}°$ W.

```
  23°        A.T.S. March 20ᵈ 6ʰ  6ᵐ      Dec. at noon 0° 7' 32" S.        H. var.
   4         Long.           1 32                       7 32               59·24
 ─────                      ──────                     ──────              7·63
6,0)9,2      A.T.G. March 20  7 38       Reduced Dec. 0  0  0              ─────
  ────                                                                    17772
  1·32                                                                    35544
                                                                          41468
                                                                        ─────────
                                                                        6,0)45,2·0012
                                                                             7 32
```

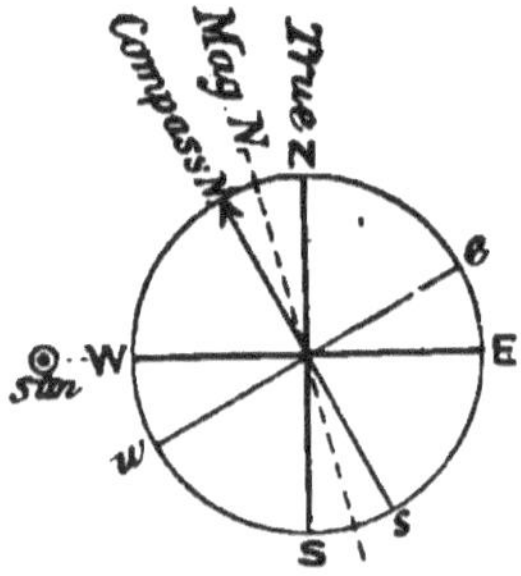

In this case the Sun's dec. is 0, and therefore it sets true West.

True bearing W.   0 0
Compass bearing W. 25 30 N.
                  ─────────
Error of Compass 25 30 W. (True reads to left)
Variation   19 15 W.
                  ─────────
Deviation   6 15 W. (Compass N. is W. of mag. N.)

FIG. 67.

AMPLITUDES.

Explanation.—(1) The illustration shows that the N. point of the Compass lies to the West of true N., because the Sun bears to the N. of the Compass West point.

(2) The Deviation is W. because the N. point of the Compass lies to the West of Magnetic N.

General Rule.—(3) In all cases the *true* bearing will be found to *read* to the *left* of the Compass bearing when the Compass Error is W., and *reads* to the *right* when the Error is E. The figure shows the *fact* without the aid of any Rule.

2. 1899, November 18th, at $5^h\ 17^m$ A.M. in lat. 28° 10′ S., long. 75° E., the Sun was observed to bear by compass E. ¾ S. Required the true amplitude and the error of the compass; also the deviation, the variation being 14° W.

Formula.—
Sine of True Amp. = Sine of Dec. × Sec of Lat.

NOTE.—For proof of Formula, see Appendix.

Calculation.—

				H. var.
A.T S. November	17^d	17^h	17^m	36·59
Long.		5	0	12·3
A.T.G. November	17^d	12	17	10977
				43908

6,0)450·057

7 30

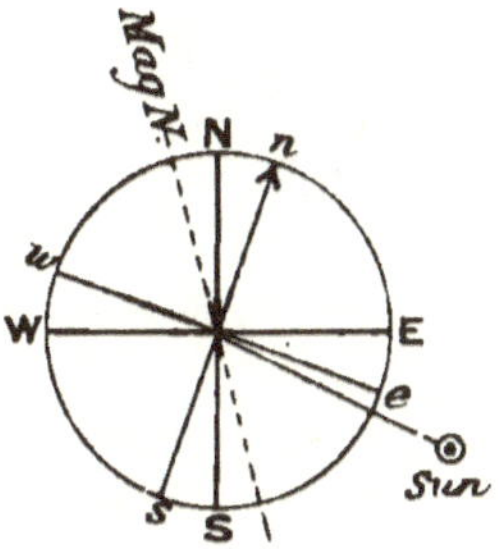

Dec. noon 17th 19° 0′ 43″ S.
Cor.   +7 30

Reduced Dec. 19 8 13   ... sine 9·515566
Lat. 28 10 0   ... sec 10·054739

True Amp. E. 21° 50′ S. ... ... sine 9·570305
Compass bearing E. 8 26 S.

Error of Compass 13 24 E. (True reads to right)
Variation 14 0 W.

Deviation 27 24 E. (Compass N. is E. of mag. N.)

FIG. 68.

Explanation.—The *true* here *reads* to the *right*, because the Sun's position is farther from the true East point than it is from the Compass E. point, and this is because the North point of the Compass lies to the East of true N.

The Deviation is E., because the North point of the Compass is to the East of Magnetic N.

3. 1899, December 1st, $8^h\ 50^m$ P.M. at ship, in lat. 59° 3′ S. long. 32° 15′ W., the sun bore by compass S. ¼ E. Required the true amplitude and error of compass; and, supposing the variation to be 5½° E., required the deviation.

TEXT-BOOK ON NAVIGATION.

A.T.S. December 1ᵈ 8ʰ 50ᵐ   21° 49' 38"·7 S.   23"·26
             2 9   4 15·9   11
             ─────   ─────
A.T.G. December 1ᵈ 10 59   21 53 54·6   6,0)25,5·86
                                                4 15·4
                             Dec. 21° 54'   ... sine 9·571695
                             Lat. 59   3   ... sec 10·288792

                             True amp. W. 46° 29' ½ S. sine 9·860487
                             Comp. bearing W. 95 37 ½ S.

                             Error of compass 49   8 E. (True reads to
                             Variation   5 30 E.   the right)

                             Deviation 43 38 E. (Compass N. is
                                                E. of mag. N.)

                             Explanation.—In this case the true
                             bearing reads to the right, because the
FIG. 69.   angle between Compass W. and Sun
                             is greater than the angle between true
W. and Sun, on account of the North point of the Compass
lying to East of true N.

4. 1899, June 10th A.M. at ship, in lat. 55° 30' N., long. 48° W., when
the correct M.T.G. was 9ᵈ 19ʰ 9ᵐ, the Sun's bearing by compass was E.N.E.
Required the true amplitude and compass error, and deviation, supposing the
variation to be 46° 30' W.

                                                            H. var.
M.T.G. June 9ᵈ 19ʰ 9ᵐ   Dec. 10th 23°   1' 28" N.   11·46
                                           −56   4·85
         4 51   ─────
                             Red. Dec. 23   0 32 N.   5730
                                                            9168
                             Dec. 23° 0'½ sine 9·592027   4584
                             Lat. 55 30   sec 10·246872   ─────
                                                           55·5810
                             True amp. E. 43 38 N. sine 9·838899
                             Compass E. 22 30 N.

                             Error of C. 21   8 W.
                             Var.   46 30 W.

                             Dev. of C. 25 22 E.

FIG. 70.

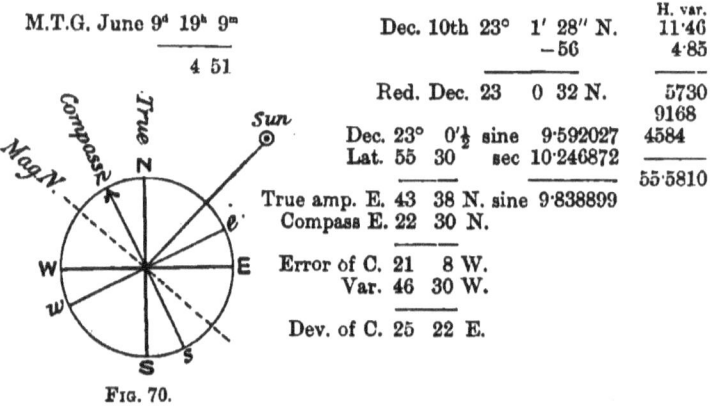

EXERCISES.

1. 1899, October 25th, 4ʰ 32ᵐ 30ˢ A.M. A.T.S., in lat. 59° 55' S., long.
170° 45' W., the Sun's amplitude by compass was E. ¾ N. Required the
true amplitude and error of the compass; and, supposing the variation to be
20° 0' E., required the deviation of the compass for the direction of the ship's
head.

2. 1899, December 23rd, 9ʰ 36ᵐ 45ˢ P.M. A.T.S., in lat. 61° 52' S., long.

152° 45' E., the Sun's amplitude by compass was S. by E. ⅜ E. Required the true amplitude and error of the compass; and, supposing the variation to be 16° 0' E., required the deviation of the compass for the direction of the ship's head.

3. 1899, June 27ᵈ 1ʰ 4ᵐ 37ˢ M.T.G., in lat. 52° 12' N., long. 139° 15 W., the Sun's amplitude by compass was N. ¾ W. Required the true amplitude and error of the compass; and, supposing the variation to be 24° 30' E., required the deviation of the compass.

4. 1899, July 19th, 6ʰ 0ᵐ A.M. A.T.S., in lat. 0° 0', long. 90° 0' W., the Sun's amplitude by compass was E.N.E. Required the true amplitude and error of the compass; and, supposing the variation to be 8° 30' E., required the deviation of the compass.

5. 1899, November 25th, 8ʰ 11ᵐ 12ˢ P.M. A.T.S., in lat. 55° 0' S., long. 122° 45' E., the Sun's amplitude by compass W. ¼ N. Required the true amplitude and error of the compass; and, supposing the variation to be 7° 30' W., required the deviation of the compass.

6. 1899, October 19ᵈ 17ʰ 34ᵐ 55ˢ M.T.G., in lat. 48° 55' N., long. 170° 30' E., the Sun's amplitude by compass W.S.W. ¼ W. Required the true amplitude and error of the compass; and, supposing the variation to be 9° 0' E., required the deviation of the compass.

7. 1899, December 28th, 9ʰ 12ᵐ 53ˢ A.M. A.T.S., in lat. 60° 0' N., long. 0° 0', the Sun's amplitude by compass S.W. by S. ¼ S. Required the true amplitude and error of the compass; and, supposing the variation to be 22° 0' W., required the deviation of the compass.

8. 1899, January 1st, 10ʰ 42ᵐ 36ˢ P.M. A.T.S., in lat. 65° 45' N., long. 180° 0' E., the Sun's amplitude by compass S. by E. ¼ E. Required the true amplitude and error of the compass; and, supposing the variation to be 26° 0' E., required the deviation of the compass.

9. 1899, November 12th, at 7ʰ 42ᵐ A.M. A.T.S., in lat. 53° 30' N.; long. 7° 15' E., the Sun's compass bearing was S.E. ¾ E. Required the true amplitude and compass error; and if the variation was 17° W., find the deviation.

10. 1899, December 31st, 0ʰ 56ᵐ M.T.G., in lat. 59° 55' S., long. 150° 30' W., the Sun's compass bearing was S. ¼ W. Required the true amplitude and compass error; and if the variation was 18° 10' E., find the deviation.

11. 1899, February 7th, at 5ʰ 48ᵐ A.M. A.T.S., in lat. 10° 20' S., long. 1° 30' W., the Sun's compass bearing was E. Required the true amplitude and compass error; and if the variation was 22° 20' W., find the deviation.

12. 1899, November 7th, at 4ʰ 21ᵐ A.M. A.T.S., in lat. 55° 12' S., long. 0° 15' W., the Sun's compass bearing was S.E. by E. Required the true amplitude and compass error; and if the variation was 19° W., find the deviation.

13. 1899, October 12ᵈ 6ʰ 44ᵐ M.T.G., in lat. 41° 0' S., long. 162° 30' E., the Sun's compass bearing was E. by N. ¼ N. Required the true amplitude and compass error; and if the variation was 14° 30' E., find the deviation.

14. 1899, December 20th, at 3ʰ 15ᵐ P.M. A.T.S., in lat. 57° 20' N., long. 131° 45' W., the Sun's compass bearing was S. by E. Required the true amplitude and compass error; and if the variation was 26° E., find the deviation.

15. 1899, January 1st, at 2ʰ 45ᵐ A.M. A.T.S., in lat. 60° S., long. 41° 15' E., the Sun's compass bearing was S. 18° 19' E. Required the true amplitude and compass error; and if the variation was 37° 30' W., find the deviation.

CHAPTER XII.

AZIMUTHS.

ART. 45.—The Azimuth of a Celestial Object—that is, its bearing reckoned from true North or South—is the angle at the Zenith between the Meridian of the observer and a vertical circle passing through the object, and is measured on the Horizon.

The Data required in the Calculation are the **True Altitude**, **Latitude**, and **Polar Distance**.

Example.—
1899, December 17th A.M. at ship, in lat. 41° 22' N.; long. 58° 15' W., the observed altitude of the Sun's lower limb was 20° 35' 40''; index error, −1' 30''; height of eye, 21 ft.; bearing by compass, S. by W. ¼ W.; correct M.T.G., 17ᵈ 2ʰ 3ᵐ 0ˢ. Required the true azimuth and the error of the compass; also the deviation, supposing the variation to be 20° W.

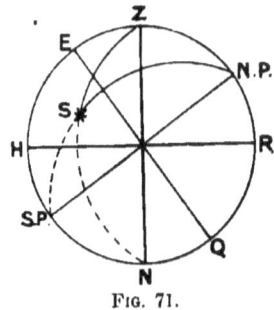
Fig. 71.

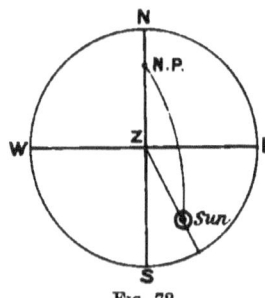
Fig. 72.

```
         M.T.G. Dec. 17ᵈ 2ʰ 3ᵐ    Sun's Dec. 23° 22' 15'' S.    5''·34
                                              11               2·05
Obs. Alt. Sun's Lower Limb 20° 35' 40''      ─────────        ──────
  Index Correction         −1  30            23 22 26          2670
                          ─────────          90  0  0          1068
                          20 34 10          ─────────        ──────
            Dip           − 4 29            113 22 26        10·9470
                          ─────────
                          20 29 41
         Ref.—Par.        − 2 23
                          ─────────
                          20 27 18
            S.D.          +16 17
                          ─────────
                          20 43 35
```

AZIMUTHS.

```
    T.A.  20 43 35    ...    ...  sec 10·029050
    Lat.  41 22  0    ...    ...  sec 10·124652
    P.D. 113 22 26
         ─────────
         2)175 28  1
         ─────────
    Half sum 87 44  0              cos 8·597152
(P.D. -½ sum) Rem. 25 38 26        cos 9·954978
                                   ──────────
                                   2)18·705832
                                   ──────────
              13° 1¼   ...    ...  sine 9·352916
                2

              S. 26  3 E.
              S. 14  4 W.
              ─────────
              Error 40  7 W.
              Var.  20  0 W.
              ─────────
              Dev.  20  7 W.
```

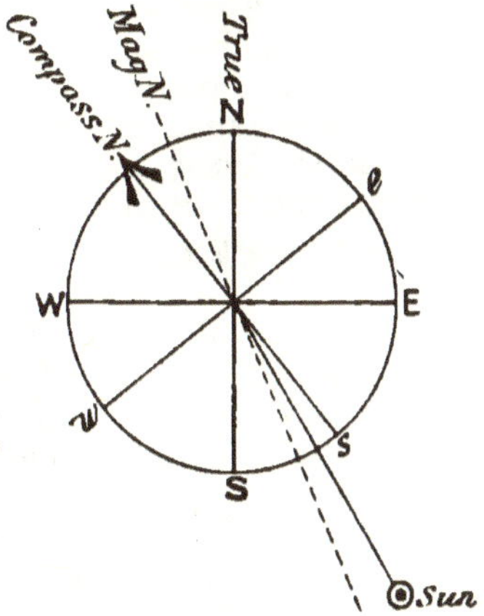

Naming the Azimuth.—When computed as above, it should be named from N. when the lat. is S., and from S. when the lat. is N., and towards E. or W. according as it is A.M. or P.M. at ship.

FIG. 73.

EXERCISES.—ALTAZIMUTHS.

1. 1899, October 31st, 2ʰ 0ᵐ 57ˢ P.M. M.T.S., in lat. 41° 44′ N., long. 124° 12′ 30″ W ; the Sun's bearing by compass, S. 36° 33′ 45″ W. ; alt. of the Sun's lower limb, 25° 35′ 10″ ; index error, −4′ 30″ ; height of eye, 25 ft. Required the Sun's true azimuth and error of the compass ; and, supposing the variation to be 18° 30′ E., required the deviation of the compass for the direction of the ship's head.

2. 1899, November 27th, 9ʰ 56ᵐ 56ˢ A.M. M.T.S., in lat. 23 27′ N., long. 122° 47′ 45″ E. ; the Sun's bearing by compass S.E. ; alt. of the Sun's lower limb, 37° 44′ 15″ ; index error, +2′ 30″ ; height of eye, 21 ft. Required the Sun's true azimuth and error of the compass ; and, supposing the variation to be 2° 15′ W., required the deviation of the compass for the direction of the ship's head.

3. 1899, September 22ᵈ 18ʰ 29ᵐ 38ˢ M.T.G., in lat. 49° 40′ S., long. 135° 24′ E. ; the Sun's bearing by compass, N. 61° 52′ 30″ W. ; alt. by observation of the Sun's upper limb, 22° 18′ 50″ ; height of eye, 20 ft. Required the Sun's true azimuth and error of the compass ; and, supposing the variation to be 0, required the deviation of the compass.

4. 1899, March 20th, at 4ʰ 4ᵐ P.M. M.T.S., in lat. 56° 20′ S. ; long. 63° 42′ W., the obs. alt. of Sun's upper limb, 16° 54′ 30″ ; index correction, +1′ 1″ ; height of eye, 27 ft. The Sun's comp. bearing W. ¼ S. Required the true azimuth, and comp. error ; and, supposing the variation to be 22° E., what is the deviation ?

5. 1899, September 1st A.M. at ship, in lat. 44° 18 N., long. 10° 15' W., when the G.M.T. was 31ᵈ 21ʰ 22ᵐ 37ˢ; the obs. alt. of the Sun's upper limb, 34° 21' 30"; index correction, +4' 15"; height of eye, 23 ft.; Sun's comp. bearing S.E. ¼ E. Required the true azimuth and comp. error; and, supposing the variation to be 25° W., what is the deviation?

6. 1899, December 10th, at 2ʰ 12ᵐ 36ˢ P.M. M.T.S., in lat. 20° 15' N., long. 161° 15' W.; the obs. alt. of Sun's lower limb, 34° 49' 20"; index correction, +1' 16"; height of eye, 28 ft. The Sun's comp. bearing S. 39° W. Required the true azimuth and comp. error; and, supposing the variation to be 10° E., what is the deviation?

ART. 46.—**Time Azimuths of the Sun by the Tables.**—When the Apparent Time at Ship, the Lat., and the Sun's Dec. are known, the Sun's True Azimuth can be found by Inspection in Burdwood's or Davis's Tables.

The same Tables will give the true Azimuth of a Star, a Planet, or the Moon, provided the Hour Angle and Declination are within the limits of the Tables.

Example.—Sun's Azimuth (see preceding Examples).

		Eq. of time.	H. var.
M.T.G. Dec.	17ᵈ 2ʰ 3ᵐ 0ˢ	3ᵐ 43ˢ·63	1·224
	+3 46	2·51	2·05
A.T.G. ,,	17ᵈ 2 6 46	3 46·14	6120
Long.	3 53 0		2448
A.T.S.	16ᵈ 22 13 46		2·50920

or 10 13 46 A.M.

A.T.S. 10ʰ 14ᵐ A.M.
Sun's Dec. 23° 22' S.
Lat. 41 22 N.

In Burdwood's Tables, with Lat. 41° and Dec. 23° (of contrary name) opposite 10ʰ 12ᵐ A.M., the Sun's True Azimuth is found to be 153° 22'; but when greater accuracy is required, corrections may be made for the remaining 2ᵐ and the minutes of Dec. and Lat., thus:

		Diff. for 4ᵐ 55
Azimuth by table, N.	153° 22' E.	,, 2ᵐ 27
Cor. for 2ᵐ	+27	
,, 22' Dec.	+9	Diff. for 1° Dec. 24
,, 22' Lat.	+3	22
True Azimuth, N.	154 1 E.	48
		48
or S.	25 59 E.	6,0)52,8
Compass S.	14 4 W.	9
Error of Compass	40 3 W.	Diff. for 1° Lat. +9
Variation	20 0 W.	22
		6,0)19,8
Deviation	20 3 W.	3

NOTE.—The Tables direct the Azimuth to be named the *same* as latitude.

AZIMUTHS.

EXERCISES.—(a) TIME AZIMUTHS OF THE SUN.

1. 1899, August 11^d 20^h 16^m 20^s M.T.G., in lat. 0° 0', long. 12° 50' W.; Sun's bearing by compass, E. ¾ N. Find the true azimuth and error of the compass by the "Time Azimuth Tables;" and, supposing the variation to be 20° 30' W., required the deviation for the direction of the ship's head.

2. 1899, September 3rd, 7^h 57^m A.M. M.T.S. in lat. 13° 29' N., long. 60° 30' E., the Sun's bearing by compass was E. by N. ¾ N. Find the true azimuth and error of the compass by the "Time Azimuth Tables;" and, supposing the variation to be 1° 10' W., required the deviation.

3. 1899, October 16th, 5^h 8^m 25^s P.M. A.T.S., in lat. 37° 17' S., long. 0° 45' E., the Sun's bearing by compass was N. 70° 18' 45" W. Find the true azimuth and error of the compass by the "Time Azimuth Tables;" and supposing the variation to be 19° 41' W., required the deviation.

4. 1899, November 10^d 15^h 52^m 0^s M.T.G., in lat. 52° 11' S., long. 169° 15' W.; the Sun's bearing by compass, W. 8° 26' 45" S. Find the true azimuth and error of the compass by the "Time Azimuth Tables;" and, supposing the variation to be 5° 31' W., what is the deviation.

5. 1899, December 24^d 19^h 33^m M.T.S., long. of Greenwich, lat. 22° 10' S., Sun's bearing by compass E. 15° 21' S. Find the true azimuth and error of the compass by the "Time Azimuth Tables;" and, supposing the variation to be 27° W., required the deviation for the direction of the ship's head.

6. 1899, November 29th, the ship in lat. 31° 30' 40" S., long. 5° 30' E., the correct mean time at Greenwich being 28^d 18^h 22^m; the Sun's bearing by compass, E. ¼ N. Required the true azimuth and error of the compass by the "Time Azimuth Tables;" and, supposing the variation to be 35° E., required the deviation for the direction of the ship's head.

7. 1899, Sepember 1st at ship, when the correct M.T.G. was 31^d 18^h 24^m 35^s in lat. 44° 23' S., long. 145° 17' E., the Sun's compass bearing was W. ½ N. Required the true azimuth, compass error and deviation by the "Tables," the variation being 10° 15' E.

8. 1899, September 23rd at ship, in lat. 0°, long. 150° W., when the correct M.T.G. was 22^d 18^h 30^s, the Sun's compass bearing West. Required the true azimuth by the "Tables;" also the compass error and deviation, the variation being 9° E.

9. 1899, October 7th A.M., the M.T.S. being 7^h 40^m in lat. 0°, long. 0° 15' W., the Sun's compass bearing S.E. by E. ¼ E. Required the true azimuth by the "Tables;" also the compass error and deviation, the variation being 20° W.

10. 1899, November 20th, the A.T.S. being 9^h 37^m A.M., in lat. 59° 42' N., long. 150° 45' W., the Sun's compass bearing S. 3° W. Required the true azimuth by the "Tables;" also compass error and deviation, the variation being 28° E.

11. 1899, December 3rd, at 4^h 7^m P.M. M.T. at ship, in lat. 35° 25' S., long. 8° 14' E., the Sun's compass bearing due West. Required the true azimuth by the "Tables;" also the compass error and deviation, the variation being 28° W.

12. 1899, September 1st at ship, in lat. 51° 25' S., long. 174° 25' E., the correct G.M.T. being 31^d 15^h 54^m 42^s, when the Sun's compass bearing was W. 8° 26' N. Required the true azimuth by the "Tables;" also the compass error and deviation, the variation being 30° E.

ART. 47.—Time Azimuth of a Star by the Tables.

Example.—1899, December 12th, 4^h 5^m A.M. mean time at ship, in lat. 5° 31' S., long. 32° 45' W., the bearing by compass of the Star Sirius was S. 56° 15' W. Required its true azimuth and error of the compass; also the deviation, supposing the variation to be 12° W.

TEXT-BOOK ON NAVIGATION.

M.T.S. Dec.	11ᵈ 16ʰ 5ᵐ		S.T. at noon	17ʰ 19ᵐ 56ˢ
Long.	2 11		Accel.	2 57 ⎫ 3 ⎭
M.T.G. Dec.	11ᵈ 18 16		M.T.S.	16 5 0

```
                    R.A.M.    9 27 56
                    R.A. Sirius 6 40 46

                    H.A. Sirius 2 47 10 W.
        H.A.    2ʰ 47ᵐ
        Dec.   16  35 S.
        Lat.    5  32 S.
```

In Davis's Tables, with Lat. 5° and Dec. 16° (of the same name), opposite 2ʰ 48ᵐ the azimuth is found to be 71° 44′.

Azimuth by Tables S. 71° 44′ W.   Change for 4ᵐ 4)16
  Cor. for 1ᵐ   − 4 ⎫   ,,   1ᵐ   4
   ,, 35′ Dec.   − 51 ⎬ = − 22
   ,, 32′ Lat. + 33 ⎭   Change for 1° 87′
                                            Dec. 35

Star's true Azimuth S. 71 22 W.   435
Compass bearing   S. 56 15 W.   261
                                          ─────
                                         6,0)304,5
Error of Compass 15 7 E.   ─────
Variation   12 0 W.   51

                                       Change for 1° Lat. 62′
Deviation   27 7 E.   32
                                             ───
                                             124
                                             186
                                           ─────
                                          6,0)198,4
                                             ───
                                              33

NOTE.—The Star's H.A. must always be found in the right-hand (or P.M.) column in the Tables.

EXERCISES.—(b) TIME AZIMUTHS OF STARS.

1. 1899, August 1st, 11ʰ 30ᵐ 30ˢ P.M. M.T.S. in lat. 60° N., long. 180° E., Markab bore by compass S. 56° 45′ E. ; variation, 15° 5′ E. Required the true azimuth and error of the compass, also the deviation for the direction of the ship's head.

2. 1899, August 17th, A.T.S. 3ʰ 20ᵐ A.M., in lat. 45° 30′ S., long. 128° 0′ E., the Star α Arietis bore by compass N.E. by N. Required the true azimuth and error of the compass by the "Time Azimuth Tables ;" and, supposing the variation to be 0, required the deviation of the compass.

3. 1899, September 28th, 10ʰ 15ᵐ P.M. A.T.S., in lat. 37° 20′ N., long. 142° 10′ W., the Star Altair bore by compass S.W. by W. ¼ W. Required the true azimuth and error of the compass by the "Time Azimuth Tables ;" and, supposing the variation to be 15° 20′ E., required the deviation for the direction of the ship's head.

4. 1899, October 14th, 2ʰ 19ᵐ A.M. M.T.S., in lat. 52° 49′ N., long. 21° 17′ W., the Star Sirius bore by compass S.E. Find the true azimuth by the "Time Azimuth Tables ;" and, supposing the variation to be 33° 15′ W., required the deviation of the compass for the direction of the ship's head.

5. 1899, November 12ᵈ 11ʰ 49ᵐ 20ˢ M.T.G., in lat. 31° 51′ S., long. 49° 50′ W., the Star α Arietis bore by compass N.E. ¼ N. Required the true

azimuth by the "Time Azimuth Tables;" and, supposing the variation to be 2° 30' E., required the deviation.

6. 1899, December 13^d 19^h 56^m 0^s M.T.G., in lat. 53° 20' N., long. 6° 30' W., the Star Arcturus bore by compass South. Required the true azimuth and error of the compass; and, supposing the variation to be 19° 30' W., required the deviation of the compass for the direction of the ship's head.

7. 1899, January 12th, in lat. 50° N., long. 50° W., when the M.T.S. was 6^h 30^m P.M., the compass bearing of Aldebaran was S. 57° 15' E. Required the true azimuth by the "Tables," compass error, and deviation, the variation being 32° W.

8. 1899, March 19th, in lat. 36° N., long. 28° E., when the M.T.G. was 18^d 10^h 35^m, the compass bearing of Spica was S. 8° 26' E. Required the true azimuth by the "Tables," compass error, and deviation, the variation being 6° 30' W.

9. 1899, June 29th, the M.T.S. being 4^h 32^m 19^s A.M., in lat. 44° 25' S., long. 65° E., the compass bearing of Markab was N. by E. ¾ E. Required the true azimuth by the "Tables," compass error, and deviation, the variation being 29° 45' W.

10. 1899, July 18th, when the M.T.G. was 17^d 16^h 37^m 40^s in lat. 37° 50' N., long. 50° W., the compass bearing of α Arietis was East. Required the true azimuth by the "Tables," compass error, and deviation, the variation being 14° 45' W.

11. 1899, September 21st, when the M.T.G. was noon in lat. 19° 20' S., long. 175° E., the compass bearing of Algenib was N. 11° 15' E. Required the true azimuth by the "Tables," compass error, and deviation, the variation being 10° 15' E.

12. 1899, December 20th, at 5^h 31^m 20^s A.M. M.T. at ship, in lat. 45° 20' N., long. 33° 25' W., the compass bearing of Regulus was W. by S. ¼ S. Required the true azimuth by the "Tables," compass error, and deviation, the variation being 33° 10' W.

CHAPTER XIII.

CHRONOMETERS AND LONGITUDE.

ART. 48. **Rating Chronometers.**—The ship's chronometer is relied on at sea to give the Mean Time at Greenwich (M.T.G.), its error on leaving port and its rate of gaining or losing per day being known.

An "Error" and "Rate" are usually supplied when the chronometer leaves the shop, but as the Rate is liable to be affected by changes of temperature and other causes during a voyage, it is advisable to ascertain the correct Error when opportunity offers at ports the ship may visit.

Rating a chronometer means finding the rate at which it gains or loses per day by dividing the sum or difference of the errors on different dates by the number of days elapsed.

Examples.—

1. A chronometer was found to be $4^m\ 30^s$ fast of mean noon at Greenwich on January 15th, 1897, and on March 30th, 1897, it was $7^m\ 35^s$ fast of mean noon at Greenwich. Required the rate.

```
            m.  s.
    Fast    4   30  on January 15.
    Fast    7   35  on March 30.
            ─────
            3   5 gained in 74 days.
               60
            ─────
         74)185(2ˢ·5 gain per day.
            148
            ─────
            370
            370
            ───
```

2. A chronometer was $14^m\ 25^s$ fast of mean noon at Greenwich on September 1st, 1897, and at mean noon at Greenwich on December 4th, 1897, it was $13^m\ 14^s\cdot5$ fast. Required the rate.

```
                m.  s.
        Fast 14  25 on September 1.
        Fast 13 14·5 on December 4.
        ─────────────────
          1 10·5 lost in 94 days.
         60
        ─────
        94)70·5(·75 losing daily.
           658
           ───
           470
           470
           ───
```

3. A chronometer was 1ᵐ 42ˢ slow of M.T.G. at mean noon on January 1st, 1897, and on March 31st at mean noon Greenwich time it was 6ᵐ 9ˢ slow. Required the rate.

```
                  m.  s.
          Slow 1  42 on January 1.
          Slow 6   9 on March 31.
          ─────────────────
           4  27 lost in 89 days.
          60
          ───
          89)267(3ˢ losing daily.
             267
             ───
```

4. On leaving Liverpool, January 19th, 1896, the ship's chronometer was 4ᵐ 2ˢ·5 slow of mean noon at Greenwich, and at Sydney on May 28th it was found to be 3ᵐ 10ˢ·5 fast of mean noon at Greenwich. Required the rate.

```
                    m.   s.
           Slow 4   2·5 on January 20.
           Slow 3  10·5 on May 28.
           ─────────────────
              52·0 gained in 130 days.
           130)52·0(·4 gaining daily.
              52·0
              ────
```

5. At New York on February 27th, 1896, a ship's chronometer was 18ˢ slow at Mean Noon Greenwich Time, and at Southampton on March 6th, at Greenwich Mean Noon it was 14ˢ fast. Required the rate.

```
                       s.
           Slow 18 on February 27.
           Fast 14 on March 6.
           ─────────────────
              32 gained in 8 days.
           8)32
           ────
              4 gaining daily.
```

6. A chronometer was fast of Mean Noon at Greenwich 2ᵐ 17ˢ on January 14th, 1897, and on June 1st it was 3ᵐ 55ˢ·5 slow of Mean Noon at Greenwich. Required the rate.

```
                    m.  s.
           Fast 2  17   on January 14.
           Slow 3  55·5 on June 1.
           ─────────────────
             6  12·5 lost in 138 days.
            60
           ─────
           138)372·5(2ˢ·7 losing daily.
               276
               ───
                965
                966
                ───
```

EXERCISES IN RATING CHRONOMETERS.

Find the chronometer "Rate" from the following errors at Mean Noon at Greenwich:—

No.	1st date.	Error.	2nd date.	Error.
		h. m. s.		h. m. s.
1	Aug. 28th, 1899	7 2·1 fast	Dec. 9th, 1899	2 24 fast
2	Nov. 20th, ,,	56 5 ,,	,, 30th, ,,	55 35 ,,
3	Sept. 30th, 1898	1 21 19 slow	Jan. 28th, ,,	1 10 43 slow
4	Oct. 29th, ,,	16 33 fast	,, 1st, ,,	15 45 fast
5	Dec. 29th, ,,	6 28·7 slow	April 17th, ..	5 56 slow
6	Jan. 31st, 1899	1 7 27 fast	May 15th, ,,	56 57 fast
7	April 12th, ,,	25 19 slow	July 1st, ,,	11 55 slow
8	Feb. 15th, ,,	13 0·3 ,.	Aug. 1st, ,,	10 30 ,,
9	Mar. 30th, ,,	1 41·1 fast	June 1st, ,,	2 25·2 fast
10	April 1st, ,,	10 24 ,,	Aug. 15th, ,,	11 32 ,,
11	Feb. 27th, ,,	0 19·5 ,,	July 31st, ,,	0 57·5 slow
12	Dec. 27th, 1898	Correct.	Feb. 20th, ,,	4 57 fast
13	,, 8th, ,,	1 4 slow	Jan. 1st, ,,	0 47·2 slow
14	,, 20th, ,,	1 2 55 fast	Mar. 30th, ,,	59 30 fast
15	Mar. 31st, ,,	5 45·5 slow	June 14th, ,,	4 45·5 slow
16	June 1st, 1899	1 8·8 ,,	Sept. 30th, ,,	0 40·1 fast
17	Sept. 11th, ,,	7 53 ,,	Oct. 11th, ,,	7 5 slow
18	June 16th, ,,	14 26·0 fast	Sept. 30th, ,,	3 18·4 ,,

ART. 49. **Longitude by Chronometer.**—The time by chronometer when corrected for its original Error and accumulated Rate gives the time at Greenwich, whilst the time at ship can be calculated from an observed Altitude of a Celestial Object and the known Latitude. The difference converted into *Arc* is the Longitude.

The Longitude is E. when Greenwich time is *less* than ship time, and W. when *greater*.

The following are required in the Formula :—
 (a) The true Altitude.
 (b) The Polar Distance.
 (c) The Latitude.

The **Polar Distance** is measured from the elevated Pole, which is the N. Pole in N. Lat., and the S. Pole in S. Lat. It is therefore 90° − Dec. when the Dec. is of the *same* name as the Lat., and 90° + Dec. when of a *different* name.

Examples.—

1. 1899, January 17th P.M. at ship, in lat. 46° 30′ S., when a chronometer showed 17ᵈ 7ʰ 36ᵐ 30ˢ, its error on M.T.G. at noon on November 20th, 1898, being 10ᵐ 24ˢ fast, and its rate 3ˢ·5 losing; the observed Altitude of the Sun's lower limb was 37° 1′ 50″ ; height of eye, 16 ft. ; index correction, +3′ 20″. Required the longitude.

CHRONOMETERS AND LONGITUDE.

	h. m. s.	d.	
Chron. 17ᵈ	7 36 30	58·3	
Fast	−10 24	3·5	
			Days elapsed.
	7 26 6	2915	Nov. 10 24)7ʰ·4
Acc. Rate	+3 24	1749	Dec. 31
			Jan. 17 ·3
M.T.G. 17ᵈ	7 29 30	6,0)20,4·05	Total 58
		3ᵐ 24ˢ	

		Dec.	H. var.	Eq. of T.	H. var.
			h.	m. s.	s.
Obs. Alt. of Sun's lower limb	37° 1′ 50″	20° 43′ 39″ S.	29·74	10 20·46	·822
I.C.	+3 20	−3 43	7·5	6·16	7·5
	37 5 10	20 39 56	14870	+10 26·62	4110
Dip	−3 55	90 0 0	20818		5754
	37 1 15	P.D. 69 20 4	6,0)22,3·050		6·1650
Cor.	−1 9		3′ 43″		
	37 0 6				
S.D.	+16 17				
True Alt.	37 16 23				
P. Dist.	69 20 4		cosec 10·028884		
Lat.	46 30 0		sec 10·162188		
Sum 2)	153 6 27				
Half sum	76 33 13		cosine 9·366489		
Half sum−Alt.	39 16 50		sine 9·801484		
	h. m. s.				
Hour Angle	3 48 30	...	log 9·359045 (Tab. XXXI. Norie)		
A.T.S. 17ᵈ	3 48 30				
Eq. of T.	+10 27				
M.T.S. 17	3 58 57				
M.T.G. 17	7 29 30				
Diff.	3 30 33				
	60				
4)	210 33 0				
Long.	52 38 15 W.				

2. 1899, February 24th, about 8ʰ 20ᵐ A.M. at ship, in lat. 56° 48′ N., and long. by account 74° 15′ W., the observed Altitude of the Sun's upper limb was 9° 17′ 45″; index correction, −6′ 30″; height of eye, 20 ft.; the chronometer showed 1ʰ 12ᵐ 20ˢ·5, and it was found to be 13ᵐ 35ˢ slow of mean noon at Greenwich on October 30th, 1898, and on December 24th it was 10ᵐ 55ˢ·5 slow of mean noon at Greenwich. Required the longitude.

TEXT-BOOK ON NAVIGATION.

```
r calculation.                                  Rate.
T. 23ᵈ 21ʰ 50ᵐ                    h. m.  s.       m.  s.              Acc. rate.
    3 25    Chron. February 24ᵈ  1 12 20·5  Slow  13 35                 2·9
   ——————                          10 55·5   ,,   10 55·5                62
   24  1 15
                                 ———————         ————————              ———
ys elapsed.                      1 23 16·0  Gained 2 39·5                58
 ————                              2 59·8         60                    174
 Oct.  1
 Nov. 30
 Dec. 24          M.T.G. 24ᵈ     1 20 16·2        55)159·5(2ˢ·9 gaining  6,0)17,9·8
  ——                                              110           daily   ————
  55                                              ————                  2 59·8
 Dec.  7
 Jan. 31                                          495
 Feb. 24                                          495
  ——
  62                                               Dec.                 H. var.
         Obs. Alt. of Sun's U.L. 9° 17' 45"       9° 25' 38" S.          55·42
                           I.C.  – 6 30             1 14                  1·33
                                 ————————         ————————              ———————
                                 9 11 15          9 24 24 S.             16626
                           Dip.  – 4 23             90                   16626
                                 ————————         ————————              5542
                                 9  6 52          99 24 24 S.
                           Cor.  – 5 35                                 6,0)7,3·7086
                                 ————————                               ——————————
                                 9  1 17                                 1 13·7
                           S.D.  –16 16          Eq. of T.               H. Var.
                                 ————————          m.    s.
                       True Alt. 8 45  1         13 24·59                ·383
                                                    ·50                  1·3
                                                 ————————               ————
                                                 +13 24·09               1149
                                                                          383
                                                                         ————
                                                                         ·4979
                     A.  8° 45'  1"
                     P. 99 24 24     ...    ...   cosec 10·005880
                     L. 56 48  0     ...    ...     sec 10·261566
                                     ————————
                     2)164 57 25
                        ————————
                        82 28 42     ...    ...  cosine  9·116944
Fig. 74.                73 43 41     ...    ...    sine  9·982246
                                   h.  m.  s.
                Sun's H.A.         3  50  38  (E. of Mer.)  log. 9·366436
                                  24
                                  ————————
                A.T.S. 23ᵈ        20   9  22
                Eq. of T.         +13 24
                                  ————————
                M.T.S. 23ᵈ        20 22 46
                M.T.G. 24ᵈ         1 20 16
                                  ————————
                                   4 57 30
                                  60
                                  ————————
                                4)297 30  0
                                  ————————
                                  74° 22' 30" W. long.
```

CHRONOMETERS AND LONGITUDE.

NOTES.—(a) When the Chronometer Time is not dated (as above), it is necessary to find the Greenwich date from the estimated time at ship and longitude.

(b) The latitude at time of observation is found at sea from the latitude at noon by allowing for the difference of latitude made during the interval between the time of observation and noon, the course and distance, or "run," being known.

3. 1899, June 10th P.M. at ship, in lat. 40° 25′ S., the observed Alt. of the Star α′ Crucis west of Meridian was 60° 11′ ; index correction, +7′ 31″ ; height of eye, 16 ft. ; a chron. which was 4ᵐ 36ˢ slow on mean noon at Greenwich on April 20 showed 10ᵈ 2ʰ 33ᵐ 52ˢ, its daily rate being 1ˢ·8 losing. Required the longitude.

```
            Chron. 10ᵈ 2ʰ 33ᵐ 52ˢ              1ˢ·8              Days elapsed.
                       +4  36                   51                 Apr. 10
                      ─────────                 ──                 May  31
                       2 38  28                 18                 June 10
                       +1  32                   90                 ──
                      ─────────              ─────────             51
       M.T.G. June 10ᵈ 2 40   0              (6,0)9,1·8
                                                1 32

     Star's R.A.         Star's dec.            R.A.M.S.
     12ʰ 21ᵐ 2ˢ         62° 32′ 21″ S.       S.T. 5ʰ 14ᵐ 29ˢ·6
                            90                accel.    19·7⎫
                        ─────────                        6·6⎭
                   P.D. 27  27  39           ─────────────
                                             5 14  55·9
   Observed Alt. Star 60° 11′  0″
               I.C.    +7  31
                      ─────────
                       60 18 31
                Dip    −3  55
                      ─────────
                       60 14 36
                Ref.      −33
                      ─────────
            T.A. 60 14  3
            Lat. 40 25  0   ...            sec   10·118416
            P.D. 27 27 39   ...            cosec 10·336165
                 ─────────
               2)128  6 42

                  64  3 21  ...    ...   cosine 9·640973
                   3 49 18  ...    ...   sine   8·823808
                    h. m. s.
     Star's H.A.    2 14  0  W.    log (xxxi) 8·919362
     Star's R.A.   12 21  2

     R.A.M.   14 35  2
     R.A.M.S.  5 14 56

     M.T.S. 10ᵈ  9 20  6
     M.T.G. 10ᵈ  2 40  0
           ──────────────
                  6 40  6
                 60
           ──────────────
               4)400  6
           ──────────────
          Long. 100°  1′ 30″ E.              FIG. 75.
```

114   TEXT-BOOK ON NAVIGATION.

NOTE.—When the Star is E. of Meridian, its H.A. may be subtracted from 24^h, as in the case of the Sun when A.M. at ship, and then add the R.A.; or the H.A. may be *subtracted* from the R.A. to find R.A.M.

The time at ship may be found in the same way as above from the altitude of a Planet or of the Moon.

EXERCISES—SUN'S ALTITUDE.

1. 1899, October 31st P.M. at ship; lat. at noon 41° 35' 30" N.; observed Alt. Sun's lower limb, 25° 35' 8"; index error, $-4'$ 30"; height of eye, 25 ft.; time by chron., $31^d\ 10^h\ 7^m\ 33^s$, which was slow September 29th $10^m\ 46^s\cdot 75$, and losing $1^s\cdot 01$ daily; run since noon, N.W. 12 miles. Required the longitude at sights and at noon.

2. 1899, November 27th A.M. at ship; lat. at noon 23° 38' 30" N.; observed Alt. Sun's lower limb, 37° 44' 15"; index error, $+2'$ 30"; height of eye, 21 ft.; time by chron., $26^d\ 13^h\ 53^m\ 34^s\cdot 6$, which was slow G.M.N. September 10th, $3^m\ 10^s$, and on October 20th was fast $2^m\ 30^s$; run since observation E.N.E. 30 miles. Required the longitude at observation and at noon.

3. 1899, September 23rd, at about $3^h\ 38^m$ P.M., lat. by D.R. at noon 49° 40' S., long. 135° 26' E.; ship becalmed; observed Alt. of Sun's upper limb, 22° 18' 52"; height of eye, 20 ft.; time by chron., $6^h\ 27^m\ 28^s$, which was slow G.M.N. July 10th $2^m\ 10^s$, and has kept mean time since; course and distance since observation, *nil*. Required longitude at observation and at noon.

4. 1899, March 20th P.M. at ship, in lat. 56° 20' S.; the observed Alt. of Sun's upper limb, 16° 54' 30"; index correction, $+1'\ 1"$; height of eye, 27 ft.; time by chron., $20^d\ 7^h\ 12^m\ 1^s$, which was slow on January 2nd $1^h\ 2^m\ 4^s$, and on February 1st was slow $1^h\ 3^m\ 54^s$ on G.M.T. Required the longitude.

5. 1899, September 1st A.M. at ship; lat. at noon 44° 42' N.; the observed Alt. Sun's upper limb, 34° 21' 30"; index correction, $+4'\ 15"$; height of eye, 23 ft.; time by chron., $31^d\ 21^h\ 34^m\ 3^s$, which was slow April 17th 4^s, and on June 6th was fast $4^m\ 8^s$; run since sights, N.W. ¼ N. 31 miles. Find the longitude at sights and noon.

6. 1899, December 10th, about $2^h\ 15^m$ P.M. at ship; lat. at noon 20° 15' N., approx. long. 161° W.; the observed Alt of Sun's lower limb, 34° 49' 20"; index correction, $+1'\ 16"$; height of eye, 28 ft.; a chron. showed $2^h\ 1^m\ 23^s$, which was fast $59^m\ 16^s$ on August 12th, and on October 31st was fast $1^h\ 2^m\ 16^s$ on G.M.T.; run since noon, E. 30 miles. Required the longitude at sights and noon.

EXERCISES—STARS' ALTITUDES.

1. 1899, July 4th, at $7^h\ 17^m$ A.M. at ship; lat. at noon 56° 9' 24" S., long. 74° 55' W.; course since noon, E. ¼ S. true; average speed, 8 knots per hour; obs. Alt. of Canopus E. of the Meridian, 54° 9' 33"; index error, $-30"$; height of eye, 41 ft. The chron. showed 12 o'clock; it was fast of G.M. noon 38^s on April 15th, and had gained daily $\cdot 75^s$, since. Required longitude by chronometer.

2. 1899, August 28th P.M. at ship, in lat. by Mer. Alt. of α Ophiuchi in the artificial horizon on shore, 15° 55' 0" S.; long. by cross-bearings 5° 44' 30" W.; obs. Alt. of Arcturus west of the Meridian 35° 0' 0"; index error, $-4'\ 0'$; height of eye, 15 ft.; time by chron., $5^h\ 58^m\ 0^s$, which had been found slow $1^h\ 10^m\ 15^s$ on 29th June, and slow $1^h\ 3^m\ 33^s$ on August 8th. Required longitude by chronometer.

CHRONOMETERS AND LONGITUDE.

3. 1890, September 28th, at about 6 P.M. ship, in lat. 50° 17' N. by α Lyrœ, long. by D.R. 160° 21' E. ; obs. Alt. of α Cygni, 68° 29' 56" E. of the Meridian ; height of eye, 22 ft.; time by chron., 7^h 25^m 21'·2, which was slow 0^m 2'·4 on June 29th, and fast 25'·6 on August 3rd. Required the longitude by chronometer.

4. 1899, October 10th, at about 5^h 30^m A.M. M.T.S., in Lat. 35° 47' 30" S., supposed long. 110° 45' E. ; obs. Alt. of Achernar, 35° 40' 0" W. of the Meridian ; index error, $-5'$ 9" ; height of eye, 21 ft.; time by chron., 10^h 20^m 20', which had been found fast 1^m 3' on May 12th, and was 40' slow August 20th. Required longitude by chronometer.

5. 1899, November 27th A.M. at ship, in lat. by Pole Star 53° 15' N.; obs. Alt. of Castor, 42° 43' 0" W. of the Meridian ; index error, $-1'$ 37" ; height of eye, 20 ft. ; time by chron., 26^d 19^h 17^m 58'·5, which had been found slow for M.G. noon 6^m 57' on August 21st, and on September 20th was 6^m 37'·5 slow. Required longitude by chronometer.

6. 1899, December 14th, at about 4^h 45^m P.M. M.T.S. ; position by D.R. at noon, lat. 51° 25' N., long. 11° 5' 45" W. ; course and distance since, S. 88° E. 56 miles ; observed Alt. of Vega, 49° 33' 0" W. of the Meridian ; index error, $+3'$ 21" ; height of eye, 19 ft. ; time by chron., 5^h 20^m 44', which was fast of G.M. noon 2^m 38'·1 on November 10th, and on November 27th was correct. Required the longitude.

7. 1899, February 3rd, approx. M.T. ship, 1^h 8^m A.M., in lat. 20° 30' N., long. by account 31° W. ; the observed Alt. of Capella W. of Mer. was 26° 29' ; index correction, $-1'$ 7" ; height of eye, 24 ft. ; time by chron., 3^h 55^m 30'·5, which was fast 43^m 19' on January 13th, and gaining daily 2'·5. Find the longitude.

8. 1899, April 5th P.M. at ship, in lat. 35° 40' S., long. by account 34° 15' E. ; about 2^h 45^m after Procyon was on the Mer., its observed Alt. was 33° 21' 30" ; index correction, $+1'$ 14" ; height of eye, 29 ft. ; time by chron., 6^h 48^m 20', which was slow 16^m 19'·5 on February 9th, and on March 1st was slow 17^m 29'·5. Find the longitude.

9. 1899, May 1st P.M. at ship; approx. M.T. ship, 6^h 30^m, in lat. 18° 14' N., long. by account 91° E. ; the observed Alt. of Rigel W. of Mer. was 25° 52' 20" ; index correction, $-1'$ 12" ; height of eye, 32 ft. ; time by chron., 12^h 0^m 0', which was slow 24^m 54' on January 11th, and on March 2nd was slow 24^m 34'. Find the longitude.

10. 1899, June 9th, at about 1^h 20^m A.M. M.T., ship in lat. 35° 15' S., long. by account 109° 30' W. ; the observed Alt. of Fomalhaut E. of Mer. was 35° 44' 30" ; index correction, $-2'$ 5" ; height of eye, 27 ft. ; time by chron., 12^h 0^m 0', which was slow 40^m 11' on April 4th, and gaining 3' daily. Find the longitude.

11. 1899, October 18th, at about 2^h 30^m A.M. M.T., ship in lat. 15° 14' S., long. by account 50° E. ; the observed Alt. of Sirius E. of Mer. was 53° 53' 10" ; index correction, $+2'$ 13" ; height of eye, 22 ft. ; time by chron., 11^h 3^m 22', which was fast 4^m 13'·7 on July 24th, and on August 29th was 2^m 11'·3 fast. Find the longitude.

12. 1899, December 14th P.M. at ship, in lat. 40° N., long. by account 14° 45' W. ; about 3^h 45^m before Pollux came to the meridian, its observed Alt. was 41° 43' 30" ; index correction, $-1'$ 9" ; height of eye, 32 ft. ; time by chron., 10^h 18^m 55', which was slow 1^h 5^m 32' on October 9th, and on November 14th was slow 59^m 14'. Find the longitude.

ART. 50. **Longitude and Azimuth.**—As the data used in the computation of *Apparent Time at Ship* and the *Azimuth* are the same, and the Formulæ very similar, it is most convenient to find the Azimuth at the same time as the Longitude by chronometer.

TEXT-BOOK ON NAVIGATION.

Example.—

Sun's T.A.	8° 45' 0"	...	...	...	...	sec 10·005084
,, P.D.	99 24 24	...	cosec 10·005880			
Lat.	56 48 0	...	sec 10·261566	...	sec 10·261566	

2)164 57 24

Half sum	82 28 42	...	cos 9·116944	...	cos 9·116944
Rem. for Time	73 43 42	...	sine 9·982246		
,, Azth.	16 55 42	...	... ...	...	cos 9·980762

Sun's H.A. 3 50 38 ... log 9·366436

2)19·364356

NOTE.—For Longitude see p. 112.

Half Az. 28° 45' ... sine 9·682178
              2

Sun's True Az. 57 30

EXERCISES.

1. 1899, July 1st A.M. at ship; lat. 56° 5' S. at noon, the observed Alt. Sun's upper limb was 9° 29' 45"; index error, −1' 30"; height of eye, 17 ft.; time by chron., 1ᵈ 3ʰ 8ᵐ 47ˢ, which was fast of Greenwich mean noon on January 1st, 2ᵐ 10ˢ, and on May 1st was slow 1ᵐ 20ˢ; run between sights and noon, W.S.W. (true) 13 miles. Required the longitude at sights and at noon; and, supposing the Sun's bearing by compass at the same time to be N. ¼ W., and the variation to be 22° 45' E., required the true azimuth and deviation for the direction of the ship's head.

2. 1899, August 7th A.M. at ship; lat. at noon 39° 31' 40" N., the observed Alt. of the Sun's lower limb was 53° 53' 5"; index correction, +3' 0"; height of eye, 29 ft.; Sun's bearing, S. 30° E. by compass; time by chron., 7ᵈ 0ʰ 0ᵐ 9ˢ, which was slow 10ˢ on January 9th, and on 29th April was correct for M.T.G.; course and distance since observation, W. 25 miles. Required the longitude at sights and at noon; and if the variation is 28° 15' W., find the true azimuth and deviation of the compass for the direction of the ship's head.

3. 1899, September 23rd P.M. at ship; lat. at noon 40° 18' S., the observed Alt. of the Sun's lower limb was 42° 55' 41"; height of eye, 7 ft.; time by chron., 22ᵈ 15ʰ 53ᵐ 29ˢ, which was fast 16ᵐ 20ˢ for G.M.N. on the 4th May, and on August 2nd was fast 10ᵐ 47ˢ; run between noon and observation, W.N.W. 21 miles; the Sun's bearing at the same time was N.W. by compass, and the variation 10° 45' E. Required the true azimuth and deviation of the compass for the direction of the ship's head.

4. 1899, September 22nd, at about 8ʰ 50ᵐ A.M., when the chron. showed 0ʰ 20ᵐ 32ˢ, the observed Alt. of the Sun's upper limb was 30° 59' 25"; index error, 2' 00" off the arc; height of eye, 30 ft.; Sun's bearing by compass, S.E. ¼ E.; variation, 11° 30' W.; lat. at observation, 42° 16' 00" N.; long. by account, 70° 40' W.; on June 16th the chronometer was slow 1ʰ 14ᵐ 30ˢ for G.M.N., and on August 5th it was 1ʰ 11ᵐ 8ˢ slow. Required longitude, true azimuth, and deviation for the direction of the ship's head.

5. 1899, December 25th, at about 1ʰ 15ᵐ P.M., the observed Alt. of the Sun's lower limb was 19° 24' 14"; index error, +5' 30"; height of eye, 22 ft.; the Sun's bearing by compass, S.; time by chron., 9ʰ 9ᵐ 44ˢ, which was fast of G.M.N. on 31st October 28ˢ, and losing 5ˢ daily; the lat. at observation was 45° 5' 30" N. long. by D.R., 60° 15' E. Required the longitude by chronometer; required also the true azimuth and error of the

compass, and, supposing the variation to be 1° 30' E., find the deviation for the direction of the ship's head.

6. 1900, January 1st P.M. at ship; lat. 44° 30' 30" N. at noon, the observed Alt. of Sun's lower limb was 20° 52' 37"; index error, +5' 30"; height of eye, 10 ft.; Sun's bearing S. ¼ E. by compass; time by chron., January 1^d 0^h 0^m 0^s, which was correct for G.M.N. on January 6th, 1899, and has been losing 1' daily since; course and distance since noon, S.W. true, 12 miles. Find longitude at noon and observation, and if the variation was 10° 45' W., find the true azimuth and deviation for the direction of the ship's head.

7. 1899, April 16th P.M. at ship; lat. at noon 30° 1' N.; the observed Alt. of the Sun's upper limb, 30° 43' 40"; index correction, − 5' 40"; height of eye, 29 ft.; time by chron., 15^d 19^h 18^m 56^s, which was slow 9^m 21^s·5 on February 9th, and on March 17th was slow 3^m 19^s·5 on G.M.T.; the Sun's compass bearing was W. ½ N.; variation, 2° W. Required longitude at sights and noon, if run since noon be N.E. ½ N. 31 miles; also compass error and deviation.

8. 1899, August 4th A.M. at ship; lat. at noon 15° 22' N.; the observed Alt. of the Sun's lower limb, 22° 31' 30"; index correction, −1' 30"; height of eye, 25 ft.; bearing by compass E. ¼ N.; time by chron., 3^d 16^h 42^m 11^s, which was slow 9^m 44^s·2 on April 20th, and on June 3rd was slow 10^m 15^s on G.M.T. Required longitude at sights and noon; run since sights W. 43 miles; also compass error and deviation, the variation being 5° W.

9. 1899, September 23rd P.M. at ship, in lat. 23° 28' S.; the observed Alt. of the Sun's upper limb, 41° 32' 15"; index correction, − 2' 30"; height of eye, 34 ft.; bearing by compass, W.N.W.; time by chron., 22^d 15^h 58^m 53^s, which was slow on April 10th 21^m 41^s, and on July 19th was slow 18^m 14^s on G.M.T. Required the longitude, compass error, and deviation, the variation being 6° 30' E.

10. 1899, October 17th A.M. at ship; lat. at noon 0° 30' S.; the observed Alt. of the Sun's lower limb, 34° 37' 40"; index correction, −1' 18"; height of eye, 17 ft.; bearing by compass, E. by S. ¼ S.; time by chron., 17^d 0^h 20^m 4^s, which was fast 58^m 22^s on July 5th, and on July 30th was fast 1^h 2^m 37^s on G.M.T. Required longitude at sights and noon; run since sights S.S.W. ¼ W. 34 miles; also find compass error and deviation, the variation, 0°.

11. 1899, November 28th P.M. at ship in lat. 50° 20' N.; the observed Alt. of the Sun's lower limb, 12° 12' 40"; index correction, −1' 5"; height of eye, 31 ft.; bearing by compass, S.W. ¼ W.; time by chron., 28^d 2^h 16^m 9^s, which was fast 30^m 6^s on July 26th, and on September 14th was fast 20^m 5^s on G.M.T. Required the longitude, compass error, and deviation, the variation being 20° W.

12. 1899, December 24th, about 3^h 20^m P.M. at ship; lat. at noon 4° 51' N., long. by account, 176° W.; the observed Alt. of the Sun's lower limb, 33° 39' 30"; index correction, +1' 27"; height of eye, 25 ft.; a chron. showed 3^h 4^m 17^s·4, which was slow 2^m 15^s·3 on December 9th, and on December 24th was correct for G.M. noon. Required the longitude at sights and noon, the run since noon being S. by E. ⅜ E., 34 miles; also, if the Sun's compass bearing was S.S.W. ¼ W. and variation 10° E., find the compass error and deviation.

CHAPTER XIV.

SUMNER'S METHOD BY PROJECTION.

ART. 51. At any given moment the Sun is vertical over some place on the Earth's surface, and if a circle be drawn with this point as the centre, the Sun would have the same altitude from all points on the circumference of this circle.

Advantage is taken of this fact to find the Ship's Position by Projection on a chart.

Two points on the circumference of the circle can be found from the Sun's altitude, and two assumed latitudes; the projection on the chart of the arc of the circle of equal altitudes passing through these two positions would pass through the actual position of the ship.

After an interval of time another altitude is observed, and with the same assumed latitudes, two other points on a new circle of position are found, and projected on the chart in the same way. The intersection of the two projected lines gives the ship's position.

In practice, the longitudes computed are marked on Mercator's Chart on the parallels of assumed latitudes, and connected by straight lines called *Lines of Position*. The point of intersection of these lines gives the latitude and longitude of the ship.

The Sun's true bearing is at right angles to the line of position at each observation—towards the east if A.M., and towards the west if P.M.

If the ship changes place during the interval, the first line of position has to be removed (parallel to itself) according to the course and distance sailed.

Example.—1899, May 15th A.M. at ship, and position uncertain, the Altitude of Sun's lower limb was observed to be 19° 27′ 45″ when the M.T.G. by chron. was 15^d 2^h 52^m 38^s; again, when A.M. at ship and the M.T.G. 15^d 7^h 12^m 52^s, the observed Altitude of Sun's lower limb was 56° 18′ 10″; index correction, +2′ 4″; height of eye, 31 ft.; the true course and distance of the ship during the interval, S.W. 36 miles. Required the "line of position" and Sun's true azimuth at each observation, and the ship's position at the second observation, the assumed latitudes being 49° N. and $49\frac{1}{2}$° N.

SUMNER'S METHOD BY PROJECTION.

First observation, M.T.G. 15ᵈ 2ʰ 52ᵐ 38ˢ.

```
  Sun's dec.         H. var.      Eq. of T.      H. var.      Sun's alt.
18° 53'  1" N.       35·38       3ᵐ 48ˢ·76       ·019        19° 27' 45"
      1 43            2·9                         2·9            +2  4
     ─────           ─────                       ─────         ─────────
 18 54 44            31842                        171        19 29 49
 90  0  0             7076                         38          -5 27
 ─────────           ──────                       ────        ─────────
 71  5 16           10,2·602                     ·0551       19 24 22
                      1 43                                     -2 33
                                                              ─────────
                                                             19 21 49
                                                             +15 50
                                                             ─────────
                                                       T.A. 19 37 39
```

Second observation, M.T.G. 15ᵈ 7ʰ 12ᵐ 52ˢ.

```
  Sun's dec.         H. var.      Eq. of T.      H. var.      Sun's alt.
18° 53'  1" N.       35·38       3ᵐ 48ˢ·76       ·019        56° 18' 10"
      4 15            7·2           ·14           7·2            +2  4
     ─────           ─────        ──────────     ─────         ─────────
 18 57 16             7076       3  48·62         038         56 20 14
 90  0  0            24766                        133          -5 27
 ─────────          ───────                      ─────        ─────────
 71  2 44           254·736                      ·1368        56 14 47
                      4 15                                     -33
                                                              ─────────
                                                             56 14 14
                                                             +15 50
                                                             ─────────
                                                       T.A. 56 30  4
```

```
A. 19° 37' 39"                      A. 56 30  4
L. 49   0   0      49 30  0         L. 49  0  0       49 30  0
P. 71   5  16      ────────         P. 71  2 44       ────────
   ─────────                           ─────────
2)139 42 55                         2)176 32 48

    69 51 27         70  6 27           88 16 24         88 31 24
    50 13 48         50 28 48           31 46 20         32  1 20
Secs   10·183057    10·187456         10·183057        10·187456
Cosecs 10·024101    10·024101         10·024211        10·024211
Cosines 9·537008     9·531806          8·479020         8·411112
Sines   9·885711     9·887281          9·721434         9 724479
        ─────────    ─────────         ─────────        ─────────
        9·629877     9·630644          8·407722         8·347258

H.A.'s 5 26 10       5 26 31           1 13 37          1  8 37
       24            24                24               24
       ─────────    ─────────         ─────────        ─────────
  14ᵈ 18 33 50   14ᵈ 18 33 29      14ᵈ 22 46 23     14ᵈ 22 51 23
      -3 49           -3 49              -3 49             -3 49
      ─────────      ─────────         ─────────        ─────────
  14 18 30  1     14 18 29 40       14 22 42 34     14 22 47 34
  15  2 52 38     15  2 52 38       15  7 12 52     15  7 12 52
      ─────────      ─────────         ─────────        ─────────
       8 22 37        8 22 58          8 30 18          8 25 18
       60             60               60               60
      ─────────      ─────────         ─────────        ─────────
   4)502 37  0    4)502 58  0      4)510 18  0     4)505 18  0
      ─────────      ─────────         ─────────        ─────────
   (1) 125 39 15 W.  (2) 125 44 30 W.  (3) 127 34 30 W.  (4) 126 19 30 W.
```

Projection.

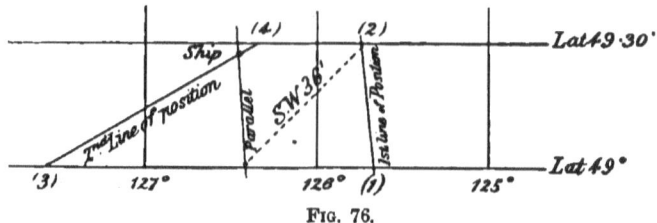

FIG. 76.

Explanation.—Mark the longitudes (1) and (2) on their respective latitudes. The line joining (1) and (2) is the First Line of Position. Next, set off 36' on a S.W. course from any point in the first line, and draw a line parallel to that line. This is the corrected position of the first line in consequence of the ship's change of place during the interval.
Then mark the longitudes (3) and (4), and draw the Second Line of Position. The intersection of this with the parallel to first line is the position of the ship.

Answers.—
First observation : Direction of line of position, N. 6° W. and S. 6° E.
    Sun's true azimuth, N. 84° E.
Second observation : Direction of line of position, N. 59° E. and S. 59° W.
    Sun's true azimuth, S. 31° E.
    Ship's position, lat. 49° 26½' N., long. 126° 27' W.

EXERCISES IN SUMNER'S METHOD.

1. 1899, September 23rd, being A.M. at ship when a chronometer showed 22ᵈ 15ʰ 52ᵐ 39ˢ M.T.G., the observed Altitude of the Sun's lower limb was 32° 13' 6", and again being P.M. at ship when the chronometer showed 22ᵈ 20ʰ 52ᵐ 40ˢ M.T.G., the observed Altitude of the Sun's lower limb was 32° 13' 13", the ship having made 21 miles on a N. 60° E. course during the interval; height of eye, 28 ft.; index error, −2' 4". Required the line of bearing when the first Altitude was taken, and the Sun's true azimuth ; also the position of the ship by Sumner's method when the second Altitude was observed, assuming the latitudes 47° 20' and 47° 40' N.

2. 1900, January 1st, being A.M. at ship when a chronometer showed 31ᵈ 10ʰ 25ᵐ 9ˢ M.T.G., the observed Altitude of the Sun's lower limb was 14° 14' 35", and again being P.M. at ship when the chronometer showed 31ᵈ 14ʰ 25ᵐ 34ˢ; the observed Altitude of the Sun's lower limb was 11° 17' 0", the ship having made 34 miles on a S. 69° E. course during the interval ; height of eye, 27 ft.; index error, *nil*. Required the line of bearing when the first Altitude was taken, and the Sun's true azimuth; also the position of the ship by Sumner's method when the second Altitude was observed, assuming the latitudes 49° and 49° 30' N.

3. 1899, June 23rd, being A.M. at ship when a chronometer showed 22ᵈ 22ʰ 40ᵐ 10ˢ M.T.G., the observed Altitude of the Sun's lower limb was 59° 4' 0", and again being P.M. at ship when the chronometer showed 23ᵈ 2ʰ 59ᵐ 55ˢ M.T.G.; the observed Altitude of the Sun's lower limb was 46° 19' 40" ; the ship having made 39 miles on a S. 83° W. course during the

interval; height of eye, 26 ft.; index error, −3′ 15″. Required the line of bearing when the first Altitude was taken, and the Sun's true azimuth; also the position of the ship by Sumner's method when the second Altitude was observed, assuming the latitudes 50° 0′ and 51° 0′ N.

4. 1889, February 21st A.M. at ship; at sea and uncertain of my position, when a chronometer showed $20^d\ 22^h\ 14^m\ 35^s$ G.M.T., the observed Altitude of the Sun's lower limb was 11° 22′ 30″; and again the same morning, when a chronometer showed $21^d\ 2^h\ 44^m\ 33^s$ G.M.T., the observed Altitude of the Sun's lower limb was 47° 17′ 20″; index correction, + 2′ 30″; height of eye, 27 ft.; the ship made 36 miles on a true N.N.W. ¾ W. course in the interval. Required the line of position when the first Altitude was observed, and the true bearing of the sun and ship's position at the second observation by Sumner's method by projection, assuming latitudes 50° 10′ and 50° 45′ S.

5. 1899, June 14th P.M.; at sea and uncertain of my position, when a chronometer showed $14^d\ 13^h\ 19^m\ 13^s$ G.M.T., the observed Altitude of the Sun's lower limb was 59° 32′ 0″; and again P.M., when a chronometer showed $14^d\ 17^h\ 0^m\ 15^s$ G.M.T., the observed Altitude of the Sun's lower limb was 26° 49′ 10″; index correction, −2′ 4″; height of eye, 29 ft.; the ship's true course and distance in the interval, S.W. ¼ S. 37 miles. Required the line of position, Sun's true bearing, and position of the ship at second observation by Sumner's method, by projection, assuming latitudes 48° and 48° 30′ N.

6. 1899, August 24th A.M.; at sea and uncertain of my position, when a chronometer showed $23^d\ 19^h\ 37^m\ 44^s$ G.M.T., the observed Altitude of the Sun's upper limb was 24° 33′ 50″; and again A.M. on the same day, when a chronometer showed $23^d\ 23^h\ 1^m\ 41^s$ G.M.T., the observed Altitude of the Sun's upper limb was 49° 9′ 20″; index correction, +1′ 15″; height of eye, 27 ft.; the ship's true course and distance in the interval, W. by N. 27 miles. Required the line of position and Sun's true bearing at the first observation, and position of the ship at the second observation by Sumner's method by projection, assuming latitudes 49° 50′ and 50° 20′ N.

7. 1899, October 11th A.M.; at sea and uncertain of my position, when a chronometer showed $10^d\ 11^h\ 8^m\ 51^s$ G.M.T., the observed Altitude of the Sun's lower limb was 24° 1′ 20″; and again P.M. on the same day, when a chronometer showed $10^d\ 15^h\ 32^m\ 56^s$ G.M.T., the observed Altitude of the Sun's lower limb was 30° 56′ 10″; index correction, −2′ 13″; height of eye, 42 ft.; the ship's true course and distance in the interval, N. 78° E. 36 miles. Required the lines of position and Sun's true bearings at the first and second observations, and position of the ship at the second observation, by Sumner's method by projection, assuming latitudes 47° and 48° N.

CHAPTER XV.

THE CHART.

CHARTS used by Navigators are all constructed on Mercator's principle (except those of Harbours, Bays, etc.). Their characteristic is that the spaces between the Parallels of Latitude increase towards the Poles, whilst the Meridians of Longitude are also parallel instead of converging as they do on a globe. The reason for this construction is explained in Part II.

Latitudes are marked on the *Graduated Meridians* at the sides of the chart, and longitudes on the *Graduated Parallels* at the top and bottom.

Soundings in fathoms for Mean Low Water of Ordinary Spring Tides are given in small figures, and Roman numerals at various places give the Time of High Water, "Full and Change." The nature of the bottom is also shown by small letters and abbreviations.

The **Compasses** on *Channel Charts* are usually *magnetic*, but on *Ocean Charts true* compasses are the rule. In the Admiralty sheets for the channels, the magnetic N. point has a *fleur de lis*, and the true N. a star—the difference between the two N. points being the variation, which is subject to an annual change.

Variation Charts are charts of the oceans or of the globe with lines of equal variation drawn on them at intervals of one or two degrees. From these charts the variation at ship can be found, provided the latitude and longitude are known.

When out of sight of land, the ship's position is fixed by the latitude and longitude found from observation of heavenly bodies. Near the coasts the position can be obtained from *soundings* and from *bearings* of lights and other known objects.

The following practical problems are in the Board of Trade Examination for Mates and Masters :—

THE CHART.

Ship's head by compass.	Card A. Deviation.	Card B. Deviation.
N.	11 E.	36 W.
N. by E.	16 E.	36 W.
N.N.E.	21 E.	30 W.
N.E. by N.	27 E.	27 W.
N.E.	31 E.	24 W.
N.E. by E.	31 E.	21 W.
E.N.E.	28 E.	17 W.
E. by N.	26 E.	12 W.
E.	24 E.	6 W.
E. by S.	20 E.	2 E.
E.S.E.	15 E.	6 E.
S.E. by E.	9 E.	12 E.
S.E.	4 E.	17 E.
S.E. by S.	1 E.	22 E.
S.S.E.	2 W.	27 E.
S. by E.	5 W.	32 E.
S.	8 W.	36 E.
S. by W.	11 W.	38 E.
S.S.W.	14 W.	37 E.
S.W. by S.	18 W.	35 E.
S.W.	21 W.	32 E.
S.W. by W.	24 W.	27 E.
W.S.W.	27 W.	20 E.
W. by S.	30 W.	12 E.
W.	26 W.	5 E.
W. by N.	23 W.	3 W.
W.N.W.	21 W.	10 W.
N.W. by W.	19 W.	16 W.
N.W.	16 W.	22 W.
N.W. by N.	11 W.	27 W.
N.N.W.	4 W.	31 W.
N. by W.	4 E.	34 W.

EXAMPLE.

Using Deviation Card B.
(a) Find the course to steer by Compass, from A (5′ S. of Kinsale Head) to B (Saltees, L. V.); also the deviation, variation, and distance.

Ans. Course S. 88° E.
      Dist. 75 miles.
      Var. 21° W.
      Dev. 5° W.

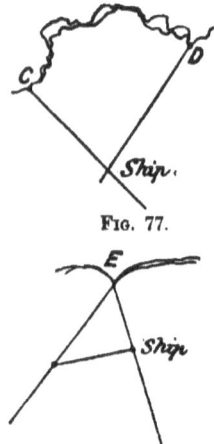

Fig. 77.

Fig. 78.

(b) With the ship's head on the above Compass course, a point C (Daunt's Rock) bore by Compass N.W., and D (Ballycottin) bore N.E. by N. ¼ N. by the same Compass. Find the ship's position.

Ans. Lat. 51° 40′ N.
Long. 8 0 W.

(c) With the ship's head as above, a point E. (Mine Head) bore by Compass N.E. by N., and after continuing 11 miles on the above-named course, it bore N. by W. ¼ W. Find the ship's position and distance from E. at the second bearing.

Ans. Lat. 51° 50½′ N.
Long. 1 21 W.
Distance from E. 12 miles.

(d) Find the course from A to B in a current which set W.S.W. 1½ mile per hour, the vessel going 12 miles per hour; also the distance made good towards B in 4 hours.

Ans. Course to Steer, E.
Dist. made good, 42 miles.

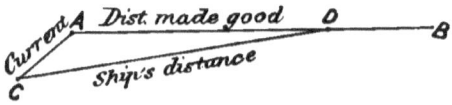

Fig. 79.

Explanation.—(a) Chart Course from A to B, E. ¼ N. (= N. 87° E.) magnetic.

From Card B.

	Comp. co.	Dev.	Corr. mag.
	E.	6° W.	N. 84° E.
	E. by S.	2 E.	S. 77 E.
(Diff. of N. 84° E. and N. 87° E.	(Diff. of dev.)	8	19 (Diff. of courses)
		3	
		19)24	
		1° to be subtracted from 6° W.	

Hence the required deviation is 5° W., which, applied to the magnetic course N. 87° E., gives the course to steer by compass S. 88° E.

The distance is found from the graduated Meridian by taking one-half *above* and the same *below* the middle latitude.

Note 1.—When the chart course is *true*, the *variation* must first be applied to find the magnetic course. The bearings must also be corrected for variation.

NOTE 2.—The deviation can generally be found by inspection, as the difference of deviation between two adjacent points is usually small.

(b) Here the compass bearings must be corrected for the deviation found above for ship's head.

Compass bearings N. 45° W.   N. 30° E.
Dev. 5 W.   5 W.
Corr. mag. bearings N. 50 W.   N. 25 E.

which, being drawn on the chart from the two objects, give the ship's position by their intersection.

(c) Correct the compass bearings as before.

Compass bearings N. 34° E.   N. 17° W.
Dev. 5 W.   5 W.
N. 29 E.   N. 22 W.

Draw lines opposite to these bearings from the given point of land; then lay the parallel rulers on A and B, and, sliding them across the two lines, find where the distance run exactly measures the space between the lines. This gives the ship's track and the positions at both observations.

(d) First draw a line from A in the direction of the set of the current, and mark on it the drift in four hours AC. From C set off with a pair of dividers the distance the ship would travel in four hours, and mark the point where it cuts AB in D. The line CD gives the course to sail in the current, and AD is the distance made good in four hours.

By Chart the Magnetic Course is E. ¼ N. = N. 84° E.

By dev. card B.
Comp. co.   Dev.   Mag. co.
E.   6° W.   N. 84° E.

Here the deviation is found to be 6° W. without interpolation, and therefore the course to steer is N. 84° E. + 6°, *i.e.* N. 90° E. or due E.

EXERCISES.

1. *Using Card A.*

(a) Find the course to steer by Compass from A (New Quay, S. Wales) to B (Rockabill Light); also give the deviation, variation, and distance.

(b) With the ship's head on the above Compass course, a point C (Cardigan Bay, L.V.) bore by Compass N. 66° W., and D (St. Tudwall) bore N. 29° E. by the same Compass: find the ship's position.

(c) With the ship's head as above, a point E (Codling, L.V.) bore by Compass N. 66° W., and after continuing 13½ miles on the same course it bore S. 55° W.: find the ship's position and distance from E.

(d) Find the course from A to B in a current which set W.N.W. (corr. mag.) 4 miles per hour, the vessel going 12 miles per hour; also the distance made good towards B in 5 hours.

2. *Using Card A.*

(a) Find the course to steer by Compass from A (Aberystwith) to B (Bardsey Island); also give the deviation, variation, and distance.

(b) With the ship's head on the above Compass course, a point (Aberdovey buoy), bore by Compass S. 74° E., and Sarn-y-Bwch (buoy) bore N. 44° E. by the same Compass: find the ship's position.

(c) With the ship's head as above, a point (St. Tudwall) bore by Compass N. 27° E., and after continuing 8 miles on the same course it bore N. 66° E.: find the ship's position and distance from St. Tudwall.

(d) Find the course from Aberystwith to Bardsey Island in a current which set S.W. (corr. mag.) 3 miles per hour, the vessel going 8 miles per hour; also the distance made good towards Bardsey in 3 hours.

3. *Using Card A.*

(a) Find the course to steer by Compass, the variation being 10° E., from A, lat. 51° 0′ N., long. 179° 0′ W., to B, lat. 49° 30′ N., long. 178° 0′ E.; also give the deviation, variation, and distance.

(b) With the ship's head on the above Compass course, a point C, lat. 51° 0′ N., long. 179° 30′ W., bore by Compass N.E. by N., and D, lat. 50° 45′ N., long. 180° 0′, bore N. 50° W. by the same Compass: find the ship's position.

(c) With the ship's head as above, a point D, lat. 50° 45′ N., long. 180° 0′, bore by Compass S. by W., and after continuing 45 miles on the same course it bore E. by S.: find the ship's position and distance from D.

(d) Find the course from A to B in a current which set north (true) 3 miles per hour, the vessel going 10 miles per hour; also the distance made good towards B in 6 hours.

4. *Using Card B.*

(a) Find the course to steer by Compass from A (Bardsey Island) to B (Fishguard); also give the deviation, variation, and distance.

(b) With the ship's head on the above Compass course, a point C (New Quay) bore by Compass S. 57° E., and D (Cardigan L.V.) bore S. 30° W. by the same Compass: find the ship's position.

(c) With the ship's head as above, a point E (Cardigan I.) bore by Compass S. 43° E., and after continuing 8 miles on the same course it bore S. 88° E.: find the ship's position and distance from Cardigan I.

(d) Find the course from A (Bardsey) to B (Fishguard) in a current which set (corr. mag.) N. 4 miles per hour, the vessel going 10 miles per hour; also the distance made good towards Bardsey in 4 hours.

5. *Using Card A.*

(a) Find the course to steer by Compass from A (Ardrossan) to B (Ailsa Craig); also give the deviation, variation, and distance.

(b) With the ship's head on the above Compass course, a point C (King's Cross Point) bore by Compass N. 7° W., and D (Pladda Lighthouse) bore N. 60° W. by the same Compass: find the ship's position.

(c) With the ship's head as above, a point (Pladda Light) bore by Compass N. 19° W., and after continuing 5 miles on the same course it bore N. 28° E.: find the ship's position and distance from Pladda Light.

(d) Find the course from A to B in a current which set (corr. mag.) E.S.E. 2 miles per hour, the vessel going 6 miles per hour; also the distance made good towards B in 4 hours.

6. *Using Deviation Card B.*

(a) Find the course to steer by Compass from Ramsey to Whitehaven; also give the deviation, variation, and distance.

(b) With the ship's head on the above Compass course, a point (Ayre Point) bore by Compass N. 52° W., and Bahama, L.V., bore S. 63° W. by the same Compass: find the ship's position.

THE CHART.

(c) With the ship's head as above, a point (St. Bee's Head) bore by Compass S. 80° E., and after continuing 8¼ miles on the same course it bore S. 26° W. : find the ship's position and distance from St. Bees Head.

(d) Find the course from Ramsey to Whitehaven in a current which set N. (corr. mag.) 2 miles per hour, the vessel going 8 miles per hour ; also the distance made good towards destination in 3 hours.

7. With Card B.

(a) Find the course to steer by Compass from A (Chicken Rock) to B (Sanda Island); also give the deviation, variation, and distance.

(b) With the ship's head on the above Compass course, a point C (Port Patrick) bore by Compass S. 11° E., and D (Corsewall Point) bore S. 76° E. by the same Compass : find the ship's position.

(c) With the ship's head as above, a point E (Maidens) bore by Compass N. 18° W., and after continuing 20 miles on the above-named course it bore S. 78° W. : find the ship's position and distance from E.

(d) Find the course from A to B in a current which set E.N.E. (corr. mag.) 3 miles per hour, the vessel going 10 miles per hour ; also the distance she makes good towards B in 4 hours.

8. With Card B.

(a) Find the course to steer by Compass from A (off Tuskar, in 52° N., and 6° 16′ W.) to B (52° 30′ N. and 5° 44′ W.); also the deviation, variation and distance.

(b) With the ship's head on the above Compass course, a point C (Tuskar) bore by Compass N. 75° W., and D (Lucifer, L.V.) bore N. 11° W. by the same Compass : find the ship's position.

(c) With the ship's head as above, a point E (Blackwater, L.V.) bore by Compass N. 6½° E., and after continuing 10 miles on the above-named course it bore N. 45° W. : find the ship's position and distance from Blackwater, L.V.

(d) Find the course from A to B in a current which set N.E. (corr. mag.) 2 miles per hour, the vessel going 6 miles per hour ; also the distance made good towards B in 3 hours.

9. With Card B.

(a) Find the course by Compass from Round Island (Scilly) to Smalls ; also the deviation, variation, and distance.

(b) With the ship's head on the above Compass course, a point (Sevenstones, L.V.) bore by Compass S. 16° W., and Round Island bore S. 55½° W. by the same Compass : find the ship's position.

(c) With the ship's head as above, a point E (lat. 51° N., long. 5° 22′ W.) bore by Compass S. 76° E., and after continuing 23 miles on the above-named course it bore S. 25° E. : find the ship's position and distance from E.

(d) Find the course from Round Island to Smalls in a current which set W. 21° N. (corr. mag.) 15 miles, the vessel sailing 30 miles in the same time ; and the distance made good towards Smalls.

10. With Deviation Card B.

(a) Find the course to steer by Compass from A (Kish, L.V.) to B (Langness Point) ; also the deviation, variation, and distance.

(b) With the ship's head on the above Compass course, a point C (Howth Light) bore by Compass N. 75° W., and Rockabill Light bore N. 31° W. by the same Compass : find the ship's position.

(c) With the ship's head as above, a point E (Chicken Rock) bore by Compass N. 44° E., and after continuing 9 miles on the above-named course it bore N. 23° W. : find the ship's position and distance from E.

(d) Find the course from A to B in a current which set S. (corr. mag.) 3 miles per hour, the vessel going 9 miles per hour ; also the distance made good towards B in 3 hours.

CHAPTER XVI.

USE OF NAPIER'S DIAGRAM.

In the list of questions proposed to candidates for Masters Certificates in the Board of Trade Regulations, Nos. 11, 12, 13, and 14 refer to the use of Napier's diagram.

Examples.—

No. 11. Having taken the following compass bearings of a distant object, find its *Correct Magnetic Bearing*, and thence the deviation.
Correct magnetic bearing required : S. 8° E.

Ship's head by standard compass.	Bearing of distant object.	Deviation required.	Ship's head by standard compass.	Bearing of distant object.	Deviation required.
N.	S. 4° E.	4° W.	S.	S. 13° E.	5° E.
N.E.	S.	8° W.	S.W.	S. 22° E.	14° E.
E.	S. 4° W.	12° W.	W.	S. 20° E.	12° E.
S.E.	S. 1° W.	9° W.	N.W.	S. 11° E.	3° E.

No. 12. With the deviation as above, construct a curve of deviation on a Napier's Diagram, and give the courses you would steer by the Standard Compass to make the following correct magnetic courses :—

Magnetic courses : S.S.W. W.N.W. N.N.E. E.NE. S.S.E.
Compass courses required : S. 14° W. N. 77° W. N. 29¼° E. N. 79° E. S. 20°½ E.

No. 13. Supposing you have steered the following courses by the Standard Compass, find the correct magnetic courses from the above curve of deviations :—

Compass courses : W.S.W. N.N.W. E.N.E. S.S.E.
Magnetic courses required : S. 82° W. N. 23° W. N. 57° E. S. 25° E.

No. 14. You have taken the following bearings of two distant objects by your Standard Compass as above, with the ship's head at W. ½ S. by compass : find the correct magnetic bearings.

Bearings by compass : W. by S. N. ¾ W.
Magnetic bearings required : N. 89° W. N. 3° E.

Explanation.—In No. 11 the *Correct Magnetic Bearing* is the *mean* of the eight observed bearings.

(a) When the bearings all read one way from S. or N., the *mean* is found by *adding* the eight bearings together and dividing the *sum* by 8.

(b) When the bearings read different ways, viz. some towards E. and some towards W., then the *mean* is found by summing up the bearings of the *same Name* separately (as in the above example), and dividing the *difference of the totals* by 8.

(c) The deviations are the differences between the correct magnetic bearing and each of the eight bearings by Compass, it being E. dev. when the correct magnetic reads to the right of the compass bearing, and W. when to the left.

Napier's Diagram and Curve of Deviations.

N.B.—The *dotted* lines represent *Compass Courses*, and the *plain lines* correct *Magnetic Courses*.

The deviations found for the eight equidistant courses are taken with a pair of dividers from the middle line, which is graduated like a compass card, and set off from the graduated line on the dotted lines for the given courses—to the left for W. dev., and to the right for E. dev. A curve is then drawn, as in the diagram. The deviation for any course (compass or magnetic) may now be found by simply measuring the part of the line (dotted or plain) intercepted by the curve.

For No. 12, place one foot of the dividers on the given course, and extend the other on the *plain* line to the curve. Keeping one foot on the middle line, turn the other *up* or *down* according as the line inclines *upwards* or *downwards*, and read off the required compass course.

For No. 13, proceed exactly in the same way as in No. 12, except that the *dotted* lines only are to be used.

NOTE.—The advantage of the diagram and curve is that it affords a ready *mechanical* way of changing magnetic courses to compass courses, and *vice versâ*, and so avoids the risk of error in applying deviation by adding or subtracting.

For No. 14, find the *Deviation* for the given direction of ship's head, by measuring the part of the *dotted* line intercepted by the curve, and apply it in the usual way to the bearings by compass, viz. E. to right, and W. to left.

130   TEXT-BOOK ON NAVIGATION.

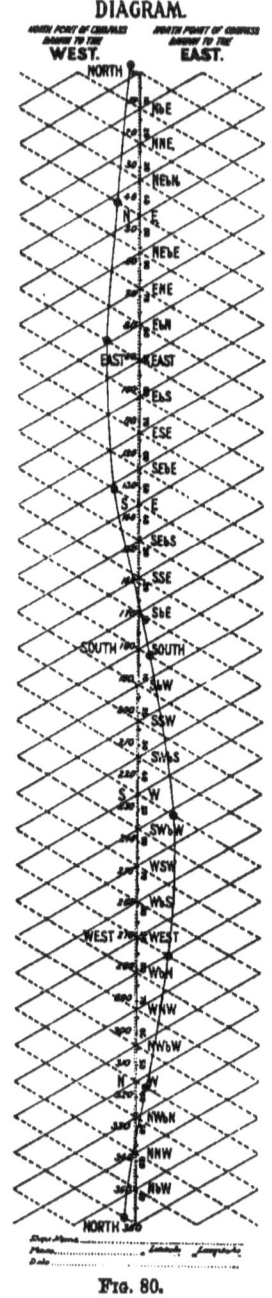

FIG. 80.

USE OF NAPIER'S DIAGRAM.

EXERCISES ON NAPIER'S DIAGRAM AND CURVE.

1. (a) Correct magnetic bearing required :—

Ship's head by compass.	Bearing of distant object.	Deviation.	Ship's head by compass.	Bearing of distant object.	Deviation.
N.	N. 83° E.		S.	N. 50° E.	
N.E.	S. 86° E.		S.W.	N. 34° E.	
E.	E.		W.	N. 40° E.	
S.E.	N. 75° E.		N.W.	N. 61° E.	

Find the correct magnetic bearing and the deviations.

(b) Correct magnetic courses : N.E. ¼ E. E. by N. S.E. by S. ½ S. N. ½ W.
  Required courses to steer.

(c) Courses steered : N.W. ¾ W. S. E. by S. N. by E.
  Required correct magnetic courses.

(d) Ship's head by compass : W.S.W.
  Bearings of distant object : S.S.W. and N.E.
  Required the correct magnetic bearing.

2. (a) Correct magnetic bearing required :—

Ship's head by compass.	Bearing of distant object.	Deviation.	Ship's head by compass.	Bearing of distant object.	Deviation.
N.	S. 85° W.		S.	N. 73° W.	
N.E.	S. 79° W.		S.W.	N. 68° W.	
E.	S. 88° W.		W.	N. 74° W.	
S.E.	N. 76° W.		N.W.	N. 79° W.	

Required the correct magnetic bearing and the deviations.

(b) Correct magnetic courses : E. ½ N. E.S.E. S. ¾ W. N.W.
  Required courses to steer.

(c) Courses steered : N. S.E. ¾ S. S.W. ½ W. N. 35° W.
  Required correct magnetic courses.

(d) Ship's head S. 65° E. Bearings by compass, W. 10° S. N. 40° E.
  Required the deviation and the correct magnetic bearings.

3. (a) Correct magnetic :—

Ship's head by compass.	Bearing of distant object.	Deviation.	Ship's head by compass.	Bearing of distant object.	Deviation.
N.	S. 4° W.		S.	S. 34° W.	
N.E.	S. 1° E.		S.W.	S. 32° W.	
E.	S.		W.	S. 11° W.	
S.E.	S. 22° W.		N.W.	S. 9° W.	

(b) Correct magnetic courses : N ½ W. W. by S. ¾ S. S. 70° E.
  E. 24° N.
  Courses to steer by compass.

(c) Courses steered by compass: W.N.W. ¼ N.   N.E. ¼ N.   S. ¾ E.
N. 40° W.
Correct magnetic courses.

(d) Ship's head at W. by N. by compass.
Bearings by compass (of distant objects): S.W. ¾ W.   N.W. by W. ½ W.
Correct magnetic bearings.

4. (a) Correct magnetic.

Ship's head by compass.	Bearing of distant object.	Deviation.	Ship's head by compass.	Bearing of distant object.	Deviation.
N.	N. 32° W.		S.	N. 4° E.	
N.E.	N. 27° W.		S.W.	N. 4° E.	
E.	N. 6° E.		W.	N. 10° W.	
S.E.	N. 2° E.		N.W.	N. 15° W.	

(b) Correct magnetic courses: S. 50° W.   N.   E. 5° S.   E. by N.
Courses to steer by compass.

(c) Courses steered by compass: W. by S.   N. by E. ½ E.   S.E. ¾ S.
S. by W.

Correct magnetic courses.
(d) Ship's head at N.N.W. by compass.
Bearings by compass (of distant objects): W.N.W.   S.S.W.
Correct magnetic bearings.

5. (a) In the following table give the correct magnetic bearing, and thence the deviation.

Correct magnetic :—

Ship's head by standard compass.	Bearing of distant object by standard compass.	Deviation required.	Ship's head by standard compass.	Bearing of distant object by standard compass.	Deviation required.
N.	N. 62° E.		S.	E. 15° S.	
N.E.	E. 24° N.		S.W.	S. 81° E.	
E.	E. 7° N.		W.	E. 5° N.	
S.E.	S. 77° E.		N.W.	N. 67° E.	

(b) With the deviations as above, give the courses you would steer by the Standard Compass to make the following courses correct magnetic :—
Magnetic courses: N.E. ½ E.   E. ¼ S.   S. ½ W.   N. ½ W.
Compass   ,,

(c) Supposing you have steered the following courses by the Standard Compass, find the correct magnetic courses.
Compass courses: N.E. ½ N.   S.E. ½ S.   W. ¼ N.   N.W. ½ N.
Magnetic.

(d) You have taken the following bearings of two distant objects by your Standard Compass as above: with the ship's head at N. ¼ W., find the bearings correct magnetic.
Compass bearings: S.S.E. ¼ E.   E. ½ N.
Correct magnetic.

USE OF NAPIER'S DIAGRAM.

6. (a) In the following table give the correct magnetic bearing of the distant object, and thence the deviation:—

Correct magnetic:—

Ship's head by standard compass.	Bearing of distant object by standard compass.	Deviation required.	Ship's head by standard compass.	Bearing of distant object by standard compass.	Deviation required.
N.	S. 32° 0' E.		S.	S. 6° 50' W.	
N.E.	S. 4° 45' E.		S.W.	S. 26° 0' E.	
E.	S. 16° 0' W.		W.	S. 43° 0' E.	
S.E.	S. 21° 30' W.		N.W.	S. 44° 35' E.	

(b) With the deviations as above, give the courses you would steer by the Standard Compass to make the following courses correct magnetic:—
Magnetic: N. by E. ¾ E.   E. ½ N.   S.S.W. ¼ W.   N.W. ½ W.
Courses to steer.

(c) Supposing you have steered the following courses by the Standard Compass, find the correct magnetic courses.
Standard: N.N.E. ¾ E.   E. by N. ½ N.   S.S.E ¼ E.   N.W. ¼ N.
Magnetic.

(d) You have taken the following bearings of two distant objects by your Standard Compass as above: with the ship's head at S.E. ½ S., find the bearings correct magnetic:—
Compass bearings: N.E. ½ E.   N. by W. ¼ W.
Correct magnetic.

CHAPTER XVII.

GREAT CIRCLE SAILING.

ART. 52.—A circle is really a plane surface bounded by a line called the circumference, but it is common to speak of this line as the circle.

A great circle of the globe is one which divides it in halves. Hence the Equator is a great circle, and as all great circles bisect one another, every other great circle must intersect the Equator in two opposite points (Fig. 81). A meridian of longitude cuts the Equator at right angles, but other great circles cross it more or less obliquely, and go to equal distances north and south. The two points on the Great Circle which are farthest from the Equator are called the Vertexes, and the meridian of longitude which passes through the Vertex is called the Meridian of Vertex.

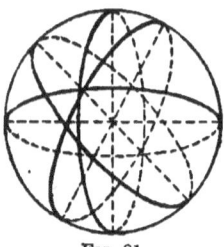

FIG. 81.

Now, the shortest distance between any two places on the globe is the arc of the Great Circle passing through them. Therefore, when this route is practicable, the navigator would save time in a sailing vessel, and coals as well as time in a steamship, by following the Great Circle track.

Illustration.—Let A be the starting-point, and B the destination, AVB the arc of the Great Circle passing through the two places, P the north pole, PA and PB the meridians of longitude through A and B, and V the highest latitude reached on the parallel LL_1 and meridian of vertex PVM; then—

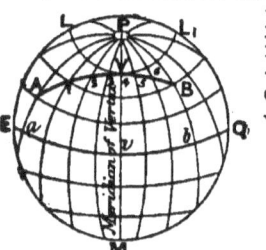

FIG. 82.

The $\angle$ APB is the Difference of Longitude of A and B.
The $\angle$ PAB is the Initial Course.
The $\angle$ PBA „ Final Course.
AB „ Distance on the Great Circle.

vV is the Latitude of Vertex.
The $\angle$ APV (or arc av) is the Longitude of A from Vertex.

GREAT CIRCLE SAILING.

The $\angle$ BPV (or arc bv) is the Longitude of B from Vertex.
PA and PB, Co-latitudes of A and B.
PV, Co-latitude of Vertex.
1, 2, 3, 4, 5, 6, points on the Great Circle.

N.B.—The Meridian of Vertex is at right angles to the Great Circle.

Inspection of the figure shows that the course from A would first be northerly until the Vertex is reached, where it would be due east, and then southerly to B. From B to A these directions would be reversed.

In proceeding along the Great Circle the course would change at each instant, but as this would not be possible in navigation, Great Circle sailing in practice consists in fixing the latitude and longitude of a number of points on the circle, and then shaping courses from point to point by Mercator's sailing. The more points chosen the nearer would the ship's track approximate to the Great Circle.

The starting-point and destination might be on the same side or on different sides of the Equator, and the track might cross the meridian of vertex or might not. Hence four different cases are distinguished—

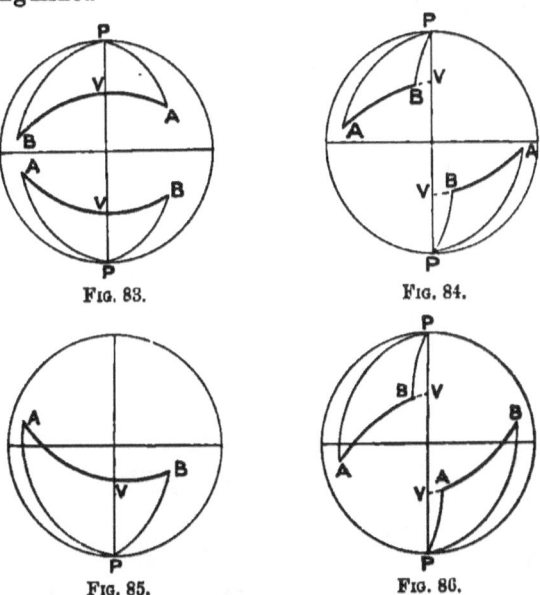

FIG. 83.   FIG. 84.

FIG. 85.   FIG. 86.

1. Vertex but not Equator " in " (Fig. 83).
2. Equator but not Vertex " in " (Fig. 86).
3. Both Equator and Vertex " in " (Fig. 85).
4. Both Equator and Vertex " out " (Fig. 84).

When both latitudes are of the same name, the Vertex is on the same side of the Equator as both, but when of different names it is on the same side as the greater.

In every case the place having the greater latitude is nearer to the Vertex than the other (see figures).

The co-latitudes PA and PB and the distance AB are three sides of a spherical triangle, of which two sides (the co-latitudes) and the included angle (difference of longitude) are known. From these data all the other details can be found by the principles of spherical trigonometry.

These details are—
(1) The initial and final courses.
· (2) The distance on the Great Circle.
(3) The latitude and longitude of Vertex.
(4) The longitude from Vertex of A and B.
(5) The latitudes and longitudes of points on the Great Circle.
(6) The courses and distances by Mercator from point to point.

Naming the Courses.—When the Vertex is "out," it can be seen, from Figs. 84 and 86, that the courses are all named as in Mercator's sailing, but when it is "in" they change from northerly to southerly, or southerly to northerly when the Vertex is passed (Figs. 83 and 85). Hence in all cases it is necessary to know whether the Vertex is "in" or "out." This can be decided when the initial and final courses have been computed; for if they are on the same side of 90° the vertex is "in," but if one is greater and the other less than 90° it is "out," and, as already stated, it is in every case on the same side of the Equator as the greater latitude.

It is seen, from Figs. 84 and 86, that when the Vertex is "out," the course is *towards* the vertex when sailing from a lower to a higher latitude, and *from* the vertex when sailing from a higher to a lower latitude.

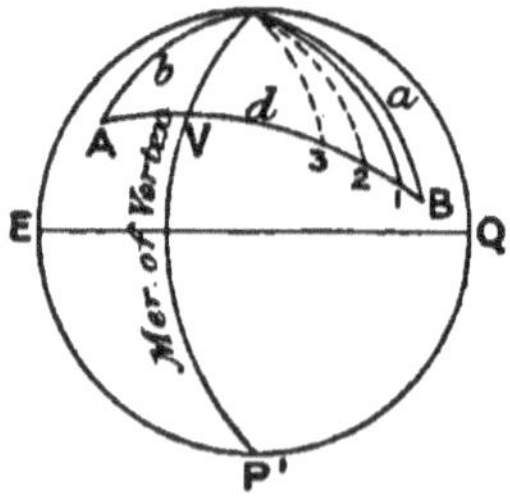

Example 1.—VERTEX ' IN."
Required the initial and final courses and the distance on the Great Circle from Cape San Lucas to Yokohama; also the position of the Vertex and the latitudes and longitudes of points on the Great Circle differing 5° in longitude.

Cape San Lucas, lat. 22° 52′ N., long. 109° 53′ W.
Yokohama, ,, 35° 26′ N., ,, 139° 39′ E.

Fig. 87.

Let A be the point having the greater latitude, and B the point having the lesser latitude, and the angle at P the difference of longitude (D). PVP$_1$ is the meridian of vertex; the angle at B is the initial course from Cape San Lucas, and the angle at A the final course; also the angle

APV is the "longitude from Vertex" of A, and BPV the long. of B from vertex.

To find the angles A and B (the courses)—

$$\text{Tan } \tfrac{1}{2}(A + B) = \frac{\cos \tfrac{1}{2}(a - b)}{\cos \tfrac{1}{2}(a + b)} \cdot \cot \tfrac{1}{2}D \quad \ldots \ldots \quad (1)$$

$$\text{Tan } \tfrac{1}{2}(A - B) = \frac{\sin \tfrac{1}{2}(a - b)}{\sin \tfrac{1}{2}(a + b)} \cdot \cot \tfrac{1}{2}D \quad \ldots \ldots \quad (2)$$

$\tfrac{1}{2}(a + b)$ 60° 51′	... sec 10·312384	...	cosec 10·058813
$\tfrac{1}{2}(a - b)$ 6 17	... cos 9·997383	...	sine 9·039197
$\tfrac{1}{2}$D 55 14	... cot 9·841457	...	cot 9·841457
$\tfrac{1}{2}$(A + B) 54 47	.... tan 10·151224		
$\tfrac{1}{2}$(A − B) 4 58	... ...	...	tan 8·939467
A 59 45			
B 49 49			

	109° 53′	W.
	139 39	E.
	249 32	
	360 0	
D	110 28	
$\tfrac{1}{2}$D	55 14	
	22 52	
	90 0	
(a)	67 8	
	35 26	
	90	
(b)	54 34	
a	67 8	
a+b	121 42	
a−b	12 34	
$\tfrac{1}{2}$(a+b)	60 51	
$\tfrac{1}{2}$(a−b)	6 17	

The greater angle being opposite to the greater side, the initial course is N. 49° 49′ W., and final S. 59° 45′ W.

To find the Distance—

$$\text{Cos } \tfrac{1}{2}d = \cos \tfrac{1}{2}(a + b) \cdot \sec \tfrac{1}{2}(A + B) \cdot \sin \tfrac{1}{2} D$$

$\tfrac{1}{2}(a + b)$ 60° 51′	... ...	cos 9·687616
$\tfrac{1}{2}$(A + B) 54 47	... ...	sec 10·239073
$\tfrac{1}{2}$D 55 14	... ...	sine 9·914598
$\tfrac{1}{2}d$ 46 4		cos 9·841287
2		

Distance 92 8 = 5528 geographical miles

To find the Latitude and Longitude of Vertex.—In the figure the angle at A (final course) and b (co-latitude of A) are known. The angle APV is the longitude of V from A, and PV is the co-latitude of vertex. AVP is a right angle; then by Napier's circular parts—

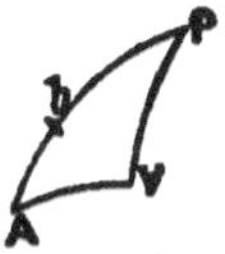

FIG. 88.

Cos lat. of V = sin A . sin b (1), and cot . long. from V = tan A . cos b (2)

A 59° 45′	... sine 9·936431	...	tan 10·234195
b 54 34	... sine 9·911046	...	cos 9·763245
Lat. of vertex 45 16 N.	... cos 9·847477		
Long. from V 45 10	... ...		cot 9·997440
Long. of A 139 39 E.			
184 49 E.			
Long. of V 175 11 W.			

To determine Points on the Great Circle.—A table may be constructed giving the longitudes of points, say at intervals of 5° from Cape San Lucas, and the corresponding longitudes from Vertex placed underneath; then the latitudes of these points may be computed from the formula—

Tan lat. = cos long. from V × tan lat. of V

138 TEXT-BOOK ON NAVIGATION.

109° 53' W.
175 11 W.
─────────
65 18

115 0 W.
175 11 W.
─────────
60 11

	C. San Lucas.	1.	2.	3.	4.	5.	
Long. from Greenwich	109° 53' W.	115° W.	120° W.	125° W.	130° W.	135° W.	Etc.
Long. from V.	65° 18'	60° 11'	55° 11'	50° 11'	45° 11'	40° 11'	Etc.
Latitudes	22° 52' N.	26° 39' N.	29° 57' N.	32° 53' N.	35° 26' N.	37° 38' N.	Etc.

FIG. 89.

In this figure, which is part of Fig. 87, the angle at P is the longitude of 1 from Vertex, PV is the co-latitude of V, and P1 the co-latitude of point 1.

By circular parts, $\cos P = \tan PV \times \cot P1$

$$\therefore \cot P1 = \frac{\cos P}{\tan PV} = \cos P \times \cot PV$$

i.e. tan lat. of $1 = \cos$ long. from V $\times$ tan lat. of V

Computation of latitude of points 1, 2, 3, etc.—

Long. from V, 60° 11'   ...   cos 9·696554
Lat. of V, 45° 16'   ...   tan 10·004043
─────────
Lat. of point 1, 26° 39'   ...   tan 9·700597

The other latitudes are found by substituting the cosines of the successive longitudes from vertex in the above computation.

The courses and distances by Mercator from point to point may be found in the ordinary way by calculation or by chart.

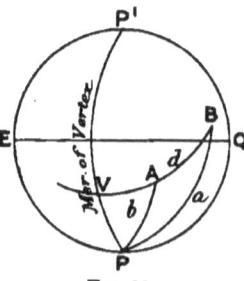

FIG. 90.

Remark.—Should a vessel be driven far from the Great Circle track by stress of weather, the navigator should strike a *new* Great Circle from the *actual position* of the ship, and not work back to the original one.

Example 2.—EQUATOR "IN."

Required the initial and final courses, the distance, and the position of the Vertex on the Great Circle from Mauritius to Cape Negrais.

Mauritius, lat. 20° 9' S., long. 57° 29' E.
Cape Negrais, ,, 16 0 N., ,, 94 13 E.

16° N.
90
─────
at. G 106

To find the Courses—

$$\tan \tfrac{1}{2}(A + B) = \frac{\cos \tfrac{1}{2}(a - b)}{\cos \tfrac{1}{2}(a + b)} \cdot \cot \tfrac{1}{2} D$$

$$\tan \tfrac{1}{2}(A - B) = \frac{\sin \tfrac{1}{2}(a - b)}{\sin \tfrac{1}{2}(a + b)} \cdot \cot \tfrac{1}{2} D$$

GREAT CIRCLE SAILING. 139

$\frac{1}{2}(a + b)$ 87° 55½'	... sec 11·441199	... cosec 10·000285	20° 9'S.
$\frac{1}{2}(a - b)$ 18 4½	... cos 9·978021	... sine 9·491728	90
$\frac{1}{2}$D 18 22	... cot 10·478849	... cot 10·478849	(Co-lat. b)69 51
			(a)106 0
$\frac{1}{2}$(A + B) 89 17	... tan 11·898069		a+b 175 51
$\frac{1}{2}$(A − B) 43 6	... ...	tan 9·970862	a−b 36 9
A 132 23			$\frac{1}{2}(a+b)$ 87 55½
B 46 11			$\frac{1}{2}(a-b)$ 18 4½

Long. 57 29 E.
94 13 E.

The first course is S. 132° 23' E. or N. 47° 37' E., and final course N. 46° E., and the Vertex is "out," because one angle is greater and the other less than 90°.

2)36 44

½D 18 22

To find the Distance—

$$\cos \tfrac{1}{2}d = \cos \tfrac{1}{2}(a + b) \cdot \sec \tfrac{1}{2}(A + B) \cdot \sin \tfrac{1}{2}D$$

$\frac{1}{2}(a + b)$ 87° 55½' ... cos 8·558801
$\frac{1}{2}$(A + B) 89 17 ... sec 11·902817
$\frac{1}{2}$D 18 22 ... sine 9·498444

$\frac{1}{2}d$ 24 12 cos 9·960062
2

Distance 48 24 = 2904 geographical miles

To find Latitude and Longitude of Vertex—

$$\cos \text{lat. of } V = \sin A \sin b \quad . \quad . \quad . \quad (1)$$
$$\cot \text{long. from } V = \tan A \cos b$$

A 47° 37' ... sine 9·868440 ... tan 10·039723
b 69 51 ... sine 9·972570 ... cos 9·537163

Lat. of V 46 6 S. ... cos 9·841010

Long. from V 69 19 ... cot 9·576886
Long. of A 57 29 E.

Long. of V 11 50 W.

151° 49' E.
79 31 W.

Example 3.—Equator and Vertex "in." Required the first course and the distance on the Great Circle from Newcastle, N.S.W., to Panama; also the position of the Vertex.

Newcastle, lat. 32° 55' S., long. 151° 49' E.
Panama, ,, 8 57 N., ,, 79 31 W.

Let. A be the place farthest from the Equator (Newcastle), and B the destination, PVP₁ the meridian of Vertex, and V the Vertex; then AB is the distance, the angle at A the first course, VP the co-latitude of Vertex, and APV the longitude of A from Vertex, and the angle APB (D) is the difference of longitude.

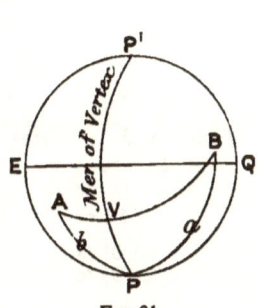

Fig. 91.

231 20
360 0

D 128 40
½D 64 20

8 57
90
a 98 57

32 55
90
b 57 5
a 98 57

a+b 156 2
a−b 41 52

$\frac{1}{2}(a+b)$ 78 1
$\frac{1}{2}(a-b)$ 20 56

140   TEXT-BOOK ON NAVIGATION.

$\frac{1}{2}(a+b)$ 78° 1' ... sec 10·682716 ... cosec 10·009569
$\frac{1}{2}(a-b)$ 20 56 ... cos 9·970345 ... sine 9·553010
$\frac{1}{2}$D 64 20 ... cot 9·681740 ... cot 9·681740

$\frac{1}{2}$(A+B) 65 10$\frac{1}{2}$ ... tan 10·334801
$\frac{1}{2}$(A−B) 9 57$\frac{1}{2}$ ... ... tan 9·244319

A 75 8
B 55 13

The first course is S. 75° 8' E., and the Vertex "in."

To find the Distance—

$\frac{1}{2}(a+b)$ 78° 1' ... ... cos 9·317284
$\frac{1}{2}$(A+B) 65 10$\frac{1}{2}$ ... ... sec 10·376908
$\frac{1}{2}$D 64 20 ... ... sine 9·954883

$\frac{1}{2}d$ 63 32   cos 9·649075
       2

Distance 127 4 = 7624 geographical miles

To find Position of Vertex—

A 75 8 ... sine 9·985213 ... tan 10·576007
b 57 5 ... sine 9·924001 ... cos 9·735135

Lat. of V 35 46 S.   cos 9·909214

Long. from V 26 2 ... ... cos 10·311142
Long. of A 151 49 E.

Long. V 177 51 E.

Example 4.—Both Equator and Vertex "Out."

Required the first course and the distance on the Great Circle from the Start to Barbadoes; also the position of the Vertex.

Start point, lat. 50° 13' N., long. 3° 39' W.
Barbadoes, ,, 13 3 N., ,, 57 37 W.

Let A be the place of higher latitude, and B the place of lower latitude, PVP$_1$ the meridian of vertex crossing the Great Circle through B and A in V (the Vertex); then BP and AP are the co-latitudes (a and b), BA is the distance, the angle at A the first course, and the angle APB the diff. longitude (D).

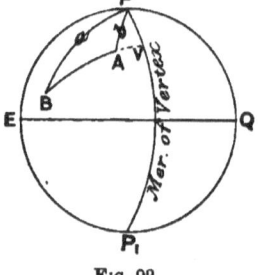

Fig. 92.

3° 39' W.
57 37 W.

D 53 58
$\frac{1}{2}$D 26 59

90 0
50 13
b 39 47
90
13 3
a 76 57
b 39 47
a+b 116 44
a−b 37 10
$\frac{1}{2}$(a+b)58 22
$\frac{1}{2}$(a−b)18 35

To find the First Course—

$\frac{1}{2}(a+b)$ 58° 22' ... sec 10·280270 ... cosec 10·069855
$\frac{1}{2}(a-b)$ 18 35 ... cos 9·976745 ... sine 9·503360
$\frac{1}{2}$D 26 59 ... cot 10·293146 ... cot 10·293146

$\frac{1}{2}$(A+B) 74 16 ... tan 10·550161
$\frac{1}{2}$(A−B) 36 19 ... ...   tan 9·866361

A 110 35
B 37 57

GREAT CIRCLE SAILING.

The first course is N. 110° 35' W., or S. 69° 25' W.
NOTE.—Vertex out, because A and B are not of the same "affection."

To find the Distance—

$\frac{1}{2}(a+b)$ 58° 22' ... ... cos 9·719730
$\frac{1}{2}(A+B)$ 74 16 ... ... sec 10·566774
$\frac{1}{2}D$ 26 59 ... ... sine 9·656799

$\frac{28\ 38\frac{1}{2}}{2}$ ... cos 9·943303

Distance 57 17 = 3437 geographical miles

To find Latitude and Longitude of Vertex—

Suppl. of A 69° 25' ... sine 9·971351 ... tan 10·425340
b 39 47 ... sine 9·806103 ... cos 9·885627

Lat. of vertex 53 12 N. ... cos 9·777454
Long. from V 26 3 E. ... ... cot 10·310967
Long. of A 3 39 W.

Long. of Vertex 22 24 E.

Remark.—From the foregoing examples, it is seen that the formulæ for the courses and distance on the Great Circle belong to oblique-angled spherical triangles, but the latitude and longitude of Vertex and the latitudes of points on the Great Circle, are found by means of Napier's "Circular Parts" for right-angled spherical triangles.

Summary of Formulæ for Great Circle sailing where the same letters always represent the same elements, viz.—

A and B, the larger and smaller courses respectively; D, the difference of longitude; d, the distance (=AB); PV, the co-latitude of Vertex; a and b, the co-latitudes of A and B; the angles APV and BPV, the longitudes from V of A and B respectively.

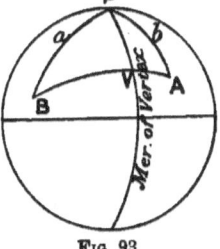

FIG. 93.

For the Courses—

$$\tan \tfrac{1}{2}(A+B) = \frac{\cos \tfrac{1}{2}(a-b)}{\cos \tfrac{1}{2}(a+b)} \cdot \cot \tfrac{1}{2}D$$

$$\tan \tfrac{1}{2}(A-B) = \frac{\sin \tfrac{1}{2}(a-b)}{\sin \tfrac{1}{2}(a+b)} \cdot \cot \tfrac{1}{2}D$$

which may be written—

$\tan \tfrac{1}{2}(A+B) = \sec \tfrac{1}{2}(a+b) \cdot \cos \tfrac{1}{2}(a-b) \cdot \cot \tfrac{1}{2}D$. (1)
and $\tan \tfrac{1}{2}(A-B) = \text{cosec } \tfrac{1}{2}(a+b) \cdot \sin \tfrac{1}{2}(a-b) \cdot \cot \tfrac{1}{2}D$ (2)

the forms used in the computation.

For the Distance—

$\cos \tfrac{1}{2}d = \cos \tfrac{1}{2}(a+b) \cdot \sec \tfrac{1}{2}(A+B) \cdot \sin \tfrac{1}{2}D$ . . (3)

For Latitude of Vertex—

$\cos \text{ lat. of } V = \sin A \cdot \sin b$ . . . . . (4)

For Longitude from Vertex—

$$\text{Cot long. from V (of A)} = \tan A \cdot \cos b \quad \ldots \quad (5)$$
$$\text{,, ,, (of B)} = \tan B \cdot \cos a \quad \ldots \quad (6)$$

For Latitude of a Point on the Great Circle—

$$\text{Tan lat. of point} = \left\{ \begin{array}{c} \cos \text{ long. from V} \\ \text{of the point} \end{array} \right\} \cdot \tan \text{lat. of V} \quad . \quad (7)$$

NOTE.—When both places are on the same parallel of latitude, the meridian of vertex would bisect the distance, making two equal right-angled spherical triangles.

ART. 53. **Composite Great Circle and Parallel Sailing.**—When the Great Circle track would take the ship into too high a latitude, then a combination of Great Circle sailing and Parallel sailing is used to obtain the shortest route.

The highest practicable latitude is made the latitude of vertex for the arcs of *two* Great Circles, one passing through the starting-point, and the other through the destination. The route would then be along the first arc to the Vertex, next on the parallel of latitude to the Vertex of the other arc, and finally along this arc to the destination.

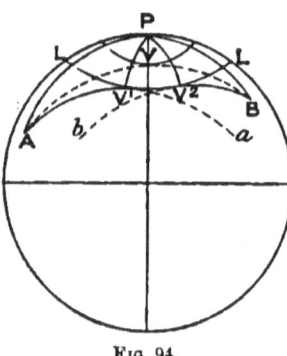

Fig. 94.

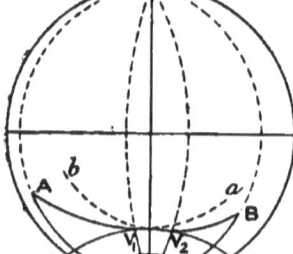

Fig. 95.

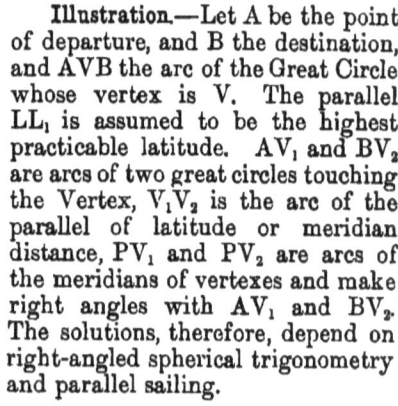

Illustration.—Let A be the point of departure, and B the destination, and AVB the arc of the Great Circle whose vertex is V. The parallel LL_1 is assumed to be the highest practicable latitude. AV_1 and BV_2 are arcs of two great circles touching the Vertex, V_1V_2 is the arc of the parallel of latitude or meridian distance, PV_1 and PV_2 are arcs of the meridians of vertexes and make right angles with AV_1 and BV_2. The solutions, therefore, depend on right-angled spherical trigonometry and parallel sailing.

Example.—Required the distance on the composite track from Rio to Cape Leeuwin, taking 55° as the limit of S. latitude; also the initial and final courses.

Lat. of Rio 23° 4' S.
Long. of Rio 43° 8' W.
Lat. of C. Leeuwin 34° 19' S.
Long. of C. Leeuwin 115° 6' E.

(1) To find the distances AV_1 and BV_2.

GREAT CIRCLE SAILING.

$$
\begin{array}{llll}
\text{Co-lat. of A } 66°\ 56' & \ldots & \cos 9\cdot593067 & \\
\text{,, \quad B } 55\ 41 & \ldots & \ldots \quad \ldots & \ldots \quad \cos 9\cdot751099 \\
\text{Co-lat. of } V_1 \text{ and } V_2\ 35\ \ 0 & \ldots & \sec 0\cdot086635 & \ldots \quad \sec 0\cdot086635 \\
\end{array}
$$

$$
\begin{array}{ll}
AV_1 = 61\ 25\tfrac{1}{2} & \ldots \quad \cos 9\cdot679702 \\
BV_2 = 46\ 30\tfrac{3}{4} & \ldots \quad \ldots \quad \ldots \quad \ldots \quad \cos 9\cdot837734 \\
\end{array}
$$

Sum 107 56
        60

6476 geographical miles

(2) To find the differences of long. APV_1 and BPV_2 or longitudes from Vertex.

$$
\begin{array}{llll}
\text{Co-lat. of A } 66°\ 56' & \ldots & \cot 9\cdot629255 & \\
\text{,, \quad B } 55\ 41 & \ldots & \ldots \quad \ldots & \ldots \quad \cot 9\cdot834154 \\
\text{,, \quad\quad } 55°\ 35\ \ 0 & \ldots & \tan 9\cdot845227 & \ldots \quad \tan 9\cdot845227 \\
\end{array}
$$

$$
\begin{array}{ll}
\text{D. long of A and } V_1\ 72\ 39 & \ldots \quad \cos 9\cdot474482 \\
\text{,, \quad B and } V_2\ 61\ 27 & \ldots \quad \ldots \quad \ldots \quad \ldots \quad \cos 9\cdot679381 \\
\end{array}
$$

Sum 134 6
Total D. long 158 14

D. long. $V_1 V_2$ 24 8 = 1448 geographical miles

(3) To find the meridian distance between V_1 and V_2.

$$
\begin{array}{llll}
\text{D. long. } 1448 & \ldots & \ldots & \log 3\cdot161068 \\
\text{Lat. } 55° & \ldots & \ldots & \cos 9\cdot758591 \\
\end{array}
$$

Mer. dist. 831   ... log $2\cdot919659$
Arcs of Great Circles 6476

Total dist. 7307

NOTE.—The distance on the Great Circle between A and B is 7131 miles.

(4) To compute the initial and final courses.

$$
\begin{array}{llll}
\text{Co-lat. of A } 66°\ 56' & \ldots & \cosec 0\cdot036189 & \\
\text{,, \quad B } 55\ 41 & \ldots & \ldots \quad \ldots & \ldots \quad \cosec 0\cdot083054 \\
\text{,, \quad\quad } 55°\ 35\ \ 0 & \ldots & \sine 9\cdot758591 & \ldots \quad \sine 9\cdot758591 \\
\end{array}
$$

Initial co. S. 38 34 E. .... sine $9\cdot794780$
Final co. N. 43 59 E. .... sine $9\cdot841645$

(5) To find the longitudes of V_1 and V_2.

Long. of A 43° 8' W.   Long. of B 115° 6' E.
D. Long. to V_1 79 39 E.   D. long. to V_2 61 27 W.

Long. of V_1 29 31 E.   Long. of V_2 53 39 E.

Various points on the arcs AV_1 and BV_2 can be fixed as in G.C.S. by assuming intervals of longitude from V_1 towards A, and from V_2 towards B.

Remark.—The student who understands Napier's Circular Parts has no need to commit the above forms to memory.

ART. 54. **Windward Great Circle Sailing.**—What is called Windward Great Circle sailing applies only to sailing vessels, and consists in putting the ship on the tack on which her course would be nearest to the Great Circle track without reference to the Rhumb or Chart course, which in some cases would differ four or five points or more from the former.

Illustration.—Let O be the starting-point, and suppose desti-

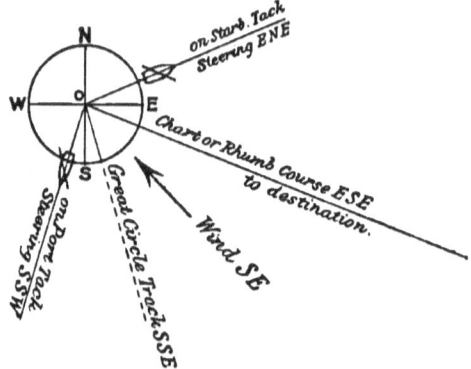

Fig. 96.

nation to bear E.S.E. by Chart, the Great Circle course from O being S.S.E.

With the wind at S.E., the course on the starboard tack would be E.N.E. and on the port tack S.S.W. It is evident that on the port tack the ship's head would only be four points off the shortest route, whilst on the starboard tack she would be steering at right angles to it. In sailing 100 miles on the former course the Traverse Table shows that she would make 70 miles towards her destination, but on the latter she would be no nearer to it, having sailed on a line at right angles to the direct route on the globe.

With the wind at S.E. by S., and steering S.W. by S. (apparently away from destination, according to chart), she would still gain 36 miles in 100 miles sailed, whilst on the starboard tack, steering E. by N., she would make good only 19 miles.

Other still more striking cases might be given, but the above are sufficient to show the advantage of the method.

All problems in Great Circle sailing, Composite sailing, and Windward Great Circle sailing can best be understood from a globe. A pair of 12-inch globes should be in every ship's equipment.

EXERCISES.—GREAT CIRCLE SAILING.

1. Find the first and final courses from Cape of Good Hope, lat. 34° 22' S., long. 18° 30' E., to Rio Janeiro, lat. 22° 55' S., long. 43° 9' W.; also find

GREAT CIRCLE SAILING.

the distance, latitude, and longitude of Vertex, and a succession of points for every five degrees of longitude.

2. Find the first and last courses, and distance from San Francisco, lat. 37° 49' N., long. 122° 29' W., to Yokohama, lat. 35° 26' N., long. 139° 39' E., the latitude and longitude of Vertex, and a succession of points for every twenty degrees of longitude.

3. Find the initial and final courses, distance on the Great Circle, latitude and longitude of Vertex, and a succession of points differing 10° in longitude.

From A, lat. 33° 51' S., long. 151° 16' E.
To B, ,, 37 47 N., ,, 122 30 W.

4. Find the initial and final courses, distance on the Great Circle, latitude and longitude of Vertex, and a succession of points differing 20° in longitude.

From A, lat. 22° 16' N., long. 114° 10' E.
To B, ,, 12 4 S., ,, 77 16 W.

5. Find the initial and final courses, distance on the Great Circle, latitude and longitude of Vertex, and a succession of points differing 5° in longitude.

From A, lat. 52° 12' N., long. 6° 12' W.
To B, ,, 13 7 N., ,, 59 25 W.

Exercises.—Composite Sailing.

1. Find the first and last courses and distance from Bahia Blanca, lat. 38° 59' S., long. 61° 39' W., to Cape Lewin, lat. 34° 31' S., long. 115° 6' E.; easting to be run on the parallel of 45° S. Find also the longitudes of points of arrival at and departure from the parallel.

2. On a composite track from Cape St. Mary, lat. 16° 40' S., long. 50° 4' E., to Monte Video, lat. 34° 43' S., long. 56° 16' W., the westing to be run on the parallel of 40° S. Required the first and last courses, total distance, and longitudes of points of arrival at and departure from the given parallel.

3. Find the first and last courses, distance, and longitudes of arrival at and departure from parallel of 50° N. when sailing on a composite track from Yokohama, lat. 35° 26' N., long. 139° 39' E., to Cape Flattery, lat. 48° 23' N., long. 124° 44' W.

4. In sailing from Port Philip, lat. 38° 18' S., long. 144° 39' E., to St. Helena, lat. 15° 55' S., long. 14° 26' W., the westing is run on the parallel of 45° S. Find the first and last courses, distance, and longitudes of points of arrival at and departure from the parallel.

5. On a composite track from Cape Agulhas, lat. 34° 50' S., long. 20° 0' E., to Albany, lat. 35° 2' S., long. 117° 54' E., the easting is run on 43° S. Find the first and last courses, distance, and longitudes of points of arrival at and departure from that parallel.

6. Find the first and last courses, distance, and longitudes of points of arrival and departure from 48° S. when sailing on a composite track from Valparaiso, lat. 33° 2' S., long. 71° 28' W., to Port Lyttleton, lat. 43° 36' S., long. 172° 50' E.

L

CHAPTER XVIII.

LONGITUDE BY LUNAR DISTANCES.

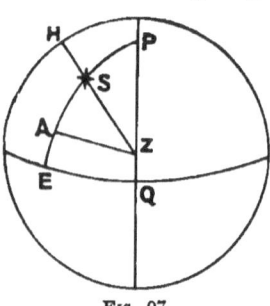

Fig. 97.

ART. 55. **Computing an Altitude.** — In some problems of Nautical Astronomy it is necessary to compute the altitude of a celestial object. Especially is this the case when setting questions, so that the altitude shall be consistent with the other data.

The data required to be known, are the hour angle of the object and the latitude and longitude of the place.

The following method has the advantage of requiring only one table of logarithms.

Example 1.—1899, March 14th, $10^h 15^m$ A.M. apparent time at ship, in lat. 51° 20' N., long. 63° 30' W. Required the true and apparent alt. of the Sun's centre.

				Dec.	H. var.
A.T.S. March 13^d 22^h 15^m		63° 30' W.		2° 29' 44" S.	59"·16
Long. 4 14		4		2 28	2·5
A.T.G. March 14^d 2 29	6,0)25,4	0		2 27 16	29580
Sun's H.A. 145		$4^h 14^m$		90	11832
				P. dist. 92 27 16	6,0)14,2900
					2,27"·9

H.A.	$1^h 45^m$	...	cos 9·952731		
Lat.	51 25	...	cot 9·901901		sin 9·893041
θ	35° 35' 6"	...	tan 9·854632	...	sec 10·089774
p − θ	56 52 10	...	...	...	cos 9·737629
True alt.	31 41 29			...	sin 9·720444
Refr. − par.	+1 25				
App. alt.	31 42 54				

Example 2.—1899, October 14^d 5^h 29^m 15^s M.T.G., in lat. 5° 40' S., long. 82° 45' E. Required the true and apparent alts. of the Star Fomalhaut, and state whether east or west of meridian.

LONGITUDE BY LUNAR DISTANCES.

M.T.G. Oct. 14^d $5^h\ 29^m\ 15^s$		82° 45′	S.T. at noon $13^h\ 31^m\ 15^s\cdot 57$
5 31 0		4	accel. 0 49·28
			4·76
M.T.S. Oct. 14 11 0 15 p.m.	331 0		·04
R.A.M.S. 13 32 10	5 31		
			13 32 9·65
R.A.M. 24 32 25	Dec. 30° 9′ 5″ S.		
R.A. Fomalhaut 22 52 9	90		
H.A. ,, 1 40 16 West	p 59 50 55		
of mer.			

h $1^h\ 40^m\ 16^s$ ... cos 9·957040
l 5° 40′ 0″ ... cot 11·003376 ... sin 8·994497

θ 83 44 55 ... tan 10·960416 sec 10·963007
p 59 50 55

$\theta - p$ 23 54 0 cos 9·961067

Star's true alt. 55 59 58 sin 9·918571
Refr. +0 38

Star's app. alt. 56 0 36

EXERCISES.

1. 1899. Find the true and appar. alts. of the Sun in lat. 48° 32′ N., long. 179° 54′ W., when the G.M.T. is June $14^d\ 13^h\ 19^m\ 13^s$.
2. 1899, August 24th A.M. at ship, the A.T.G. being $23^d\ 19^h\ 35^m\ 26^s$ in lat. 50° 25′ N., long. 0°. Find the true and appar. alts. of the Sun.
3. 1899, January 15th P.M. at ship, in lat. 14° 27′ N., long. 143° 20′ W., when the M.T.G. is $15^d\ 16^h\ 48^m\ 39^s$. Required the true and appar. alts. of Sirius.
4. 1899, February 23rd P.M. at ship, in lat. 19° 41′ N., long. 45° 23′ W., when the M.T.S. is $11^h\ 15^m\ 26^s$. Find the true and appar. alts. of Capella.
5. 1899, February 4th A.M. at ship, in lat. 47° 23′ N., long. 180°, the appar. T. ship is $9^h\ 14^m\ 15^s$ A.M. Find the true and appar. alts. of the Moon.
6. 1899, when the M.T.G. is September $17^d\ 3^h\ 1^m\ 34^s$ in lat. 48° 22′ N., long. 153° 27′ E. Find the true and appar. alts. of the Moon.
7. 1899, January 16th in lat. 50° 5′ N., long. 124° 29′ W., the M.T.G. being $16^d\ 21^h\ 33^m\ 23^s$. Find the true and appar. alts. of Mars.
8. 1899, May 19th in lat. 48° 15′ N., long. 139° 20′ W., the M.T.S. being $9^h\ 30^m\ 15^s$. Find the true and appar. alts. of Jupiter.

ART. 56. **Lunars.**—Before the invention of chronometers the "Lunar Problem" was almost the only means available at sea for finding the Greenwich time, on which the determination of longitude depends. It may still be of great use in the event of chronometers "running down" or becoming unserviceable.

In the N.A., pp. xiii. to xviii., are given for each day the Lunar Distances (east or west) of a number of conspicuous objects at intervals of three hours of M.T.G. These true distances are the angles between the centres of the Moon and the other bodies as measured at the Earth's centre (Fig. 98). The M.T.G. corresponding to any other distance may be found by interpolation, a correction for "second differences" being made on account of the "inequality" of the Moon's motion.

Let M' and S' be the apparent positions of the Moon and

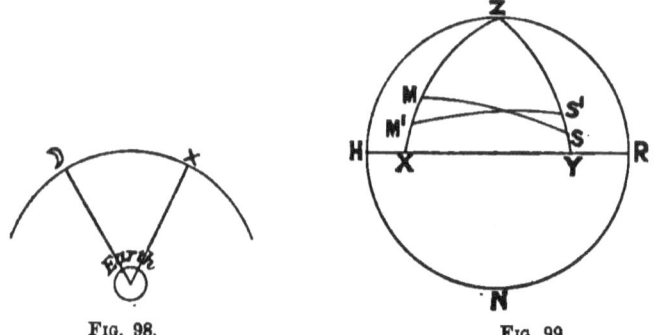

FIG. 98.   FIG. 99.

another body, and ZM, ZS the arcs of two vertical circles passing through their centres. The true position (M) of the Moon will be *higher* than the apparent position, because the Moon's parallax *exceeds* the refraction; but the true position (S) of the other object will be *lower*, because its parallax is *less* than the refraction. Therefore M'S' represents the *apparent* distance, and MS, the *true* distance.

Many methods have been devised for "clearing" the apparent lunar distance from the effects of refraction and parallax, some of which (as Thomson's) require special tables.

Examples of two of these methods are here given, namely, Borda's and Thomson's.

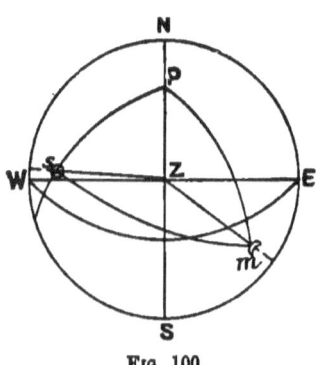

FIG. 100.

The data required for the computations are—
(1) The approximate Greenwich time.
(2) The apparent distance.
(3) The apparent and true altitudes.

One of the altitudes has usually to be computed (see Art. 55).

Example 1 (Sun and Moon).— 1899, June 17th P.M. at ship, in lat. 53° 24' 0" N., a chronometer which was supposed to be $12^m\ 48^s$ slow of M.T.G. showed $17^d\ 6^h\ 30^m$; the observed alt. of the Sun's lower limb was $14^h\ 13^m\ 30^s$; the observed distance between the nearest limbs of the Sun and Moon, 105° 26' 9"; index cor. for distance, $-30''$; height of eye, 24 ft.

LONGITUDE BY LUNAR DISTANCES. 149

Required the Moon's true and apparent altitudes, the error of chron. on M.T.G., and the longitude of the place.

```
Time by chron. June 17ᵈ 6ʰ 30ᵐ 0ˢ                      For 12ʰ  5"·96
                   Slow       12 48                              6·7
                                                                4172
              Approx. M.T.G. 17  6 42 48                        3576
Moon's elements—                                              12)39·932
  S.D.              H.P.            R.A.           Dec.         3·328
 15' 11"·21       55' 38"·43     12ʰ 36ᵐ 32ˢ    9° 33' 13"·5 S. For 12ʰ 21"·86
 Cor. +3·33          12·20              34         3 23              6·7
                                                                 15302
   15 14·54         55 50·63     12 35 58         9 29 50·5      13116
 Aug. +6·70   Red.  − 6·90                       90  0  0     12)146·462
                                                                12·20
   15 21·24         55 43·73                 p 99 29 50·5      For 1ᵐ  1·967
Sun's elements—                                                (R.A.)  17·2
  R.A.              Dec.          R.A.M.S.        S.D.
 5ʰ 42ᵐ 43·2     23° 23' 33" N.  5ʰ 42ᵐ 5'·52    15 45"·9        3934
     1  9·6             29          59·14                       13769
                                     6·90                        1967
  5 43 52·8         23 24  2          ·13                     33·8324
                       90                                     For 1ᵐ  11"·8
              P.D. 66 35 58       5 43 11·69                  (Dec.)  17·2
                                                                236
                                                                826
 Obs. alt. 14° 13' 30"     Obs. dist. 105° 26'  9"              118
 Dip       − 4 48          Index cor.   − 0 30              6,0)20,2·96

           14  8 42                 105 25 39                   3  23
 S.D.      +15 46          Moon's S.D.  15 21            For 1ʰ 10··394
                           Sun's S.D.   15 46                    6·7

 App. alt. 14 24 28                                            72158
    Cor.   − 3 31          App. dist. 105 56 46                62364
                           Moon's A.A. 26  7 19  sec 10·046792  6,0)6,9·6398
 True alt. 14 20 57        Sun's A.A.  14 24 28  sec 10·013878    1ᵐ 9·6
 Lat.  53 24  0  sec 10·224590                                For 1ʰ  4"·29
 P.D.  66 35 58  cosec 10·037275        2)146 28 33                    6·7

    2)134 20 55                         73 14 16  cos 9·459997   3003
                                        32 42 30  cos 9·925019   2574
       67 10 27  cos 9·588755   Moon's T.A. 26 55 26  cos 9·950174  28·743
       52 49 30  sine 9·901346  Sun's T.A.  14 20 57  cos 9·986236 T.A. 26° 55'
                                                                  Corr.     48

 Sun's H.A. 6ʰ 29ᵐ 50ˢ log 9·751966         2)41 16 23            A.A.  26  7
 Sun's R.A. 5 43 53                                             2)19·382096
                                                                  Corr. 47' 20"
                                        A   20 38 12                        7
 R.A.M. 12 13 43                        B   29 24 14  sin 9·691048         40
 M.'s R.A. 12 55 58                                                    48  7

                                       A+B  50  2 26  cos 9·807701
                                       A−B   8 46  2  cos 9·994895
 M.'s H.A.   22 15  cos 9·997944
   Lat.   53 24  0  cot 9·870793  sin 9·904617                  2)19·802596

     θ  36 28 13  tan 9·868737 sec 10·094654
     p  99 29 51

     p−θ  63  1 38   ...    ...   cos 9·656642 ½ dist. 52 49  0 sine 9·901298
 M.'s T.A. 26 55 26                                      2
    Cor.      48  7               sin 9·655913
                                     True dist. 105 38  0
 M.'s A.A. 26  7 19
```

TEXT-BOOK ON NAVIGATION.

To find M.T.G.—

True dist. 105° 38′ 0″
Dist. by N.A. 105 17 22 for 17ᵈ 6ʰ 0ᵐ 0ˢ p. log 3152 diff. −16

Difference   20 38   ...   p. log 9407

0 42 38 p. log 6255

17 6 42 38
Cor. for second diff. +3

M.T.G. 17 6 42 41
Chron. 17 6 30 0

Chron. slow   12 41

To find M.T.S. and longitude—

R.A.M. 12ʰ 13ᵐ 43ˢ
R.A.M.S. 5 43 12

M.T.S. 17ᵈ 6 30 31
M.T.G. 17 6 42 41

0 12 10 = long. 3° 2′ 30″ W.

THOMSON'S METHOD.

Moon's H.P.	55′ 44″	log 0·0492	log 0·0492	Table xiv.	
Sun's app. alt. 14° 24 28		log 1·0641			
Moon's app. alt. 26 7 19			log 0·8163	,, xv.	
App. dist. 105 56 46	log S. 0·9829	log T. 1·5439		,, xvi.	Thomson
				,, xvii.	
First cor.	4 45 35	log 2·0962			
Second ,,	4 52 59	... ...	log 2·4094	,, xvii.	
Third ,,	2 43			,, xviii.	

True dist. 105 38 3

Dist. by N.A. 105 17 22 for 17ᵈ 6ʰ 0ᵐ 0ˢ p. log 3152 (second diff. −16)

Diff.   20 41   ...   ... p. log 9397 Table xix.

0 42 44 p. log 6245   ,,

17 6 42 44
Cor. for second diff.   +3

M.T.G. 17 6 42 47
Chron.   6 30 0

Chron. slow   12 47

which differs only 6ˢ from the result by Borda's method of direct calculation.

LONGITUDE BY LUNAR DISTANCES.

The tables being in sequence, there is no difficulty in remembering the form.

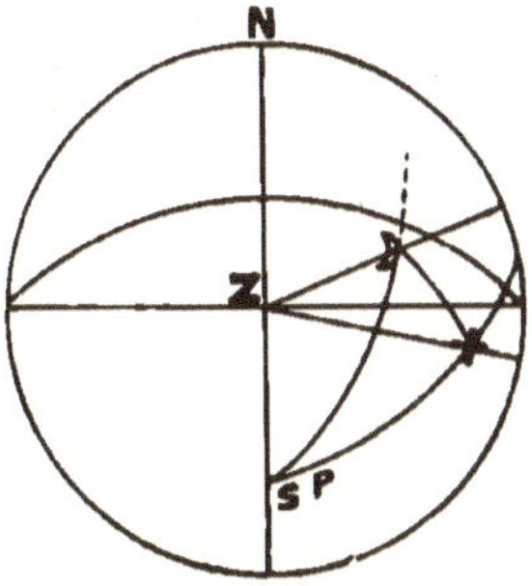

Example 2 (Star and Moon).—1899, May 30th A.M. at ship, in lat. 51° 25′ S.; time by chron., 29ᵈ 15ʰ 18ᵐ 47ˢ, which was estimated 2ᵐ slow of G.M.T. With the following observations, find true and apparent alts. of Fomalhaut, and the error of chron. for G.M.T., and longitude :—

```
   Obs. alt. of Moon's      Obs. dist. between
   L.L.E. of mer.          Fomalhaut and
                           Moon's N.L.
      26° 24′ 10″           30° 11′ 20″
I.C.   − 2  11         I.C.  + 2  53
Height of eye,     30 ft.
```

FIG. 101.

```
               Moon's R.A.     Var.     Moon's dec.    Var.
Chron. 29ᵈ 15ʰ 18ᵐ 47ˢ  21ʰ 2ᵐ 7ˢ·89   2·308    13° 2′ 58″ S.   12·46
Slow            2  0              48·01    20·8        4 19         20·8
                                  ─────              ─────
G.M.T. 29  15  20  47   21  2  55·9     48·006   12 58 39 S.   259·168
```

```
                          Moon's S.D.         Moon's H.P.
Sid. T.  4ʰ 27ᵐ 10ˢ·94    16′ 18″·09          59′ 43″·48
         2  27·84     Cor.    − ·68      Cor.   − 2·49
Accel. {    3·28      Aug.    + 7·8      Red.   − 7·1
              ·12             ──────             ──────
                              16 25·2            59 32·9
R.A.M.S. 4  29  42·18
```

```
   Obs. alt. 26° 24′ 10 ′         R.A. Star 22ʰ 52ᵐ 6ˢ·57
        I.C.  − 2  11             Dec. Star 30° 9′ 4″·5 S.
              ─────────
              26  21  59
   Dip.       − 5  22
              ─────────
              26  16  37
   S.D.      + 16  25
              ─────────
   A.A. 26  33  2
   Cor  + 51 22
        ─────────
   T.A. 27  24  24
   Lat. 51  25   0                sec  10·011236
   P.D. 77   1  21                cosec 10·205053
        ─────────
        155  50  45
         77  55  23         ...   cos  9·320623
         50  30  59         ...   sin  9·887507
                                        ─────────
         Moon's E.H.A.   4ʰ  8ᵐ 14ˢ     9·424424
         Moon's R.A.    21   2  56
                        ─────────
         R.A.M.        16  54  42
         R.A. Star     22  52   6·6
                        ─────────
         Star's E.H.A.  5  57  24·6
```

TEXT-BOOK ON NAVIGATION.

To compute alt. of Star—

Star's H.A. $5^h\ 57^m\ 24^s{\cdot}6$ ... cos 8·053018
l 51 25   0 ... cot 9·901901 ... sin 9·893041

θ 0 30 52   tan 7·954919 ... sec 10 000018
p 59 50 55
$p - \theta$ 59 20   3   ...   ...   cos 9·707597

Star's T.A. 23 29 51 ...   ... sin 9·600656
Ref.   +2 10

Star's A.A. 23 32   1

To compute true dist. and M.T.G.—

Obs. dist. 30° 11′ 20″
I.C.   +2 53
Moon's S.D. +16 25

App. dist. 30 30 38
Moon's A.A. 26 33   2   ... sec = 10·048400
Star's A.A. 23 32   0   ... sec = 10·037712

            80 35 40

            40 17 50   cos = 9·882353
             9 47 12   cos = 9·993634

Moon's T.A. 27 24 24   cos = 9·948297
Star's T.A. 23 29 50   cos = 9·962407

         2)50 54 14
                             2)19·872803
A = 25° 27′ 7″
B = 59 44 36   sin = 9·936401

A + B = 85 11 43   cos = 8·923036
A − B = 34 17 29   cos = 9·917076

                             2)18·840112

½ T. dist. 15° 15′ 6″ ... sin = 9·420056
           2

T. dist. 30 30 12
N.A.   30 36 40   P.L 5236 $29^d\ 15^h\ 0^m\ 0^s$

          6 28   P.L. 1·4446

                P.L. ·9210   0 21 35·5
       Mean diff. P.L.'s + 358, 2nd cor.   −0 44

                             G.M.T. 29 15 20 51·5
                             Chron. 29 15 18 47

                             Slow   2   4·5

LONGITUDE BY LUNAR DISTANCES.

To find M.T.S. and longitude—

```
        R.A.M.    16ʰ 54ᵐ 42ˢ
        R.A.M.S.   4 29 42
        ─────────────────
        M.T.S. 29ᵈ 12 25  0
        M.T.G.  29 15 20 51·5
        ─────────────────
                2 55 51·5
        Long.   43° 57' 52" W.
```

Thomson's Method.

Moon's H.P.	0° 59' 33"	log 0·0204	log 0·0204	Table xiv.
Star's A.A.	23 32 1	log 0·8587		,, xv.
Moon's A.A.	26 33 2		log 0·8097	
App. dist.	30 30 38	log S. 0·7056	log T. 0·7703	,, xvi. } Thomson.
First cor.	4 13 10	log 1·5847		xvii.
Second ,,	5 45 10	... ...	log 1·6004	,, xvii.
Third ,,	1 13	... ...	... ...	,, xviii.

True dist. 30 30 11

which differs by only 1" from the result by direct method.

Remark.—Questions are sometimes set where it is not stated whether the object whose altitude is given is E. or W. of the meridian, but this may be determined by noticing in the Nautical Almanac whether the other object is E. or W. of the Moon, and, taking the H.A. in conjunction with the distance, the wrong supposition would put one of the objects too near the horizon, or even below it.

EXERCISES.

1. 1899, February 17th P.M. at ship, in lat. 30° 25' S.; time by chron., 17ᵈ 10ʰ 20ᵐ 10ˢ, which was estimated to be 5ᵐ 31ˢ fast of G.M.T. With following observations, find true and apparent alts. of the Moon, error of chron. for G.M.T., and long. of place; height of eye, 24 ft. :—

Obs. alt. of Sun's L.L.	Obs. dist. between Sun and Moon.
36° 8' 0"	96° 19' 30"
I.C. −1 17	I.C. +1 7

2. 1899, May 5th A.M. at ship, in lat. 35° 20' N.; time by chron., 4ᵈ 20ʰ 50ᵐ 30ˢ, which was estimated to be 2ᵐ 5ˢ fast of G.M.T. With following observations, find true and apparent alts. of the Sun, error of chron. for G.M.T., and long. of ship; height of eye, 26 ft. :—

Obs. alt. of Moon's L.L. W. of mer.	Obs. dist. Sun and Moon.
40° 22' 10"	55° 41' 0"
I.C. −1 9	I.C. +1 25

3. 1899, June 30th A.M. at ship, in lat. 42° 25' S.; time by chron., 29ᵈ 14ʰ 2ᵐ 25ˢ, which was supposed to be correct for G.M.T. With following observations, find true and apparent alts. of Saturn, long. of ship, and error of chron. for G.M.T.; height of eye, 29 ft. :—

Obs. alt. of Moon's L.L. E. of mer.   Obs. dist. Moon's R.L. and Saturn's centre.
38° 35' 20"   107° 53' 50"
I.C. −3 12   I.C. −2 24

4. 1899, September 15th, at about 7ʰ 15ᵐ P.M., in lat. 51° 20' S.; time by chron., 15ᵈ 18ʰ 20ᵐ 30ˢ, which was estimated to be 10ˢ slow of G.M.T. With following observations, find error of chron. on G.M.T., and long. of ship; height of eye, 39 ft. :—

Obs. alt. of Moon's L.L.   Obs. alt. of centre of Mars.
48° 1' 40"   14° 44' 25"
I.C. −2 11   I.C. −2 11

Observed distance between Moon's N.L. and Mars' centre, 100° 43' 20".

5. 1999, August 20th A.M. at ship, in lat. 29° 56' N., long. by D.R. 32° 33' E.; when the chron. showed 19ᵈ 12ʰ 12ᵐ 54ˢ, supposed to be correct for M.T.G., an alt. of the Moon's lower limb was observed, 21° 24' 52" W.; no index error; dip, 14 ft. At the same time the distance between the Moon's near limb and α Arietis was observed, 86° 35' 50". Required the alt. of Arietes, the error of the chron., and longitude.

6. 1899, September 16th P.M. at ship, in lat. 43° 24' S., supposed long. 65° 45' E.; chron. showed 15ᵈ 23ʰ 40ᵐ 18ˢ, supposed to be 10ᵐ fast of M.T.G.; obs. alt. of Sun's lower limb was 19° 25' 5"; index error, +2'; height of eye, 20 ft. The obs. distance between the Sun and Moon's near limbs was 137° 51' 29"; index error, +2' 10". Required the Moon's true and apparent altitudes, the error of chron., and the longitude.

7. 1899, December 21st A.M. at ship, in lat. 23° 27' 30" N.; time by chron., 20ᵈ 11ʰ 54ᵐ 54ˢ, supposed of; obs. alt. Moon's lower limb, 26° 32' 31" W. of the meridian; obs. distance of the Moon's near limb and Jupiter's centre, 105° 45' 16"; no index error; height of eye, 28 ft. Required the true and apparent altitude of Jupiter, the error of the chron. on G.M.T., and the longitude.

CHAPTER XIX.

LATITUDE BY DOUBLE ALTITUDES.

ART. 57. **Double Altitudes.**—This is a problem for determining the latitude from—
(*a*) Two different altitudes of the same object (Sun, Moon, or Planet).
(*b*) Simultaneous altitudes of two different objects.
The data required for the computation are—
(1) The polar angle.
(2) The two altitudes.
(3) The declination.

Definition.—The **polar angle** is the angle at the Pole between the hour circles passing through the two positions of the object or objects.

(1) For the Sun, the polar angle is the interval between the apparent times of observation; therefore, if mean times are given, the equation of time must be applied to each, or the change of equation of time applied to the mean time interval.

(2) A Star's polar angle is the interval of sidereal time between the observations.

(3) When simultaneous altitudes of two stars are observed, the polar angle is the difference of their right ascensions.

(4) For the Moon or a Planet, the polar angle is the sidereal interval corrected for the change of right ascension in the elapsed mean time.

Correction for "Run."—Both altitudes are supposed to be observed at the same place; therefore, if a ship changes her position during the interval, a correction depending on the course and distance run and the bearing of the object must be applied to one of the altitudes, in order to make it what it would be at the place where the other altitude is observed.

Thus, if the ship's course is *the same* as the bearing of the object at the first observation, that is, *directly towards* it, and the distance run ten miles, the first altitude would have to be *increased* by 10′; but if the course is *opposite* to the bearing, that is, *directly away from* the object, the first altitude will have to be *decreased* by 10'; whilst if the course were at right angles to the bearing, there would be no correction.

When the course is *obliquely towards or from* the object, the angle between the bearing and the course must be found. With this angle as a course in the Traverse Table and distance

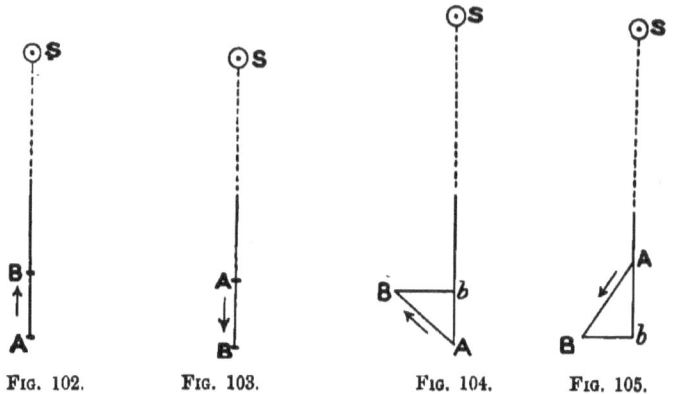

FIG. 102.   FIG. 103.   FIG. 104.   FIG. 105.

run in the Distance column, the correction will be found in the Latitude column, and will be *added* to or *subtracted* from the first altitudes according as the course is *towards* or *from* the Sun.

Illustration.—Let A and B be the ship's positions at the first and second observations respectively, and AS the direction of the object—

Fig. 102. Course directly towards the object; correction = AB +
" 103.   "   "   from   "   ;   "   = AB −
" 104.   "   obliquely towards "   ;   "   = Ab +
" 105.   "   "   from   "   ;   "   = Ab −
" 106.   "   at right angles to bearing;   "   = 0

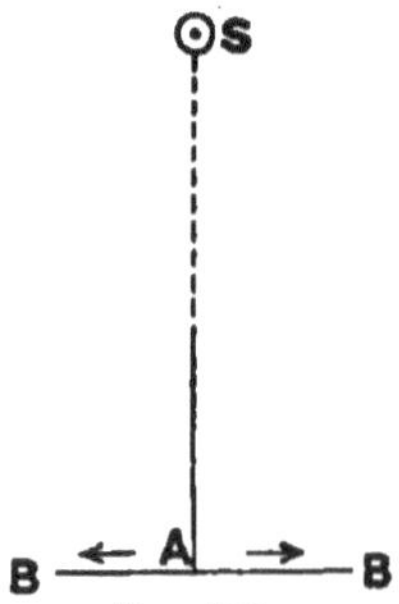

In Figs. 104 and 105 the angle at A is the course, and AB the distance in the Traverse Table, and Ab the latitude, since Bb is a perpendicular upon AS.

When the bearing is given at the *first* observation, the computation gives the latitude at the *second* observation, and *vice versâ*.

Therefore, if the bearing at the *second* observation is given, the "correction for run" must be applied to the *second* altitude, *reversing* the rule for applying it; that is, if the run is *towards* the object, *subtract* the correction from the *second* altitude, and if *from* the object,

FIG. 106.

LATITUDE BY DOUBLE ALTITUDES.

add the correction. The latitude found is the latitude at the *first* observation.

Of the various methods for solving the problem, the best known is "Ivory's method," but the latitude found requires a correction for the change of declination in half the elapsed time. This correction may be more than 2′ in extreme cases. The mode of computing and the rule for applying the correction are given at the end of the following example.

Example 1.—DOUBLE ALTITUDE OF THE SUN.

1899, March 10th A.M. at ship, in N. lat. and 30° W. long., the following altitudes of the Sun's lower limb were observed with a sextant whose index correction is +1′ 11″; height of eye, 20 feet :—

  First A.T.S. 8ʰ 10ᵐ 45ˢ A.M.
  Second ,, 1 2 10 P.M.
  Obs. alt. Sun's L.L. 17° 20′ 40″
  ,, ,, 34 20 30
  Bearing S. 57° E.

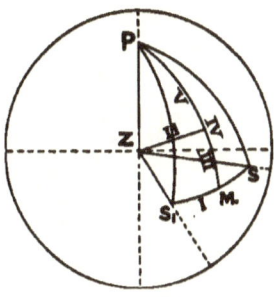

FIG. 107.

The ship's run during the interval was N. 11° E. 39 miles. Required the latitude at the second observation.

Ivory's Method.

  First A.T.S. 9ᵈ 20ʰ 10ᵐ 45ˢ   Sun's dec. 4° 4′ 8″·4 S.   58″·8
  Second ,,   10 1 2 10   Cor.   − 35·3   ·6
                                                            ─────
  Interval   4 51 25   4 3 33·1   35 28

  ½ interval   2 25 42·5   H. var. 58″·8
  Middle time 10 2 36 27·5   ½ interval 2·4
  Long.   2 0 0   ─────
                 ───────────   2352
  A.T.G. 10   0 36 27·5   1176
                                                            ─────
                                  Change of dec. in ½ int. 141·12

Correction for run—
  Sun's bearing S. 57° E.
                180
               ─────
               N. 123 E.
  Course N. 11 E.
               ─────
  Angle 112
               ─────
  Supplement 68

In Traverse Table the course 68° and distance 39 give 14′·6 = 14′ 36″ in Lat. col., which is the correction to be *subtracted* from first alt., because the angle *exceeds* 90° (see Art. 57).

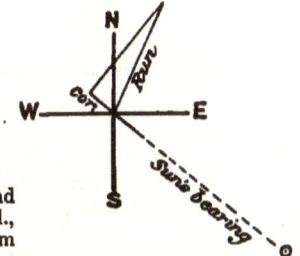

FIG. 107a.

158 TEXT-BOOK ON NAVIGATION.

```
First obs. alt. 17° 20' 40''           Second obs. alt. 34° 20' 30''
        I.C.    +1 11                         I.C.    +1 11
               ────────                              ────────
                17 21 51                              34 21 41
        Dip        4 23                       Dip        4 23
               ────────                              ────────
                17 17 28                              34 17 18
        Ref. par.  2 54                       Ref. par.  1 16
               ────────                              ────────
                17 14 34                              34 16  2
        S.D.      16  7                       S.D.      16  7
               ────────                              ────────
                17 30 41                    Second true alt. 34 32 9
        Cor. for run −14 36                 First      ,,    17 16 5
               ────────                                    ────────
        First true alt. 17 16  5            Diff.      17 16 4
        Second    ,,    34 32  9                       ────────
                     ────────               ·½ diff.    8 38 2
        Sum          51 48 14
                     ────────
               ½ sum 25 54  7
```

½ int. 2ʰ 25ᵐ 42ˢ·5   cosec 0·226362
Dec.   4   3   33   sec   0·001090   ...   ...   sine 8·849951

Arc 1, 36 19 16   cosec 0·227452   sec 0·093821   sec 0·093821
½ sum alts. 25 54 7   cos 9·954022   sine 9·640314
½ diff. ,,   8 38 2   sine 9·176439   cos 9·995050

Arc 2, 13 10 43   sine 9·357913   sec 0·011590*

Arc 3, 56 35 51   cos 9·740775
Arc 4, 95 2 25 = 180° − 84° 57' 35''   ...   ...   cos 8·943772

Arc 5, 38 26 34   sec 0·106111
Arc 2,   sec 0·011590*

Approx. }
  lat. } 49 41 39   cosec 0·117701
Cor.   1 24 N.

True lat. 49 43 3 N.

Correction for change of declination—
Ch. of dec. 141''·1   ...   log 2·149527
Approx. lat. 49° 42'   ...   sec 0·187752
½ int.   ...   cosec 0·226362
Arc 2   ...   sine 9·357913

Cor. 83''·49   log 1·921554

1' 23''

Rule for applying the correction of latitude—
(a) When *second* altitude is the *greater*,
    Allow it N. when the Sun is going N.
       ,, S.   ,,   ,,   S
(b) When first altitude is the greater,
    Reverse the above rule.

LATITUDE BY DOUBLE ALTITUDES.

Example 2.—DOUBLE ALTITUDE OF A STAR.

1899, October 26th A.M., in lat. by account 30° N., long. 47° 15′ E., the following altitudes of the Star Capella were observed:—
(1) M.T.S. 26th 0ʰ 14ᵐ 45ˢ A.M. Obs. alt. 55° 55′ 50″
(2) ,, 26th 4 34 45 A.M. ,, 65 1 24 Bearing N.E.
height of eye, 29 feet; index cor. −5′ 22″; run of ship during the interval between observations, E.S.E. 48 miles. Required the latitude at the time of first observation.

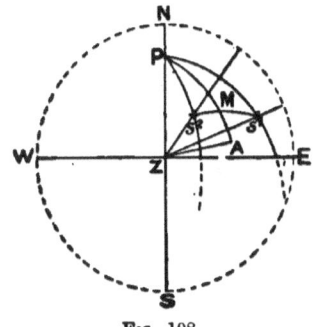
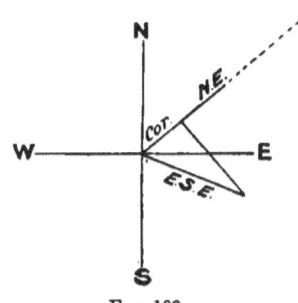

FIG. 108. FIG. 109.

```
          First M.T.S. 25ᵈ 12ʰ 14ᵐ 45ˢ
          Second  ,,   25  16 34 45
               Interval  4 20  0
                                39·4
                Accel.           3·3
                        ─────────────
          Sid. interval  4 20 42·7
             ½ int.      2 10 21·3
```

Correction for ship's run—
The angle between the star's bearing and the ship's course is 6 points. With 6 points as a course, and dist. 48′, the Traverse Table gives diff. lat. 18′·4 = 18′ 24″, which is the correction to be subtracted from the second alt. Fig. 109 (see Art. 57).

Star's dec. from N.A. 45° 53′ 42″ N.—

First obs. alt. 55° 55′ 50″	Second obs. alt. 65° 1′ 24″
I.C. −5 22	I.C. −5 22
55 50 28	64 56 2
Dip −5 16	Dip −5 16
55 45 12	64 50 46
Ref. −0 39	Ref. −0 27
First true alt. 55 44 33	64 50 19
Second ,, 64 31 55	Cor. for run −18 24
Sum 2)120 16 28	Second true alt. 64 31 55
	First ,, 55 44 33
½ sum 60 8 14	
	Difference 2)8 47 22
	½ diff. 4 23 41

160   TEXT-BOOK ON NAVIGATION.

½ int. 2ʰ 10ᵐ 21ˢ	cosec 10·268731		
Dec. 45° 53' 42"	sec 10·157406	... ...	sine 9·856164

Arc 1, 22 0 55	cosec 10·426137	sec 10·032881	sec 10·032881
½ sum 60 8 14	cos 9·697164	sine 9·938130	
½ diff. 4 23 41	sine 8·884383	cos 9·998721	

Arc 2, 5 50 32	sine 9·007684	sec 10·002261	
Arc 3, 20 21 24	... ...	cos 9·971993	
Arc 4, 39 14 11	... ...	... ...	cos 9·889045

Arc 5, 59 35 35	sec 10·295731
Arc 2, 5 50 32	sec 10·002261

Lat. 30 13 58 N. cosec 10·297992

NOTE.—In this example the correction for run is applied to the second altitude, because the bearing is given when second altitude was observed. See rule given with Example 1.

Remarks.—Inspection of the above computation will show—
(1) When Dec. is 0, Arc 4 is 90°, because sin 0 = 0, and therefore cos arc 4 is 0.
(2) When diff. of alts. is 0, arc 2 = 0, and arc 5 = co-lat.
(3) Arc 4 = 90° when polar dist. = 90°, and is greater or less than 90° according as the polar dist. is greater or less than 90°. Therefore, when lat. and dec. are of different names, the supplement must be taken for arc 4.
(4) Arc 5 is always the *difference* of 3 and 4 when lat. and dec. are of *different* names, and generally the *sum* when lat. and dec. are of the *same* name. But there may be a doubt as to whether they should be added or subtracted in some cases, which, however, can be decided by computing the H.A. at middle time from the formula sin H_1 = sin arc 2 × sec. lat., using the lat. first found, and if this agrees with the known H.A. at middle time, the lat. is correct; if not, subtract 3 from 4.

EXERCISES.

1. 1899, May 10th, in lat. by account 20° 38' 30" N., long. 115° 45' W. With the following observations, find the latitude by double altitudes:—

A.T.S. 10ʰ 58ᵐ 30ˢ A.M. Obs. alt. Sun's L.L. 75° 0' 19" Index error, +0' 30"
 ,, 3 17 30 P.M.   ,,   ,,   43 16 50   Dip, 25 ft.

Sun's bearing at second observation, W. ½ S. ; run during the interval, N. ½ W. 5 miles per hour. Required the latitude at first observation.

2. 1899, June 23rd, lat. N., mer. of Greenwich. With the following observations, find the latitude by double altitudes:—

M.T.G. 22ᵈ 22ʰ 40ᵐ 10ˢ Obs. alt. Sun's L.L. 59° 3' 59" Index error, −3' 15"
 ,, 23 2 59 55   ,,   ,,   46 19 41   Height of eye, 26 ft.

Sun's bearing at first observation, S. 38½° E ; run during the interval, W. by S. 9 knots per hour. Required the latitude at second observation.

LATITUDE BY DOUBLE ALTITUDES.

3. 1900, January 1st, in lat. N., long. 180° E. With the following observations, find the latitude by double altitudes at second observation:—

Chron. dec. 31ᵈ 10ʰ 15ᵐ 40ˢ } slow 10ᵐ 20ˢ on M.T.G.
„ 31 14 15 14
Alt. Sun's L.L. 14° 14' 35"   Index error, 0
Alt. 11 17 1   Dip, 27 ft.

Sun's true bearing at the first observation, S.E.; run, E. 40 miles during the interval.

4. 1899, September 23rd, lat. N., long. 82° 30' E. With the following observations of the Sun's lower limb, find the latitude by double altitudes at the second observation:—

A.T.S. 9ʰ 30ᵐ 10ˢ A.M.   Obs. alt. 32° 13' 6"   Index error, −2' 4'
„ 2 30 20 P.M.   „ 32 13 13   Height of eye, 28 ft.

Sun's bearing at first observation, S. 46° E.; course and dist. during interval, N. 60° E.; rate, 5·8 knots per hour.

5. 1899, January 20th, in lat. by D.R. 12° 20' N., long. 70° 39' W. With the following observations of Planet Venus' lower limb, find the latitude by double altitudes at the second observation:—

M.T.G. 20ᵈ 2ʰ 42ᵐ 36ˢ   Obs. alt. 56° 37' 13"   Index error, −2' 10"
„ 20 6 28 6   „ 13 37 31   Height of eye, 20 ft.

Bearing of Venus at first observation, S. 26° W.; run, E. 10 knots per hour.

6. 1899, June 23rd P.M. at ship, in lat. N., long. 30° E. With the following observations of Jupiter's centre, find the latitude by double altitudes at the second observation:—

M.T.S. 5ʰ 47ᵐ 0ˢ P.M.   Obs. alt. 36° 33' 23"   Index error, 0.
„ 9 47 0 P.M.   „ 36 24 35   Height of eye, 19 ft.

Bearing of Jupiter at first observation, S. 38° E.; run during interval, N. 54° W. 25'.

7. 1899, November 27th A.M. at ship, in lat. N., long. 29° 15' E. With the following observations of Sirius, find the latitude by double altitudes at the second observation:—

M.T.S. 15ʰ 45ᵐ 20ˢ   Obs. alt. 41° 26' 31"   Index error, +2' 10"
„ 19 0 0   „ 8 39 54   Height of eye, 23 ft.

Bearing of Sirius at first observation, S.S.W. ½ W.; run, N.N.E. ½ E. 8 knots per hour.

8. 1899, April 26th at ship, in lat. by account 33° N., long. 178° 35' E. With following observations, find latitude at time of taking the second observation:—

Time by chron. 25ᵈ 10ʰ 44ᵐ 15ˢ   Obs. alt. } 59° 16' 50"   Bearing, S.E. by E.
„ 25 16 4 32   Sun's L.L.
„   36 45 40

The chron. was fast 23ᵐ 19ˢ of G.M.T.; the ship's true course and distance in the interval, N.N.E. 40 miles; index cor. −2' 9"; height of eye, 28 ft.

9. 1899, July 2nd at ship, lat. S., long. 50° 27' W. Find lat. at second observation from following:—

A.T.S. 2ᵈ 9ʰ 40ᵐ 25ˢ A.M.   Obs. alt. Sun's L.L.   9° 38' 10"   Bearing, N.E. by N.
„ 2 2 19 45 P.M.   „   „   10 19 50

The ship's true course, N.E. by N., 9 knots per hour during the interval; index cor. −3' 8"; height of eye, 33 ft.

10. 1899, October 11th, in lat. 47° 30' N. by account, long. 148° 23' E. With following observations, find latitude at second observation:—

M.T.G. 10ᵈ 11ʰ 8ᵐ 51ˢ   Obs. alt. Sun's L.L. 24° 1' 20"   Bearing, S. 46° E.
„ 10 15 32 56   „   „   30 56 10

M

162   TEXT-BOOK ON NAVIGATION.

The ship's true course and distance in the interval, N. 78° E. 36 miles; index cor. −2′ 13″; height of eye, 42 ft.

11. 1899, January 15th, in lat. N., long. 143° 20′ W. With following observations of Sirius, find latitude at first observation :—

M.T.G. 15ᵈ 16ʰ 48ᵐ 39ˢ   Obs. alt. Sirius, 26° 42′ 50″
 ,,   15 20 8 44   ,,   ,,   58 42 50   Bearing S. 11° E.

The ship's true course and distance in the interval, S.W. 24 miles; index cor. +4′ 21″; height of eye, 42 ft.

12. 1899, January 16th, in lat. N., long. 124° 29′ W. With following observations of Mars, find latitude at second observation :—

M.T.G. 16ᵈ 21ʰ 33ᵐ 23ˢ   Obs. alt. Mars' ⎫
 ,,   16 23 58 35   centre   ⎬ 62° 34′ 40″ Bearing, S.S.W. ¼ W.
                          ,,   ⎭ 44 20 50

The ship's true course and distance in the interval, S.E. by E. ¾ E. 24 miles; index cor. +3′ 19″; height of eye, 19 ft.

ART. 58. Simultaneous Altitudes.

Example.—1899, July 18th P.M. at ship, in long. 179° 45′ W.; chron. shows 18ᵈ 21ʰ 3ᵐ 49ˢ M.T.G.; observed alt. of Altair east of meridian, 36° 29′ 40″; observed alt. of Spica west of meridian, 15° 6′ 30″; height of eye, 30 ft.; index cor. +4′ 15″. Required the ship's position by Sumner's method of projection; also the true azimuths of the stars, assuming lats. 47° 40′ N. and 48° 20′ N.

	Altair.		Spica.	
	h. m. s.	h. m. s.	h. m. s.	h. m. s.
M.T.G.	18 21 3 49	S.T. at noon 7 44 18·79	R.A. 19 48 55·3	13 19 54·9
	11 59 0	Accel. 3 26·99	Dec. 8° 36′ 17″ N.	10° 38′ 12″ S.
		·62	90	90
M.T.S.	18 9 4 49			
	R.A.M.S.	7 47 46·40	P.D. 81 23 43	100 38 12
	M.T.S.	9 4 49		
	R.A.M.	16 52 35·4		

179° 45′
    4
─────
6,0)71,9. 0
─────
 11 59

Altair.		Spica.	
Obs. alt. 36° 29′ 40″		15° 6′ 30″	
Index cor. +4 15		Index cor. +4 15	
36 33 55		15 10 45	
Dip −5 22		Dip −5 22	
36 28 33		15 5 23	
Ref. −1 17		Ref. −3 30	
True alt. 36 27 16		T.A. 15 1 53	
Lat. 47 40 0	48° 20′ 0″	Lat. 47 40 0	48° 20′ 0″
P.D. 81 23 43		P.D. 100 38 12	
2)165 30 59		2)163 20 5	
82 45 30	83 5 30	81 40 3	82 0 3
46 18 14	46 38 14	66 38 10	66 58 10
Secants 10·171699	10·177312	10·171699	10·177312
Cosecs 10·004916	10·004916	10·007527	10·007527
Cosines 9·100559	9·080198	9·161121	9·143510
Sines 9·859147	9·861547	9·962845	9·963928
9·136321	9·123973	9·303192	9·292277

LATITUDE BY DOUBLE ALTITUDES.

	h. m. s.	h. m. s.		h. m. s.	h. m. s.
H.A. (Altair)	2 53 42 E.	2 51 8 E.	(Spica)	3 33 5 W.	3 30 13 W.
R.A. ,,	19 48 55	19 48 55	,,	13 19 55	13 19 55
R.A.M.	16 55 13	16 57 47		16 53 0	16 50 8
R.A.M.S.	7 47 46	7 47 46		7 47 46	7 47 46
M.T.S.	18ᵈ 9 7 27	18ᵈ 9 10 1		18ᵈ 9 5 14	18ᵈ 9 2 22
M.T.G.	18 21 3 49	18 21 3 49		18 21 3 49	18 21 3 49
Diff.	11 56 22	11 53 48		11 58 35	12 1 27
	60	60		60	60
	4)716 22 0	4)713 48 0		4)718 35 0	4)721 27 0
Longs.	179° 5′ 30″ W.	178° 27′ 0″ W.	179° 38′ 45″ W.	180° 21′ 45″ W.
				360 0 0
				179 38 15 E.

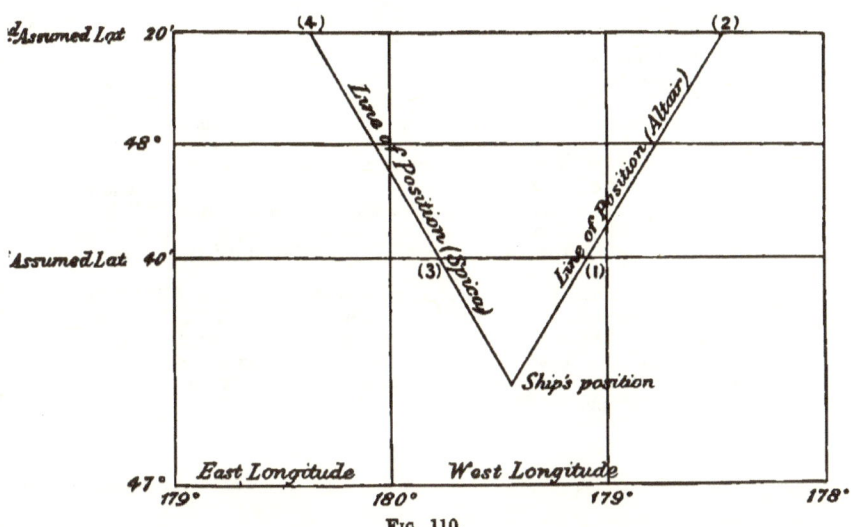

Fig. 110.

NOTE.—The half-sum and remainder for the second assumed latitude are most readily found by adding half the difference of the assumed latitudes to the half-sum and remainder for the first assumed latitude.

The longitudes being projected on a chart, two lines of position are obtained, which by their intersection give the ship's position. The true azimuths of the stars are 90° from the lines of position, Altair to the S. and E., and Spica to the S. and W.

The projection gives—

Ship's position, lat. 47° 28′ N.; long. 179° 21′ W.
Altair's line of position, S. 40° W.; azimuth, S. 50° E.
Spica's ,, S. 44° E. ,, S. 46° W.

EXERCISES.

1. 1899, December 14th P.M. at ship, lat. 46° 17′ N., long. 29° 45′ W. by D.R.; time by chron. 14ᵈ 6ʰ 29ᵐ M.T.G.; observed alt. of Venus west of meridian, 9° 14′ 30″; at the same time α Arietis was observed east of the meridian, 37° 13′ 0″; index error, −1′ 10″; height of eye, 17 ft. Find ship's position and the true bearing of the stars by Sumner's method by projection on the chart, assuming lats. 46° 30′ N. and 47° 0′ N.

2. 1899, December 26th A.M. at ship, when the chron. indicated 25ᵈ 10ʰ 10ᵐ 24ˢ M.T.G., the observed alt. of Sun's lower limb was 15° 29′ 30″, and the observed alt. of the Moon's lower limb was 13° 15′ 10″ W. of meridian, taken at the same time; index error, +2′ 11″; height of eye, 24 ft. Required the position of the ship, the true azimuth of the objects by Sumner's method by projection on the chart, assuming lats. 46° 0′ N. and 46° 30′ N.

3. 1899, April 15th P.M. at ship, when the chron. indicated 15ᵈ 7ʰ 15ᵐ 54ˢ M.T.G.; the observed alt. of Mars west of the meridian was 55° 43′ 10″ and the observed alt. of Jupiter taken at the same time was 17° 24′ 30″ east of the meridian; index error, 0; height of eye, 23 ft. Required the ship's position and the true azimuths of the planets by Sumner's method by projection on the chart, assuming lats. 38° 0′ N. and 38° 20′ N.

4. 1899, April 5th A.M. at ship, when a chron. showed 4ᵈ 21ʰ 23ᵐ 35ˢ G.M.T., the observed alt. of Venus' centre east of meridian was 16° 22′ 10″, and at the same time the observed alt. of Saturn's centre west of meridian was 19° 9′ 10″; height of eye, 25 ft.; index cor. to both, +3′ 13″. Required the position of the ship and true bearings of the planets by Sumner's method by projection, assuming lats. 46° and 46° 40′ N.

5. 1899, July 18th P.M. at ship, when a chron. showed 18ᵈ 21ʰ 3ᵐ 49ˢ G.M.T., the observed alt. of Altair east of meridian was 36° 29′ 40″, and at the same instant the observed alt. of Spica west of meridian was 15° 6′ 30″; height of eye, 30 ft.; index cor. to both alts. +4′ 15″. Required the position of the ship and true bearings of the stars by Sumner's method by projection, assuming lats. 47° 40′ and 48° 20′ N.

6. 1899, December 9th P.M. at ship, when a chron. showed 9ᵈ 9ʰ 53ᵐ 20ˢ M.T.G., the observed alt. of Capella east of meridian was 37° 27′ 25″, and at the same time the observed alt. of Altair west of meridian was 26° 31′ 55″; height of eye, 17 ft.; index cor. to both, −2′ 14″. Required the position of the ship and true bearings of the stars by Sumner's method by projection, assuming lats. 48° and 48° 30′ N.

CHAPTER XX.

FINDING ERROR OF CHRONOMETER.

Art. 59. Error of Chronometer by a Single Altitude of the Sun or a Star.—For this purpose it is necessary to know the correct latitude and longitude of the place of observation. The altitude would therefore be usually observed in port, using an artificial horizon.

The computation of the hour angle is exactly the same as in the problem of finding the longitude by chronometer. Thence the mean time at place and mean time at Greenwich are found, and the error of chronometer determined.

Example 1.—1899, May 15th P.M. at place, in lat. 47° 34′ N., and long. 52° 40′ 50″ W. ; approx. M.T.G. by chron. 15^d 7^h 40^m 36^s ; observed alt. by artificial horizon of the Sun's lower limb, 62° 54′ 30″ ; index cor. + 1′ 4″. Required the error of chron. on A.T.S., M.T.S. and M.T.G.

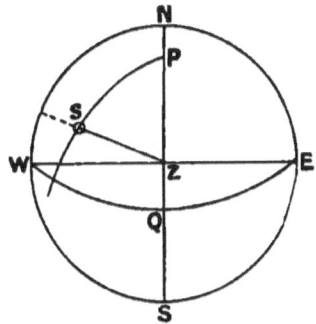

Fig. 111.

Approx. M.T.G., May 15^d 7^h 40^m 30^s.

Sun's dec.	Var. 1 hour.	Eq. of time.	Var. 1 hour.		Alt.
18° 53′ 1″·1 N.	35″·38	3^m 48^s·76	·019		62 54 30
+4 32·4	7·7	− ·15	7·7	I.C.	+1 4
18 57 33·5	24766	3 48·61	133		2)62 55 34
90	24766		133		31 27 47
P.D. 71 2 26·5 N.	6,0)27,2·426		·1463	Cor.	−1 26
	4′ 32″·4				31 26 21
				S.D.	+15 50
				T.A.	31 42 11

166   TEXT-BOOK ON NAVIGATION.

```
         a   31° 42' 11''
         l   47  34   0     ...   ...   sec   10·170869
         p   71   2  27     ...   ...   cosec 10·024223
            ─────────────
         2)150 18 38
            ─────────────
             75   9 19     ...   ...   cos   9·408581
             43  27  8     ...   ...   sine  9·837430
H.A. and A.T.S.   4ʰ 13ᵐ 36ˢ (1)  ...   log   9·441103
Eq. of time      − 3  49
                ──────────
M.T.S. 15ᵈ  4   9  47  (2)
Long.       3  30  43
           ──────────
M.T.G. 15ᵈ  7  40  30  (3)
Chron.      7  40  36
           ──────────
(3)  0   0   6  fast of M.T.G.
(2)  3  30  49     ,,    M.T.S.
(1)  3  27   0     ,,    A.T.S.
```

Example 2.—1899, March 23rd, 0ʰ 15ᵐ A.M., approx. A.T.S. in lat. 10° 19' N., long. 99° 52' 15'' E.; observed alt. of Star Antares (east of meridian) by artificial horizon, 39° 18' 50''; index cor. +3' 8''; time shown by chron. 5ʰ 14ᵐ 9ˢ. Required the error of chron. on M.T.G. and A.T.S.

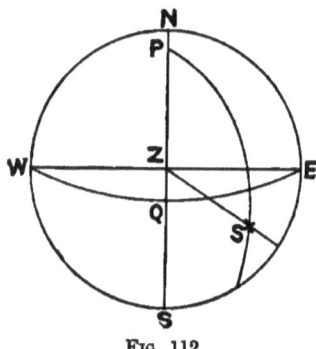

FIG. 112.

```
                                                          R.A.M.S.
Approx. M.T.S. 22ᵈ 12ʰ 15ᵐ 0ˢ,   99° 52' 15'' E.  S.T. at noon 23 59  5·23
Long.            6  39 29         4                          ⎧ 49·28
                ──────────       ─────           Accel.      ⎨  5·75
Approx. M.T.G. 22  5 35 31   6,0)39,9 29  0                  ⎩   ·08
                                 ──────────                   ──────
                                 6ʰ 39ᵐ 29ˢ                   0  0  0·34

       R.A. Antares 16ʰ 23ᵐ 15ˢ·43      Dec. Antares 26° 12' 34'' S.
                                                     90  0   0
                                                    ────────────
                                        P. distance 116 12 34
```

FINDING ERROR OF CHRONOMETER. 167

```
Obs.'alt.  39° 18' 50"
I.C.        +3   8                    Eq. of time.      Var. 1 hour.
          ─────────                   7ᵐ 0ˢ·97            ·76
          2)39 21 58                   -4·26             5·6
          ─────────                   ─────────         ─────
            19 40 59                   6 56·71           456
Ref.        -2 38                                        380
          ─────────                                     ─────
True alt.   19 38 21                                    4·256

  Lat.   10 19  0    ...   ...   sec 10·007079
  P.D.  116 12 34    ...   ...   cosec 10·047119

        2)146  9 55

          73  4 57    ...         cos 9·463878
          53 26 36    ...         sin 9·904860

H.A. Star  4  7 46 E.  ...        log 9·422936
R.A. Star 16 43 15·4
          ─────────               M.T.S. 12ʰ 15ᵐ 29ˢ·1
R.A.M. 12 15 29·4                 Eq. of time   6 56 ·7
R.A.M.S.  0  0  0·3               ─────────────────────
          ─────────               A.T.S. 12  8  32·4
M.T.S. 12 15 29·1                 Chron. 5  14   9
Long.   6 39 29                   ─────────────────────
          ─────────               Error of chron. 6 54 23·4 slow of A.T S.
M.T.G.  5 36  0·1                                 12  0   0
Chron.  5 14  9                                   ─────────────
          ─────────                               or 5  5 36·6 apparently fast.
Error of chron.  21 51·1 slow of M.T.G.
```

EXERCISES.

1. 1899, on May 5th, at about 10ʰ 45ᵐ A.M. mean time at ship, in lat. 35° 20' N., long. 29° 25' E., the observed alt. of the Sun's lower limb taken with an artificial horizon was 130° 0' 30"; index cor. -1' 4"; the chron. showed 8ʰ 50ᵐ 30ˢ. Required its error on M.T.G. and M.T.S.

2. 1899, on August 30th, at about 9ʰ 3ᵐ A.M. mean time at ship, in lat. 40° 20' N., long. 25° 30' W., the observed alt. of the Sun's lower limb in an artificial horizon was 79° 10' 14"; index cor. -3' 12". Time by a hack chron. 10ʰ 30ᵐ, which was estimated slow of the standard chron. 15ᵐ 15ˢ. Required the error of the standard on M.T.G. and M.T.S.

3. 1899, on January 7th, when the apparent time at ship was 3ʰ 44ᵐ P.M., in lat. 40° 15' S., long. 100° 40' E., a chron. showed 9ʰ 10ᵐ 12ˢ; the observed alt. of Sun's lower limb in an artificial horizon was 79° 2' 30"; index cor. -1' 22". Required the error of the chron. on A.T.S. and M.T.G.

4. 1899, on February 21st, when the A.T.S. was 10ʰ 48ᵐ A.M. in lat. 50° 25' S., long. 55° 20' W., a chron. showed 3ʰ 5ᵐ 26ˢ; the observed alt. of Sun's lower limb was 47° 17' 20"; index cor. -2' 30"; height of eye, 27 ft. Required the error of the chronometer on A.T.S. and M.T.G.

5. 1899, on March 20th, when the A.T.S. was 5ʰ 18ᵐ P.M. in lat. 24° 12' N., long. 45° 30' W., a chron. showed 8ʰ 30ᵐ 6ˢ; the observed alt. of Sun's lower limb was 8° 56' 20"; index cor. -2' 12"; height of eye, 32 ft. Required the error of the chronometer on A.T.S. and M.T.G.

6. 1899, on April 26th P.M. at ship, when the M.T.S. was approximately 3ʰ 36ᵐ in lat. 33° 20' N., long. 178° 35' E., a chron. showed 3ʰ 41ᵐ 35ˢ; the

observed alt. of Sun's lower limb taken in an artificial horizon was 73° 18' 50"; index cor. −2' 10". Required the error of the chronometer on M.T.G.

7. 1899, on May 15th, at about $10^h 47^m$ A.M. mean time at ship, in lat. 49° 26' N., long. 126° 30' W., when a chron. showed $7^h 10^m 14^s$, the observed alt. of the Sun's lower limb taken in an artificial horizon was 112° 27' 30"; index cor. +2' 4". Find the error of the chron. on G.M.T. and A.T.S.

8. 1899, on June 14th, approximate mean time at ship 5^h P.M., in lat. 48° 32' N., long. 179° 56' W., when a chron. showed $4^h 48^m 39^s$, the observed alt. of the Sun's lower limb was 26° 49' 10"; index cor. −2' 4"; height of eye, 29 ft. Required the error of the chron. on M.T.G.

9. 1899, on April 21st, at about $8^h 30^m$ P.M. mean time at ship, in lat. 25° 10' S., long. 125° 10' W., when a chron. showed $4^h 45^m$, the observed alt. of Pollux taken in an artificial horizon was 46° 22' 30" W. of mer.; index cor. +4' 18". Required the error of the chron. on M.T.G.

10. 1899, in lat. 40° 20' S., long. 43° 35' W., at about midnight on July 25th, when a chron. showed $2^h 25^m$, the observed alt. of Antares taken in an artificial horizon was 93° 30' 30" W. of mer.; index cor. −2' 20". Required the error of the chron. on M.T.S. and M.T.G.

11. 1899, in lat. 19° 41' S., long. 45° 23' W., at about $11^h 15^m$ P.M. on February 23rd, the observed alt. of Spica E. of mer. was 25° 59' 30"; index cor. +1' 30"; height of eye, 26 ft., when a chron. showed $2^h 14^m 36^s$. Find the error of the chron. on M.T.G. and M.T.S.

12. 1899, in lat. 48° 17' N., long. 50° 40' W., at about $6^h 30^m$ P.M. on December 9th, the observed alt. of Capella E. of mer. in an artificial horizon was 74° 44' 30"; index cor. −2' 12", when a chron. showed 10^h. Find the error of the chron. on M.T.G.

ART. 60. **Error of Chronometer by Equal Altitudes.** (*a*) *Equal Altitudes of the Sun.*—If the times by chronometer when the Sun has equal altitudes east and west of the meridian be noted, the mean of these times, with a correction on account of the change of declination during the interval, gives the time shown by the same chronometer when the Sun's centre crosses the meridian, that is, *apparent noon*. This affords a means of finding the error of the chronometer.

The observation can best be made on shore with an artificial horizon, the latitude and longitude of the place of observation being accurately known.

If there were no change of declination, it is obvious that the hour angles E. and W. would be equal, and the mean of the times would be what the chronometer showed at apparent noon. But as the Sun's declination changes, the middle time would fall on one side or the other of apparent noon by a small interval called the "Equation of Equal Altitudes."

Thus, when the Sun's polar distance increases, the east H.A. would be greater than the west H.A., because it would take less time to fall to the same altitude, and the middle time would be a few seconds *before* apparent noon. Again, when the polar distance decreases, the west H.A. would be the greater, because it would take a longer time for the Sun to attain the same altitude, and so the middle time would then be a few seconds *after* apparent noon.

FINDING ERROR OF CHRONOMETER.

The object of the calculation is therefore to find the correction to be applied to the *middle time* to find the exact time by

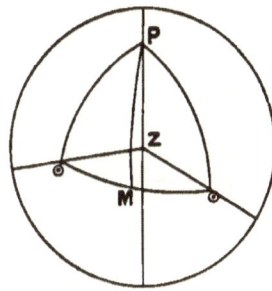

Fig. 113.

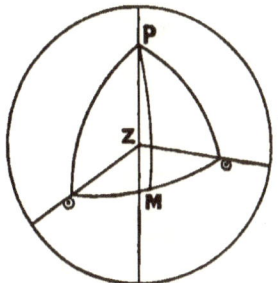

Fig. 114.

chronometer when the Sun's centre was on the meridian, and thence the error of the chronometer.

The working out of the formula gives the correction in two parts.

The first part is marked + or − according as the Sun's *polar distance* is *increasing* or *decreasing*.

The second part is marked + or − according as the *declination* is *increasing* or *decreasing*.

The *algebraical sum* of the two parts is the *Equation of Equal Altitudes* with the proper sign.

The elements required for the calculation are, the *Time Interval*, the *Sun's Declination*, and the *Equation of Time*, with the latitude and longitude of the place of observation.

(b) *Equal Altitudes of a Star.*— The intervals from the time when a star crosses the meridian to the times when it has equal altitude east and west of the meridian are exactly equal, because its declination does not change. Hence the middle time by chronometer would be the time of meridian passage, and as the time of meridian passage can be found independently, the error of the chronometer can be determined.

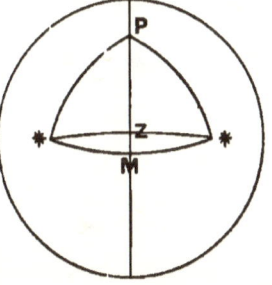

Fig. 115.

Example 1 (the Sun).—1899, January 12th, in lat 30° 26′ S., long. 73° 24′ W., the following times were noted when the Sun had equal alts. Required the error of the watch on A.T.S. and M.T.G. :—

73° 24′
4
─────
6,0)29,3 36
─────
4 53 36

170   TEXT-BOOK ON NAVIGATION.

	A.M. 10ʰ 26ᵐ 19ˢ	P.M. 1ʰ 24ᵐ 12ˢ
A.T.S. Jan. 12ᵈ	0 0 0	First obs. 10 26 19 A.M.
Long.	4 53 36	Second obs. 1 24 12 P.M.

A.T.G. Jan. 12ᵈ   4 53 36   Int.   2 57 53
                                 ½ int. 1 28 56·5

                              Middle time 11 55 15·5

Dec.	H. var.	Eq. of T.	H. var.		
21° 38′ 1″ S.	24·66	+8ᵐ 33ˢ·23	·964	Dec. 11th 21° 47′ 40″·6
−2 1	4·9	4·72	4·9	Dec. 13th 21 27 56·7
21 36 0	22194	8 37 95	8676	Change in 2 days	19 43·9
	9864		3856		60
	6,0)12,0·834		4·7236		1183·9
	2′ 1″				

        Int.   2ʰ 57ᵐ 53ˢ ...   log A, 7·7357   ... log B, 7·7021
        Lat.   30° 26′   ...   tan 9·7690
        Dec.   21 36   ...   .   tan 9·5976
 Change in 2 days   1184″   ...   log 3·0734   log 3·0734

                 +3·785   log 0·5781
                 −2·361   ...   . 0·3731

 Eq. of equal alts. +1·424

 Middle time by watch 11ʰ 55ᵐ 15ˢ·5   A.T.G. 12ᵈ 4ʰ 53ʰ 36ˢ
              Cor.   1·424   Eq. of time   +8 37·95

 Watch showed at app. noon 11 55 16·924   M.T.G.   5 2 13·95
                           12 0 0   Watch   11 55 16·92

       Watch slow of A.T.S. 0 4 43·076 Slow of M.T.G. 5 6 57·03

Example 2 (a Star).—1899, May 19th, in lat. 35° 20′ S., long. 32° E., the following times by chronometer were noted when the Star Spica had equal altitudes:—

Star E. of mer.	Star W. of mer.
Chron. 5ʰ 38ᵐ 40ˢ	9ʰ 12ᵐ 38ˢ

Required the error of chronometer on A.T.S., M.T.S., and M.T.G.

		Eq. of T.	H. var.
May 19ᵈ 0ʰ 0ᵐ 0ˢ (mean noon)		3ᵐ 42ˢ·54	·109
Long. 2 8 0		·23	2·1
May 18 21 52 0 M.T.G.		3 42 77	109
			218
			·2289

FINDING ERROR OF CHRONOMETER.   171

To compute the time when the star is on the meridian—

S.T. at noon (18ᵈ)	3ʰ 43ᵐ 48ˢ·81			First obs.	5ʰ 38ᵐ 40ˢ	
Accel.	3 26·99			Second obs.	9 12 38	
	8·54					
				Int. 2)3 33 58		
R.A.M.S.	3 47 24·34					
R.A. of Star	13 19 55·47			½ int.	1 46 59	
Sid. int.	9 32 31·13			Middle time by chron. }	7 25 39	
Solar int.	8 58 31·53					
	31 54·76			A.T.S.	9ʰ 34ᵐ 40ˢ·10	
	30·91			Chron.	7 25 39	
	·13					
				Chron. slow	2 9 1·10	
M.T.S.	9 30 57·33					
Eq. of T.	3 42·77			M.T.S.	9 30 57·33	
				Chron.	7 25 39	
A.T.S.	9 34 40·10					
Long.	2 8 0			Chron. slow	2 5 18·33	
M.T.G.	7 22 57·33					
Mid. T. by chron.	7 25 39					
Chron. fast of M.T.G.	2 41·27	(1)				
Chron. slow of M.T.S.	2 5 18·33	(2)				
Chron. slow of A.T.S.	2 9 1·10	(3)				

EXERCISES.

1. 1889, January 12th, in lat. 30° 26′ S., long. 73° 24′ W., the following times were noted when Sun had equal altitudes. Required the error of chron. for A.T.S. and M.T.G. :—

   A.M. 10ʰ 26ᵐ 19ˢ   P.M. 1ʰ 24ᵐ 12ˢ

2. 1899, March 21st, in lat. 35° 39′ N., long 139° 43′ 30″ E., at the following times the Sun had equal alts. :—

  East mer. 6ʰ 17ᵐ 45ˢ West mer. 0ʰ 25ᵐ 41ˢ

Find the error of chron. on A.T.S. and M.T.G.

3. 1899, April 2nd, in lat. 17° 56′ N., long. 76° 51′ W., the Sun had equal alts. at following times :—

  East mer. 3ʰ 15ᵐ 27ˢ West mer. 8ʰ 19ᵐ 13ˢ

Find the error of chron. on A.T.S., M.T.S., and M.T.G.

4. 1899, May 26th, in lat. 32° 3′ S., long. 115° 45′ 30″ E., the Sun had equal alts. at following times :—

  East mer. 2ʰ 38ᵐ 57ˢ West mer. 6ʰ 56ᵐ 14ˢ

Required the error of chron on A.T.S., A.T.G., and M.T.G.

5. 1899, June 14th, in lat. 21° 40′ N., long. 21° 18′ W., at the following times the Sun had equal altitudes :—

  A.M. 10ʰ 51ᵐ 19ˢ P.M. 3ʰ 36ᵐ 55ˢ

Required the error of chron. on A.T.S., A.T.G., and M.T.G.

172   TEXT-BOOK ON NAVIGATION.

6. 1899, May 19th, in lat. 35° 20′ S., long. 32° E., the following times by chron. were noted when Spica had equal alts. :—

   East mer. 5ʰ 38ᵐ 40ˢ   West mer. 9ʰ 12ᵐ 38ˢ

Find the error of chron. on A.T.S., M.T.S., and M.T.G.

7. 1899, February 10th, in lat. 37° 50′ S., long. 144° 58′ 30″ E., the following times were noted when α Argûs had equal alts. :—

   East mer. 8ʰ 26ᵐ 24ˢ   West mer. 2ʰ 29ᵐ 40ˢ

Find the error of chron. on M.T.S., M.T.G., and A.T.G.

8. 1899, January 28th, in lat. 45° 30′ N., long. 73° 34′ 45″ W., the following times were noted when Arcturus had equal alts. :—

   East mer. 6ʰ 10ᵐ 19ˢ   West mer. 3ʰ 10ᵐ 41ˢ

Find the error of chron. on A.T.S. and M.T.G.

9. 1899, April 30th, in lat. 46° 20′ N., long. 122° 40′ W., the following times were noted when Arcturus had equal alts. :—

   East mer. 1ʰ 30ᵐ 52ˢ   West mer. 7ʰ 53ᵐ 21ˢ

Find the error of chron. on M.T.S., M.T.G., and A.T.G.

CHAPTER XXI.

CONSTRUCTION OF CHARTS.

ART. 61. **Example.**—Let it be required to construct a chart on Mercator's projection, embracing 43° to 46° N. lat., and 20° to 23° W. long., on a scale of $1\frac{1}{4}$ inch ($1\frac{1}{4}''$) to a degree of longitude.

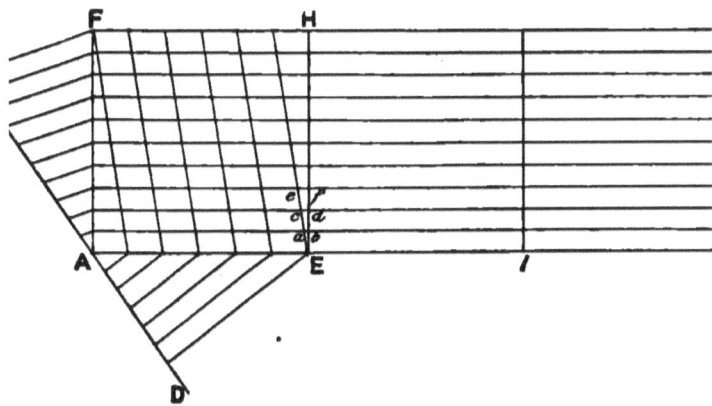

FIG. 116.

(a) **To construct a Suitable Diagonal Scale.**—Draw a horizontal line, AB, and divide it into equal parts $1\frac{1}{4}$ inch in length; next, draw a line, CD, through A, making any angle with AB. Measure off six equal parts on AB, and 10 equal parts on AC. Join DE and CF, and draw parallel lines to DE from the division marks on AD to meet the horizontal line AB, and other parallel lines from the division marks on AC to meet the vertical line AF (or make AF of indefinite length and set off ten equal parts on it). This divides AE into six equal parts, and AF into ten equal parts.

Now draw lines parallel to AB (as in the figure) through the division marks on AF. Also divide FH into six equal parts by marking the intersections of lines parallel to AF from the

174   TEXT-BOOK ON NAVIGATION.

division marks of AE, and draw lines diagonally as in the figure.

The diagonal scale is then complete, and may be used for the construction of the chart. One division of AE = 10′, and one move upwards gives 1′.

(b) **To construct the Chart.**—Draw a horizontal line for the

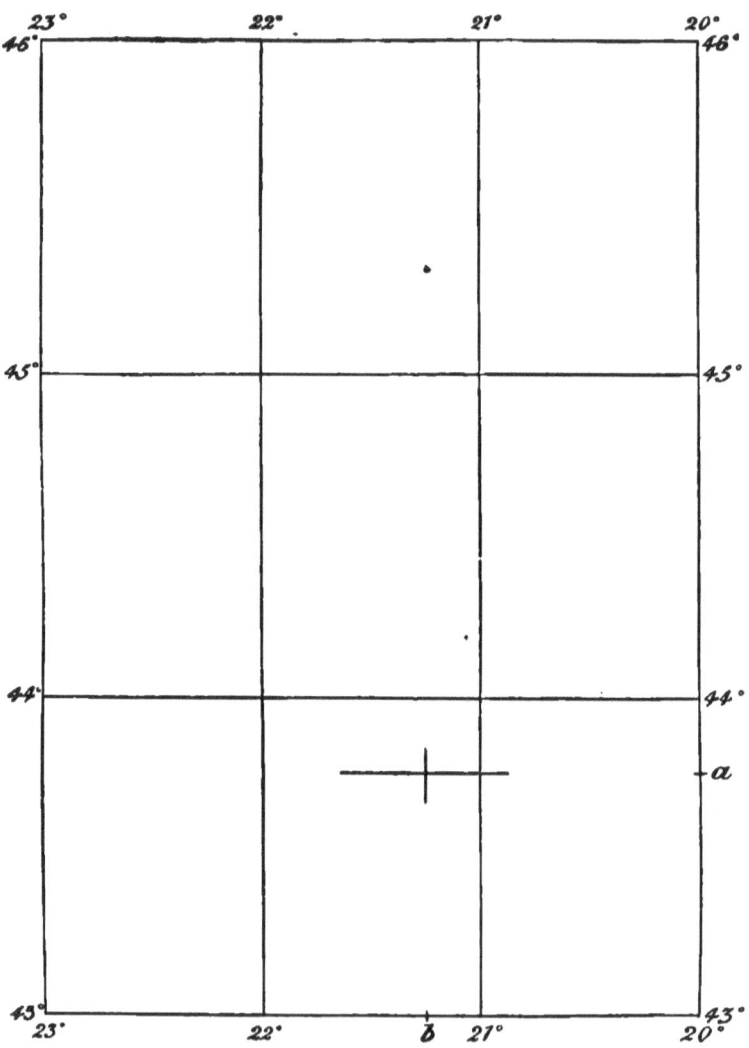

Fig. 117.

CONSTRUCTION OF CHARTS. 175

parallel of 43° N. lat., and divide it into parts each = 1¼ in Mark the divisions 20°, 21°, 22°, 23° of W. longitude, and draw vertical lines of indefinite length for meridians of longitude. Now turn to the table of "meridional parts" and find the difference between the meridional parts for 43° and 44°, which will be found to be 83.

Place one point of a pair of dividers on 1 on the line EB, and extend the other point to the second division to the left of E; this gives 80', and to get the additional 3', carry the right-hand point of the dividers up to the third horizontal line, extending the other point to the intersection of this line with the diagonal line. There are now 83' between the points of the dividers, and this, being set off on a meridian from the parallel 43°, gives the point through which to draw the parallel 44°. Proceed in the same way for 45° and 46°, and mark the degrees of latitude on the outside lines, which then become graduated meridians; the top and bottom lines are graduated parallels. Both these may be subdivided into single minutes. The minutes of longitude are all equal, but the minutes of latitude are not strictly so, especially in high latitudes. For high latitudes the larger subdivisions, say 10' apart, may be marked off in the same way as the degrees, and for practical purposes these may be subdivided into ten equal parts for single minutes.

A position in a given latitude and longitude may be marked on this chart by means of the diagonal scale, instead of using 1' subdivisions of the graduated meridians and parallels.

Example.—It is required to mark on the chart a position in 43° 45' N. lat. and 21° 14' W. long.

Mer. parts for 43° = 2863; for 43° 45' = 2925; difference, 62

Take 62 from the diagonal scale, and set it off on the graduated meridian at a; then take 14 and set off from 21° at b on the graduated parallel; draw a horizontal line through a, and a vertical line through b: the intersection of these lines is the position required.

EXERCISES IN CONSTRUCTION OF MERCATOR'S CHARTS.

1. Construct a Mercator's chart on a scale $2''\cdot 4$ to a degree of longitude, extending from lat. 56° N. to 57° 30' N., and from long. 25° W. to 28° W. A ship from lat. 57° 8' N., long. 25° 10' W., sailed as follows by compass: S.W. by S. 50 miles, dev. 6° 10' W.; W.N.W. 53 miles, dev. 6° 50' W.; N.E. 40 miles, dev. 10° E.; N. by W. 52 miles, dev. 1° 10' E. The variation is two points W. Find latitude and longitude in.

2. Construct a Mercator's chart, scale $0\cdot 8''$ to 1° long., extending from 63° 30' S. to 65° 30' S., and from 120° to 123° E., putting in parallels at each ½°. A ship leaves A, lat. 65° 7' S., long. 120° 52' E., and sails the following true courses and distances: N.E. 47 miles; N.N.W. 63 miles; S.W. by S. 42 miles; E. 27 miles. Find latitude and longitude in. The variation being 8° 30' W., what is the magnetic bearing of A from B?

3. Construct a Mercator's chart to extend from lat. 54° 30' N. to 56° N.,

and long. 30° W. to 33° W., scale 1"·3 to 1° long. A point of land in lat. 54° 32' N., long. 32° 56' W. bears S.W. from the ship by compass, distant 7 miles; ship's head, E. ¾ N.; dev. 9° 55' E.; variation, 18° W. Ship sailed as follows by compass: E. ¾ N. 90 miles, dev. 9° 55' E.; N.N.W. ¼ W. 40 miles, dev. 2° 20' W.; W.S.W. 45 miles, dev. 7° 50' W.; N.E. ¾ E. 75 miles, dev. 10° 35' E. Find latitude and longitude in, true course, and distance made good.

4. Construct a Mercator's chart to extend from lat. 54° N. to 55° 30' N., and long. 3° to 6° W., scale 1"·75 = 1° long. Insert following positions: Abbey Head, lat. 54° 47' N., long. 3° 58' W.; Little Ross, 54° 46' N., 4° 5' W.; Burial Island, 54° 29' N., 5° 25' W.; South Rock, 54° 22' N., 5° 26' W. At the ship, Abbey Head bore N. 46° E., and Little Ross, N. 37° W.; she then steamed S. 63° W. 12 knots for 4 hours, when her position was fixed by cross-bearings of Burial Island W. 8° N., and South Rock S. 23° W. Find true position of ship, set and drift of current.

PART II.

CHAPTER XXII.

SOLUTION OF PLANE TRIANGLES.

ART. 62.—**Trigonometrical Ratios.**

The Trigonometrical Ratios are known as sines, tangents, secants, cosines, cotangents, and cosecants.

They are best understood by first considering the geometrical relations of lines to the circle (see Fig. 118).

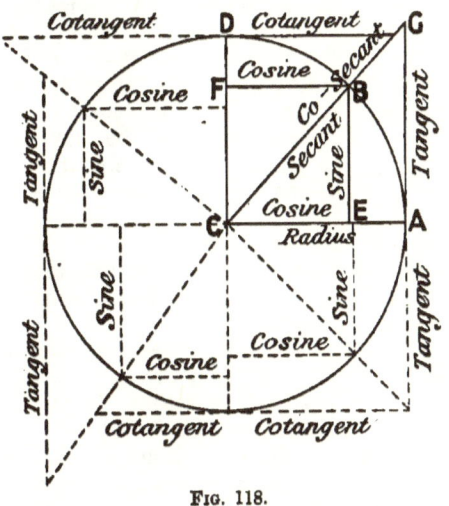

FIG. 118.

Geometrical definitions.—

(1) The *sine* is a perpendicular from one end of an arc upon the radius passing through the other end, and falls within the circle.

(2) The *tangent* is a perpendicular at the end of the radius, and touches the circle.

(3) The *secant* is a line from the centre meeting the tangent, and cuts the circumference.

(4) The cosine, cotangent, and cosecant are seen to be the sine, tangent, and secant respectively of the complement of the arc AB, namely, the arc BD. CE is cosine because = FB. Of course the arc AB is the measure of the angle ACB, and BD measures the complement angle BCD.

When the angle or arc exceeds 90°, as ACB_1, ACB_2, ACB_3, the same definitions apply.

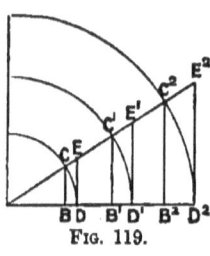

Fig. 119.

But it is evident that the *circle* might be drawn from the same centre C, and with *any* length of radius, which would appear to give the sines, tangents, etc., different values, because the lines would have different lengths (Fig. 119). What does *not* vary, however, is the *ratios* which these lines bear to the *radius*, whether the circle be small or large, and these *ratios* are the real values of the sine, etc., the radius of the circle being 1.

A glance at Fig. 118 will show—

(1) *When the angle decreases* the sine decreases, and becomes 0 when the angle is 0; the tangent does the same, whilst the secant decreases to 1 or radius.

(2) *When the angle increases* the sine increases, until it becomes equal to radius or 1 when the angle is 90°, whilst the tangent and secant increase to an infinite length (because they become parallel lines) when the angle is 90°.

In the same manner, the variations in the values of the cosines, cotangents, and cosecants can be traced, when it will be found that cosine of 0° is equal to radius or 1, and decreases to 0 as the angle increases to 90°; the cotangent is infinite for 0° and 0 for 90°, and the cosec. is infinite for 0° and 1 for 90°. The sign ∞ stands for infinity.

Summary.—

For 0° $\begin{cases} \text{sine} = 0; & \tan = 0; & \sec = 1 \\ \text{cosine} = 1; & \cot = \infty; & \csc = \infty \end{cases}$

For 90° $\begin{cases} \text{sine} = 1; & \tan = \infty; & \sec = \infty \\ \text{cosine} = 0; & \cot = 0; & \csc = 1 \end{cases}$

Sines which are *above* the horizontal diameter are called $+$, and *below* $-$. Hence the sines of angles between 0 and 180° are $+$, and the sines of angles between 180° and 360° are $-$ (see Fig. 118).

Also cosines are $+$ or $-$ according as they are to the *right* or *left* of the vertical diameter. Hence the cosines of angles between 0 and 90° and between 270° and 360° are $+$, whilst cosines of angles between 90° and 270° are $-$ (see Fig. 118).

SOLUTION OF PLANE TRIANGLES.   179

The signs (+ and −) of the tangents, cotangents, secants, and cosecants are derived from the above, and *not* from their positions in the figure. They are easily determined from the following relations—

$$\text{Tan} = \frac{\text{sine}}{\cos}, \cot = \frac{\cos}{\text{sine}}, \sec = \frac{1}{\cos}, \text{cosec} = \frac{1}{\text{sine}}$$

Therefore when sine and cos. are both + or both −, the tan. and cotan. are +; but when one is + and the other −, the tan. and cotan. are −.

The sec. has the same sign as the cosine.
The cosec. has the same sign as the sine.

In navigation problems angles above 180° are not used; consequently it is only necessary to remember that up to 90° the ratios (sines, etc.) are all +. Between 90° and 180° sines and cosecs. are +, but tangents, secants, cosines, and cotangents are −.

It is also seen from Fig. 118 that the sines, tangents, etc., of an angle above 90° are equal to the sines, tangents, etc., of the supplement (what it wants of 180°), but with a − sign, except the cosecant and sine.

Other relations known as *Reciprocals* are useful to remember. They are as follow: (a) sines and cosecants, (b) cosines and secants, (c) tangents and cotangents, and are so called because—

(a) Sine $= \dfrac{1}{\text{cosec}}$, and cosec $= \dfrac{1}{\text{sine}}$

(b) Secant $= \dfrac{1}{\text{cosine}}$, and cosine $= \dfrac{1}{\text{sec}}$

(c) Tan $= \dfrac{1}{\text{cotan}}$, and cotan $= \dfrac{1}{\text{tan}}$

The definitions of Fig. 118 may be applied to express the ratios of the sides of a right-angled triangle to each other.

(1) The ratio of BC to AC is exactly similar to that of EB to CB (Fig. 120) (radius), and is therefore the sine of the angle A.

The ratio of AB to AC is like that of CE to CB (Fig. 118), and is the cosine of the angle A, which may be expressed—

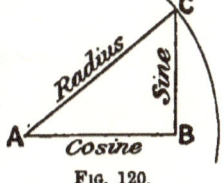

FIG. 120.

$$\frac{BC}{AC} = \text{sine A}; \quad \frac{AB}{AC} = \cos A$$

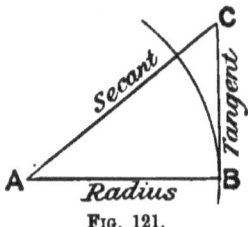

Fig. 121.

(2) The ratio of BC to AB is the same as that of AG to AC (radius) in Fig. 118, and is therefore the tangent of the angle A.

The ratio of AC to AB is like that of CG to AC (radius) in Fig. 118, and is the secant of A;

$$i.e. \ \frac{CB}{AB} = \tan A \ ; \ \frac{AC}{AB} = \sec A$$

(3) This figure is the same as the preceding one, but with C as the centre of the circle, and CB as radius; then—

$$\frac{AB}{BC} = \tan C \ ; \ \frac{AC}{BC} = \sec C$$

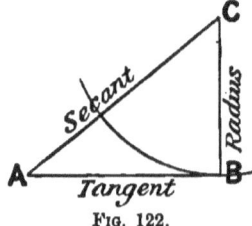

Fig. 122.

Now, as the angles A and C are complementary to each other (their sum being equal to 90°), what is the sine of one is the cosine of the other, the tangent of one is the cotangent of the other, and the secant of one is the cosecant of the other.

Calling the sides of the right-angled triangle by the general names hypotenuse, base, and perpendicular, the above may be summarized as follows:—

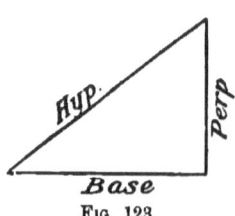

Fig. 123.

(1) Ratios of the other sides to the hypotenuse—

$$\frac{\text{perp.}}{\text{hyp.}} = \text{sine A and cosine C}$$
$$\frac{\text{base}}{\text{hyp.}} = \cos \text{A and sine C}$$

see Fig. 120.

(2) Ratios of the other sides to the base—

$$\frac{\text{perp.}}{\text{base}} = \tan \text{A and cotan C}$$
$$\frac{\text{hyp.}}{\text{base}} = \sec \text{A and cosec C}$$

see Fig. 121.

(3) Ratios of the other sides to the perpendicular—

$$\frac{\text{base}}{\text{perp.}} = \tan \text{C and cotan A}$$
$$\frac{\text{hyp.}}{\text{perp.}} = \sec \text{C and cosec A}$$

see Fig. 122.

NOTE.—It will be noticed that the sine and tangent stand *opposite* to the angle, whilst the cosine and secant are *adjacent*.

SOLUTION OF PLANE TRIANGLES.

Remark.—By means of these ratios all cases in plane right-angled triangles may be solved, including the problems in plane sailing, parallel sailing, middle-latitude sailing, and traverse sailing.

ART. 63. Solution of Right-angled Triangles.—Of the five parts exclusive of the right angle, viz. the other two angles and the three sides, any two being given, one of which is a side, the other three can be found.

Example 1.—Given the hypotenuse AC = 270, and the angle A = 36° 40', required the angle C and the sides AB and BC.

(1) The angle C = 90° − 36° 40' = 53° 20'

(2) $\dfrac{BC}{AC}$ = sine A

∴ BC = AC × sine A
A = 36° 40'   ...   log sine 9·776090
AC = 270 ...   ...   log 2·431364

BC = 161·2   log 2·207454

(3) $\dfrac{AB}{AC}$ = cos A

∴ AB = AC × cos A
A = 36° 40'   ...   log cos 9·904241
AC = 270 ...   ...   log 2·431364

AB = 216·6   log 2·335605

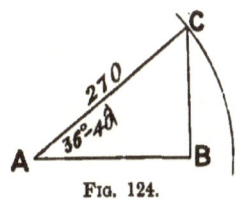

FIG. 124.

Example 2.—Given the side AB = 46·7, and the angle C = 29° 45', required the angle A, and the sides BC and AC.

(1) The angle A = 90° − 29° 45' = 60° 15'

(2) $\dfrac{BC}{AB}$ = tan A

∴ BC = AB × tan A
A = 60° 15'   ...   log tan 10·242948
AB = 46·7   ...   log 1·669317

BC = 81·71   log 1·912265

(3) $\dfrac{AC}{AB}$ = sec A

∴ AC = AB × sec A
A = 60° 15'   ...   log sec 10·304329
AB = 46·7   ...   log 1·669317

AC = 94·1   ...   ...   log 1·973646

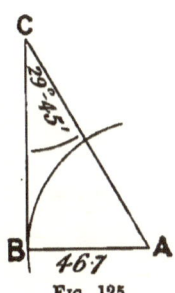

FIG. 125.

This might also be solved by the ratios

$$\dfrac{CB}{AB} = \text{cotan } C$$

and $\dfrac{AC}{AB}$ = cosec C

182   TEXT-BOOK ON NAVIGATION.

Example 3.—Given BC = 1342, and the angle A = 55° 15′, required the angle C and the sides AB and AC.

(1) The angle C = 90° − 55° 15′ = 34° 45′.

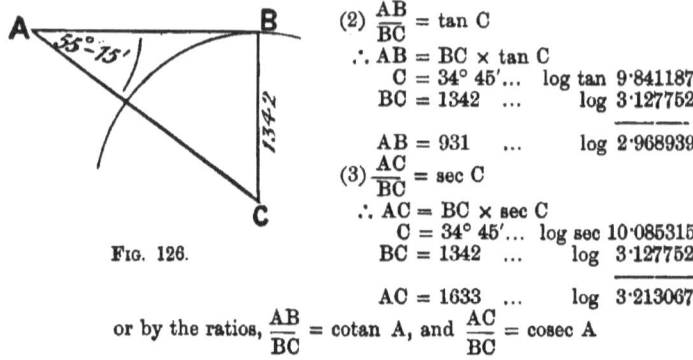

Fig. 126.

(2) $\dfrac{AB}{BC} = \tan C$

∴ AB = BC × tan C
    C = 34° 45′...   log tan 9·841187
    BC = 1342 ...   log 3·127752

    AB = 931 ...   log 2·968939

(3) $\dfrac{AC}{BC} = \sec C$

∴ AC = BC × sec C
    C = 34° 45′...   log sec 10·085315
    BC = 1342 ...   log 3·127752

    AC = 1633 ...   log 3·213067

or by the ratios, $\dfrac{AB}{BC} = \cotan A$, and $\dfrac{AC}{BC} = \cosec A$

Note.—Any side may be taken as radius, but it simplifies the solution to take the given side as radius, and use the ratios of the other two sides to the given side.

Example 4.—Given two sides AB = 368 and BC = 215, required the angles and third side.

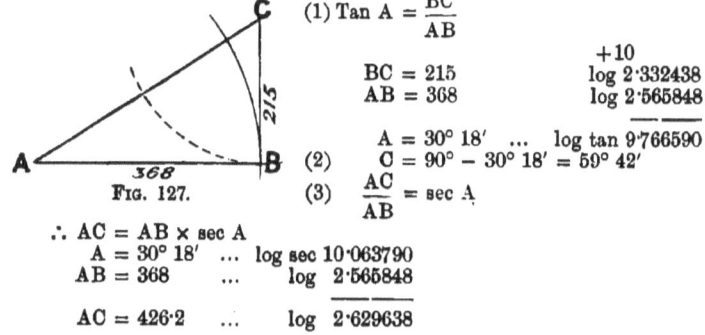

Fig. 127.

(1) $\tan A = \dfrac{BC}{AB}$

                          +10
BC = 215   log 2·332438
AB = 368   log 2·565848

A = 30° 18′ ...   log tan 9·766590
(2)   C = 90° − 30° 18′ = 59° 42′
(3) $\dfrac{AC}{AB} = \sec A$

∴ AC = AB × sec A
   A = 30° 18′ ...   log sec 10·063790
   AB = 368   ...   log 2·565848

   AC = 426·2 ...   log 2·629638

Or BC might be taken as radius, with centre C; then—

(1) $\tan C = \dfrac{AB}{BC}$
(2)   A = 90° − C
(3) $\dfrac{AC}{BC} = \sec C$

∴ AC = BC × sec C

Example 5.—Given the base AB = 162, and hypotenuse AC = 245, required the angles and the perpendicular.

SOLUTION OF PLANE TRIANGLES.

(1) $\text{Sec } A = \dfrac{AC}{AB}$

$$\begin{array}{ll} & +10 \\ AC = 245 \quad \ldots & \log 2\cdot 389166 \\ AB = 162 \quad \ldots & \log 2\cdot 209515 \\ \hline A = 48°\ 36' \quad \ldots & \log \sec 10\cdot 179651 \end{array}$$

(2) $C = 90° - 48°\ 36' = 41°\ 24'$

(3) $\dfrac{BC}{AB} = \tan A$

∴ $BC = AB \times \tan A$

$$\begin{array}{ll} A = 48°\ 36' \quad \ldots & \log \tan 10\cdot 054719 \\ AB = 162 \quad \ldots & \log\ \ 2\cdot 209515 \\ \hline BC = 183\cdot 8 \quad \ldots & \log\ \ 2\cdot 264234 \end{array}$$

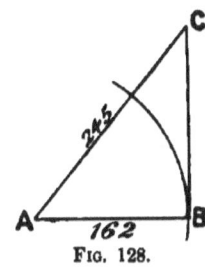

Fig. 128.

Or AC may be taken as radius; then—

(1) $\text{Cosine } A = \dfrac{AB}{AC}$

(2) $C = 90° - A$

(3) $\dfrac{BC}{AC} = \text{sine } A$

∴ $BC = AC \times \sin A$

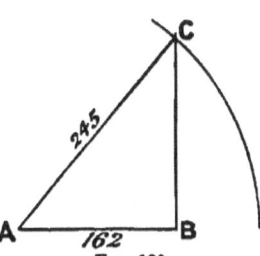

Fig. 129.

Example 6.—Given hypotenuse AC = 17·8, and perpendicular BC = 15·5, required the angles and the base.

(1) $\text{Sec } C = \dfrac{AC}{BC}$

$$\begin{array}{ll} & +10 \\ AC = 17\cdot 8 \quad \ldots & \log 1\cdot 250420 \\ BC = 15\cdot 5 \quad \ldots & \log 1\cdot 190332 \\ \hline C = 29°\ 27' \quad \ldots & \log \sec 10\cdot 060088 \end{array}$$

(2) $A = 90° - 29°\ 27' = 60°\ 33'$

(3) $\dfrac{AB}{BC} = \tan C$

∴ $AB = BC \times \tan C$

$$\begin{array}{ll} C = 29°\ 27' \quad \ldots & \log \tan 9\cdot 751757 \\ BC = 15\cdot 5 \quad \ldots & \log\ \ 1\cdot 190332 \\ \hline AB = 8\cdot 752 \quad \ldots & \log\ \ 0\cdot 942089 \end{array}$$

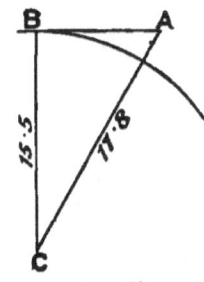

Fig. 130.

Or AC may be taken as radius; then—

(1) $\text{Sine } A = \dfrac{BC}{AC}$

(2) $C = 90° - A$

(3) $\dfrac{AB}{AC} = \sin C, \text{ or } = \cos A$

∴ $AB = AC \times \text{sine } C, \text{ or } = AC \times \cos A$

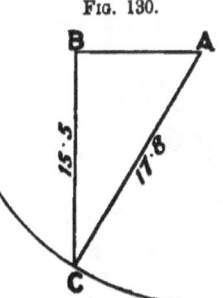

Fig. 131.

EXERCISES.—RIGHT-ANGLED TRIANGLES (PLANE).

1. In triangle ABC, AB=503, AC=357, C=90°. Find A, B, and BC.
2. In triangle PQR, PQ=371, QR=204, Q=90°. Find P, R, and PR.
3. In triangle OPB, OP=125·7, PB=695, O=90°. Find P, B, and OB.
4. In triangle ABC, A=64° 41', C=90°, AB=1037. Find AC and BC.
5. In triangle ABC, B=34° 17', C=90°, BC=905. Find AB and AC.
6. In triangle ABC, AB=147, BC=116, C=90°. Find A, B, and AC.
7. In triangle PQR, P=33° 14', PQ=4372, Q=90°. Find PR and RQ.
8. In triangle ABC, C=90°, AC=706, BC=981. Find A, B, and AB.
9. At 145 ft. from the base of a flagstaff its angle of elevation is observed to be 32° 16'. Find the height.
10. The difference of latitude made good is 236·5 miles south, the departure 215·7 miles west. Find the course, and distance sailed.
11. The length of the shadow cast by a vertical post is 39 ft. 6 in. when the sun has an elevation of 21° 32'. Find the height of the post.
12. From a ship a promontory was observed to be $3\frac{1}{2}$ points on the bow, and after making 7 miles on her course, it was found to be abeam. What distance was the vessel from the point at the two bearings?
13. A lighthouse, whose top is 200 ft. above the sea-level, marks a danger situated at three cables' length from the shore. Required the vertical "danger angle."
14. From a reef awash at low water, the angle of elevation to the summit of a hill distant 2050 yards in a horizontal direction is 1° 12'; height of the eye, 6 ft.; tidal rise, 21 ft. Required the height of the hill above high water.

ART. 64.—Solution of Oblique-angled Plane Triangles.

An oblique-angled triangle has six parts, namely, three sides and three angles.

The sum of any two sides is greater than the third side.

The sum of the three angles is equal to two right angles, or 180°.

The greater angle has the greater side opposite.

Any three parts being given, one being a side, the other three can be found.

The solutions comprise *three cases*, according to what three parts are given.

I. Given two angles and a side opposite to one of them, or two sides and an angle opposite to one of them, to find the other three parts.

II. Given two sides and the included angle, to find remaining angles and third side.

III. Given the three sides, to find the angles.

CASE I.—Formula: *The sides are proportional to the sines of the opposite angles.*

Example 1.—Let the capital letters A, B, C represent the angles, and the small letters a, b, c the sides opposite to them respectively.
  Given C = 118° 30', B = 21° 22', and b = 67·5. Required A, a, and c.
  To find A—
  Since the sum of the three angles = 180°, the third angle A = 180° − (118° 30' + 21° 22') = 40° 8' . . . . (1)

SOLUTION OF PLANE TRIANGLES.

To find a—
$$a : b :: \sin A : \sin B$$
$$\text{or } \frac{a}{b} = \frac{\sin A}{\sin B}$$
$$\therefore a = \frac{b \times \sin A}{\sin B}$$

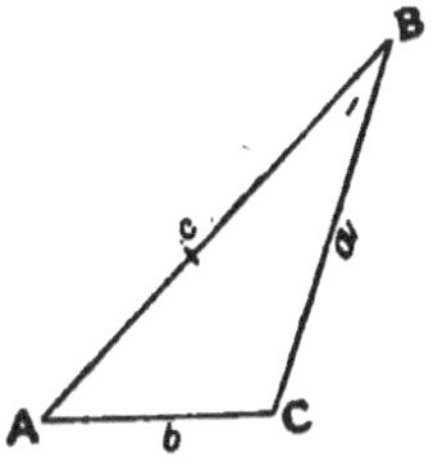

b 67·5	... ...	log 1·829304
A 40° 8′	... ...	sin 9·818103
	Sum of logs	11·647407
B 21° 22′	... ...	sin 9·561501
$a = 121·9$	... ...	log 2·085906 . (2)

Fig. 132.

To find c—
$$c : b :: \sin C : \sin B$$
$$\text{or } \frac{c}{b} = \frac{\sin C}{\sin B}$$
$$c = \frac{b \times \sin C}{\sin B}$$

b 67·5	... ...	log 1·829304
C 118° 30′	... ...	sin 9·943899
		11·773203
B 21° 22′	... ...	sin 9·561501
$c = 162·8$	... ...	log 2·211702 . . . . . (3)

c may also be found from the formula—
$$\frac{c}{a} = \frac{\sin C}{\sin A}$$
$$c = \frac{a \times \sin C}{\sin A}$$

Example 2.—Given $a = 350$, $c = 367$, and C (opposite to the greater side) = 54° 30′. Required b, B, and A.

To find C—
$$\text{Sin } C : \sin A :: c : a$$
$$\text{or } \frac{\sin C}{\sin A} = \frac{c}{a}$$
$$\therefore \sin C = \frac{\sin A \times c}{a}$$

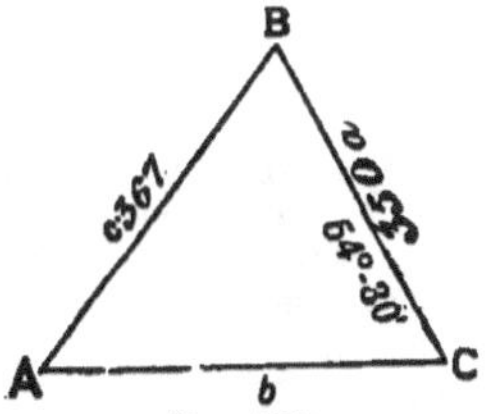

A 54° 30′	... ...	sin 9·910686
c 350	... ...	log 2·544068
	Sum of logs	12·454754
a 367	... ...	log 2·564666
A = 50° 56′	...	sin 9·890088 . . . . (1)
C = 54 30		
105 26		
180 0		
B = 74 34	. . . . (2)

Fig. 133.

To find b—

$$b : c :: \sin B : \sin C$$
$$\text{or } \frac{b}{c} = \frac{\sin B}{\sin C}$$
$$b = \frac{c \times \sin B}{\sin C}$$

c 367	...	log 2·564666
B 74° 34'	...	sin 9·984050
		12·548716
C 54° 30'		sin 9·910686
$b = 434\cdot5$ ...	...	log 2·638030 . . . . . (3)

Example 3.—Given $b = 1134$, $a = 875$, and A (opposite to the shorter side) 35° 20'. Required the other angles and third side.

This case is ambiguous, because the side a may belong to two different triangles, ABC and AB_1C, to both of which the side b and the angle A is common, but the angle at B_1 is acute, and the angle at B of the triangle ABC is obtuse, and the third angles at C are different.

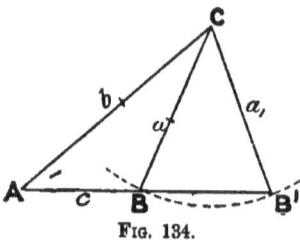

FIG. 134.

To find B and B_1—

$$\sin B : \sin A :: b : a$$
$$\text{or } \frac{\sin B}{\sin A} = \frac{b}{a}$$
$$\sin B = \frac{b \times \sin A}{a}$$

b 1134	...	log 3·054613	
A 35° 20'	...	sin 9·762177	
		12·816790	
a 875		log 2·942008	
$B_1 = 48° 33'$	...	sin 9·874782	(1)
180 0			
B = 131 27			

C in the larger triangle $AB_1C = 180° - (35° 20' + 48° 33') = 96° 7'$ }
C in the smaller triangle $ABC = 180° - (35° 20' + 131° 27') = 13° 13'$ } · (2)

The third side c will have two values according as the larger or smaller value of C is used, and may be found from the formula—

$$c : a :: \sin C : \sin A \; .. \; (3)$$

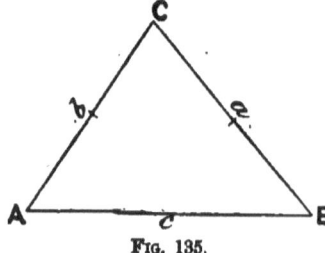

FIG. 135.

CASE II.—Given two sides a and b and the included angle C, to find A, B, and c.

The formula for finding the angles A and B is—

SOLUTION OF PLANE TRIANGLES.

(1) $\tan \frac{1}{2}(A - B) : \tan \frac{1}{2}(A + B) :: (a - b) : (a + b)$

or $\dfrac{\tan \frac{1}{2}(A - B)}{\tan \frac{1}{2}(A + B)} = \dfrac{a - b}{a + b}$

$\therefore \tan \frac{1}{2}(A - B) = \dfrac{\tan \frac{1}{2}(A + B) \times (a - b)}{(a + b)} = \dfrac{a - b}{a + b} \times \cot \frac{1}{2} C$

from which half the difference of the angles is found. The $\frac{1}{2}$ sum of these angles is also known by subtracting C from 180, and dividing the remainder by 2.

The greater angle = $\frac{1}{2}$ sum + $\frac{1}{2}$ diff.
the lesser angle = $\frac{1}{2}$ sum − $\frac{1}{2}$ diff.

(the greater angle being opposite to the greater side.)

(2) The third side c is found as in Case I.

$$\dfrac{c}{a} = \dfrac{\sin C}{\sin A}$$

$$\therefore c = \dfrac{a \times \sin C}{\sin A}$$

or $\dfrac{c}{b} = \dfrac{\sin C}{\sin B}$

and $c = \dfrac{b \times \sin C}{\sin B}$

Example.
Given $a = 145\cdot5$, $b = 98\cdot1$, and C = 87° 20′, find A, B, and c.

```
    a  145·5                    180°  0′
    b   98·1                  C  87  20
                                 _____
(a + b) 243·6        (A + B) 2)92  40
(a − b)  47·4                  _____
                     ½(A + B)  46  20
```

To find A and B.

$\frac{1}{2}(A + B) = $ 46° 20′ ...   ... tan 10·020220
$(a - b) = $ 47   4   ...   ... log 1·675778
                                      _____
                                   sum 11·695998
$(a + b) = $ 243   6   ...   ... log 2·386677
                                      _____
$\frac{1}{2}(A − B) = $ 11   31   ...   ... tan 9·309321
$\frac{1}{2}(A + B) = $ 46   20

A = sum 57 51
B = diff. 34 49

To find c.

$c : a :: \sin C : \sin A$

or $\dfrac{c}{a} = \dfrac{\sin C}{\sin A}$

therefore $c = \dfrac{a \times \sin C}{\sin A}$

TEXT-BOOK ON NAVIGATION.

$$a = 145\cdot5 \quad \ldots \quad \ldots \quad \log 2\cdot162863$$
$$C = 87° 20' \quad \ldots \quad \ldots \quad \sin 9\cdot999529$$
$$c = 145\cdot3 \quad \ldots \quad \ldots \quad \log 2\cdot162392$$

or $c : b :: \sin C : \sin B$

$$\frac{c}{b} = \frac{\sin C}{\sin B}$$

and $c = \dfrac{b \times \sin C}{\sin B}$

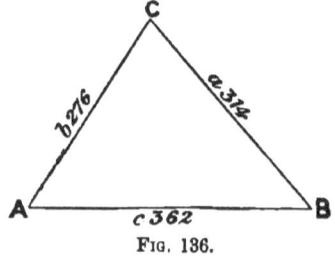
Fig. 136.

CASE III.—Given the three sides a, b, c, to find the angles A, B, C.

The formula for finding an angle (A) is—

$$\cos \tfrac{1}{2}A = \sqrt{\frac{s(s-a)}{bc}}$$

where $s = \tfrac{1}{2}(a + b + c)$

a 314	s 476	... log 2·677607	b 276	... log 2·440909
b 276	$s - a$ 162	... log 2·209515	c 362	... log 2·558709
c 362		Sum 4·887122	bc	... sum 4·999618
2)952	bc	... log 4·999618		
$s = 476$		2)19·887504		
$s - a = 162$				
	$\tfrac{1}{2}$A = 28° 32'	... cos 9·943752		
	2			
	A = 57 4·			

B and C may be found as in Case I. Thus—

$$\frac{\sin B}{\sin A} = \frac{b}{a}$$

$$\sin B = \frac{b \times \sin A}{a}$$

b 276	... ...	log 2·440909
A 57° 4'	... ...	sin 9·923919
		12·364828
a 314		log 2·496930
B = 47° 32$\tfrac{1}{2}$'	... ...	sine 9·867898
A = 57 4		
Sum 104 36$\tfrac{1}{2}$		
180 0		
C = 75 23$\tfrac{1}{2}$		

SOLUTION OF PLANE TRIANGLES.

This case may also be solved by dropping a perpendicular, CD, on the base, making two segments, AD and DB, which can be found from the formula—

$$\frac{\text{Difference of segments of base}}{\text{difference of sides}} = \frac{\text{difference of sides}}{\text{sum of segments}}$$

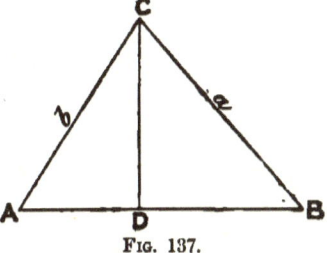

Fig. 137.

The difference of segments and the sum of segments (= base) being known—

The greater segment = ½ sum + ½ difference
The lesser   „   = ½ sum − ½ difference

NOTE.—The greater segment is always next to the greater side, and by making the longest side the base, etc., the perpendicular falls within the base.

The angles A and B can be computed by the rules for right-angled triangles.

```
                  a 314
                  b 276
                  ─────
              Sum 590   ...   ...   log 2·770852
              Diff.  38  ...   ...   log 1·579784
                                    ────────────
                                         4·350636
              Base 362  ...   ...   log 2·558709
                                    ────────────
        Diff. of segments 61·93 ... ... log 1·791927

              ½ diff.  30·96
              ½ sum   181·00

         Greater segm. DB = 211·96
         Lesser segm.  AD = 150·04
```

To find A—
$$\cos A = \frac{AD}{AC} = \frac{150·04}{276}$$

To find B—
$$\cos B = \frac{DB}{BC} = \frac{211·96}{314}$$

```
  150·04  ...  log 2·176206        211·96  ...  log 2·326254
     276  ...  log 2·440909           314  ...  log 2·496930
               ───────────                       ───────────
  A = 57° 4'  ...  cos 9·735297   B = 47° 32½'  ...  cos 9·829324
     C = 180° − (A + B) = 180° − (57° 4' + 47° 32½') = 75° 23½'
```

TEXT-BOOK ON NAVIGATION.

EXERCISES.—OBLIQUE-ANGLED PLANE TRIANGLES.

If ABC be any triangle, complete the solution in the following cases :—
1. $a = 1056$, $b = 359$, $c = 1267$.
2. $a = 364$, $b = 217$, $c = 494$.
3. $a = 96$, $b = 80$, $c = 63$.
4. $A = 37° 54'$, $B = 74° 25'$, $a = 104·6$.
5. $B = 14° 20'$, $C = 101° 13'$, $b = 296$.
6. $A = 97° 34'$, $C = 50° 16'$, $c = 16$.
7. $a = 6954$, $b = 4865$, $C = 39° 40'$.
8. $a = 267$, $c = 341$, $B = 68° 46'$.
9. $b = 61$, $c = 49$, $A = 97° 18'$.
10. $a = 865$, $b = 742$, $A = 52° 26'$.
11. $b = 469$, $c = 348$, $C = 39° 4'$.
12. $a = 1065$, $c = 897$, $A = 49° 36'$.
13. $a = 587$, $b = 925$, $A = 21° 23'$.
14. $a = 7$, $b = 8$, $c = 9$.
15. $a = 600$, $b = 500$, $c = 1000$.
16. $a = 50$, $b = 60$, $c = 70$.
17. $A = 18° 20'$, $B = 150°$, $b = 10,000$.
18. $A = 78° 40'$, $C = 39° 55'$, $c = 135$.
19. $b = 237·09$, $c = 130·96$, $A = 57° 59'$.
20. $c = 2265·4$, $a = 1779$, $B = 58° 17'$.
21. $a = 1300$, $b = 500$, $A - B = 30° 22'$.
22. $a = 1586·6$, $b = 5374·5$, $A = 15° 11'$.
23. $a = 200$, $b = 300$, $A = 35° 15'$.
24. The angular elevation of a tower at a place A due south of it is 45°, and at a place B, due west of A, the elevation is 30° : AB = 300 ft. Find height of tower.
25. AB is a line 250 ft. long, in the same horizontal plane as the foot D of a tower CD ; the angles DAB and DBA are respectively 61° 23' and 47° 14' ; the angle of elevation CAD is 34° 50'. Find height of tower.

CHAPTER XXIII.

SOLUTION OF SPHERICAL TRIANGLES.

ART. 65.—A spherical triangle is a portion of the surface of a sphere bounded by arcs of three great circles, and, like a plane triangle, has three sides and three angles. The sides, being arcs of circles, are measured like angles, in degrees, minutes, and seconds (° ′ ″).

When one of the angles is 90°, the triangle is *right angled;* when one side is 90°, it is *quadrantal;* and when no side or angle is 90°, it is an oblique-angled triangle.

Unlike a plane triangle, a spherical triangle may have two or three right angles, or one, two, or three sides of 90°, but no angle or side can be as much as 180°.

The sum of any two sides is greater than the third side.

The sum of the three angles of a spherical triangle is always more than 180°, but less than 540°, and the sum of the three sides is less than 360°.

The greater angle is opposite to the greater side, as in plane triangles.

The angles at the base of an isosceles triangle are equal, as in plane triangles.

ART. 66. **Solution of Right-angled Spherical Triangles.**— For the purpose of simplifying the solution of right-angled triangles, "Napier's Circular Parts" and "Napier's Rules" have been devised.

Napier's Circular Parts.—These are five in number (the right angle being omitted), namely, the two sides including the right angle, and the complements of the third side and remaining angles.

Let the capital letters A, B, c represent the angles, and the small letters a, b, c the sides. The circular parts are then a, b, and the complements of A, B, c.

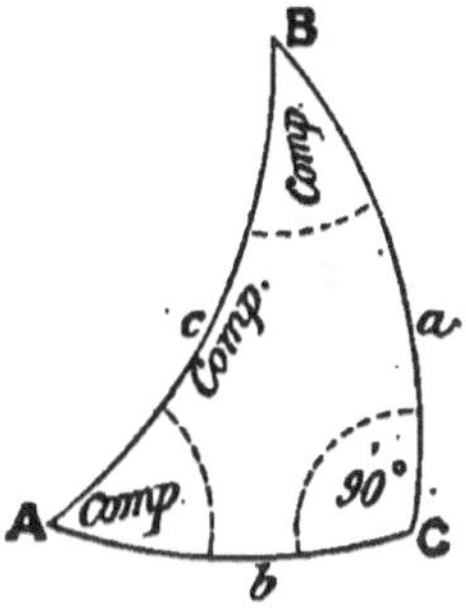

Any one of the five may be a middle part, and will have two "adjacent parts" and two "opposite parts." Reference to the figure will give the following summary:—

FIG. 138.

Middle part.	Adjacent parts.	Opposite parts.
A	b and comp. of c.	a and comp. of B.
c	Comp. of A and comp. of B.	a and b.
B	a and comp. of c.	b and comp. of A.
a	b and comp. of B.	Comp. of A and comp. of c.
b	a and comp. of A.	Comp. of c and comp. of B.

When any two are given, the other three can be found by means of Napier's Rules.

Napier's Rules.—

I. Sine of middle part = product of tangents of adjacent parts.
II. Sine of middle part = product of cosines of opposite parts.

The problem of finding an unknown side or angle from two given parts may be varied in thirty different ways, but they may all be reduced to three cases.

CASE I.—*To find a side or angle when two "adjacents" are given.*—Suppose A and B given, to find c. Here complement of c is middle part, and complements of A and B adjacent parts; then by Rule I.—

Sine of comp. of c = tan of comp. of A × tan of comp. of B
that is, cos c = cot A × cot B.

Example 1.—Given A = 64° 30′, B = 56° 25′, to find c.
A 64° 30′ ... ... cot 9·678496
B 56 25 ... ... cot 9·822154
c = 71 32 ... ... cos 9·500650

Example 2.—Given a = 46° 20′, c = 113° 40′, to find B.
Sine (comp. of B) = tan A × tan (comp. of c)
that is, cos B = tan a × cot c
a 46° 20′ ... ... tan 10·020220
c 113 40 ... ... cot 9·641747 (−)
  62 40 ... ... cos 9·661967 (−)
 180   0
B = 117 20

CASE II.—*To find a side or angle when two "opposites" are given.*—Suppose A and c given, to find a. Here a is middle part to which comp. of A and comp. of C are opposite parts; then by Rule II.—

Sine a = cos (comp. of A) × cos (comp. of c)
that is, sin a = sine A × sin c

Example 1.—Given A = 69° 30′, c = 75° 50′, to find a.
A 69° 30′ ... ... sin 9·971588
c 75 50 ... ... sin 9·986587

a = 65 15½ ... ... sin 9·958175

SOLUTION OF SPHERICAL TRIANGLES.

Example 2.—Given $a = 48°\ 27'$, $B = 82°\ 15'$ to find A.
 Sine (comp. of A) = cos a × cos (comp. of B)
 that is, cos A = cos a × sin B

a 48° 27' ... ... cos 9·821693
B 82 15 ... ... sine 9·996015

A = 48 55 ... ... cos 9·817708

CASE III.—*To find a side or angle when one "adjacent" and one "opposite" are given.*

Remark.—One of the three must be a middle part, to which the other two are either adjacents or opposites.

Suppose A and b given, to find a. Here b is middle part, and a and comp. of A adjacent parts; then by Rule I.—

$$\text{Sin } b = \tan a \times \tan (\text{comp. of A})$$
that is, sin b = tan a × cot A

and by transposing, $\cot A = \dfrac{\sin b}{\tan a}$

which brings the unknown term on one side of the equation, and the given terms on the other.

Example 1.—Given $a = 44°\ 36'$, $b = 79°\ 24'$, to find A.

Using above equation, $\cot A = \dfrac{\sin b}{\tan a}$

b 79° 24' ...   sin 9·992525
a 44 36 ...   tan 9·993936

A = 45 5½' ... ... cot 9·998589

Example 2.—Given $b = 61°\ 50'$, $B = 83°\ 44'$, to find c. Here, with b as middle part, comps. of B and c are opposite parts; then by Rule II.—

Sin b = cos (comp. of B) × cos (comp. of c)
that is, sin b = sin B × sin c

and transposing, $\sin c = \dfrac{\sin b}{\sin B}$

b 61° 50' ...   ... sin 9·945261
B 83 44 ...   ... sin 9·997397

c = 62 29 ...   ... sin 9·947864

NOTE.—In a right-angled spherical triangle an angle and opposite side are of the same affection.

EXERCISES.—RIGHT-ANGLED SPHERICAL TRIANGLES.

Solve the triangles in following cases. C = 90°.
 1. $a = 51°\ 36'$, $b = 65°\ 12'$.
 2. A = 78° 11', B = 39° 49'.
 3. $c = 84°\ 13'$, $a = 61°\ 44'$.
 4. $c = 79°\ 27'$, A = 51° 30'.

O

5. $a = 32° 51'$, $B = 56° 17'$.
6. $a = 54° 37'$, $A = 67° 53'$.
7. $a = 70° 29'$, $b = 102° 38'$.
8. $a = 110° 17'$, $b = 98° 46'$.
9. $A = 114° 51'$, $B = 67° 32'$.
10. $A = 92° 31'$, $B = 118° 44'$.
11. $c = 107° 19'$, $a = 69° 8'$.
12. $c = 88° 36'$, $b = 101° 25'$.
13. $b = 97° 20'$, $c = 94° 13'$.
14. $c = 113° 10'$, $B = 74° 32'$.
15. $c = 83° 12'$, $B = 98° 15'$.
16. $b = 95° 52'$, $A = 76° 13'$.
17. $a = 106° 27'$, $B = 94° 8'$.
18. $b = 114° 22'$, $B = 108° 19'$.

ART. 67. **Solution of Oblique-angled Spherical Triangles.**—As this is not a treatise on Spherical Trigonometry, the only cases here given are those which bear upon the problems of Navigation and Nautical Astronomy, so that the student may be able to deduce and understand the rules for the solution of these problems.

The "*cases*" and the formulæ belonging to them have a close analogy to those of Oblique-angled Plane Trigonometry, and are therefore the more easily remembered.

CASE I.—Given the three sides a, b, and c, to find the three angles A, B, and C.

The fundamental formula expressing the relation of the angles to the sides is—

$$\cos A = \frac{\cos a - \cos b . \cos c}{\sin b . \sin c} \quad \ldots \ldots (1)$$

$$\text{or } \cos B = \frac{\cos b - \cos a . \cos c}{\sin a . \sin c} \quad \ldots \ldots (2)$$

$$\text{or } \cos C = \frac{\cos c - \cos a . \cos b}{\sin a . \sin b} \quad \ldots \ldots (3)$$

It is seen that (2) and (3) are obtained from (1) by merely permuting the letters.

But as these formulæ are not in a form suitable for calculation by logarithms, the following, which are derived from them, are used:—

$$\sin \tfrac{1}{2}A = \sqrt{\frac{\sin (s - b) . \sin (s - c)}{\sin b . \sin c}} \quad . \ (1)$$

$$\cos \tfrac{1}{2}A = \sqrt{\frac{\sin s . \sin (s - a)}{\sin b . \sin c}} \quad \ldots \ (2)$$

$$\tan \tfrac{1}{2}A = \sqrt{\frac{\sin (s - b) . \sin (s - c)}{\sin s . \sin (s - a)}} \quad \ldots \ (3)$$

where s stands for $\tfrac{1}{2}(a + b + c)$.

SOLUTION OF SPHERICAL TRIANGLES.

The formulæ for finding B and C may be written down from the above by changing the letters.

CASE II.—Given two sides a and b and the included angle C, to find the angles A and B.

$$\operatorname{Tan} \tfrac{1}{2}(A + B) = \frac{\cos \tfrac{1}{2}(a - b)}{\cos \tfrac{1}{2}(a + b)} \cdot \cot \tfrac{1}{2}C \quad . \quad . \quad (1)$$

$$\operatorname{Tan} \tfrac{1}{2}(A - B) = \frac{\sin \tfrac{1}{2}(a - b)}{\sin \tfrac{1}{2}(a + b)} \cdot \cot \tfrac{1}{2}C \quad . \quad (2)$$

from which half the sum and half the difference of A and B are obtained.

The greater angle A = ½ sum + ½ difference
The lesser angle B = ½ sum − ½ difference

The third side c may be found from the formula—

$$\cos \tfrac{1}{2}c = \frac{\cos \tfrac{1}{2}(a + b)}{\cos \tfrac{1}{2}(A + B)} \cdot \sin \tfrac{1}{2}C \quad . \quad . \quad . \quad (3)$$

Or c may be obtained independently from the formula—

$$\sin \frac{C}{2} = \sqrt{\sin \left(\frac{a+b}{2} + \theta\right) \cdot \sin \left(\frac{a+b}{2} - \theta\right)}$$

$$\text{where } \sin^2 \theta = \sin a \cdot \sin b \cdot \cos^2 \frac{C}{2} \quad . \quad . \quad (4)$$

CASE III.—Given two angles and a side opposite to one of them, to find the side opposite to the other; or, given two sides and an angle opposite to one of them, to find the angle opposite to the other.

Rule.—The sines of the sides are proportional to the sines of the opposite angles.

Given A, B, and a, to find b.

$$\text{Sin } b : \sin a :: \sin B : \sin A$$
$$\text{or } \frac{\sin b}{\sin a} = \frac{\sin B}{\sin A}$$
$$\text{therefore } \sin b = \frac{\sin a \cdot \sin B}{\sin A} \quad . \quad . \quad (1)$$

Given a, b, and A, to find B.

$$\text{Sin B} : \sin A :: \sin b : \sin a$$
$$\text{or } \frac{\sin B}{\sin A} = \frac{\sin b}{\sin a}$$
$$\text{therefore } \sin B = \frac{\sin A \cdot \sin b}{\sin a} \quad . \quad . \quad (2)$$

Similarly for other angles and sides.

From these formulæ only one side or one angle can be found, so they will not give a complete solution of the triangle; but the third angle or side can be found from the following:—

$$\cot \tfrac{1}{2}C = \frac{\cos \tfrac{1}{2}(a+b)}{\cos \tfrac{1}{2}(a-b)} \cdot \tan \tfrac{1}{2}(A+B) \quad \ldots \quad (3)$$

$$\tan \tfrac{1}{2}c = \frac{\cos \tfrac{1}{2}(A+B)}{\cos \tfrac{1}{2}(A-B)} \cdot \tan \tfrac{1}{2}(a+b) \quad \ldots \quad (4)$$

or c may be found from the sine rule as above after C is known.

NUMERICAL SOLUTIONS.

Example 1.—Given $a = 84° 30'$, $b = 108° 44'$, $c = 72° 20'$, to find A, B, and C.

Using the formula—

$$\cos \tfrac{1}{2}A = \sqrt{\frac{\sin s \cdot \sin (s-a)}{\sin b \cdot \sin c}}$$

$$= \sqrt{\sin s \cdot \sin (s-a) \cdot \operatorname{cosec} b \cdot \operatorname{cosec} c}$$

(the cosecs being reciprocals of the sines).

```
          a  84° 30'
          b 108  44                cosec 10·023639
          c  72  20                cosec 10·020981
            ───────
         2)265  34

    s =    132  47     ...         sin   9·865653
(s − a) =   48  17     ...         sin   9·872998
                                       ───────────
                                   2)19·783271

   ½A =     38  49     ...         cos   9·891635
             2
           ───────
   A  =     77  38
```

To find B by the "Sine rule"—

$$\frac{\sin B}{\sin A} = \frac{\sin b}{\sin a}$$

$$\text{and } \sin B = \frac{\sin A \cdot \sin b}{\sin a} = \sin A \cdot \sin b \cdot \operatorname{cosec} a$$

```
   A  77° 38'            sin    9·989804
   b 108  44             sin    9·976361
   a  84  30             cosec 10·002004
                              ──────────
       68  20             sin   9·968169
      180   0
      ──────
  B = 111  40  (the supplement, because opposite to greater side)
```

SOLUTION OF SPHERICAL TRIANGLES.

To find C by the "Sine rule"—

$$\frac{\sin C}{\sin A} = \frac{\sin c}{\sin a}$$

and $\sin C = \dfrac{\sin A \cdot \sin c}{\sin a} = \sin A \cdot \sin c \cdot \operatorname{cosec} a$

A 77° 38'	... ...	sin 9·989804
c 72 20	... ...	sin 9·979019
a 84 30	... ...	cosec 10·002004
C = 69 14	... ...	sin 9·970827

Without the "Sine rule" A, B, and C can be most conveniently found from the formula (3), namely—

$$\tan \tfrac{1}{2}A = \sqrt{\frac{\sin(s-b)\cdot\sin(s-c)}{\sin s \cdot \sin(s-a)}}$$

$$\tan \tfrac{1}{2}B = \sqrt{\frac{\sin(s-a)\cdot\sin(s-c)}{\sin s \cdot \sin(s-b)}}$$

$$\tan \tfrac{1}{2}C = \sqrt{\frac{\sin(s-a)\cdot\sin(s-b)}{\sin s \cdot \sin(s-c)}}$$

because the logs need only to be taken out once, thus—

```
      a  84° 30'
      b 108  44
      c  72  20
      ─────────
     2)265  34
      ─────────
      s = 132  47
  (s - a) =  48  17
  (s - b) =  24   3
  (s - c) =  60  27
```

To find A—

(s − b) 24° 3'	...	sin 9·610164	s 132° 47'	...	sin 9·865653
(s − c) 60 27	...	sin 9·939482	(s − a) 48 17	...	sin 9·872998
		19·549646			9·738651
		9·738651			
		2)19·810995			

$\tfrac{1}{2}A = 38° 49'$ ... tan 9·905497
$A = 77\ 38$

To find B—

(s − a) 48° 17'	...	sin 9·872998	s 132° 47'	...	sin 9·865653
(s − c) 60 27	...	sin 9·939482	(s − b) 24 3	...	sin 9·610164
		19·812480			19·475817
		19·475817			
		2)20·336663			

$\tfrac{1}{2}B = 55° 50'$ ... tan 10·168331
$\phantom{\tfrac{1}{2}B =\ } 2$

$B = 111\ 40$

To find C—

$(s-a)$ 48° 17'	...	sin 9·872998		s 132° 47'	...	sin 9·865653
$(s-b)$ 24 3	...	sin 9·610164		$(s-c)$ 60 27	...	sin 9·939482
		19·483162				19·805135
		19·805135				

$$2\overline{)19 \cdot 678027}$$

$\tfrac{1}{2}$C = 34° 37'   ... tan 9·839013

C = 69 14

Example 2.—Given two sides $a = 88° 20'$, $b = 60° 50'$, and included angle $C = 75° 30'$, to find A, B, and C.

$$\operatorname{Tan} \tfrac{1}{2}(A + B) = \frac{\cos \tfrac{1}{2}(a - b)}{\cos \tfrac{1}{2}(a + b)} \cdot \cot \tfrac{1}{2}C \quad \ldots \ldots (1)$$

$$\operatorname{Tan} \tfrac{1}{2}(A - B) = \frac{\sin \tfrac{1}{2}(a - b)}{\sin \tfrac{1}{2}(a + b)} \cdot \cot \tfrac{1}{2}C \quad \ldots \ldots (2)$$

NOTE.—Instead of dividing by cosine in equation (1), and sine in equation (2), the computation is simplified by multiplying by secant and cosecant, and adding the logs.

$a = $ 88° 20'
$b = $ 60 50

$a + b = $ 149 10	$\tfrac{1}{2}(a + b) = $ 74° 35'	sec 10·575385		cosec 10·015915
$a - b = $ 27 30	$\tfrac{1}{2}(a - b) = $ 13 45	cos 9·987372		sin 9·376003
C = 75 30	$\tfrac{1}{2}$C = 37 46	cot 10·111100		cot 10·111100
	$\tfrac{1}{2}(A + B)$ 78 2	tan 10·673857		
	$\tfrac{1}{2}(A - B)$ 17 40	...	...	tan 9·503018

A = 95 42
B = 60 22

To find c—

$$\cos \tfrac{1}{2}c = \frac{\cos \tfrac{1}{2}(a + b)}{\cos \tfrac{1}{2}(A + B)} \cdot \sin \tfrac{1}{2}C$$

$\tfrac{1}{2}$C 37° 45'	...	sin 9·786906
$\tfrac{1}{2}(a + b)$ 74 35	...	cos 9·424615
$\tfrac{1}{2}(A + B)$ 78 2	...	sec 10·683311
$\tfrac{1}{2}c = $ 38 17	...	cos 9·894832

$c = $ 76 34

Example 3.—(a) Given two angles $A = 67° 30'$, $B = 55° 50'$, and an opposite side $a = 78° 10'$, to find b.

$$\frac{\sin b}{\sin a} = \frac{\sin B}{\sin A}$$

$$\sin b = \frac{\sin a \cdot \sin B}{\sin A}$$

$$= \sin a \cdot \sin B \cdot \operatorname{cosec} A$$

SOLUTION OF SPHERICAL TRIANGLES.

$$
\begin{array}{lll}
a\ 78°\ 10' & \ldots & \sin\ 9\cdot990671 \\
B\ 55\ 50 & \ldots & \sin\ 9\cdot917719 \\
A\ 67\ 30 & \ldots & \operatorname{cosec}\ 10\cdot034385 \\
\hline
b = 61\ 14 & \ldots & \sin\ 9\cdot942775
\end{array}
$$

(b) Given two sides, $a = 75°\ 54'$, $b = 63°\ 25'$, and an opposite angle, $A = 72°\ 10'$, to find B.

$$\frac{\sin B}{\sin A} = \frac{\sin b}{\sin a}$$

$$\sin B = \frac{\sin A \cdot \sin b}{\sin a}$$

$$= \sin A \cdot \sin b \cdot \operatorname{cosec} a$$

$$
\begin{array}{lll}
A\ 72°\ 10' & \ldots & \sin\ 9\cdot978615 \\
b\ 63\ 25 & \ldots & \sin\ 9\cdot951476 \\
a\ 75\ 54 & \ldots & \operatorname{cosec}\ 10\cdot013286 \\
\hline
B = 61\ 28 & \ldots\ \sin & 9\cdot943377
\end{array}
$$

EXERCISES.—OBLIQUE-ANGLED SPHERICAL TRIANGLES.

1. If $a = 82°\ 26'$, $b = 92°\ 2'$, $c = 104°\ 28'$, find A, B, and C.
2. If $a = 140°$, $b = 70°$, $c = 80°$, find A, B, and C.
3. If $a = 113°\ 2'$, $b = 82°\ 40'$, and $C = 138°\ 50'$, find A, B, and c.
4. If $b = 60°$, $c = 84°\ 22'$, and $A = 95°$, find B, C, and a.
5. If $a = 115°\ 20'$, $c = 84°\ 42'$, and $B = 65°\ 30'$, find A, C, and b.
6. If $A = 75°$, $B = 15°$, and $c = 60°$, find a, b, and C.
7. If $A = 137°\ 22'$, $B = 60°$, and $c = 81°$, find a, b, and C.
8. Find A, when $B = 88°\ 52$, and $b = 73°\ 50'$, $a = 120°$.
9. If $A = 136°$, $a = 155°\ 55'$, $b = 144°\ 45'$, solve.
10. If $a = 90°$, $b = 75°\ 29'$, $c = 84°\ 13'$, solve.
11. If $b = 90°$, $A = 100°\ 20'$, $B = 82°\ 50'$, solve.
12. If $c = 90°$, $A = 63°\ 32'$, $B = 98°\ 17'$, solve.

CHAPTER XXIV.

APPLICATION OF FORMULÆ TO PROOF OF RULES.

A. NAVIGATION.

ART. 68. **Plane and Traverse Sailings.**—For short distances, the surface of the Earth may be regarded as a plane. In plane and traverse sailings no notice is taken of the curvature, and all cases may be solved by the formulæ for right-angled triangles.

Thus in sailing from A to B, the hypotenuse represents the distance, the perpendicular the D. lat., and the base the departure, whilst the angle which AB makes with AC (lying true N. and S.) is the course.

In the Traverse Table the values of D. lat. and Dep. in miles are given for every degree of course up to 89°, and all distances up to 300 miles, being computed from the formulæ—

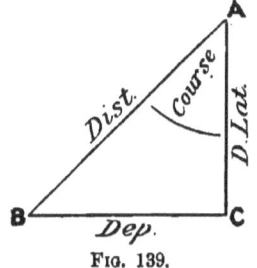

FIG. 139.

$$\frac{\text{D. lat.}}{\text{dist.}} = \cos \text{course}$$

$$\frac{\text{dep.}}{\text{dist.}} = \sin \text{course}$$

whence D. lat. = dist. × cos course . . . . (1)
and dep. = dist. × sin course . . . . (2)

When D. lat. and dep. are given, the course and distance made good may be found from the formulæ—

$$\text{Tan course} = \frac{\text{dep.}}{\text{D. lat.}} \quad \ldots \ldots \ldots (3)$$

distance = D. lat. × sec course . . . . (4)

ART. 69. **Parallel Sailing.**—Let P be the pole, PE and PQ

APPLICATION OF FORMULÆ TO PROOF OF RULES. 201

meridians of longitude, EQ an arc of the Equator, DL an arc of a parallel of latitude, and C the centre of the Earth.

Then QL or ED will represent the lat. of the parallel DL, EQ is the diff. long., and DL the departure.

Join CL, and draw FL perpendicular to PC.

It is evident that FL is a radius of the small circle of which DL is a part, and CQ a radius of the great circle of which EQ is a part; that is, a radius of the Earth. And since the arcs are proportional to the radii of the two circles—

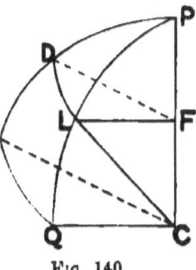

Fig. 140.

Arc DL : arc EQ :: rad. FL : rad. CQ
therefore DL : EQ :: FL : CL

because CL is also a radius of the Earth;

$$\text{or } \frac{DL}{EQ} = \frac{FL}{CL} = \sin FCL, \text{ or } \cos QCL$$

because CFL is a plane right-angled triangle, and QCL is the complement of FCL.

Now, the angle QCL is measured by the arc QL, which is the latitude of the parallel DL—

$$\text{therefore } \frac{\text{dep.}}{\text{D. long.}} = \cos \text{lat.} \qquad (1)$$

from which are derived the formulæ—

$$\text{Dep.} = \text{D. long.} \times \cos \text{lat.} \qquad . \qquad . \qquad (2)$$

$$\text{D. long.} = \frac{\text{dep.}}{\cos \text{lat.}} \text{ or dep.} \times \sec \text{lat.} \qquad (3)$$

Now, the formula $\frac{\text{dep.}}{\text{D. long.}} = \cos \text{lat.}$ expresses the relation between the base, hypotenuse, and angle at the base of a right-angled plane triangle. Such a triangle may therefore be used to derive the formulæ of parallel sailing, by calling the base departure, the hypotenuse diff. of longitude, and the angle at the base latitude.

Then $\frac{\text{dep.}}{\text{D. long.}} = \cos \text{lat.}$ . . . . (·1)

and $\frac{\text{D. long.}}{\text{Dep.}} = \sec \text{lat.}$ . . . . (2)

From (1) dep. = D. long. × cos lat. . (3)
from (2) D. long. = dep. × sec lat. . (4)

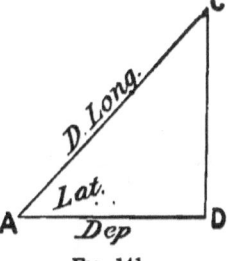

Fig. 141.

which are the same as those derived from the sphere. For exercises, see Art. 9.

ART. 70. **Middle-latitude Sailing.**—This is a combination of plane and parallel sailing, the dep. being common to both, but the latitude used is the "middle lat.," or mean of the two latitudes, for in this case the ship does not sail on a parallel of latitude.

It is evident that the formulæ for right-angled triangles will solve every case, but this method is only made use of in modern navigation for the purpose of finding the D. long. in the "day's work,": the formula employed being—

D. long = dep. × sec of middle lat.

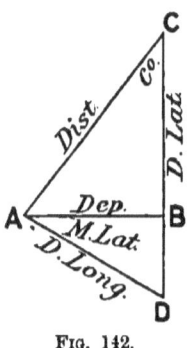

FIG. 142.

The *mean* is not the true "middle lat.," being too small, but is sufficiently near for short distances.

ART. 71. **Mercator's Sailing.**—This is represented by two similar triangles, DEC and ABC, the angle at C being the course, which is common to both triangles. The sides of the smaller triangle represent the elements of plane sailing. The extended perpendicular CB represents the meridional D. lat. found from the meridional parts for the two latitudes; then the base AB will be the D. long. in its correct proportion to the departure. The object aimed at in Mercator's method is to find the rhumb course from point to point. For this purpose, the diff. lat. is extended by means of the meridional parts so as to bear the same proportion to the diff. long. as the true diff. lat. bears to the departure. In the right-angled triangle ABC—

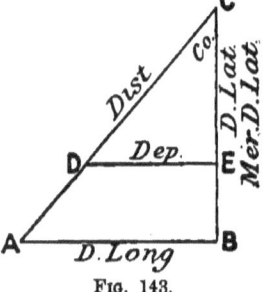

FIG. 143.

$$\text{Tan course} = \frac{\text{D. long.}}{\text{mer. D. lat.}} \quad \ldots \ldots (1)$$

and in the triangle DEC—

$$\frac{\text{dist.}}{\text{D. lat.}} = \sec \text{ course}$$

therefore dist. = D. lat. × sec course . . . (2)

These are the only formulæ required for computing the true rhumb course and the distance on that course. Evidently this distance is longer than the distance on a great circle, but the method has the advantage of leading to the destination without change of course. For exercises, see Art. 10.

ART. 72. **Great Circle Sailing.**—Let AB be the arc of a great circle from the starting-point (A) to the destination (B), and CA, CB arcs of meridians of longitude; then the angles A and B are the *initial* and *final* courses, the angle C is the difference of longitude, and CA, CB the co-latitudes of A and B.

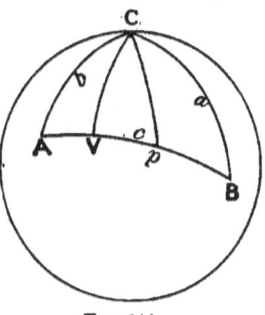

Also, let the arc CV be drawn at right angles to AB; then V is the vertex, and CV the co-latitude of vertex, putting the letters a, b, c for the sides respectively opposite the angles A, B, C.

By Case II., spherical trigonometry—

FIG. 144.

$$\text{Tan } \tfrac{1}{2}(A + B) = \frac{\cos \tfrac{1}{2}(a - b)}{\cos \tfrac{1}{2}(a + b)} \cdot \cot \tfrac{1}{2}C.$$

$$\text{Tan } \tfrac{1}{2}(A - B) = \frac{\sin \tfrac{1}{2}(a - b)}{\sin \tfrac{1}{2}(a + b)} \cdot \cot \tfrac{1}{2}C$$

The greater angle (A) $= \tfrac{1}{2}(A + B) + \tfrac{1}{2}(A - B)$
„ lesser „ (B) $= \tfrac{1}{2}(A + B) - \tfrac{1}{2}(A - B)$

The initial course A is obviously between N. and E., and the final course B between S. and E.

The distance or third side AB may be found from formula (3), Case II., Art. 67, namely—

$$\cos \tfrac{1}{2}C = \frac{\cos \tfrac{1}{2}(a + b)}{\cos \tfrac{1}{2}(A + B)} \cdot \sin \tfrac{1}{2}C$$

$$= \cos \tfrac{1}{2}(a + b) \cdot \sec \tfrac{1}{2}(A + B) \cdot \sin \tfrac{1}{2}C$$

i.e. $\cos \tfrac{1}{2}$ dist. $= \cos \tfrac{1}{2}$ sum of co-lats. $\times \sec \tfrac{1}{2}$ sum of first and last courses $\times \sin \tfrac{1}{2}$ diff. long.

The latitude of vertex, longitude from vertex, and the latitudes of points on the great circle are found by applying the rules for the solution of right-angled spherical triangles.

1. *To find "Latitude of Vertex."*—In the right-angled triangle ACV are known the angle A and the side b.

Taking CV (co-latitude of V) as middle part, the opposite parts are comp. of a and comp. of b;

therefore $\sin CV = \cos(\text{comp. of A}) \times \cos(\text{comp. of } b)$
$= \sin A \times \sin b$

that is, cos lat. of V = sine of initial course × cos lat. of A

2. *To find "Longitude from Vertex" of* A, *viz. the angle* ACV.—In the same triangle, taking comp. of ACV as middle part, CV and the comp. of b are adjacent parts;

therefore sin (comp. of ACV) = tan CV × tan (comp. of b)
or cos ACV = tan CV × cot b

that is, cos "long. from vertex" = cot lat. of vertex × tan lat. of A

Similarly, the latitude of V and "longitude from V" of B can be found from the triangle BCV.

The longitude of the vertex can be found by applying the "longitude from vertex" to the longitude of A or B.

3. *To find the Latitude of a Point (p) whose "Longitude from Vertex" is assumed, viz. the angle pCV.*—In the right-angled triangle pVC, the side CV and the angle pCV being known—

sin (comp. of pCV) = tan CV × tan (comp. of Cp)
or cos pCV = tan CV × cot Cp

transposing, cot $Cp = \dfrac{\cos p\text{CV}}{\tan \text{CV}} = \cos p\text{CV} \times \cot \text{CV}$

that is, tan lat. of p = cos assumed long. from V × tan lat. of V.

Similarly, the latitudes of other points on the great circle track can be found. The courses and distances from point to point are calculated as in Mercator's sailing.

ART. 73. **Current Sailing.**

1. Given the course and rate of sailing of a ship, and the "set" and "drift" of a current during the same interval of time, to find the course and distance made good.

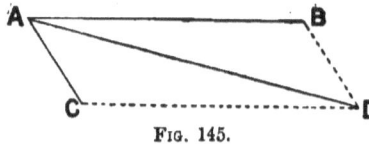

FIG. 145.

Let AB represent the course and the distance sailed, and AC the "set" and "drift" of the current. Draw CD parallel to AB, and BD parallel to AC, and join AD; then AD will represent the course and distance made good during the time elapsed.

Here the angle BAC is known, and therefore ABD is also known, because it is equal to 180° − BAC. Also the sides AB and BD of the triangle ABD are known (because BD = AC). Hence two sides and included angle of the oblique-angled triangle ABD are known, from which the other angles and the third side AD can be computed. From the given direction of AB and the angle BAD the course of AD may be found.

Example.—A ship steers E. by N. at 10 miles per hour in a current setting S. by W. 3 miles per hour. Find the course and distance made good.

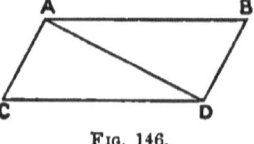

FIG. 146.

The angle BAC = 10 points, therefore ACB = 6 points (or 67° 30'), because the two together are equal to two right angles. Applying the formula for two sides and included angle—

APPLICATION OF FORMULÆ TO PROOF OF RULES.

$$\text{Tan } \tfrac{1}{2}(A - B) = \frac{(a - b) \tan \tfrac{1}{2}(A + B)}{(a + b)}$$

where A represents the angle ADC, and B the angle CAD.

	180° 0'		$\tfrac{1}{2}$(A + B) 56° 15'	...	tan 10·175107
	67 30		(a − b) 7	...	log 0·845098
(A + B)	112 30				11·020205
			(a + b) 13		log 1·113943
$\tfrac{1}{2}$(A + B)	56 15				
a	10		$\tfrac{1}{2}$(A − B) 38 52		tan 9·906262
b	3		$\tfrac{1}{2}$(A + B) 56 15		

a + b	13			
a − b	7	∠ CAD 17 23	{(being the lesser angle opposite to lesser side)

Course of AC N. 78 45 E.

N. 96 8 E.
180 0

Course made good S. 83 52 E.

To find distance made good (AD).

$$\frac{AD}{CD} = \frac{\sin ACD}{\sin CAD}$$

$$\text{therefore } AD = \frac{CD \times \sin ACD}{\sin CAD}$$

ACD 67° 30'	...	...	sin 9·965615
CD 3	...	...	log 0·477121
			10·442736
CAD 17° 23'	...	...	sin 9·475327
AD 9·3	...	...	log 0·967409

2. Given the course to a given point, the set and drift of the current, and the rate of sailing of the ship; to find the course to steer and the distance made good towards the point in a given time.

Let A be the place of the ship, and B the destination, and AC the set and drift of the current. From C set off CD equal to the distance the ship would sail in the given time, cutting AB in the point D. Then CD gives the course to steer, and AD is the distance made good towards B.

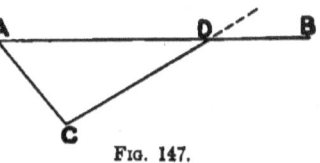

Fig. 147.

In the triangle ACD, the two sides AC and CD are given, and the angle at A opposite to the latter, from which the other angles and remaining side can be found by the rule of sines (Case I.).

TEXT-BOOK ON NAVIGATION.

Example.—Given the direct course from A to B, E. by S., and the set and drift of the current S.S.E. 4 miles per hour, the ship's rate of sailing being 7 miles per hour, required the course to steer in order to counteract the effect of the current, and the distance made good towards B in 2 hours?

Draw AC in a S.S.E. direction, and set off on it 8 miles (drift in 2 hours), and from C set off 14 miles (2 hours' sailing), cutting AB in the point D, and

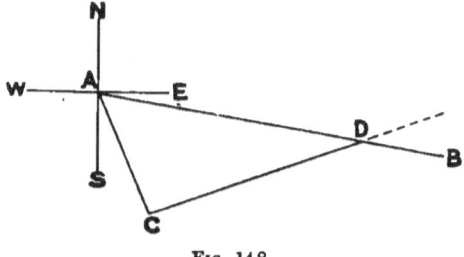

Fig. 148.

join CD. The line CD will give the course to steer, and AD is the distance made good.

Here AC = 8, CD = 14, and the angle CAD (opposite to CD) = 56° 15'.

$$\text{Then } \frac{\sin D}{\sin A} = \frac{AC}{CD}$$

$$\text{and } \sin D = \frac{AC \times \sin A}{CD}$$

A 56° 15'	...		sine	9·919846
AC	8	...	log	0·903090
				10·822936
CD	14	... ...	log	1·146128
D	28° 22'		sine	7·676808

Course from A to B   S. 78 45 E.

S. 107 7 E.
180 0

Course to steer N. 72 53 E.

To find AD.

A = 65° 15'
D = 18 22
―――
84 37
180 0
―――
C = 95 23

$$\frac{AD}{CD} = \frac{\sin C}{\sin A}$$

$$\therefore AD = \frac{CD \times \sin C}{\sin A}$$

C 95° 23'	...	sin	9·998080
CD 14	...	log	1·146128
			11·144208
A 56 15		sin	9·919846
Dist. made good 16·8		log	1·224362

APPLICATION OF FORMULÆ TO PROOF OF RULES. 207

3. The ship's position by observation and by dead reckoning being known, the set and drift of the current may be found. (Solved by right-angled triangles.)

Example.—At noon on a certain day, according to dead reckoning from the previous noon, a ship was in lat. 45° 30′ N., long. 28° 15′ W.; but by observation she was found to be in lat. 46° 15′ N., long. 27° 22′ W. Required the set and drift of the current.

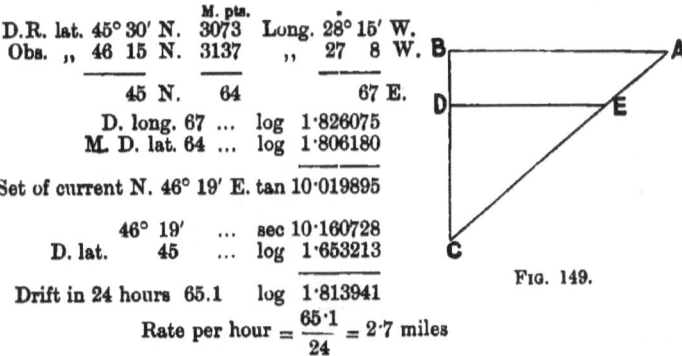

FIG. 149.

```
                        M. pts.
D.R. lat. 45° 30′ N.    3073    Long. 28° 15′ W.
Obs.  ,,  46 15 N.      3137     ,,   27  8  W.
          ─────────              ──────────────
          45 N.          64            67 E.
          D. long. 67 ...  log 1·826075
          M. D. lat. 64 ...  log 1·806180
                              ─────────
Set of current N. 46° 19′ E.  tan 10·019895

          46° 19′      ...  sec 10·160728
D. lat.      45        ...  log  1·653213
                              ─────────
Drift in 24 hours 65.1      log  1·813941

Rate per hour = 65·1/24 = 2·7 miles
```

NOTE.—This is a case of finding a course and distance by Mercator, and is solved by the formulæ for right-angled triangles.

Examples 1 and 3 can be solved approximately by means of the Traverse Table.

Example 1.—

Courses.	Dist.	N.	S.	E.	W.
E. by N.	10	2·0	—	9·8	—
S. by W.	3	—	2·9	—	0·6
		2·0	2·9	9·8	0·6
			2·0	0·6	
			0·9	9·2	

Course and distance made good, S. 84° E. 9·3 miles.

Example 3.—D. lat. 64 and dep. 67 give course 46° . . . . . (1)
Course 46°, with D. lat. 45, give dist. 65 . . . . (2)

Questions in current sailing may be varied in many ways, but all of them can be solved on the principles of the three foregoing examples.

B. Nautical Astronomy.

Art. 74. The Amplitude.—In the projection, on the plane of the horizon, N., E., S., W. are the cardinal points, P is the elevated pole (in this case the north pole), Z is the zenith, and O an object on the horizon.

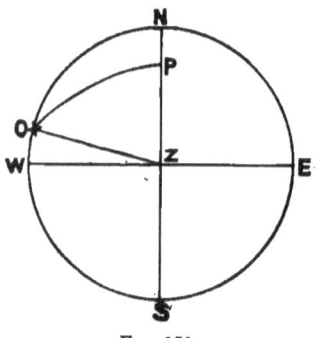

Fig. 150.

The arc WO or the angle ·WZO is the amplitude, PO is the polar distance, and PZ the co-latitude.

The triangle PZO is quadrantal, because ZO = 90°.

The circular parts are therefore the angles at Z and O, and the complements of PO, PZ, and the angle ZPO; and taking complement PO for middle part, the angle PZO and complement of PZ are opposite parts; therefore—

Sin (comp. of PO) = cos ∠ PZO × cos (comp. of PZ)
or cos PO = cos ∠ PZO × sin PZ
that is, sin dec. = sin amp. × cos lat.

whence sin amp. = $\dfrac{\text{sin dec.}}{\text{cos lat.}}$

= sin dec. × sec lat.

which is the well-known formula for finding the amplitude.

The equator cuts the horizon in the east and west points; therefore an object whose dec. is 0 rises true East and sets true West. When its dec. is north, it rises and sets towards the north; and when its dec. is south, it rises and sets towards the south. Hence the rule for naming the amplitude the *same* as the *declination*.

Art. 75. To find the Time of Rising or Setting of a Celestial Object.—The angle ZPO is the hour angle of the object when setting.

Taking (comp. of ZPO) as middle part, the complements of PO and PZ are adjacent parts; then—

sin (comp. of ZPO) = tan (comp. of PO) × tan (comp. of PZ)
that is, cos ZPO = cot PO × cot PZ
or cos H.A. = tan dec. × tan lat.

This formula gives the H.A. of the object when rising or setting, and thence the time.

N.B.—When lat. and dec. are of the *same* name, the *Supplement* must be taken for the H.A.

APPLICATION OF FORMULÆ TO PROOF OF RULES.

ART. 76. The Altitude Azimuth.
—Let S be the object, PS its polar distance (p), PZ the co-lat. ($90° - l$), ZS the zenith distance ($90° - a$). The angle PZS is the azimuth from N., and its supplement SZX is the azimuth from S.

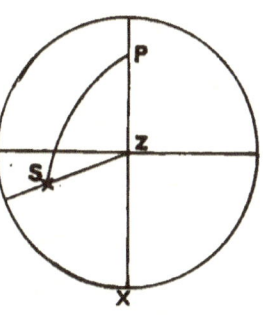

Fig. 151.

PZS is a spherical triangle, of which the three sides are given to find an angle PZS.

The general formula is—

$$\cos A = \frac{\cos a - \cos b . \cos c}{\sin b . \sin c}$$

whence in the figure—

$$\cos PZS \text{ or } -\cos SZX = \frac{\cos PS - \cos ZS . \cos PZ}{\sin ZS . \sin PZ}$$

Putting the letters Z, a, l, p for the azimuth, altitude, latitude, and polar distance respectively—

$$-\cos Z = \frac{\cos p - \cos (90° - a) . \cos (90° - l)}{\sin (90° - a) . \sin (90° - l)}$$

$$= \frac{\cos p - \sin a . \sin l}{\cos a . \cos l}$$

adding 1 to each side of the equation—

$$1 - \cos Z = 1 + \frac{\cos p - \sin a . \sin l}{\cos a . \cos l}$$

$$\therefore 2 \sin^2 \frac{Z}{2} = \frac{(\cos a . \cos l - \sin a . \sin l) + \cos p}{\cos a . \cos l}$$

$$= \frac{\cos (a + l) + \cos p}{\cos a . \cos l}$$

$$\sin^2 \frac{Z}{2} = \frac{\cos \tfrac{1}{2}(a + l + p) \cos (a + l - p)}{\cos a . \cos l}$$

$$\therefore \sin^2 \frac{Z}{2} = \frac{\cos s . \cos (s - p)}{\cos a . \cos l}$$

where $s = \tfrac{1}{2}(a + l + p)$

i.e. $\sin^2 \dfrac{Z}{2} = \cos s . \cos (s - p) . \sec a . \sec l$

or $\sin \dfrac{Z}{2} = \sqrt{\cos s . \cos (s - p) . \sec a . \sec l}$

from which is derived the rule for computing the azimuth of the *opposite name* from the *latitude*, namely, add together the log secants of the altitude and latitude, and the log cosines of the half-sum and remainder; divide by two to get the square root, which gives the log sine of half the azimuth.

P

ART. 77. **Time Azimuth.**—Here the hour angle ZPS, the polar distance PS, and the co-latitude PZ are given to find the angle PZS, which is the azimuth of the *same name* as the *latitude*.

This is a case of two sides and the included angle being given to find one of the angles at the base. From the general formulæ—

$$\operatorname{Tan} \tfrac{1}{2}(A + B) = \frac{\cos \tfrac{1}{2}(a - b)}{\cos \tfrac{1}{2}(a + b)} \cdot \cot \tfrac{1}{2}C$$

$$\operatorname{Tan} \tfrac{1}{2}(A - B) = \frac{\sin \tfrac{1}{2}(a - b)}{\sin \tfrac{1}{2}(a + b)} \cdot \cot \tfrac{1}{2}C$$

by substituting the Z and S in the figure for A and B, and the letters h, p, l' for the H.A., polar distance, and co-latitude respectively, we have—

$$\operatorname{Tan} \tfrac{1}{2}(Z + S) = \frac{\cos \tfrac{1}{2}(p - l')}{\cos \tfrac{1}{2}(p + l')} \cot \tfrac{1}{2}h$$

$$\operatorname{Tan} \tfrac{1}{2}(Z - S) = \frac{\sin \tfrac{1}{2}(p - l')}{\sin \tfrac{1}{2}(p + l')} \cot \tfrac{1}{2}h$$

therefore, $\tan \tfrac{1}{2}(Z + S) = \sec \tfrac{1}{2}(p + l') \cdot \cos \tfrac{1}{2}(p - l') \cdot \cot \tfrac{1}{2}h$
and $\tan \tfrac{1}{2}(Z - S) = \operatorname{cosec} \tfrac{1}{2}(p + l') \cdot \sin \tfrac{1}{2}(p - l') \cdot \cot \tfrac{1}{2}h$

whence the following form for computation:—

$\tfrac{1}{2}(p + l')$	... sec		cosec
$\tfrac{1}{2}(p - l')$	... cos		sine
$\tfrac{1}{2}h$	... cot		cot
$\tfrac{1}{2}(Z + S)$	... tan	...	
$\tfrac{1}{2}(Z - S)$	... ...	... ...	tan

The *sum* of these angles is the *greater* of the two angles Z and S, and the *difference* is the *lesser* angle; and since the greater angle is opposite to the greater side, the azimuth Z is the greater angle when p is greater than l', and the smaller angle when p is less than l'.

NOTE.—When $\tfrac{1}{2}(p + l')$ exceeds 90°, the sec is negative, and therefore tan $\tfrac{1}{2}(Z + S)$ is negative, in which case the supplement must be taken for $\tfrac{1}{2}(Z + S)$.

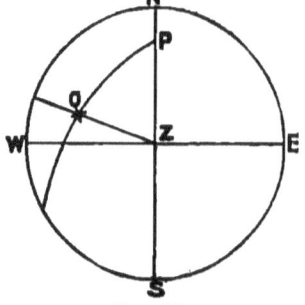

FIG. 152.

ART. 78. **To find the Hour Angle of a Celestial Object, the Altitude, Latitude, and Polar Distance being known.**—Let O be the object, then PO is its polar distance (p), ZO its zenith distance ($= 90° - a$), and PZ the co-latitude ($= 90° - l$); the angle ZPO is the hour angle. Therefore, in the spherical triangle PZO, three sides are given to find an angle ZPO.

APPLICATION OF FORMULÆ TO PROOF OF RULES. 211

The general formula for finding an angle is—
$$\cos A = \frac{\cos a - \cos b . \cos c}{\sin b . \sin c}$$
which, applied to the figure, gives—
$$\cos ZPS = \frac{\cos ZO - \cos PO . \cos PZ}{\sin PO . \sin PZ}$$

Let the letters a, l, p, and h stand for altitude, latitude, polar distance, and hour angle respectively;

then
$$\cos h = \frac{\cos (90° - a) - \cos p . \cos (90 - l)}{\sin p . \sin (90° - l)}$$
$$= \frac{\sin a - \cos p . \sin l}{\sin p . \cos l}$$

Subtracting each side of the equation from 1—
$$1 - \cos h = 1 - \frac{\sin a - \cos p . \sin l}{\sin p . \cos l}$$
$$i.e.\ 2 \sin^2 \frac{h}{2} = \frac{\sin p . \cos l + \cos p . \sin l - \sin a}{\sin p . \cos l}$$
$$= \frac{\sin (p + l) - \sin a}{\sin p . \cos l}$$
$$= \frac{2 \cos \tfrac{1}{2}(p + l + a) . \sin \tfrac{1}{2}(p + l - a)}{\sin p . \cos l}$$

Now let $s = \tfrac{1}{2}(p + l + a)$, then $(s - a) = \tfrac{1}{2}(p + l - a)$;

therefore
$$\sin^2 \frac{h}{2} = \frac{\cos s . \sin (s - a)}{\sin p . \cos l}$$
$$= \operatorname{cosec} p . \sec l . \cos s . \sin (s - a)$$

and
$$\sin \frac{h}{2} = \sqrt{\operatorname{cosec} p . \sec l . \cos s . \sin (s - a)}$$

which is the formula used for computing the hour angle in the *longitude by chronometer* and other time problems.

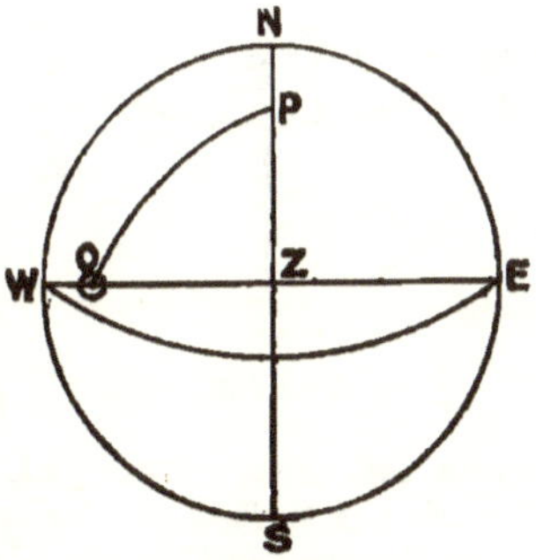

ART. 79. **To find the Time when the Sun or other Celestial Object is on the Prime Vertical.**—Let O be an object on the prime vertical; then PO is its polar distance, PZ is the co-latitude, and the angle P is the hour angle.

PZO is a spherical triangle, right angled at Z, of which the circular parts are PZ and ZO, and the complements of PO and the angles P and O.

FIG. 153.

Taking complement of P as middle part, PZ and the complement of PO are adjacent parts;

then sin (comp. of P) = tan PZ × tan (comp. of PO)
or cos P = tan PZ × cot PO
that is, cos H.A. = cot lat. × tan dec.

In the case of the Sun, the H.A. is the interval from apparent noon.

For any other object, add its R.A. to its H.A. west to find the R.A. of meridian, from which subtract the sun's R.A. to find the time.

ART. 80. **Latitude by Ex-Meridian Altitude. Direct Method.**—

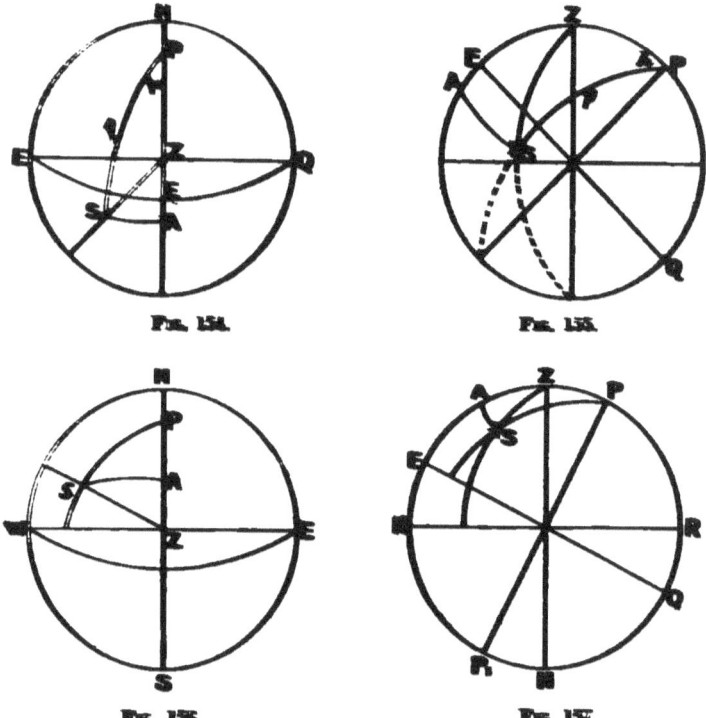

Fig. 154.   Fig. 155.

Fig. 156.   Fig. 157.

Given the hour angle (h), the declination (d), and the true altitude (a), to find the latitude.

In the figures S is the object, PS the polar distance, ZS the zenith distance, the angle SPA(h) the hour angle, and SA an arc of a great circle at right angles to the meridian, EQ the equator.

Then EZ is the latitude, and is equal to the sum or difference of the arcs ZA and EA.

In the right-angled spherical triangle PSA—

$$\cos h = \cot p \cdot \tan PA$$
$$= \tan \delta \cdot \tan PA$$
$$\therefore \tan PA = \cos h \cdot \cot \delta$$

But EA (arc 1) is the complement of PA, because PE = 90°;

$$\therefore \cot \text{arc } 1 = \cos h \cdot \cot \delta \quad \ldots \ldots (1)$$

Again, in the same triangle—

$$\cos p = \cos PA \cdot \cos SA$$
i.e. $\sin \delta = \sin \text{arc } 1 \cdot \cos SA$
$$\therefore \cos SA = \sin \delta \cdot \csc \text{arc } 1 \quad (2)$$

Also in the right-angled triangle SZA—

$$\cos SZ = \cos SA \cdot \cos ZA$$
i.e. $\sin a = \sin \delta \cdot \csc \text{arc } 1 \cdot \cos ZA$

by substituting the value of $\cos SA$ in (2)—

$$\therefore \cos ZA \text{ (arc 2)} = \frac{\sin a}{\sin \delta \cdot \csc \text{arc } 1}$$

or $\cos \text{arc } 2 = \sin a \cdot \csc \delta \cdot \sin \text{arc } 1 \ldots \ldots (3)$

Arc 1 is named the same as the dec., and arc 2 the same as the zenith distance, and the sum or difference, according as they are of the same or different names, gives the latitude as in the meridian altitude problem.

For computation the formulæ may be arranged as follows:—

h — — — ... cos
δ — — — ... cot   ... cosec
arc 1 — — — ... cot   ... sine
a — — —   ... sine
arc 2 — — —   cos
lat. — — —

NOTE.—The same letters apply to the figures in both projections.

N.B.—When S is on the equinoctial (i.e. dec. = 0), then the perpendicular SA coincides with the equinoctial EQ (Fig. 19), and arc 1 disappears.

The lat. XZ is then found from the right-angled triangle ZSE, in which ES measures the H.A., and ZS = co-alt. or zenith distance.

To find the Latitude.—In the right-angled triangle PZA—
$$\cos PZ = \cos PA \cdot \cos ZA$$
that is, sin lat. = cos arc 5 . cos arc 2
and therefore cosec lat. = sec arc 5 . sec arc 2

The foregoing formulæ are arranged for computation as follows:—

	° ′ ″			
½ pol. angle — — —		cosec		
Dec. — — —		sec		sin
Arc 1, — — —		cosec	sec	sec
½ sum alts. — — —		cos	sin	
½ diff. ,, — — —		sin	cos	
Arc 2, — — —		sin	*sec	
Arc 3, — — —			cos	
Arc 4, — — —				cos
Arc 5, — — —		sec		
Arc 2, — — —		*sec		
Lat. — — —		cosec		

The Ambiguous Case.—When the sum of arcs 3 and 4 is less than 90°, it may be doubtful whether they should be *added* or *subtracted* to find arc 5. But if the lat. be first found by adding, and the lat. by account is not given, the correctness of the answer may be tested by computing the hour angle at the middle time, and seeing if it agrees with the known hour angle. If not, then *subtract* 3 from 4, and again compute the latitude.

To compute the Hour Angle (H_1) at the Middle Time.—In the right-angled triangle PZA—
$$\sin AZ = \sin PZ \cdot \sin ZPA$$
$$\therefore \sin ZPA = \frac{\sin AZ}{\sin PZ} = \sin AZ \cdot \operatorname{cosec} PZ$$

i.e. sin H.A. at mid. time = sin arc 2 . sec lat.

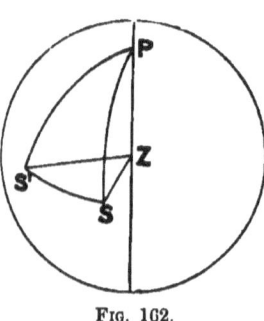

Fig. 162.

ART. 82. **Direct Method.**—In the triangle PSS_1 are given, the two sides PS and PS^1 and the included angle at P, from which can be found the third side SS_1 and the angles PSS and PSS_1.

Then in the triangle ZSS_1 with the three sides given, the angle ZS_1S can be found. Also the angle PS_1Z = $PS_1S - ZS_1S$.

Lastly, in the triangle PZS_1 the two sides PS_1 and ZS_1 and the in-

APPLICATION OF FORMULÆ TO PROOF OF RULES. 217

cluded angle PS_1Z are known, from which can be found the co-lat. PZ, and thence the latitude.*

ART. 83. Computing an Altitude.—Given the hour angle and declination of an object, and the lat. of observer, to find the altitude.

In the figure, S is the object, PS the polar distance, the angle

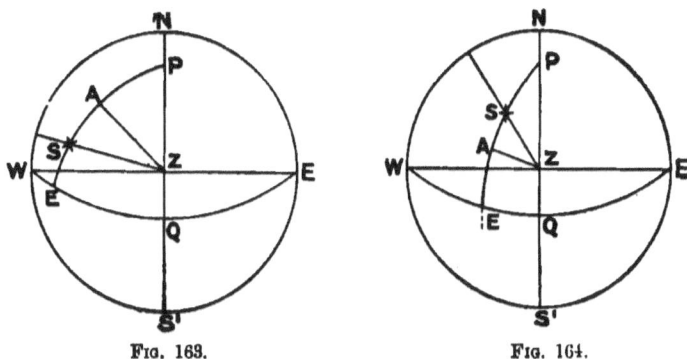

FIG. 163.   FIG. 164.

ZPS the hour angle, PZ the co-lat., and ZA an arc of a great circle perpendicular to PS.

Let $PA = \theta$;

then $SA = (p \sim \theta)$

In the right-angled spherical triangle PAZ—

$\cos h = \cot PZ \cdot \tan PA$
$= \tan l \cdot \tan \theta$
$\therefore \tan \theta = \cos h \cdot \cot l$ . . . . . . (1)

In the same triangle—

$\cos PZ = \cos PA' \cdot \cos AZ$
$i.e.\ \sin l = \cos \theta \cdot \cos AZ$
$\therefore \cos AZ = \sin l \cdot \sec \theta$ . . (2)

Also in the triangle SAZ—

$\cos SZ = \cos AZ \cdot \cos AS$
$i.e.\ \sin a = \sin l \cdot \sec \theta \cdot \cos (p \sim \theta)$

by substituting the value of cos AZ from (2).

NOTE.—$(p \sim \theta)$ signifies the difference between p and θ, because one or the other may be the greater according as the perpendicular ZA falls with in or without the triangle PZS.

---

* The latitude found by this method requires no correction for change of declination, because the actual polar distances are used, but it is subject to the same ambiguity as Ivory's method when SS_1 produced passes near the zenith. In the latter case the angle PS_1Z may be either the difference or the sum of PS_1S and ZS_1S.

For computation the following arrangement is convenient (l = lat., a = alt., p = polar distance, h = hour angle) :—

h	...	cos	
l	...	cot	sin
θ	...	tan	sec
p			
$p \sim \theta$			cos
a	... ...		sin

Remark.—This method has the merit of requiring only one table for all the logarithms.

ART. 84. The Lunar.—Let ZX and ZY be two vertical circles passing through the apparent positions M' and S' of the Moon and another celestial object (Sun, Star, or Planet); the true positions M and S, as seen from the Earth's centre, will be on the same vertical circles, and differ from the apparent positions by the effects of refraction and parallax. The Moon's parallax exceeds the refraction, but the parallax of any other body is less than the refraction. Hence the Moon's apparent place is lower, and that of the other body higher, than the true positions.

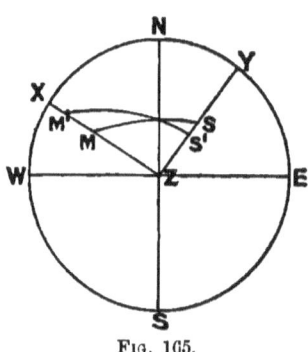

FIG. 165.

M'S' and MS are arcs of two great circles, and represent the apparent and true distances.

XM' and XM are the apparent and true altitudes of the Moon, and YS' and YS the apparent and true altitudes of the other body.

The angle Z is common to the two triangles ZM'S' and ZMS.

The problem is to find the true distance MS, the apparent distance M'S', and the true and apparent altitudes being given.

In the spherical triangle ZM'S' are given the three sides, ZM', ZS', and M'S', to find the angle Z. Then in the triangle ZMS, the two sides ZM and ZS and the included angle Z being known, the third side MS or true distance can be found.

The problem can therefore be solved by simply applying the formulæ of Cases I. and II. of oblique-angled spherical trigonometry, but as this involves two separate computations, several methods have been devised for *clearing the distance* by a single logarithmic computation. The method most generally

APPLICATION OF FORMULÆ TO PROOF OF RULES. 219

used is known as *Borda's*, which requires only one table of logarithms. It is obtained as follows:—

Applying the general formula, $\cos A = \dfrac{\cos a - \cos b \cdot \cos c}{\sin b \cdot \sin c}$

In the triangle ZM'S', $\cos Z = \dfrac{\cos M'S' - \cos ZM' \cdot \cos ZS'}{\sin ZM' \cdot \sin ZS'}$. (1)

In the triangle ZMS, $\cos Z = \dfrac{\cos MS - \cos ZM \cdot \cos ZS}{\sin ZM \cdot \sin ZS}$. (2)

Let m and m' represent the Moon's true and apparent altitudes;
s and s' the true and apparent altitudes of the other body;
d and d' the true and apparent distances;

Then equation (1) becomes—
$$\cos Z = \dfrac{\cos d' - \sin m' \cdot \sin s'}{\cos m' \cdot \cos s'}$$
and equation (2) becomes—
$$\cos Z = \dfrac{\cos d - \sin m \cdot \sin s}{\cos m \cdot \cos s}$$
by putting altitudes for zenith distances

$$\therefore \dfrac{\cos d - \sin m \cdot \sin s}{\cos m \cdot \cos s} = \dfrac{\cos d' - \sin m' \cdot \sin s'}{\cos m' \cdot \cos s'}$$

Adding 1 to each side—

$$1 + \dfrac{\cos d - \sin m \cdot \sin s}{\cos m \cdot \cos s} = 1 + \dfrac{\cos d' - \sin m' \cdot \sin s'}{\cos m' \cdot \cos s'}$$

$$\therefore \dfrac{\cos d + \cos m \cdot \cos s - \sin m \cdot \sin s}{\cos m \cdot \cos s}$$
$$= \dfrac{\cos d' + \cos m' \cdot \cos s' - \sin m' \cdot \sin s'}{\cos m' \cdot \cos s'}$$

that is—

$$\dfrac{\cos d + \cos (m + s)}{\cos m \cdot \cos s} = \dfrac{\cos d' + \cos (m' + s')}{\cos m' \cdot \cos s'}$$

$$\therefore \cos d + \cos (m + s) = \dfrac{\cos m \cdot \cos s}{\cos m' \cdot \cos s'} \cdot \{\cos d' + \cos (m' + s')\}$$

but $\cos d = 1 - 2 \sin^2 \dfrac{d}{2}$, and $\cos (m + s) = 2 \cos^2 \dfrac{m + s}{2} - 1$

$$\therefore 1 - 2 \sin^2 \dfrac{d}{2} + 2 \cos^2 \dfrac{m + s}{2} - 1 = - 2 \sin^2 \dfrac{d}{2} + 2 \cos^2 \dfrac{m + s}{2}$$

$$= \dfrac{\cos m \cdot \cos s}{\cos m' \cdot \cos s'} \cdot \{\cos d' + \cos (m' + s')\}$$

$$= \dfrac{\cos m \cdot \cos s}{\cos m' \cdot \cos s'} \cdot 2 \cos \dfrac{m' + s' + d'}{2} \cdot \cos \dfrac{m' + s' - d'}{2}$$

Dividing both sides by 2, and putting $\sin^2 \theta$ for
$$\frac{\cos m . \cos s}{\cos m' . \cos s'} . \cos \frac{m' + s' + d'}{2} . \cos \frac{m' + s' - d'}{2}$$
we have—
$$-\sin^2 \frac{d}{2} + \cos^2 \frac{m+s}{2} = \sin^2 \theta$$
$$\therefore \sin^2 \frac{d}{2} = \cos^2 \frac{m+s}{2} - \sin^2 \theta$$
$$= \tfrac{1}{2}\{1 + \cos(m+s)\} - \tfrac{1}{2}(1 - \cos 2\theta)$$
$$= \tfrac{1}{2}\{\cos(m+s) + \cos 2\theta\}$$
$$= \tfrac{1}{2}\left\{ 2 \cos\left(\frac{m+s}{2} + \theta\right) . \cos\left(\frac{m+s}{2} - \theta\right)\right\}$$
$$\therefore \sin \frac{d}{2} = \sqrt{\cos\left(\frac{m+s}{2} + \theta\right) . \cos\left(\frac{m+s}{2} - \theta\right)}$$

FORM FOR COMPUTATION.

	°	′	″	
App. dist. (d')	—	—	—	
Moon's ,, alt. (m')	—	—	—	... sec
Sun's or Star's ,, ,, (s')	—	—	—	... sec
$\tfrac{1}{2}$ sum $= \left(\dfrac{m' + s' + d'}{2}\right)$	2)—	—	—	
	—	—	—	... cos
$\tfrac{1}{2}$ sum $- d' = \left(\dfrac{m' + s' - d'}{2}\right)$	—	—	—	... cos
Moon's true alt. (m)	—	—	—	... cos
Sun's or Star's ,, (s)	—	—	—	... cos
$(m + s) =$	2)—	—	—	
$\left(\dfrac{m+s}{2}\right)$	—	—	—	
(θ)	—	—	—	2)— — ... sin
sum $\left(\dfrac{m+s}{2} + \theta\right)$	—	—	—	... cos
diff. $\left(\dfrac{m+s}{2} - \theta\right)$	—	—	—	... cos
$\dfrac{d}{2}$	—	—	—	2)— — ... sin
$d =$	2			

Remark.—It will be noticed that the first four logs are the same as in the alt.-azimuth problems, and that all other logs taken out are cosines, whilst the angles sought correspond to sines.

APPLICATION OF FORMULÆ TO PROOF OF RULES.

ART. 85. The "Danger Angle."—When sailing near the land where there are outlying dangers, a safe passage may be secured by calculating the angle subtended at the safe distance by the height of the top of a lighthouse or cliff above sea-level. Setting this angle on the sextant, the navigator so shapes his course that this vertical angle is not exceeded, allowance being also made for the dip of the land horizon (Tab. viii. Norie or Rapier).

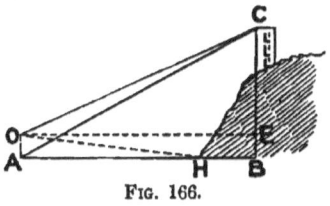

FIG. 166.

Referring to the figure, the height BC and the safe distance AB being known (by the chart)—

$$\text{Tan} \angle \text{BAC} = \frac{CB}{AB}$$

Let AO be the height of observer's eye above sea-level, OE a horizontal line, and H the base of the cliff or land horizon; then ∠ EOH is the dip, EB = OA, and CE = CB − EB.

Now, $\tan \angle \text{COE} = \frac{CE}{OE}$, and the dip being added to ∠ COE, the danger angle COH is found.

Example.—The height of the top of a lighthouse which stands inland 500 feet is 725 feet, and, owing to the existence of a sunken rock, it is not safe to pass nearer the land than 1¼ mile. Required the greatest vertical angle for safe navigation which the top of the light-house should make with the base of the cliff on which it stands, the height of eye being 20 ft.

Referring to the figure—

CE = 725 − 20 = 705 ft., AH = 6600 ft.
and AB or OE = 6600 × 500 = 7100 ft.

$$\text{Now, tan COE} = \frac{CE}{OE} = \frac{705}{7100}$$

705	log 2·848189
7100	log 3·851258
	tan 8·996931

∠ CDE = 5° 40′
Dip   10

∠ COH 5 − 50 the danger angle required

To find the Width of a River or Channel.—Let C be a prominent object on the opposite bank, and A and B two stations whose distance apart is known, B being chosen so that the ∠ ABC is a right angle. The horizontal angle at A, subtended by B and C, can be measured by means of a sextant or prismatic compass.

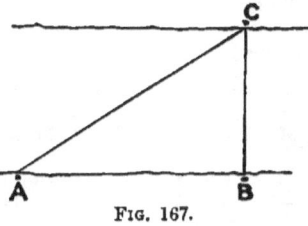

FIG. 167.

Then $\dfrac{CB}{AB} = \tan A$

and $CD = AB \times \tan A$

Example.—Given the distance AB = 600 yards, and the angle CAB = 42° 30′, required the width (CB) of the river.

A 42° 30′	tan 9·962052
AB 600	log 2·778151
BC = 549·8	log 2·740203

therefore the width of the river is 550 yards nearly.

Remark.—Many other problems in heights and distances connected with navigation depend on the rules for the solution of right-angled plane triangles.

MISCELLANEOUS PROBLEMS.

1. Two ports in lat. 38° S. and 50° S. are on the same meridian; a vessel leaves the former, and, after sailing due W. 300 miles, steers due S. until the second parallel is reached. How far is she then distant from the second port?

2. Two places, A and B, are in the same latitude, and in longs. 10° W. and 20° 20′ W. respectively; a ship, sailing 10 knots, took exactly 48 hours to pass from one to the other. Find the latitude.

3. A vessel sailed from lat. 35° 20′ N., long. 30° W., 250 miles due N., 250 miles due W., 250 miles due S., and 250 miles due E. Required her latitude and longitude in.

4. A ship which sails 8 knots per hour, within 6 points of the wind, has to reach a port 120 miles S.S.W. of her, wind S.W. How long will she take?

5. A ship, after steaming S.E. by E. 12 knots per hour for 23 hours, from a position lat. 25° 36′ S., long. 62° 35′ E., found herself in lat. 27° 25′ S., long. 66° 45′ E. Find set and drift of current.

6. A ship has to reach a port 30 miles N.N.W. of her; the wind is N. by E., and she can sail 7 knots within 6 points of the wind, making $\frac{1}{4}$ point leeway; she starts on the starboard tack, wishing to make her port in two tacks. After what interval must she go about?

7. An observer in a boat finds that the angle of elevation of the top of a lighthouse standing upon a cliff is 32°; after the boat has been rowed 100 yards directly towards the lighthouse, the elevation of the same point is 71°. Find the distance of the boat from the foot of the cliff at both observations; and, if at both points of observation the angle subtended by the lighthouse is found to have the same value, find the heights of cliff and lighthouse.

8. The upper part of a mast is 30 feet long, and subtends, at a point 150 feet in a horizontal line from the base, the same angle as that subtended at the same point by a man 6 feet high standing at the base. Find height of the mast.

9. The meridional parts for lat. 38° 25′ are 2500·08. Calculate those for 38° 31′.

10. The meridional parts for lat. 57° 46′ are 4267·97. Calculate those for 57° 50′.

11. A ship sails from Rio de Janeiro, lat. 22° 54′ S., long. 43° 6′ W., to Madeira, lat. 32° 43′ N., long. 16° 40′ W. Find longitude of point where the track crosses the equator.

12. Find the latitude of the point of "Maximum Separation in Latitude" in sailing from lat. 40° 30' N., long. 120° E., to lat. 45° N., long. 172° E.

13. The vernier of a sextant, the limb of which is divided to 10', has 40 divisions, coinciding with 119 divisions on the limb. With what accuracy may the sextant be read off?

14. The limb of a sextant is divided to 15', and 59 divisions on the limb are formed into 60 divisions on the vernier. With what accuracy may the instrument be read?

15. If every inch on a barometer scale be divided into 10 parts, show how to form a vernier such that the reading of the instrument may be taken to the five-thousandth part of an inch.

16. What condition must hold in order that a star may never set in a given latitude? What stars of the first, second, or third magnitudes never set in lat. 52° 30' N. or 46° S.?

17. Find the time when altitude of sun on the prime vertical is 18° 25', his declination being 12° 16' N.

18. What is the length of the day in lat. 50° 25' N., when the Sun's declination is 18° 47' N.?

19. In lat. 53° 24' N., find the difference between the longest and shortest days.

20. In what latitude is the shortest day just half the longest?

21. A steamer is being steered S.W. ½ W., at the rate of 12 miles per hour, in a known current setting W.N.W. 2½ miles per hour. Required the course and distance made good per hour.

22. Given the course and distance to a port, N. 75° W. 196 miles; a steady current is setting S. 15° W. 1·9 mile per hour, and the vessel steams 10 knots. What course should be steered to counteract the current, and what progress is made towards the port in 12 hours?

23. A ship's position is, by dead reckoning, lat. 48° 21' S., long. 43° 28' W.; and by observation, lat. 47° 165', long. 41° 3' W. Assuming the discrepancy to be due to a current, what was its set and drift during the time elapsed since the previous observations?

24. The true course to destination is N. 33° W., and distance 75 miles; but after steaming 5 hours on that course at 14 miles per hour, it was found to bear N. by W. (true), and distant by cross-bearings 10 miles. What was the effect of the current?

25. How long should it take a steamer to reach her destination, bearing W.N.W. 85 miles distant, if steering the proper course to counteract the effect of a steady current setting S. 20° W. at the rate of 2¾ miles per hour, the ship's rate of steaming being 13 knots?

26. A departure is taken at 6 P.M., from a position determined by cross-bearings in lat. 51° 18' N., long. 9° 42' W.; her course is set W. ½ S., and after steaming at the rate of 14 knots per hour until noon the following day, the position is found by observation to be lat. 51° 5' N., long. 16° 20'. Required the set and drift of the current.

CHAPTER XXV.

INSTRUMENTS.

ART. 86. **Principle of the Sextant.**—The principle of the sextant is based on the optical fact that the angle of incidence of a ray of light upon a mirror is equal to the angle of reflection.

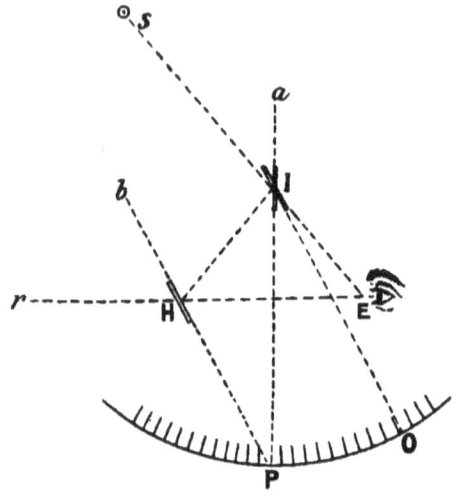

FIG. 168.

Let I and H represent the index and horizon glasses of a sextant. Light from S falling on I is reflected at an equal angle to H, and thence reflected to the eye at E, so that the object S would be seen by reflection in the direction Er, and if r were the horizon, the object's image would be seen on the horizon.

IO represents the index bar when the mirrors are parallel, and IP when it has been moved to P. Produce PI to a, PH to b, and EH to r.

Now, since $\angle$ HPI = $\angle$ S$I a$ = $\angle$ EIP (Euc. I. 15), the $\angle$ HIE s bisected by IP, making $\angle$ HIP = $\angle$ EIP.

INSTRUMENTS.

Also, since ∠ bHI = ∠ EHP = ∠ bHr (Euc. I. 15), the ∠ rHI is bisected by bH, making ∠ rHb = ∠ bHI.

The ∠ rHI = ∠ HIE + ∠ HEI (Euc. I. 32); therefore ∠ rHb = ½ ∠ HIE + ½ ∠ HEI = ∠ HIP + ½ ∠ HEI.

Again, ∠ rHb (= bHI) = ∠ HIP + ∠ HPI; therefore HIP + HPI = HIP + ½ HEI.

Taking away the common angle HIP, there remains ∠ HPI = ½ ∠ HEI; and as ∠ OIP = ∠ HPI, because IO and HP are parallel, it follows that the angle through which the index bar IP has moved from zero is half the angle described by the image of the object. For this reason an arc of ½° on the limb is read as 1°, and an arc of 60° will measure 120°.

ART. 87. **Principle of the Vernier, or Nonius.**—The value of the smallest division on an ordinary sextant is 10′, but the

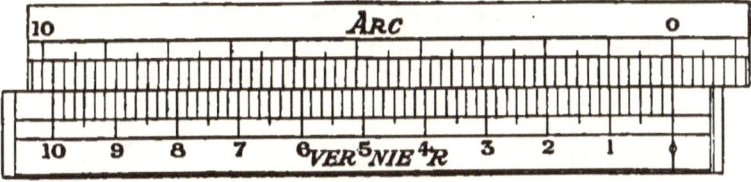

FIG. 169.

vernier attached to the index bar subdivides into single minutes and 10, 20, 30, etc., seconds.

This is accomplished by taking 59 divisions of the limb for the length of the vernier, and dividing it into 60 equal parts;

Thus, 60V = 59L

$$\therefore 1V = \frac{59}{60} L = 1 - \frac{1}{60}$$

that is, one division on the vernier is $\frac{1}{60}$ of 10′ = 10″ shorter than one on the limb. Six of these vernier divisions are marked 1, and the whole length reads 10′.

When the pointer or 0 on the vernier coincides with a cut on the limb, the next vernier cut falls short of the one above it on the limb by 10″, the second one by 20″, the third by 30″, etc. So that if the index is moved forward until the first, second, third, etc., cut coincides with the one above it on the limb, the pointer will have moved 10″, 20″, 30″, etc., and so on to the end of the vernier scale. When the 10′ cut coincides, the pointer will have moved a whole division of the limb, *i.e.* 10′. Therefore, to find how far the pointer or 0 has moved from the last limb division, it is only necessary to see what cut on the vernier coincides with a cut on the limb.

The same rule holds good when the reading is "off" the arc, only in this case both scales must be read from left to right.

Q

CHAPTER XXVI.

PROJECTIONS.

ART. 88.—The representation of the circles of a sphere on a flat surface is called a *Projection*. The boundary of this surface is a circle of the same diameter as the sphere, and is called the *Primitive Circle*, whose plane is called the *Plane of Projection*. The point where the eye is supposed to be placed is the *Projecting Point*.

I. *Stereographic Projection* supposes the eye to be at some point on the surface of the sphere, and looking at the circles on the concave surface of the opposite hemisphere. Great circles will appear as arcs of circles on the plane of projection, except those in the same plane as the eye, and passing through the projecting point, which will appear to be straight lines.

This is the kind of projection used for illustrating the problems of nautical astronomy. The projection may be either on the plane of the meridian or the plane of the horizon.

(a) *Projection on the Plane of the Meridian* (Figs. 170 and

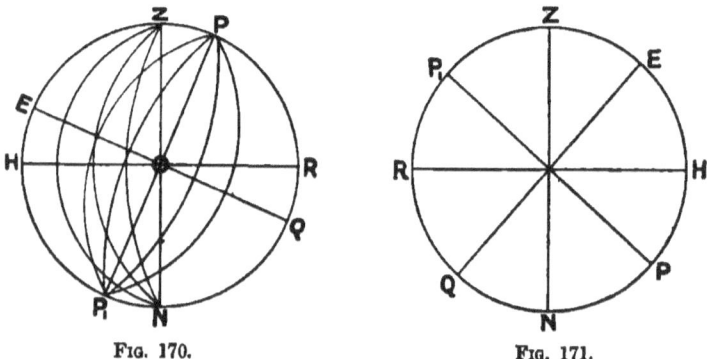

FIG. 170.   FIG. 171.

171).—The primitive circle represents the celestial meridian, the projecting point O being the east or west point of the horizon; Z and N are the zenith and nadir; P and P_1, the north and south poles respectively; HR (at right angles to ZN) is the horizon; and EQ (at right angles to PP_1) is the equinoctial. The elevation of the upper pole P above the horizon, viz. the arc RP, is equal to the latitude of the observer's position. The arc EZ also equals the latitude. Supposing R to be the north point of the horizon, then P is the north pole (Fig. 170), and the latitude is north. For south latitude, the south pole P_1 must be above the horizon (Fig. 172).

Hour circles are represented by arcs from P to P_1, and vertical or altitude circles by arcs from Z to N.

(b) *Projection on the Plane of the Horizon* (Fig. 172).—Here the *primitive circle* is the horizon, N., E., S., W. being the cardinal points. The zenith (Z) is the projecting point, hence all *vertical circles* appear as straight lines passing through Z, NZS being the *celestial meridian*, and WZE the *prime vertical*. P is the north pole, at a distance from N. equal to the latitude. The arc ZQ is also equal to the latitude, and PZ is the co-latitude; WQE is the equator, cutting the horizon in the E. and W. points; Pp_1, Pp_2, etc., are hour circles.

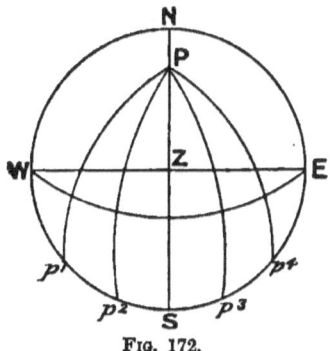

FIG. 172.

The projection on the plane of the horizon has some advantages over that on the plane of the meridian for illustrating most of the problems of nautical astronomy, and is therefore more commonly employed.

(c) *Projection on the Plane of the Equinoctial.*—Here the

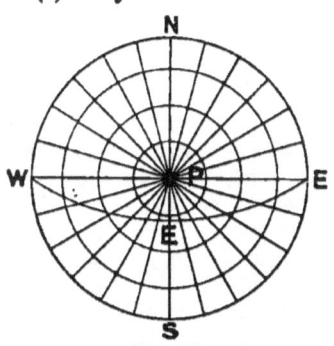

FIG. 173.

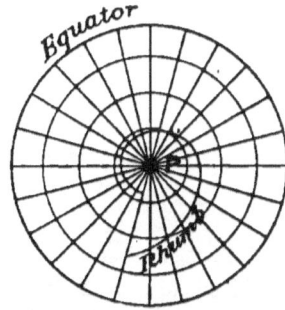

FIG. 174.

primitive is the equinoctial, and the pole P the projecting point. The concentric circles are parallels of declination. Hour circles are represented by straight lines, since their planes pass through the projecting point. NPS is the celestial meridian, and EPW the six-o'clock hour circle; ECW is the ecliptic.

A projection of the Earth on the plane of the equator shows the meridians of longitude as straight lines, and the parallels of latitude as concentric circles whose common centre is the pole. This is useful for showing the course of a rhumb line, also for mapping the circumpolar regions.

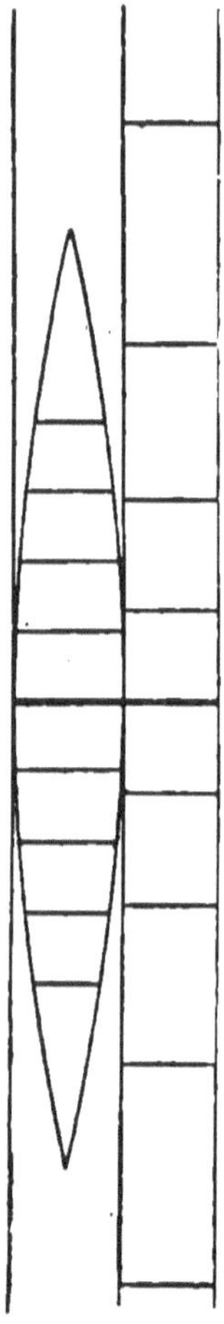

II. *Orthographic Projection* supposes the spectator to be at such a distance that the diameter of the sphere bears an infinitely small proportion to that distance. Circles whose planes pass through the projecting point appear as straight lines, and all other great circles as ellipses. This projection is often used in astronomy.

III. *Gnomonic Projection* supposes the eye to be at the centre of the sphere and looking at the concave surface. It is used in astronomy and dialling.

IV. *Mercator's Projection.*—This is the kind of projection used in the construction of Navigating Charts, and represents the Earth's surface as a plane. The meridians of longitude are drawn as parallel straight lines, and the divisions of latitude are expanded in the same proportion as the meridian distances are lengthened. The effect of this is to make the rhumb course a straight line, which is of great advantage to the navigator.

The distance from the equator of any given parallel of latitude, and the amount of expansion of a degree of latitude, can be found from the table of *Meridional Parts;* for example, the parallel of 30° is 1800 geographical miles from the equator on the globe, but the meridional parts for 30° = 1888, which means that on a Mercator's chart drawn to the same scale, this parallel would be 1888 geographical miles from the equator. Again, the meridional parts for 31° = 1958, and 1958 − 1888 = 70 miles, which would be the length of the expanded degree from 30° to 31°.

FIG. 175.

The meridional parts for any given latitude may be calculated by adding together the secants of all minutes of latitude up to the given latitude; thus—

$$\sec 0' + \sec 1' + \sec 2' + \sec 3', \text{etc.}$$

or they may be calculated independently from the formula—

log of mer. parts = $3·8984895 + \log (\log \cot \frac{1}{2}$ co-lat. $- 10)$

PART III.

CHAPTER XXVII.

LAWS OF STORMS.

THE facts gathered from observations, and known as the *Laws of Storms*, apply more particularly to the violent Hurricanes, Cyclones, and Typhoons (which may be all classed under the name "Cyclones") met with between the parallels of 10° and 35° of latitude; but also in a modified degree to the strong winds, gales, and storms of higher latitudes.

These facts may be summarized as follows:—

1. The tropical Cyclones are never met with in the belt between 10° N. and 10° S. lat., but outside of 10° they occur in the following regions:—

N. Atlantic: The western part near the West Indies.
N. Indian Ocean: The Bay of Bengal and Arabian Sea.
N. Pacific: The China and Java Seas.
S. Indian Ocean: The Western part, passing near Mauritius, and hence called "Mauritius Hurricanes."
S. Pacific: Eastern part.

2. Cyclones (as the name implies) have a rotatory motion round a centre, and a progressive motion, varying from three or four miles an hour to twelve or more.

Near their origin they cover only a small area, but afterwards expand to as much as a thousand or more miles in diameter.

3. The barometer stands lowest at the centre (often 28 inches or less), and gradually increases in height towards the circumference, whilst the wind force is generally greater the nearer the centre.

4. The bearing of the centre from the ship may be inferred approximately by counting from the wind direction 8 or 10 points to the *Right* in the N. Hemisphere, and the same number of points to the *Left* in the S. Hemisphere.

5. The path of the centre, or *Line of Progression*, divides the Storm Area into two semicircles, and by observing the change of wind, it may be determined whether the ship is

on the right or left hand side of the Line of Progression—a matter of great importance in the handling of the ship. Thus, if the wind changes to the *Right*, the ship is on the Right-hand side, but if to the left, she is in the Left-hand semicircle.

If the wind is "steady," *i.e.* changes very slightly in direction, it indicates that the Ship is *on* the Line of Progression.

6. The Direction of Rotation in the N. Hemisphere is to the Left, and in the S. Hemisphere to the Right.

FIG. 176.

7. The approach of a Cyclone is often indicated by well-marked signs, such as:—
(*a*) An oppressive stillness, with threatening sky and much lightning.
(*b*) A long "ground swell."
(*c*) A dense cloud bank on the Horizon.
(*d*) A rapid motion of the upper clouds.

8. When within the Storm Area, the *approach* of the centre is shown by—
(*a*) A rapidly falling barometer.
(*b*) Increase of wind, with heavy squalls, much lightning and rain, and general murkiness.
(*c*) Heavy and confused sea.
(*d*) Continuous "veering" or "backing" of the wind, except when *on* the Line of Progression.

The *receding* of the centre would be indicated by—
(*a*) Rising barometer.
(*b*) Wind becoming more "steady" and decreasing in force.
(*c*) Weather clearing, but the sea still confused and dangerous.

9. Vessels, especially steamships, sometimes *overtake* Hurricanes, because their speed is greater than the rate of progression of the Storm Centre. In this case the above signs of *approach* and *recession* would be reversed.

10. In entering the Central Area, which may be up to a hundred miles in diameter, the wind suddenly ceases, with glimpses of a clear sky and a general calm of perhaps hours' duration, interrupted by puffy squalls, whilst the sea is particularly confused and dangerous. After passing through the centre, the wind strikes the ship from the opposite point of the compass with renewed Hurricane force.

11. The paths of Storm Centres follow the same general directions for the same regions of the Globe. The following are the usual tracks:—
(*a*) N. Atlantic: Having their origin near the Windward Islands, they first travel about W.N.W., then more northerly, and about 25° N. to 30° N. they recurve to the N.E. towards Mid-Atlantic.

LAWS OF STORMS. 231

(b) N. Pacific: The same general direction as in the N. Atlantic.
(c) N. Indian Ocean, including Bay of Bengal and Arabian Sea: A general N. and W. direction.
(d) China Seas: Between N.W. and W.S.W.
(e) Java Sea: Between N. and W.
(f) S. Indian Ocean: Starting from about 12° South, they travel in a S.W. direction towards Mauritius, then more southerly, and recurve to the S.E. in 25° to 30° S. lat.

12. Seasons of Greatest Frequency of Cyclones.—The following Table of recorded Cyclones from Scott's Meteorology shows that Cyclones are most frequent in the N. Indian Ocean at the Change of Monsoons, and in the other Cyclone Regions during the hottest months :—

Cyclone region.	Jan.	Feb.	Mar.	April.	May.	June.	July.	Aug.	Sept.	Oct.	Nov.	Dec.	Totals of recorded observations.	Average no. per annum
West Indies (in 300 years)	5	7	11	6	5	10	42	96	80	69	17	7	355	3·55
S. Indian Ocean (39 years)	9	13	10	8	4	—	—	—	1	1	4	3	53	1·36
Bombay (25 years)	1	1	1	5	9	2	4	5	8	12	9	5	62	2·5
Bay of Bengal (139 years)	2	—	2	9	21	10	3	4	6	31	18	9	115	·8
China Seas (85 years)	5	1	5	5	11	10	22	40	58	35	16	6	214	2·5

The Table also shows that the *most* cyclone-infested regions of the globe are the W. Indies, Arabian Sea, and China Seas, and that the worst cyclone months in the various regions are—
(a) Bay of Bengal and Arabian Sea: April, May, October, and November.
(b) N. Atlantic: June to September.
(c) China and Java Seas: July to October.
(d) S. Indian Ocean: January to March.
The same rule would give for—
(e) N. Pacific: June to September.
(f) S. Pacific: January to March.

13. The Rules of Action to be taken when a ship is caught in a Cyclone can be best understood by reference to the following diagrams :—

These Rules may be summed up as follows:—
(1) If *on* the Line of Progression of an approaching Storm Centre, run or steam before the wind until clear of the track of the centre; then, in the N. Hemisphere, heave to on the Port Tack, but in the S. Hemisphere on the Starboard Tack. By so doing, the ship would always " come up " to the sea when hove to.
(2) In the Right-hand semicircle—
(a) Northern Hemisphere: Head reach or heave to on the Starboard Tack.

(b) Southern Hemisphere: Run with wind on Port Quarter, or heave to on Starboard Tack.

(3) In the Left-hand Semicircle—

(a) Northern Hemisphere: Run, having the wind on Starboard Quarter, or heave to on the Port Tack.

(b) Southern Hemisphere: Head reach or heave to on the Port Tack.

(4) When overtaking a Cyclone, it is obvious that the ship should be stopped or her course altered so that she may not approach the centre.

14. **General Remarks.**—The most dangerous position for a ship is the track of the Centre of a Cyclone, which should be avoided if possible. This can generally be done, if there is sea room, by attention to the foregoing Rules.

Observations tend to show that a Storm Area is more circular in front than in the rear; so that the Rule for finding the bearing of the centre is most reliable in the most dangerous situation.

The same Rules may be applied when Heavy Gales are encountered in higher latitudes.

It is common to give the name of "Dangerous Semicircle" to the Right-hand Semicircle in N. Lat., and the Left-hand Semicircle in S. Lat., because the direction of the air-currents is towards the Line of Progression in front of the Storm Centre.

If it is suspected that the ship is overtaking a storm, the engines should be stopped, or the vessel "hove to," and the barometer carefully watched. Should the barometer begin to rise, it is a sure sign of *overtaking;* but if it continues to fall (although less rapidly), it is a case of *meeting* a storm.

LAW OF STORMS.

Questions to be answered by candidates for Mates' and Masters' Certificates.

1. Suppose the wind in a cyclone to blow from   , state what is the probable bearing of the centre in the   hemisphere.
2. If the wind in the same cyclone changes to   , on what side of the line of progression is the ship, and what action would you take?
3. Under what circumstances would the change in the direction of the wind be the reverse of the above?
4. What are the usual indications that a ship is on the line of progression of the centre of a cyclone?
5. What are the usual indications that a ship is approaching the centre of a cyclone or receding from it?
. 6. State what are the usual tracks taken by cyclones in   , and at what season of the year do they most frequently occur in those regions?

LAW OF STORMS.
Answers.

1. In the Northern Hemisphere the centre bears about eight to ten points to the right, and in the Southern Hemisphere eight to ten points to the left from the given direction of the wind.

2. If the wind changes to the right the ship is on the right-hand side of

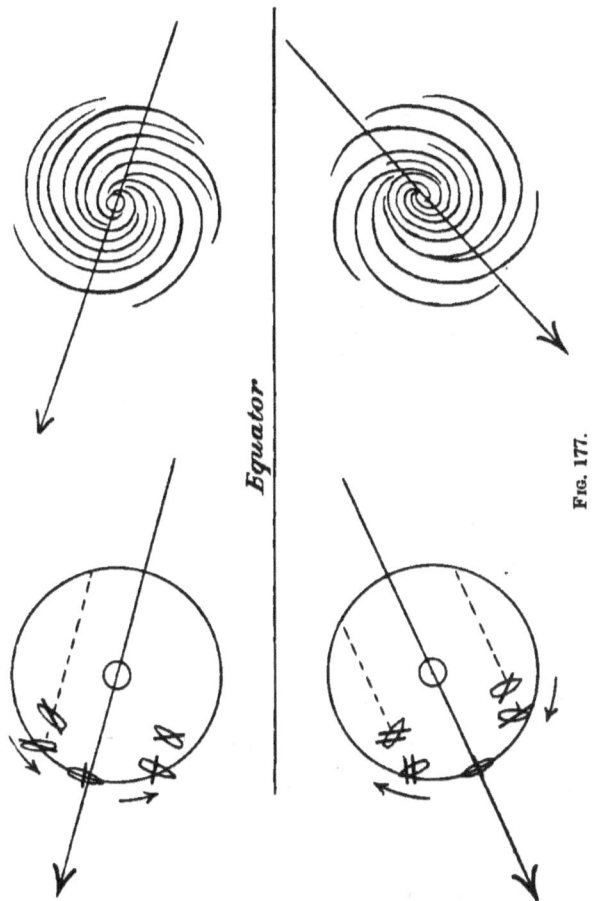

Fig. 177.

the line of progression, if to the left she is on the left-hand side. These rules apply to both hemispheres.

The action to take is as follows :—

Northern Hemisphere—

RIGHT-HAND SIDE.—If possible, sail out on the starboard tack to a safer distance from the centre, or heave to on the starboard tack.

LEFT-HAND SIDE.—If possible, run with the wind on starboard quarter until in a safer position, or heave to on the port tack.
Southern Hemisphere—
RIGHT-HAND SIDE.—If possible, run with the wind on the port quarter until in a safer position, or heave to on the starboard tack.
LEFT-HAND SIDE.—If possible, sail out on the port tack to a safer position, or heave to on the port tack.

3. The wind would change the reverse way if the ship were on the other side of the line of progression, also if she were overtaking the storm on the same side as before.

4. In advance of the centre the barometer falls more and more rapidly as the centre approaches, the wind continues to blow from the same quarter, but increases in violence, with frequent heavy squalls, a confused angry sea, a dense cloud bank, and general threatening appearances of weather.

In the central area the barometer is lowest, with a confused and dangerous sea, and calms varied by strong gusts.

In the rear of the centre the barometer rises, and the wind blows at first with great violence from the opposite quarter, afterwards moderating and the squalls becoming less frequent, the sea being heavy and confused, but the weather gradually clearing.

5. (a) When approaching the centre, not on the line of progression, there are the same barometer and weather indications as stated in the preceding answer, but the wind changes in direction as it increases in force, with heavy cross-seas.

(b) When receding from the centre the barometer rises, the wind becomes more steady, with less frequent squalls and less threatening appearance of weather, though the sea is still cross and dangerous.

6. North Atlantic (Tropical).—The cyclones, in this region, are supposed to take their rise near the Windward or Leeward Islands, and first travel about W.N.W. towards Florida, then more northerly, and on reaching the parallel of 30° often recurve to the N.E. They occur most frequently during July, August, and September.

South Atlantic (Tropical).—They travel first about W.S.W., then more southerly, and sometimes recurve to the S.E. They occur most frequently in the hottest months—December, January, and February.

South Indian Ocean (Tropical).—They are supposed to have their origin in the belt of calms, south of the Equator, often near Java, and travel first W.S.W., then more southerly towards the Mauritius, and sometimes recurve to the S.E. They occur from November to May, but most frequently in December, January, and February.

China Seas.—The storms take their origin near the Philippine Islands, and travel generally between W.N.W. and W.S.W. They occur most frequently in July, August, and September.

Bay of Bengal.—The storms in this region begin in the neighbourhood of the Andaman and Nicobar Islands, and travel between W. and N.N.W. They occur most frequently in April to June, and September to November.

Arabian Sea.—Between N. and W., and most frequent in May and October.

North Pacific (Tropical).—The cyclones of this region begin in about 10° N., and travel first W.N.W., then more northerly, and finally recurve to the N.E. July, August, and September are the worst months.

South Pacific (Tropical).—The cyclones of this region first travel about W.S.W., then more southerly, and finally to the S.E. They occur most frequently from November to April.

Outside of lat. 30°, storms in the North Atlantic and North Pacific travel to N.E., and in the South Indian Ocean and South Pacific to the S.E.

CHAPTER XXVIII.

MAGNETISM AND DEVIATION OF THE COMPASS.

For a full and scientific discussion of the deviation of the compasses in iron ships, the student is referred to Evans's "Elementary Manual," edited and revised by Captain Creak, R.N.

The purpose of the following remarks is to enable students to clearly understand the principles involved in the Board of Trade Syllabus of Questions dealing with the practical application of the laws of magnetism to the case of iron ships and their compasses.

Definition.—Magnetism is a property which may be acquired by masses of iron and steel, by virtue of which they attract or repel each other under certain circumstances.

Kinds of Magnets.—There are three kinds of magnets, viz. natural, artificial, and electro-magnets. The natural magnets are the Earth and pieces of magnetic iron ore or "lodestone." Artificial magnets are bars or "needles" of hard tempered steel, which have been magnetized by rubbing contact with other magnets, or by the influence of an electric current. Electro-magnets are elongated or horseshoe-shaped masses of soft iron wound round with copper wire, through which an electric current is passing. The strongest magnets are made in this way; but the magnetic properties disappear when the current ceases.

Degrees of Magnetism.—Magnetism may exist in three different states or degrees in steel or iron, viz. PERMANENT, SUB-PERMANENT, and TRANSIENT INDUCED.

A bar of hard-tempered steel receives magnetism slowly, but when once magnetized, it retains its magnetic strength for an indefinite length of time, and is said to have *permanent magnetism*.

Iron or steel of a less degree of hardness is more easily magnetized, but is liable to lose a considerable portion of its magnetism with lapse of time, and is therefore said to have *sub-permanent* magnetism.

Soft or malleable iron becomes instantly magnetized by

induction from any magnetic source, but loses it when the inducing cause is removed. This temporary magnetic condition is known as TRANSIENT INDUCED magnetism.

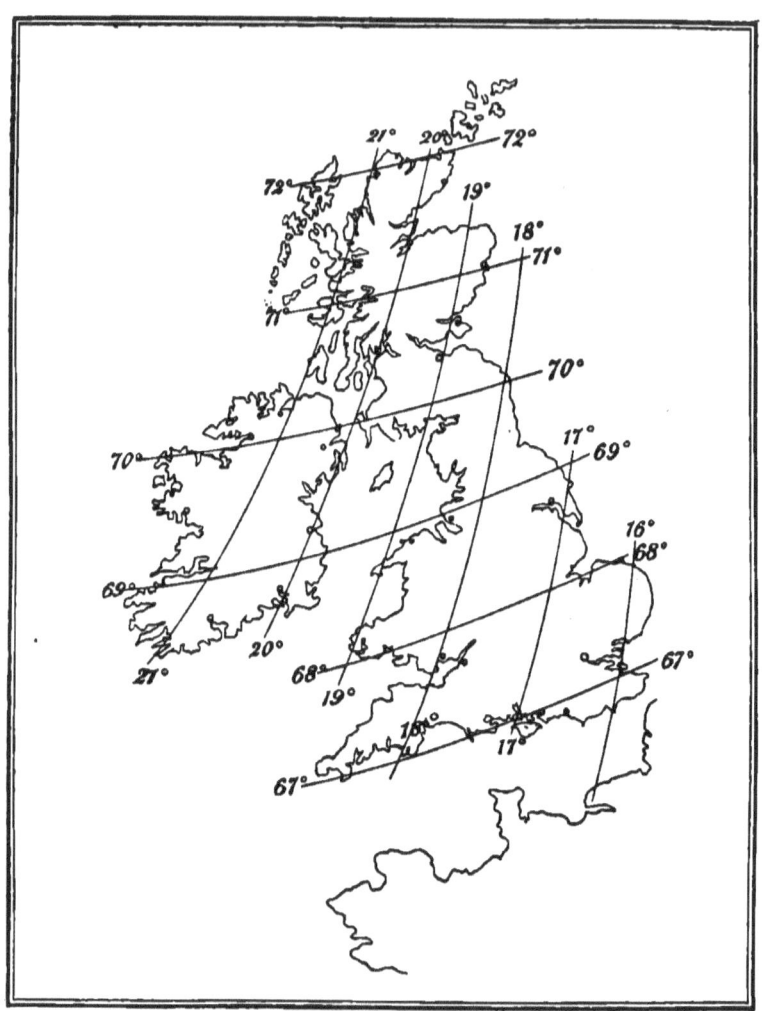

FIG. 178.—Variation chart of British Islands.

Saturation.—There is a limit to the magnetic strength which a piece of steel or iron can acquire, and when this limit is reached it is said to be *saturated*.

Magnetic Poles.—Every magnet has two POLES, or points where the greatest energy is concentrated, distinguished as RED and BLUE poles, because it is common to paint artificial magnets in these colours. The red pole of a permanent magnet is the one which would point north when the magnet is suspended at its centre and free to turn round. This end is also called the N. or *marked* end. Therefore the end of a compass-needle which points N. is the red pole.

First Law of Magnetism.—It is a well-known law of magnetism that *like poles repel each other, and unlike poles attract each other*. This can be proved on a compass-needle, for if the red pole of a magnet be brought near to the N. or marked end, it will repel it, whilst the blue pole will attract it. Similarly, the S. end will be repelled by the blue pole, and attracted by the red. It follows from this experiment that the Earth's NORTH magnetic pole is BLUE, because the red pole of the needle is attracted towards it. The needle, if floating on water and free to move, would not, however, travel towards the N. pole, because its length is as nothing compared with the distance from the Earth's pole, its blue end being as strongly repelled as the red end is attracted. It is the same with the action of the south magnetic pole. The only result is, therefore, to give the needle a north and south *direction*.

If a needle hundreds of miles in length could be supposed to be afloat in the North Atlantic, with its red pole towards north, no doubt it would travel bodily towards the N. pole of the Earth, because its red pole would then be attracted more strongly than the blue would be repelled.

Second Law.—It is another law of magnetism that the effect of a magnetic force (such as the pole of a magnet or an induced pole in soft iron) on a small magnet varies nearly in the inverse ratio of the cube of the distance ; that is, at double the distance it is one-eighth ($2 \times 2 \times 2 = 8$); at treble the distance it is only one twenty-seventh ($3 \times 3 \times 3 = 27$), etc. Also—

Third Law.—A magnet placed *end on* to a pivoted needle causes twice the deflection that it would do if *broadside on* at the same distance between the centres of the magnet and needle.

These facts have a very important bearing on the effect of the magnetic forces in a ship on her compasses, and show the advantage of placing the compasses as far as practicable from masses of iron having magnetic properties, especially when end on.

Transient Induced Magnetism.—When either pole of a magnet is brought near to one end of an iron rod, it develops two magnetic poles in the iron, the near end being of the opposite name to that of the pole applied. The process may be repeated through a series of rods end on, so that the N. point of a compass needle would be deflected by the last one in the same manner as by the magnet itself.

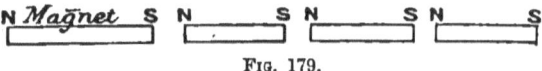

FIG. 179.

The same fact may be illustrated by the familiar experiment of causing a number of soft iron bars or nails to hang on to the end of a strong magnet; but, whether in contact or not, the effect is the same.

The above are examples of *magnetic induction* and the acquiring of *transient induced magnetism*.

Lines of Force.—The mysterious action of one magnet on

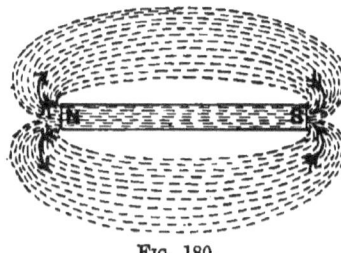

FIG. 180.

another at a distance, and the magnetization of iron with or without contact, are explained by the existence of *lines of force*, which issue from the red pole and return to the blue pole. These lines of force pass without interruption through all intervening non-magnetic substances, and it is the passing of lines of force into iron which
gives it magnetic properties. Their existence and direction can be shown by experiment (Fig. 180). If a sheet of white paper be laid on a magnet and iron filings sprinkled on the paper, the filings will arrange themselves in lines corresponding to the direction of the lines of force from one pole of the magnet to the other.

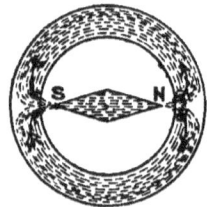

FIG. 181.

If a suspended magnetic needle were enclosed in a thick iron shell, the lines of force would follow round through the iron by preference to reach the blue pole, and the consequence would be that the needle would lose its directive power. This is the reason why compass-bowls are made of copper, and not of iron. Another reason is that poles would be induced in an iron bowl, which would oppose the Earth's directive action on the needle. Therefore a ship's compass which is closely surrounded by iron would have a weakened directive power.

The Earth's Magnetism.—The Earth itself acts as a magnet, having a blue pole towards the north and a red pole towards the south. Its action on a compass needle is merely to give a *directive* tendency, as before explained, but it is also a powerful *inducing* source. No iron can escape from its influence. Hence fireirons, knife-blades, needles, etc., are nearly always magnets with two fixed poles, as may be tested on a compass-needle, whilst bars or rods of soft iron have also induced poles from the same cause, which are not, however, fixed. A fairly strong magnet can be made by holding a bar of hardened steel in a N. and S. direction and tapping it with a hammer; the vibration quickens the process of magnetization by the Earth's induction.

A Ship's Induced Magnetism.—Even a ship becomes a huge magnet from the Earth's induction in the process of building, whilst the separate parts, such as beams, stanchions, davits, rudder and stern-posts, etc., become so many temporary magnets, liable to act separately and collectively in disturbing the compasses.

Variation.—The magnetic poles of the Earth do not coincide with the true geographical poles, the N. or blue pole being in about lat. 70° N., long. 90° W., and the S. or red pole was at one period found to be in lat. $73\frac{1}{2}$° S., long. $147\frac{1}{2}$° E. (from which it is seen they are not exactly opposite). This is the cause of the variation of the compass, which is the difference in direction between true N. and magnetic N. The amount of variation is stated on ordinary charts, and for the entire globe on special variation charts. The variation is not constant at any place, but changes at a known rate (in the United Kingdom now about 8' or 9' annually).

Previous to the year 1657 the variation was E. in this country. In that year there was no variation at London, the compass pointing true N. Then it became westerly, and attained its maximum W. variation in 1816. Since then it has constantly decreased, and will again become 0 in 1977; then, becoming E. again, it will reach its maximum in the year 2290. The explanation of this progressive change of variation is that the N. magnetic pole describes a complete circle round the true N. pole in a period of about 640 years. Twice in the course of this period the compass would point true N. at Greenwich, *i.e.* when the magnetic pole is on the meridian of Greenwich, and on the opposite meridian (Fig. 182). The annual change would evidently be greatest when the magnetic pole is at or near to the points O and O_1 in the figure, and slowest when near the points W. and E., when the direction of the needle is a tangent to the small circle. At present the magnetic poles are coming towards the meridian of

FIG. 182.

Greenwich. The same remarks apply to other places. A line in the direction magnetic N. and S. at any place is called the magnetic meridian; hence variation may be defined as the angle between the true and magnetic meridians. An irregular line round the Earth, about midway between the poles, is the magnetic equator, which cuts the true geographical equator in two nearly opposite points about longs. 170° W. and 12° W.

Magnetic Dip.—A magnetized needle, poised at its centre so as to be free to move in a vertical plane, and provided with a scale of degrees, is called a dipping-needle (Fig. 183).

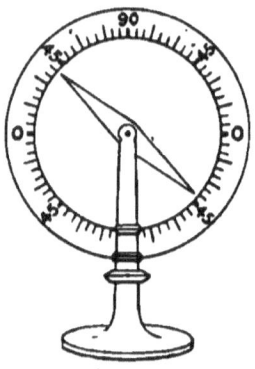

FIG. 183.

At the magnetic equator, and placed in the plane of the magnetic meridian (*i.e.* pointing N. and S. magnetic), such a needle would lie horizontal; but if carried northwards the red pole would dip more and more, until the magnetic pole is reached, where it would stand upright, red pole downwards. South of the magnetic equator the blue pole would dip, and the needle would again assume an upright position at the S. magnetic pole, blue pole downwards.

At Greenwich the dip at the present time is 68°, but in the West Atlantic and Gulf of St. Lawrence it is much greater. Like the variation, the dip at Greenwich is decreasing annually, and will be least in the year 1977, when the variation is 0.

These are the results of observation. The explanation is that in N. magnetic latitude the red pole of the needle points *direct*

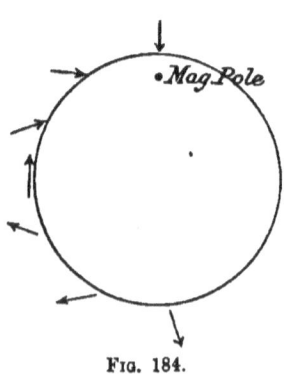

FIG. 184.

to the N. magnetic pole, and in S. magnetic latitude the blue pole to the S. magnetic pole of the Earth, notwithstanding the mass of the Earth intervening, whilst the horizontal plane at any place is a tangent to the Earth's surface (see Fig. 184).

The dip at any place decreases when the magnetic pole, in making its circuit round the true pole, is approaching the meridian of the place, and increases when it is receding. Thus in the Fig. 182, the dip at Greenwich decreases when the magnetic pole is moving from O_1 to O, and increases from

MAGNETISM AND DEVIATION OF THE COMPASS. 241

0 to O_1, being greatest when the pole is at O_1, and least when at 0. At present the dip at Greenwich is decreasing.

The position assumed by the dipping-needle may be considered the *natural* one for any magnetized needle which is free to move, but the horizontal position of a compass needle is a *forced* one. The reason why the compass card lies horizontal under all circumstances is that the point of support is above the centre of gravity of the card and needles, and the dragging down force is not sufficient to overcome their weight. It is evident from the above that an ordinary compass would be of no use for steering purposes near the magnetic poles, and is at its best on the magnetic equator.

Parallelogram of Forces.—Before considering the magnetism of iron ships and its effects on her compasses, it is necessary to understand the principle of the parallelogram of forces.

It is a law in mechanics that if two adjacent sides of a parallelogram represent two given forces in direction and amount, then the diagonal from the point of meeting of these two sides will represent the resultant (or equivalent force) both

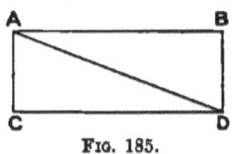

Fig. 185.

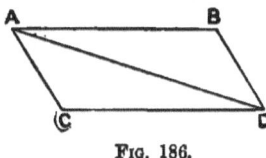

Fig. 186.

in direction and amount. Thus, if AB and AC in the parallelogram ABCD represent two given forces, then the diagonal AD will represent the joint effect or resultant. The sides of the triangle ABD represent equally well the forces and resultant.

A good illustration of this principle may be taken from the case of sailing in a current. For example, let the direction from A to B (Fig. 187) be the course steered, and the length of AB the rate of sailing per hour; also let AC represent the "set" of the current and "drift" in one hour.

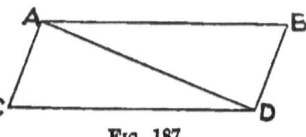

Fig. 187.

Draw BD parallel to AC, and CD parallel to AB, and join AD, then AD will be the course and distance made good, and represents the resultant or joint effect of wind or steam, and the current acting for the same length of time.

It is the same with any other kind of force, and it will be seen how the principle applies in dealing with the *magnetic* forces in a ship.

AB and AC are called components, and these can be found

R

from the resultant, as well as the resultant from the components. The finding of the components is called "resolution," and of the resultant "composition." Thus, a force represented by AD is "resolved" into components represented by AB and AC. This expression will be frequently used hereafter.

Magnetic Condition of Iron Ships.—(*a*) *Sub-permanent Magnetism.*—It has been stated that the hull of an iron ship becomes a magnet from the Earth's induction. The effect of lying for a considerable time in one direction, and the hammering to which the iron is submitted, is to fix the acquired magnetism so that it becomes sub-permanent, *i.e.* less than permanent. After launching there is some reduction of the strength of this sub-permanent magnetism, but it can often be traced throughout the whole existence of the ship, showing that a portion of it is almost of a permanent character.

If the hull be supposed to be divided into halves as she lies on the stocks by a plane at right angles to the line of dip (*q.v.*), the half towards the N. will have red polarity, and the other half blue.

Thus, if built in the United Kingdom, with head towards N., the line of dip being DP (Fig. 188) and EQ at right angles,

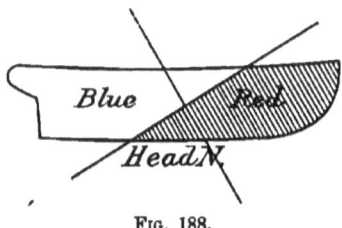

Fig. 188.

the ship would be an irregularly shaped magnet, red towards the bow, and blue towards the stern.

If built heading S., the red polarity would be at the stern, and blue at the bow.

If built heading E., the port side would be red, and the

Fig. 189.   Fig. 190.

starboard side blue (Fig. 189); but if built head W., the starboard side would be red, and the port side blue (Fig. 190).

MAGNETISM AND DEVIATION OF THE COMPASS. 243

The magnetic poles or points of greatest concentration of magnetic strength might be supposed to be about the centre of the two halves. So that in all these cases the line joining the poles would be inclined, one pole being nearer the keel than the other (unless built on the magnetic equator).

But it is only the horizontal component of the total force which causes deviation when the ship is upright. (The effect

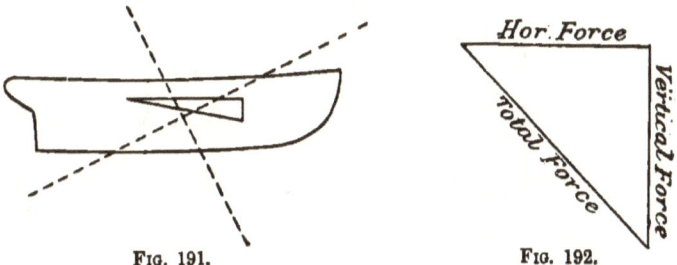

FIG. 191. FIG. 192.

of the vertical component will be considered later.) The parallelogram (or triangle) of forces will show the relation of the horizontal to the total force (Fig. 192).

It is clear that the greater the dip the less will be the horizontal component of the ship's magnetic force, and *vice versâ*. Hence, other things being equal, a ship built in a high latitude

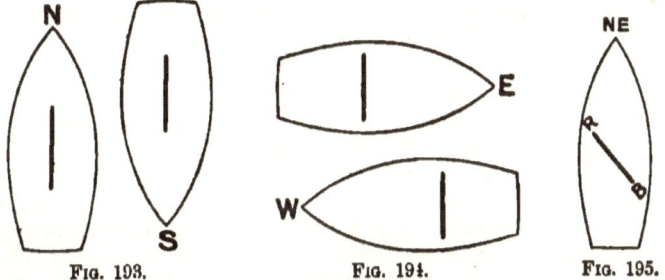

FIG. 193. FIG. 194. FIG. 195.

should cause less deviation from sub-permanent magnetism than one built in a low latitude, the ship being upright.

If a ship were built heading magnetic N. or S., the line representing the horizontal force of sub-permanent magnetism would be parallel to the keel (Fig. 193); if E. or W., it would be at right angles to the keel (Fig. 194); but if in any other direction it would lie obliquely (Fig. 195). In the latter case, the parallelogram of forces would again resolve the total horizontal force into two components—one represented by a line parallel to the keel, and the other at right angles to it.

Induction in Horizontal and Vertical Iron.—The effect of the Earth's induction on the horizontal and vertical iron of a ship may now be considered.

A bar of soft iron has the greatest magnetic strength from the Earth's induction when it lies in the line of dip, but is neutral when at right angles to it.

Thus, in N. magnetic latitude the end B of the horizontal

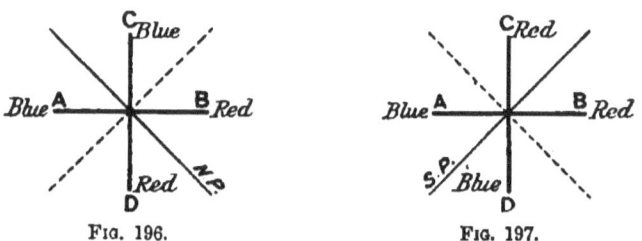

FIG. 196.   FIG. 197.

bar AB and the lower end of the vertical bar CD are red poles, because towards the N. dipping-pole (Fig. 196).

In S. magnetic latitude, the end A of the horizontal bar and the lower end of the vertical bar are blue, because towards the S. magnetic pole (Fig. 197).

A bar lying at right angles to the line of dip would be neutral.

This explains why the stern post, rudder post, and other vertical iron *below* the compass attract the N. point of the compass when north of the magnetic equator and repel it when south, whilst they have no effect when the ship is *on* the magnetic equator. Also, why the ends of horizontal beams attract and repel alternately as the ship's head turns from one quadrant to another, but are neutral when her head is N. or S. magnetic.

Another inference is that the magnetic strength of vertical iron in a ship *decreases* on going towards the magnetic equator, and *increases* towards the poles, whilst that of a horizontal beam lying N. and S. *increases* towards the equator, and *decreases* towards the poles.

Coefficients.—The term COEFFICIENTS is used to represent the components of magnetic forces in a ship. It may be assumed that these forces have a mechanical value, although a very small one, and could be measured by the same standard as other mechanical forces, but in reference to a ship's compasses they are estimated by the amount of deviation they produce when acting most effectively. This deviation itself is often called the coefficient.

The letters A, B, C, D, E name the coefficients, of which B, C, and D are the most important.

MAGNETISM AND DEVIATION OF THE COMPASS. 245

Coefficient B represents a magnetic force whose poles lie horizontally in the fore and aft midship line, and is a combination of the fore and aft component of sub-permanent magnetism and the induced magnetism of vertical iron before or abaft the compass. It is named +B if the combined effect is to attract the N. point of the compass towards the bow, and −B if towards the stern.

Coefficient C represents a force whose poles lie horizontally at right angles to the fore and aft midship line, and is the athwartship component of the sub-permanent magnetism only, for vertical iron is supposed to be evenly balanced towards the sides. It is named +C when the N. point of the compass is attracted to the starboard side, and −C when to the port side.

The horizontal force of sub-permanent magnetism is therefore represented by coefficients B and C, whose signs and relation to each other depend on the direction of ship's head during her construction, as illustrated in the following figures :—

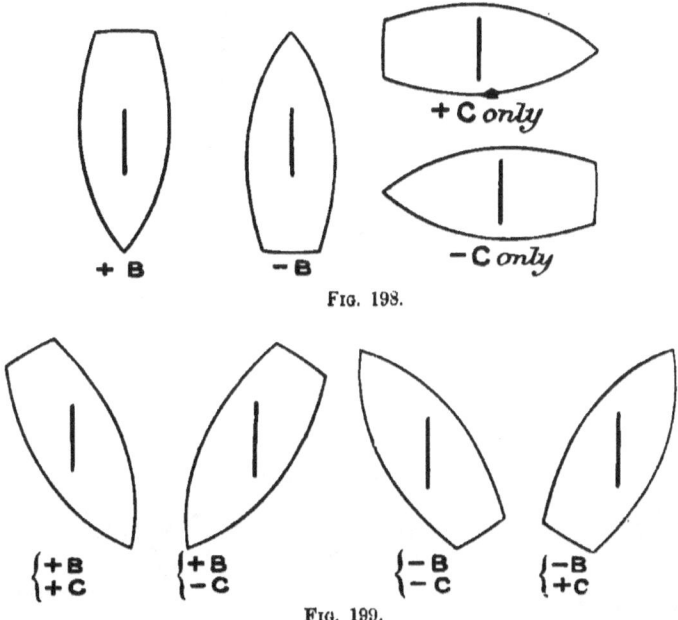

Fig. 198.

Fig. 199.

Semicircular Deviation.—The deviation due to coefficient B or coefficient C is called *semicircular*, because it is east on courses over one half of the compass, and west in the other half.

Coefficient +B (blue pole forward) causes E. deviation when

the ship's head is to the eastward, and W. deviation when to the westward; − B (blue pole aft) gives deviations of opposite name to the above. The maximum amount in both cases is found when the ship's head is E. or W. *by compass*.

Coefficient +C gives E. deviation when the ship's head is to the northward, and W. deviation when to the southward; − C gives the opposite, the maximum amount being found when heading N. or S. *by compass*. B causes *no* deviation when ship's head is N. or S. by compass, and C causes *no* deviation when her head is E. or W., because in these cases the forces act in a line with the compass needle.

The kind of deviation due to coefficients B and C is seen at a glance in the following figures, where + stands for E. deviation, and − for W. deviation.

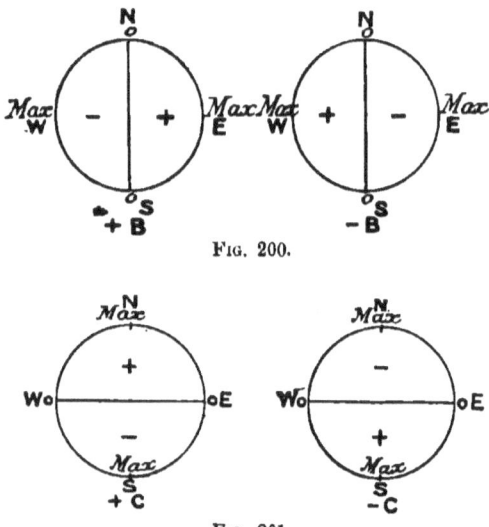

FIG. 200.

FIG. 201.

Transient Induced Magnetism in Ships.—The *transient induced* magnetism of horizontal iron is represented by coefficients D and E. Beams from side to side, and elongated iron lying fore and aft *end on* to the compass, give coefficient +D; athwartship beams *end on* and fore-and-aft iron extending beyond the compass both ways, gives − D.

Horizontal stringers from starboard quarter towards port bow give +E, and when extending from port quarter towards starboard bow they give − E. When end on to the compass, they give a coefficient of the opposite sign.

Quadrantal Deviation: Coefficients D and E.—The deviations

MAGNETISM AND DEVIATION OF THE COMPASS. 247

caused by coefficients D and E are called *quadrantal*, because they change their signs with change of course from quadrant to quadrant.

Thus, +D gives E. deviation between N. and E. and between

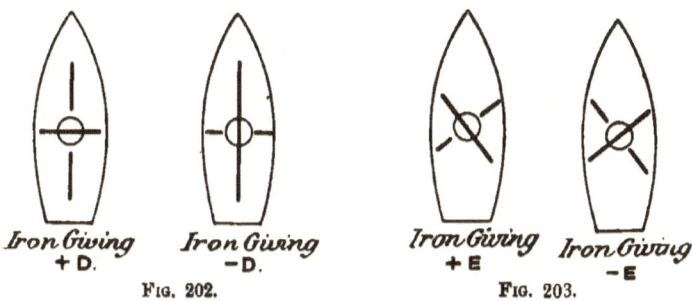

Fig. 202.   Fig. 203.

S. and W., but W. deviation between E. and S. and between W. and N. Also, +E gives E. deviation between N.W. and N.E. and between S.E. and S.W., but W. deviation between N.E. and S.E. and between S.W. and N.W. −D and −E give deviations of opposite sign.

It is an instructive exercise to prove the foregoing results by tracing the influence of one of these masses of horizontal iron on the compass as the ship's head is swung round.

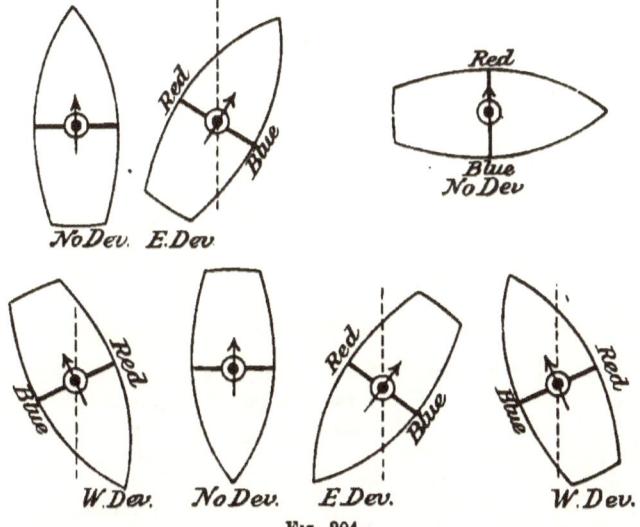

Fig. 204.

These figures show the magnetic polarity of a beam from side to side (giving coefficient +D) and the deviation produced in each quadrant. The zero points (*i.e.* points on which there is no deviation) are N., E., S., and W., and the maximum deviation is on courses halfway between these points, namely, at N.E., S.E., S.W., and N.W.

Similarly, it can be shown that the zero points for coefficient E. are N.E., S.E., S.W., and N.W., and the courses of maximum deviation N., E., S., and W.

The student should reason out for himself the effect of the various masses of horizontal iron as the ship is swung round the compass.

It is conceivable that the iron producing +D or +E might be exactly counterbalanced by iron causing −D or −E, but if not, then the balance on the side of the greater is the coefficient.

The following figures show the quadrantal deviations due to coefficients D and E, and the zero and maximum points. The sign + stands for E. deviation, and − for W. deviation.

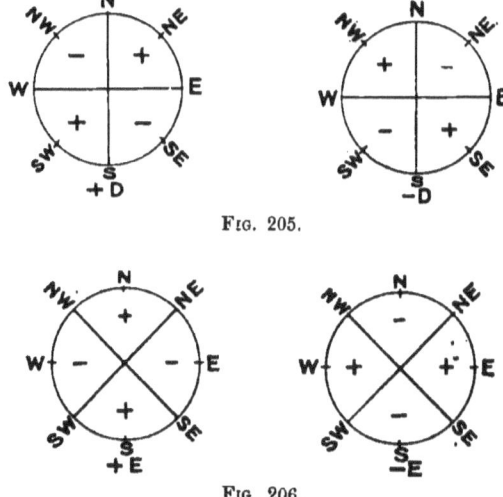

Fig. 205.

Fig. 206.

Coefficient A and its causes.—Coefficient A is of the nature of an index error, being of the same sign or name on all courses, and is often not due to magnetism at all. It may arise from any of the following causes:—

(*a*) The misplacing of the lubber line.

(*b*) The magnetic axis of the needles not coinciding in direction with the line through the N. and S. points of the card.

(c) An error in the correct magnetic bearing used in finding the deviations.

(d) Unsymmetrical arrangement of iron with respect to the compass.

(e) Retained magnetism during the process of "swinging" the ship.

The first three causes, if detected, could be removed, but the last two can scarcely be remedied. In any case the deviation from these causes is small, and is liable to be confounded with the deviation due to coefficient E, which is also very small.

The existence of a coefficient A is made evident in a table of deviations when the sum of the E. deviations is not equal to the sum of the W. deviations. When the former is greater, A is +; when the latter, it is −.

Analysis and Compensation.—The importance of making an analysis of the causes of deviation and of knowing how much is due to each is, that suitable methods of compensation may be employed, namely, to counterbalance the permanent magnetism by means of magnets, and the transient induced magnetism by iron correctors which themselves have the same kind of magnetism from the same source (the Earth), but act on the compass in the opposite sense.

Separation of Coefficients.—It is a fortunate circumstance that the effects of the various coefficients of forces can be sorted out from the observed deviations, as explained in the answer to Question 62, and a table can be made of the deviations due to each coefficient for every point of the compass (see answer to Question 62), which, being summed up algebraically, give the total deviation for each course.

The practical method of compensation is explained in answers to Question 40.

Compensation of Coefficient B.—The most troublesome coefficient to deal with is B, because it is a combination of sub-permanent and transient induced magnetism, as before explained. When on the magnetic equator, it would be easy to compensate the former with a magnet, and then, on proceeding north or south, the amount of change in the deviation when the ship's head is E. or W. could be compensated by a "Flinders bar."

But since B and C of sub-permanent magnetism stand to each other in the relationship of difference of Latitude and Departure, the ship's head whilst building being the course, it is possible to find this B if that direction is known. For this purpose the ship's head should be steadied on N. or S. by compass, and the deviation from coefficient C observed. Then, with the direction of ship's head whilst being built as a course, and C in departure column (in the Traverse Table), the required B will be found in the latitude column. This is the amount to be compensated by the fore-and-aft magnet, and the balance of observed B with the Flinders bar, either before or abaft the compass as required.

Effect of Magnetic Forces on Different Compasses.—It may be remarked that although the same coefficients of magnetic forces in a ship disturb both steering and standard compasses, yet the *measure* of the coefficients as shown by the deviations will not be the same for both, because the forces act at different distances. Consequently, observations and corrections must be made for each compass separately.

It is explained in the answer to Question 63 how this can be done for several compasses simultaneously.

Compensation good for all Latitudes.—If the compensations are properly made when a ship is ready for sea, they should stand good for all latitudes provided the sub-permanent magnetism does not change, for on this supposition the strength of the magnets and the coefficients of sub-permanent magnetism would always be in the same relation to each other, and no deviation would occur, although the directive power of the needle varies with change of latitude, whilst the horizontal force of transient induced magnetism and of the correctors would also remain in the same relation to each other.

Changes in Deviation.—But, however carefully a compass may be compensated, changes may be expected with the lapse of time, and as these changes will be chiefly in the deviation due to sub-permanent magnetism, a readjustment of the magnets will be necessary from time to time. The effect of the vibration caused by the propeller, the buffeting of heavy seas, collisions, or stranding is to considerably modify the magnetic condition of the ship when she left the builder's hands.

Retained Magnetism.—Another troublesome cause of change of deviation is what is called *retained magnetism;* that is, if a ship remains for a considerable time in one position in dock or on the same course at sea for several days, the induced magnetism from the earth is retained for a certain length of time, and causes a deviation when the course is altered. In sailor's language, the ship tends to follow her last course. For example, a steamer in crossing the Atlantic, bound for Liverpool, lays on an easterly course for several days together, and when on rounding Tuskar her course is changed to the northward, the retained red polarity on the port side (and blue on the starboard side) causes an E. deviation, which tends to set her to the eastward of her proper course. To this cause is owing, no doubt, the fact that steamers from the westward, coming up Channel in thick weather, are apt to be found too near the Welsh coast, and steamers from the south, bound up the English Channel, too near the English coast.

Uncompensated Compasses.—An uncompensated compass having large deviations on some courses and very small deviations on others is very unsatisfactory for steering purposes, for two reasons:—

(1) The deviation will be subject to greater changes, because when a ship approaches the magnetic equator the directive force of the needle increases, and therefore the coefficients of sub-permanent magnetism, which are assumed to remain nearly constant, will cause less deviation. On receding from the magnetic equator, the deviation would increase for the same reasons.

(2) The directive tendency of the needle would be much increased when a strong blue magnetism in the ship is towards the north from the compass, and correspondingly decreased when red is towards the north. The consequence would be that in some directions a considerable change of course would show only a small change in the compass course, and in other directions a small change of course would appear a large one by compass.

Horizontal Iron and Uncompensated Compass.—The deviation caused by horizontal iron would not be subject to these changes, because the horizontal component of transient induced force would vary in the same ratio as the directive force of the needle.

Diminution of Directive Force.—In an ordinary iron ship, the directive force is rarely more than six-sevenths of what it would be in a wooden ship or on shore, and in armoured war ships not more than three-fourths, according to Captain Creak, R.N., who has had very large experience of such matters. The chief cause of this is the horizontal iron, for in most cases it has a red polarity to the north of the compass. Sub-permanent magnetism weakens the directive force on *some* courses, but strengthens it on others, according as a red pole is north or south of the compass, and so the mean effect all round is *nil*. These remarks apply to uncompensated compasses.

The consequence of a diminution of directive force is *greater deviations*, for the forces which cause deviation are acting on a weakened compass.

The object aimed at by compensating a compass is to secure a uniform directive force on all courses, so that if any deviations remain they will be small in amount and regular in their changes.

Heeling Error.—So far it has been assumed that the ship is upright, but when she has a list from the pressure of the wind, or shifting of cargo, or unequal trimming of bunker coals, further deviations—sometimes very large, and known as heeling error—occur on northerly and southerly courses, but disappear on east and west courses.

The Causes of Heeling Error.—(*a*) It has been already explained that there is a *vertical* component of sub-permanent magnetism as well as a horizontal. The upper pole of this force, nearest to the steering compass aft, is blue if the ship is built head north (Fig. 207), and red if built head south (Fig. 208), and so long as the

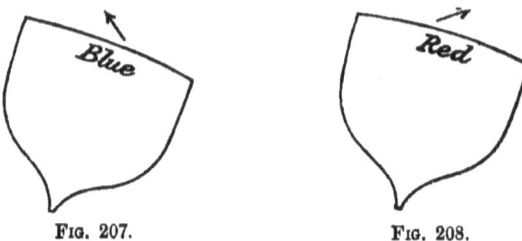

FIG. 207.   FIG. 208.

ship is upright, it only tends to *attract* the N. point of the needle *downwards* or *repel* it *upwards*, according as the upper pole is *blue* or *red*, without causing any deviation. The case is different when the ship heels, for then this upper pole is no longer in the vertical plane passing through the centre of the compass, but towards the higher side, deviating the N. point of the compass towards the higher or lower side. This effect is evidently greatest on N. and S. courses, and *nil* on E. and W. courses, because in the former case the force acts at right angles to the needle, and in the latter case in the plane of the magnetic meridian.

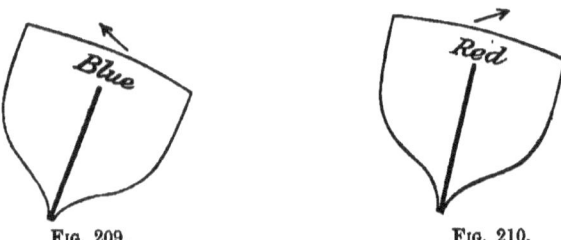

FIG. 209.   FIG. 210.

(*b*) The upper pole of vertical iron below the compass affects the compass in exactly the same way as the sub-permanent pole when the ship heels, only it must be remembered that vertical iron has the blue pole uppermost only in N. magnetic latitude, and red pole uppermost in S. magnetic latitude, being neutral on the magnetic equator. Therefore the heeling error from this cause decreases on sailing towards the magnetic equator and *vice versâ*.

(*c*) The athwartship iron, which gives coefficient D when the ship is upright, acquires a vertical component when she heels, which is of the same nature as the magnetism of vertical iron, and is subject to the same changes—the higher end being *blue* in N. magnetic latitude, and red in S. magnetic latitude.

It is evident that in some cases these three causes of heeling error would act together, and the heeling error would be large. In other cases the first would be opposed to the other two, and they would tend to neutralize each other, and give a small heeling error.

Their combined forces might be represented by a single force acting vertically under the centre of the compass, which would therefore be the resultant of—

(1) The vertical component of sub-permanent magnetism.
(2) The induced magnetism of vertical soft iron.
(3) The vertical component of the induced magnetism of transverse beams.

This resultant force is called + if the balance on the side of the greater is *blue uppermost*, and − if *red*. The total effect may be summed up by saying that a + vertical force causes a heeling error towards the *higher* side, and a − force a heeling error towards the *lower* side (Figs. 211 and 212).

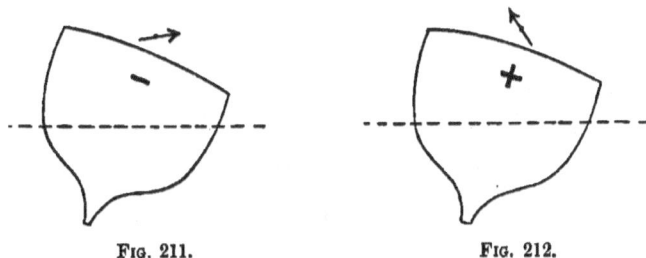

Fig. 211.   Fig. 212.

Experience shows that in the majority of sailing ships the force below the compass is + in N. latitude and − in S. latitude, with the result that N. of the magnetic equator the ship is drawn to windward on northerly courses, and to leeward on southerly courses. South of the magnetic equator the rule would generally be reversed. The same rules apply to steam ships, if higher side is substituted for windward, etc.

Navigators make a practical application of these facts by keeping "away" on northerly courses, and "closer" on southerly courses, when north of the magnetic equator, and the reverse when south of it.

Since the vertical force draws the N. point of the compass to one side or the other when the ship heels, it is called a coefficient C, although it would be more logical to distinguish it in some way from the coefficient C already discussed, due to horizontal force, say by using C_1 to represent it.[1]

Like coefficient C of horizontal sub-permanent force, the

[1] For a similar reason the part of coefficient B due to vertical iron before or abaft the compass might be distinguished as B_1.

new coefficient C_1 causes no deviation (heeling error) when the ship's head is E. or W. (magnetic), and the greatest deviation on N. and S. courses, because in the former case it acts in the plane of the magnetic meridian, and in the latter at right angles to the needle.

The heeling error is usually compensated by means of one or more magnets placed vertically under the compass, and capable of being moved up or down. There are two modes of procedure :—

(1) The ship's head being at N. or S. magnetic, she is caused to heel, say about 10°, and the error observed. Then the magnet is placed in position, and moved up or down, until the compass points correct magnetic N.

(2) The card having been removed, and the ship upright, with her head E. or W. magnetic, a dipping needle, weighted so as to lie horizontal when in the plane of the magnetic meridian on shore, is placed on the pillar in the centre of the bowl, and lying N. and S. Should the *red* end dip, showing a *blue* force below the compass, the *red* pole of the vertical magnet should be uppermost, and *vice versâ*, being moved up or down until the needle is again horizontal.

In either of these methods, what has been done is to counteract the force below the compass by a force of opposite name ; by observation of the heeling error in the first case, and in the second case by observing the vertical action of the force below the compass on the dipping needle.

In a compensated compass the vertical force of the athwartship iron (producing coefficient D) is counterbalanced for moderate angles of heeling by a vertical force developed in the correctors; which leaves only the effects of vertical iron and the sub-permanent magnetism to be dealt with.

The Flinders bar increases the heeling error, for its upper pole is of the same name as that of the vertical iron it is intended to counteract when the ship is upright. Therefore the compensating magnet has to neutralize the combined vertical forces of sub-permanent magnetism, vertical iron, and Flinders bar. But as the two latter change with change of latitude, it is evident that this mode of compensation is effective only so long as the ship remains in the same latitude, and constant shifting of the magnet would be necessary as the ship proceeds north or south, and probably a reversal of the poles on the other side of the magnetic equator.

The only efficient compensation would be to use the magnet for the part due to sub-permanent magnetism only, and a vertical soft iron bar, having its lower end *above* the compass, to counteract the vertical force of soft iron *below* the compass. Some arrangement might possibly be made for the purpose in the construction of the binnacle.

It will now be understood that the principle of compensation is to counteract the existing magnetic forces in a ship by employing counter equal forces of the same kind, to act on the compass needles.

The order of procedure recommended in the "Elementary Manual" for ships of the Royal Navy, where the conditions are pretty constant, is—
1st. Coefficient D.
2nd. Heeling error.
3rd. Coefficients B and C.

But in the Merchant Service, where cargoes change every trip, and where the compensation is often hurriedly made when ready for sea, the "Tentative Method" of adjustment is usually carried out in the following order:—
1. Coefficient C (athwartship magnet).
2. Coefficient B. (1) Flinders bar, and (2) fore and aft magnet.
3. Coefficient D (soft iron correctors).
4. Heeling error (vertical magnet).

Coefficient E is rarely compensated, although a counter E could be introduced by moving one of the D correctors forward, and the other a little aft. For + E the starboard corrector should be moved aft, and the port one as much forward; for − E, the reverse.

A table of the remaining deviations is obtained by "swinging" the ship in dock or other convenient place.

Yet, however accurately the compensation may be made, changes in the deviation, from causes already explained, will occur during a voyage, and no careful navigator will place much reliance on "deviation cards." In fact, nothing can relieve him from the obligation to find the deviation from frequent observations, which is now happily made easy by Azimuth Tables.

CHAPTER XXIX.

SYLLABUS.

BOARD OF TRADE QUESTIONS ON DEVIATION OF THE COMPASS FOR ORDINARY MASTERS.

The applicant is to answer correctly at least 12 of such of the following questions as are marked with a cross by the examiner.

The examiner's attention is specially called to the importance of questions 11, 12, 13, 14, and 39, which must be marked in all cases.

1. State briefly the essentials of an efficient compass.
2. State briefly the chief points to be considered when selecting a position for your compass on board ship, and what should be particularly guarded against.
3. What do you mean by deviation of the compass, and how is it caused?
4. Describe how you would determine the deviation of your compass: (1) by reciprocal bearings; (2) by figures on the dock walls; (3) by bearings of a distant object; (4) by the bearings of the sun or other celestial body.
5. Having determined the deviation with the ship's head on the various points of the compass, how do you know when it is easterly and when westerly?
6. Why is it necessary, in order to ascertain the deviations, to bring the ship's head in more than one direction?
7. For accuracy, what is the least number of points to which the ship's head should be brought for constructing a curve or table of deviations?
8. How would you find the deviation when sailing along a well-known coast?
9. Name some suitable objects by which you could readily obtain the deviation of the compass when sailing along the coasts of the Channel you have been accustomed to use.
10. Supposing you have no means of ascertaining the magnetic bearing of a distant object when swinging your ship for deviations, how could you find it approximately from equidistant compass bearings; and at what distance, as a rule, should the object be from the ship?
11. Having taken the following compass bearings of a distant object, find the object's magnetic bearing, and thence the deviation.

Magnetic bearing required :—

Ship's head by standard compass.	Bearing of distant object by standard compass.	Deviation required.	Ship's head by standard compass.	Bearing of distant object by standard compass.	Deviation required.
N.	S. 4° E.		S.	S. 13° E.	
N.E.	S.		S.W.	S. 23° E.	
E.	S. 4° W.		W.	S. 21° E.	
S.E.	S. 1° W.		N.W.	S. 11° E.	

12. With the deviation as above, construct a curve of deviations on a Napier's diagram, and give the courses you would steer by the standard compass to make the following courses, correct magnetic.

Magnetic courses : S.S.W. W.N.W. N.N.E. E.N.E. S.S.E.
Compass courses required.

13. Supposing you have steered the following courses by the standard compass, find the correct magnetic courses made from the above curve of deviations.

Compass courses : W.S.W. N.N.W. E.N.E. S.S.E.
Magnetic courses required.

14. You have taken the following bearings of two distant objects by your standard compass as above ; with the ship's head at W. ½ S., find the bearings, correct magnetic.

Compass bearings : W. by S. and N. ¾ W.
Magnetic bearings required.

15. Do you expect the deviation to change ? If so, state under what circumstances.
16. How often is it desirable to test the accuracy of your table of deviations ?
17. What is meant by variation of the compass ? What is it caused by ? And where can you find the variation for any given position ?
18. The earth being regarded as a magnet, which is usually termed the blue, and which the red magnetic pole ?
19. Which end of a magnet (or compass needle) is usually termed the red or "marked" end, and which the blue ?
20. What effect has the pole of one magnet of either name on the poles of another magnet ?
21. What is meant by transient induced magnetism ?
22. Which is the red and which is the blue pole of a mass of soft vertical iron, by induction, and what effect would the upper and lower ends of it have on the compass needle (a) in the northern hemisphere, (b) in the southern hemisphere, (c) on the magnetic equator ?
23. Describe what is usually termed the sub-permanent[1] magnetism of an iron ship, and state when and how it is acquired, and which is the red and which is the blue pole, and why it is called sub-permanent magnetism ?
24. Describe the meaning of the expression " coefficient A."
25. Describe the meaning of the expression " coefficient B," its signs and effects.
26. Describe the meaning of the expression " coefficient C," its signs and effects.
27. Describe the meaning of the expression " coefficient D," its signs and effects.
28. Describe the meaning of the expression " coefficient E," its signs and effects.
29. Would you expect any change to be caused in the error of your compass by the ship heeling over either from the effect of the wind or the cargo, etc.?
30. The compasses of iron ships being more or less affected by what is

---

[1] The term sub-permanent magnetism in these questions is used in the original sense as proposed by the late Sir G. B. Airy, to denote the character of the permanent magnetism of an iron ship as distinguished from the permanent magnetism of a magnetized steel bar. The terms "sub-permanent" and "permanent" throughout these questions may, therefore, be considered as synonymous.

termed the heeling error, on what courses is this error usually at its minimum, and on what courses at its maximum?

31. Describe clearly the three principal causes of the heeling error on board ship.

32. State to which side of the ship in the majority of cases is the north point of the compass drawn when ship heels over in the northern hemisphere.

33. Under what conditions (that is, as regards position of ship whilst building, and the arrangement of iron in the ship) is the north point of the compass needle usually drawn to windward or the high side of the ship in the northern hemisphere; and if not allowed for, what effect has it on the assumed position of the ship when she is steering on northerly, also on southerly courses, in the northern hemisphere?

34. Under what conditions (as in Question 33) is the north point of the compass needle usually drawn to leeward or the low side of the ship in the northern hemisphere, and, if not allowed for, what effect would it have on the assumed position of the ship, when she is steering on northerly, also on southerly courses, in the northern hemisphere?

35. The effects being as you state, on what courses would you keep away, and on what courses would you keep closer to the wind in the northern hemisphere in order to make good a given compass course (a) when north point of compass is drawn to windward or the high side of ship; and (b) when drawn to leeward or the low side?

36. Does the same rule hold good in both hemispheres with regard to the heeling error?

37. State clearly how that part of the heeling error due to the permanent part of the magnetism of the ship varies as the ship changes her position on the globe, and what is the reason of this?

38. State clearly how that part of the heeling error due to the induction in transverse iron (which was horizontal when ship was upright) and iron vertical to the ship's deck, varies as the ship changes her position on the globe.

39. Your steering compass having a large error, show by "Beall's Compass Deviascope" how you would correct it by compensating magnets and soft iron (as usually practised by compass adjusters in the Mercantile Marine) in order to reduce the error within manageable limits. Show also how the heeling error can be compensated.

40. As the coefficient B (capable of being corrected) usually consists of two parts, one due to the permanent magnetism of the ship, and the other to vertical induction in soft iron, how should each of the two parts, strictly speaking, be corrected when compensating the compass?

41. If the whole of coefficient B be corrected by a permanent magnet, as is usually done, what is likely to ensue as the ship changes her magnetic latitude?

42. Provided the needles of your compass are not so long and powerful, and so near, as to cause the soft iron correctors to become magnetized by induction, would the coefficient D if properly compensated be likely to remain so in all magnetic latitudes and both hemispheres? If so, state the reason why.

43. State at what distance, as a general rule, the magnets and soft iron correctors should be placed from the compass needles, and what will be the consequence if they are placed too near the needles.

44. Is it necessary that the magnets used for compensating coefficients B and C should be placed on the deck? If not, state where they may also be placed, and the rules to be observed in placing them into position.

45. Can the compensation of the heeling error be depended upon when the ship changes her latitude? If not, state the reason.

ANSWERS TO THE QUESTIONS ON DEVIATION OF THE COMPASS.

1. The card should be light, accurately centred and divided, and the needles well magnetized, the magnetic axis coinciding with the N. and S. points of the card. The cap should be sound, and the pivot sharp and smooth. The point of the pivot should be level with the gimbals. The bowl should be of copper, and the mounting should not allow of any lateral play.

2. The standard compass should be in the midship line, and should afford a clear view all round for taking bearings, and not less than five feet from iron of any kind, especially vertical iron and any iron which is likely to change its position, such as derricks, cowls, davits, etc. Also as far as possible from a dynamo, if electric lighting is used. Steering compasses should be so placed that the card can be well seen by the steersman, and should be as far removed as possible from the influence of iron.

3. Deviation is the amount of deflection of the N. point of a compass from the correct magnetic N. caused by the magnetic influence of iron in the ship, whether in hull, equipment, or cargo.

4. (1) By reciprocal bearings : An observer is stationed at a compass on shore in a position free from local attraction, and another observer at the standard compass on board. Then, as the ship is swung round, bearings are taken of each compass by the other observer at the same time, whilst the ship's head is steadied on the required points. The difference between the bearing observed on board and the reversed shore bearing is the deviation. (2) By figures on the dock wall : These figures at Liverpool show the true bearings of the Vauxhall Chimney from the river, so that the figure in a line with the chimney as seen from the ship shows its true bearing from the observer's position. The difference between this and the bearing by the standard compass is the total error, from which the variation may be separated, and the result is the deviation. (3) By bearings of a distant object : When the correct magnetic bearing of a distant object is known, the difference between the bearing of it by compass on board, and the correct magnetic is the deviation. (4) By the bearings of a celestial object : The true azimuth of the object being found for a given moment, either by calculation or by tables, the bearing of the object at the same time is taken by compass; the difference shows the error of compass for the actual position of ship's head, and the variation being known, the deviation may be found.

5. If the correct magnetic bearing reads to the *right* of the compass bearing, the deviation is *east;* if to the *left*, it is *west.*

6. Because the deviation changes with every change of course.

7. Eight equi-distant points, viz. N., N.E., E., S.E., S., S.W., W., N.W.

8. By comparing the bearing of two known objects when in one (or in line) with the known correct magnetic direction of one from the other obtained from a chart.

9. (*a*) English Channel : Bolt Head and Start Lighthouse, Prawle Point and Start Point, etc. (*b*) Irish Channel : Skerries and South Stack, Smalls and Bishop Lights, Wolf and Longship. (*c*) North Sea : South Foreland Lights, Whitby Lights, and Tyne Harbour Lights.

10. The ship's head should be steadied on eight equi-distant points in succession, and the bearing of a distant object observed when on each point. The mean of these may be taken as the correct magnetic. The object should be seven or eight miles distant, so that its bearing would not be sensibly affected by the swinging of the ship.

11, 12, 13, and 14 are practical.

15. Yes, the deviation may be expected to change : (*a*) From lapse of time; (*b*) By keeping the ship's head for a considerable time in an opposite

direction to that in which she was built; (c) By change of latitude; (d) By shocks or strains from heavy seas; (e) By being kept for a long time on one course, magnetism is induced which is retained on a change of course.

16. The deviation table should be tested at every opportunity.

17. Variation is the difference in direction between the magnetic north and true north, and is due to the fact that the earth's magnetic poles do not coincide with its geographical poles. The amount, together with the annual change, may be found on a magnetic chart.

18. The earth's north magnetic pole is the blue pole, and the south the red.

19. The end of a magnetized needle, freely suspended, which points north is the red pole, and the other the blue pole.

20. Poles of the same name repel each other, and poles of different names attract each other.

21. Transient induced magnetism is the magnetic condition of soft iron caused by the influence of a magnetic force in its neighbourhood or that of the earth itself.

22. North of the magnetic equator the upper end is the blue pole and the lower the red, but south of the magnetic equator this rule is reversed. (a) In the northern hemisphere the upper end would attract the north end of the compass needle, and the lower end repel it. (b) In the southern hemisphere the upper end would repel and the lower attract the north end of the needle. (c) On the magnetic equator the effect would be *nil*.

23. The sub-permanent magnetism of an iron ship is the magnetic condition of a more or less enduring character which she has when she is launched. It is acquired in the course of building from the earth's inductive influence, aided by the vibration due to hammering, riveting, etc. The red pole is that which was towards the north, and the blue towards the south when the ship was on the stocks. It is called sub-permanent magnetism to distinguish it from the magnetism of a tempered steel bar, and because it is liable to be weakened or modified by circumstances and by lapse of time.

24. Coefficient A is a deviation of the same amount and sign on all courses, usually due to a faulty or badly-placed compass.

25. Coefficient B represents a horizontal magnetic force, whose poles are in the fore and aft midship line. It is called +B when the blue pole is towards the bow, and −B when towards the stern. +B causes E. deviation on easterly courses, and W. deviation on westerly courses. −B causes W. deviation on easterly courses, and E. deviation on westerly courses. The effect is *nil* on N. and S. courses, and greatest on E. and W. courses.

26. Coefficient C represents a horizontal magnetic force, whose poles are in a line at right angles to the fore and aft midship line. It is called +C when the blue pole is towards the starboard side, and −C when towards the port side. +C causes E. deviation on northerly courses, and W. deviation on southerly courses. −C causes W. deviation on northerly courses, and E. deviation on southerly courses. The effect is *nil* on E. and W. courses, and greatest on N. and S. courses.

27. Coefficient D represents the effect of transient induced magnetism in horizontal iron in fore and aft and athwartship directions. The deviation due to this cause changes its name in successive quadrants, and is thence called quadrantal. +D causes E. deviation from N. to E. and S. to W. courses, and W. deviation from S. to E. and N. to W. −D gives W. deviation from N. to E. and S. to W. courses, and E. deviation from S. to E. and N. to W. The deviation due to this cause is *nil* on the cardinal points and greatest on N.E., S.E., S.W. and N.W.

28. Coefficient E represents the effect on the compasses of induced magnetism in horizontal iron lying in a direction oblique to the fore and aft line. The deviation due to this cause is quadrantal, and is *nil* on N.E., S.E., S.W., and N.W., and greatest on the cardinal points. +E gives E. deviation

from N.W. to N.E. and S.E. to S.W., and W. deviation from N.E. to S.E. and S.W. to N.W. —E gives W. deviation from N.W. to N.E. and S.E. to S.W., and E. deviation from N.E. to S.E. and S.W. to N.W.

29. Yes, because when the ship heels, horizontal athwartship iron acquires a vertical component, and besides, the upper poles of vertical iron and the poles of sub-permanent magnetism will then not be in the same vertical plane with the centre of the compass.

30. The heeling error is usually greatest on N. and S. courses, and least on E. and W. courses, because in the former case the force causing it acts at right angles to the compass needle, and in the latter case coincides with the direction of the needle.

31. The three principal causes of heeling error are: (1) Traverse iron acquires a vertical component. (2) The upper induced poles of vertical iron will not be in the same vertical plane with the centre of the compass. (3) The poles of the fore and aft component of the ship's sub-permanent magnetism fall to the right or left of the vertical plane, through the centre of the compass.

32. In the N. hemisphere the N. point is usually drawn towards the higher side.

33. In ships built with head towards the north, the N. point of a steering compass aft will generally be attracted towards the higher side, and, if not allowed for, will draw the ship to windward on northerly courses, and to leeward on southerly courses.

34. In ships built with head toward south, the effect of the subpermanent magnetism would be to draw the N. point of the compass to leeward, and if this exceeded the influence of the vertical and athwartship soft iron, the N. point of the compass would be drawn towards the lower side. If not allowed for, it would take the ship to leeward on northerly courses, and to windward on southerly courses.

35. (a) When the N. point is drawn to windward, I would keep away on northerly courses, and closer to the wind on southerly courses. (b) When the N. point is drawn to leeward, I would keep closer to the wind on northerly courses, and keep away on southerly courses.

36. The rule for the northern hemisphere is generally reversed in the southern, especially in ships built heading south, for the poles of vertical induction are reversed, but in ships built heading north, the same rule may still obtain, that is, when the sub-permanent vertical force exceeds the induced.

37. It *decreases* on approaching the magnetic equator, and *increases* on receding from it, because the directive force of the compass changes as the magnetic latitude changes, being greatest at the magnetic equator and *nil* at the magnetic poles, whilst the sub-permanent magnetism of the ship is nearly constant.

38. The induced vertical force becomes weaker on approaching the magnetic equator, is *nil* at the equator and the poles are reversed on crossing it. Therefore the part of the heeling error due to this cause, drawing the N. point to windward in north mag. lat., disappears at the magnetic equator, and repels the N. point to leeward in S. mag. lat.

39. Practical.

40. The part of co-efficient B due to the sub-permanent magnetism should be compensated by means of a magnet, and the part due to vertical induction in soft iron by an upright bar of soft iron before or abaft the compass in the midship line.

41. It will be found to be over-compensated or under-compensated as the ship changes her mag. lat., for the part caused by vertical induction changes with change of latitude.

42. Yes, because the horizontal inductive force of the correctors will vary in the same ratio as that of the horizontal iron which produces co-efficient D.

43. The magnets used for compensating should not be nearer the centre of compass than twice their own length, and the soft iron correctors should not be nearer than 1¼ times the length of the needles. If the magnets are too near, the compensation which was made with the ship's head on the cardinal points would not be perfect on other points. If the soft iron correctors are too near, the poles of the compass needles would have a tendency to induce opposite polarity in the correctors.

44. The magnets used for compensating B and C need not be on the deck, but may be placed in any convenient position near the compass, in or outside the binnacle, provided the following conditions are complied with : (1) The ship should be upright and on even keel. (2) The fore and aft magnets should be parallel to the deck, and bisected by a vertical plane athwartships, passing through the centre of the compass. (3) The athwartship magnets should be parallel to the deck, and placed so as to be bisected by a fore and aft vertical plane passing through the centre of the compass.

45. No, because a part of the heeling error is due to vertical induction in soft iron, which varies in force as the latitude changes. This part of the heeling error can not be compensated by means of a permanent magnet fixed in position.

QUESTIONS ON DEVIATION OF THE COMPASS FOR MASTERS (EXTRA).

(From the Board of Trade " Syllabus of Examination.")

1. Describe an artificial magnet, and how a steel bar or needle is usually magnetized.

2. Which end of the compass needle, or a magnet, is commonly termed the red, and which the blue pole ?

3. Which is the red magnetic pole of the earth, and which the blue ? and give their geographical positions.

4. What effect has the pole of one magnet of either name on the pole of the same name of another magnet ? and what would be the consequence of the pole of one magnet of either name being brought near enough to affect the pole of contrary name, if in these cases both magnets were freely suspended ?

5. By applying this law to all magnets, natural as well as artificial, describe what would be the result on a magnetic bar or needle, freely suspended, not by weight or by the nature of its mounting constrained to preserve a horizontal position ; and what would be the result, if so mounted, but free to move in every direction, the earth being regarded as a natural magnet.

6. What is the cause of the variation of the compass ?

7. What is meant by the deviation of the compass ?

8. What is meant by the term "local attraction" ? under what circumstances have ships' compasses, from recent careful investigation, been found to be affected by it ? and name some of the localities in different parts of the world where this disturbance is to be found, and consequently where increased vigilance is necessary (see Appendix, " Evans' Elementary Manual," 1888).

9. What do you understand by the term "soft" iron ; and what are its properties as regards acquiring and retaining magnetism ?

10. What do you understand by the term "hard" iron ; and what are its properties as regards acquiring and retaining magnetism ?

11. Describe the meaning of the term "horizontal force" of the earth ; where is it the greatest, and where the least, and what effect has it in respect to the increase or decrease of the directive force of the compass needle ?

12. Does the magnetic equator coincide with the geographical equator ? if not, state clearly how it is situated.

13. Where can the values of the magnetic dip, the earth's horizontal force, and the variation, be found?

14. State in what parts of the globe lying in the usual tracks of navigation the variation changes very rapidly, and what special precautions should be observed when navigating these localities; also why a "variation" chart is then very useful.

15. Why is a knowledge of the magnetic dip and the earth's horizontal force important in dealing with compass deviations?

16. Describe the meaning of the term "vertical force" of the earth; where is it the greatest and where the least?

17. Would you expect a compass to be the more seriously affected by any given disturbing force when near the magnetic equator, or near the poles? and state the reason.

18. State briefly (a) the essentials of an efficient compass; and (b) what you would consider a good arrangement of the needles (that is, whether long or short, single or double, etc.) with the view to good compensation.

19. In stowing away spare compass cards or magnets, how would you place them with regard to each other, or what might be the probable consequence?

20. State briefly the chief points to be considered when selecting a position for your compass on board ship, and what should be particularly guarded against.

21. What is meant by transient induced magnetism?

22. Which is the *red* and which the *blue* pole of a mass of soft vertical iron (or indeed of any soft iron not in a horizontal position) by induction, and what effect would the upper and lower ends of it have on a compass needle in the *Northern* hemisphere?

23. Which is the *red* and which the *blue* pole of a mass of soft vertical iron by induction, and what effect would the upper and lower ends of it have on the compass needle in the *Southern* hemisphere?

24. What effect would a bar of soft vertical iron have on the compass needle on the magnetic equator?

25. Describe what is usually termed the sub-permanent[1] magnetism of an iron ship, and state when and how it is acquired, and which is the sub-permanent red and which is the blue pole, and why it is called sub-permanent magnetism.

26. What is meant by "the composition of forces" and "the parallelogram of forces?" and show how the knowledge of these is valuable in ascertaining and compensating the sub-permanent magnetism of an iron ship.

27. Describe the nature of the coefficients B and C, plus (+) and minus (−), and the different magnetic forces they represent; also why they are said to produce semicircular deviations.

28. Can semicircular deviations be produced by any other force than the sub-permanent magnetism of the ship? If so, by what?

29. On what points, by compass bearings of the ship's head, does +B give westerly deviation, and on what points does it give easterly; also on what points does −B give westerly, and on what points easterly?

30. On what points does +C give westerly deviation, and on what points easterly; also on what points does −C give westerly, and on what points easterly deviation?

31. The value of either coefficient B or C being given, also the magnetic direction of the ship's head while she was being built, determine by the traverse tables the approximate value of the other coefficient C or B; and,

---

[1] See footnote, p. 257

the value of both these coefficients being given, determine approximately the direction by compass of the ship's head whilst being built, assuming, of course, that these coefficients resulted altogether from sub-permanent magnetism.

32. Would you expect the greatest disturbance of the needle from the effects of sub-permanent magnetism alone to take place when ship's head is in same direction as when building, or when her head is at right angles to that direction, and in what direction of the ship's head would you expect to find the least disturbance?

33. Describe quadrantal deviation, and state what coefficients represent it; also on what points of the ship's head, by compass, each of these coefficients gives the greatest amount of deviation, and why it is called quadrantal deviation.

34. On what points of the compass will each of the coefficients D and E, + and −, give easterly, and on what points westerly, deviation?

35. What conditions of the iron of a ship will produce +D and what −D?

36. State clearly, then, which end of the horizontal iron running athwartship (such as beams, etc.), and of horizontal iron running fore and aft of a ship, acquires *red* and which *blue* polarity, by induction, when ship's head is at N.E., S.E., S.W., and N.W. respectively.

37. Describe the nature of the deviation represented by coefficients +A and −A, and describe the errors in the construction of the compass, and other causes, that frequently produce it.

38. What is the object of compensating the compass by magnets, etc., and what are the general advantages of a compensated compass over an uncompensated one?

39. Before adjusting the compass of an iron ship, what is it desirable to do with the view to eliminating, as far as possible, what may be termed the unstable part of the magnetism of the ship?

40. Describe clearly the tentative method of compass adjustment (that is, the compensation of coefficients B, C, and D, with ship upright) as generally practised by compass adjusters in ships of the mercantile marine.

41. State at what distance, as a general rule, the magnets and soft iron correctors should be placed from the compass needles, and what will be the consequence if they are placed too near the needles.

42. Is it necessary that the magnets used for compensating co-efficients B and C should be placed on the deck? If not, state where they may also be placed, and the rules to be observed in placing them into position.

43. Does the B found on board ship usually arise altogether from sub-permanent magnetism, or does part of it usually arise from some other cause or causes?

44. If the part of B due to induced magnetism in vertical soft iron, as well as the part due to sub-permanent magnetism, are corrected by a magnet alone, as is generally the case, what is frequently the consequence on the ship changing her magnetic latitude and hemisphere?

45. How should each of these two parts of B then, strictly speaking, be compensated?

46. Assuming, for the sake of clearness, that your steering compass is unavoidably placed very near to the head of the stern-post (and other vertical iron at the stern), thereby causing a very large − B from induced magnetism, describe briefly any method by which the approximate position for the com-pensating vertical iron bar (Flinders' or Rundell's) could be estimated in order to reduce the error; describe also how you would proceed, in order to improve, if not to perfect, its position after observations have been made on the magnetic equator.

47. State if standard compasses, as well as steering compasses, are

generally subject to this disturbance from induced magnetism in vertical iron; also whether the attraction in all cases is found to be towards the stern; and if not, state the conditions under which it might be toward the bow, and how the compensating soft iron bar should then be placed.

48. Generally speaking, does the magnetism induced in vertical iron usually have any effect in producing the coefficient C, ship upright, or is it generally produced by sub-permanent magnetism alone? State also your reasons for saying so.

49. Provided the needles of your compass are not so long and powerful, and so near, as to cause the soft iron correctors to become magnetized by induction, would the coefficient D, if properly compensated as you have described (Ans. 40), be likely to remain so in all latitudes and both hemispheres? If so, state the reason why.

50. Under what circumstances does the character of A and E so change as to render it desirable that these coefficients should be disregarded or modified?

51. Supposing your compasses were allowed to remain uncompensated, explain clearly what would be the probable changes (ship upright) in the deviations produced separately by (1) the sub-permanent magnetism of the ship alone, (2) by the induced magnetism in vertical soft iron, (a) on reaching the equator, (b) in the southern hemisphere.

52. Assuming you were able to arrive at the proper proportions to be corrected, and were then to exactly compensate the sub-permanent magnetism of the ship by means of a permanent magnet, and the induced magnetism in vertical iron by a soft iron bar, would you expect any deviation to take place in your compass as the ship changed her latitude and hemisphere? And state your reasons for saying so.

53. Supposing the coefficient D from horizontal soft iron were allowed to remain uncompensated, would you, or would you not, expect the D to differ in name or amount on the ship changing her magnetic latitude, and hemisphere? And state the reason.

54. Describe how you would determine the deviation of your compass, (1) by reciprocal bearings, (2) by figures on the dock walls, (3) by bearings of a distant object.

55. Describe, in detail, how you would determine the deviation of your compass by the bearings of the Sun. Also by a star or planet.

56. Describe the uses to which the Napier's diagram can be applied, and its special advantages.

57. Describe clearly how the Napier's diagram is constructed.

58. For accuracy, what is the least number of points to which the ship's head should be brought for constructing a complete curve of deviations, or a complete table of deviations?

59. Nearing land, and being anxious to check your deviations on a few courses you may probably require to steer, what is the least number of points it would be necessary to steady the ship's head upon, if making use of a Napier's diagram, in order to ascertain the deviation on each of the points, say, in a quadrant of the compass? and describe clearly how you would do this at sea.

60. Supposing you have no means of ascertaining the magnetic bearing of the distant object when swinging your ship for deviations, how could you find it, approximately, from equi-distant compass bearings; and how far, as a rule, should the object be from the ship when swinging, or steaming round?

61. Having taken the following equi-distant compass bearings of a distant object, find the object's magnetic bearings, and thence the deviations:—
(a) Magnetic bearing required—

Ship's head by standard compass.	Bearing of distant object by standard compass.	Deviation.	Ship's head by standard compass.	Bearing of distant object by standard compass.	Deviation.
N.	S. 75° W.		S.	S. 34° W.	
N.E.	S. 64° W.		S.W.	S. 31° W.	
E.	S. 56° W.		W.	S. 49° W.	
S.E.	S. 50° W.		N.W.	S. 71° W.	

(b) Construct a curve of deviations on a Napier's diagram, with the deviations as above, and give the courses you would steer by the standard compass to make the following courses, correct magnetic.
Magnetic courses : N.N.W.   S.S.E.   W.N.W.   E.S.E.
Compass courses required.

(c) Supposing you have steered the following courses by the standard compass, find the correct magnetic courses made from the above curve of deviations.
Compass courses : N.N.E.   E.N.E.   S.S.W.   W.S.W.
Magnetic courses required.

(d) You have taken the following bearings of two distant objects by your standard compass as above; with the ship's head at N.E. ¼ E., find the bearings, correct magnetic.
Compass bearings : S.E. by S.   and   N.N.W.
Magnetic bearings required.

62. Assuming the deviations observed with the ship's head by compass to be as follow (or as in question 61, whichever may be given), determine the value of the coefficients A, B, C, D, and E, and from them construct a complete table of deviations (or for as many points as the Examiner may direct).

Deviation at N.   (21° W.)   S.   (20° E.)
N.E. (10° W.)   S.W. (23° E.)
E.   (2° W.)   W.   (5° E.)
S.E. (7° E.)   N.W. (17° W.)

63. When swinging your ship, if it be required to construct Deviation Tables for two or more compasses situated in different parts of the vessel, describe the process, and how you would employ the Napier's diagram for this purpose.

64. State your rule for determining whether deviation is easterly or westerly.

65. Is a knowledge of the value of the various coefficients of any advantage? If so, state why?

66. Describe (a) what is commonly known by the term "retentive" or "retained" magnetism, and how the ship acquires it when in port and at sea; (b) its effect on the compass needle whilst ship's head continues in the same direction; (c) the immediate consequence when the direction of the ship's head is altered; and (d) the special precautions to be invariably observed at sea on the alteration of the ship's course.

67. Describe a "Dumb-card" or "Pelorus," and its use (a) in compensating a compass, (b) in determining the deviation.

68. If you determine the deviation by an azimuth or an amplitude of a heavenly body, it is then combined with variation, which together is sometimes called the *correction* for the compass. State when the deviation is the difference between the variation and the *correction*, and when the sum; and when it is of the same name as that of the *correction*, and when of the contrary name.

69. In observing azimuths of heavenly bodies, the best method is by

"time azimuths," since these can be observed without an altitude when the ship is in port, or when the horizon cannot be defined from any cause. Given the Sun's declination, the hour of the day, and the latitude to find the true bearing of the Sun.[1]

70. By night, if it be desirable to observe the *correction* of the compass. Given the day of the year, and time at ship, also the latitude of the place, to determine what stars will be in good position for this purpose.

71. If your correcting magnets are so mounted that their positions can be altered, describe the process by which, on open sea, you can place the ship's head correct magnetic N. (or S.), and correct magnetic E. (or W.), and can make the correction perfect.

72. Given the name of a star, the time, the place of ship, the variation of the compass, and the bearing of the star by compass, determine the deviation, and name it east or west.

73. Would you expect any change to be caused in the error of your compass by the ship heeling over, either from the effect of the wind or the cargo?

74. Describe clearly the three principal causes of the heeling error on board an iron ship.

75. Towards which side of the ship would that part of magnetism induced in continuous transverse iron (which was horizontal while ship was upright) help to draw the north point of the needle when ship heels over, (a) in the northern hemisphere, (b) in the southern hemisphere?

76. Supposing the compass were placed between the two parts of a divided beam or other athwartship iron, towards which side of the ship would iron so situated help to draw the north point of the needle when ship heels over, (a) in the northern hemisphere, (b) in the southern hemisphere?

77. Would you expect that part of the magnetism induced in iron exactly perpendicular to the ship's deck, such as stanchions, bulkheads, etc., if below the compass, to cause any part of the heeling error when ship heels over, and if so, towards which side of the ship, (a) in the northern hemisphere, (b) in the southern hemisphere?

78. If an ordinary standard compass placed higher than the iron top sides be compensated whilst the ship is upright, what coefficient will be affected by heeling?

79. Under what conditions (that is, as regards position whilst building and the arrangement of iron in the ship) is the north point of the compass needle usually drawn to windward, or the high side of the ship, in the northern hemisphere?

80. Under what conditions, as a rule, is the north point of the compass needle usually drawn to leeward, or the low side of the ship, in the northern hemisphere?

81. State to which side of the ship, in the majority of cases, is the north point of the compass drawn when ship heels over in the northern hemisphere; and when this is the case, and it is not allowed for, what effect has it on the assumed position of the ship when she is steering on northerly, and also on southerly courses?

82. On what courses would you keep away, and on what courses would you keep closer to the wind in both the northern and southern hemispheres in order to make good a compass course, (a) when north point of compass is drawn to windward, or the high side of ship; and (b) when drawn to leeward or the low side?

---

[1] The process of finding time azimuths by the ordinary formulæ of spherical trigonometry is tedious, and since on board an iron ship these observations should be often repeated, the candidate will be allowed to use any table or graphic or linear method that will solve the problem within a half of a degree, the altitude of the heavenly body not being given.

83. If a ship is beating to windward, when she tacks, under what circumstances will the heeling error retain the same name, and under what circumstances will it take the contrary name?

84. If a ship is placed on the opposite tack by the change of wind, the ship's course being the same by compass, will the heeling error change its name?

85. In which direction of the ship's head does the heeling error attain its maximum value, and in which direction does it generally vanish?

86. Explain clearly how that part of the heeling error due to the permanent part of the magnetism of the ship varies as the ship changes her geographical position, and what is the reason of this?

87. Explain clearly how that part of the heeling error due to the induction in transverse iron (which was horizontal when ship was upright), and iron vertical to the ship's deck, varies as the ship changes her geographical position.

88. What, then, would be the probable nature of the heeling error, that is, whether to high or low side of the ship, and whether the error would be equal to the sum or difference, etc., of the forces—(1) in high north latitudes, (2) on magnetic equator, (3) in high south latitudes? Assuming the polarity of the sub-permanent magnetism of the ship under, and affecting, the compass to be as given below; the vertical induction in soft iron, of course, obeying the ordinary laws in the above geographical positions (1), (2), (3).

(a) In cases where the effect of *red* vertical sub-permanent magnetism is equal to that of the vertical induction in the soft iron of the ship.

(b) Where the effect of *red* vertical sub-permanent magnetism is greater than that of the vertical induction in the soft iron.

(c) Where the effect of *red* vertical sub-permanent magnetism is less than that of the vertical induction in the soft iron.

(d) Where the effect of *blue* vertical sub-permanent magnetism is equal to that of the vertical induction in the soft iron.

(e) Where the effect of *blue* vertical sub-permanent magnetism is greater than that of the vertical induction in the soft iron.

(f) Where the effect of *blue* vertical sub-permanent magnetism is less than that of the vertical induction in the soft iron.

89. Can the heeling error be compensated? If so, state the means to be employed, and how the compensation may be effected.

90. Can the compensation of the heeling error be depended on in every latitude? If not, state the reason.

91. Do the soft iron correctors used for compensating the coefficient +D have any effect on the compass needle when the ship heels over? and, if so, do they draw the needle towards the low or the high side of the ship, and do they counteract, or otherwise, the effect produced by the vertical induction in the soft iron, (a) in the northern hemisphere, (b) in the southern hemisphere; and what is the reason of this?

92. Given the heel, the direction of the ship's head by compass, and the heeling error observed, to find the approximate heeling error, with a greater or less given heel, and with the ship's head on some other named point of the compass, the ship's magnetic latitude being in both cases the same.

93. Describe any instrument to show the ship's heel (generally called a clinometer), and state how and where it should be fixed.

94. Should the clinometer be observed when the ship is swung to determine the deviation when the ship is upright? If so, state the reason why.

95. Would you expect the Table of Deviations supplied by the compass adjuster from observations made in swinging the ship to remain good during the voyage, or would you expect the deviations to change? If so, state under what circumstances.

96. Is it desirable that a record of your observations for deviations should

SYLLABUS.

be kept as a guide for any subsequent voyage, in case the ship should be in the same locality, or for further correction of the compass? If so, describe some suitable form for keeping such record.

97. Would you, under any circumstances, consider it a safe and proper procedure to place implicit confidence in your compasses, however skilfully they may have been adjusted? If not, what precautions is it your duty to take at all times?

ANSWERS TO THE QUESTIONS ON COMPASS DEVIATION FOR MASTERS (EXTRA).

1. An artificial magnet is a bar or needle of hard-tempered steel which has been magnetized by artificial means, and so distinguished from a loadstone or natural magnet.

Artificial magnets can be made in several ways:

(a) For small needles, it is sufficient to place either pole of a strong magnet on one end of the needle and draw it in contact to the other end, where it is taken off. If the pole used is the red pole, the end of the needle last in contact with it becomes a blue pole, and the other a red pole.

(b) The method of "separate touch." The bar to be magnetized is placed with its ends resting on opposite poles of two strong magnets; two other magnets are held, one in each hand, with opposite poles downwards, and brought together at the middle of the bar. The magnets are then inclined so as to make angles of about 30° with the bar, and are moved in contact with it towards the ends, where they are taken off. This process should be repeated several times on both sides of the bar, taking care that a red pole is moved towards the end supported by a red pole, and the blue pole towards the end resting on a blue pole.

(c) By using an electro-magnet. One half of the bar is drawn from the middle to the end over one pole of a strong electro-magnet, and the other half over the other pole.

(d) By suspending the bar inside a solenoid and passing a strong electric current through the latter.

2. The end which points north when the magnet is freely suspended is named the red pole, and the other end the blue pole.

3. The south magnetic pole of the Earth is the red, and the north the blue pole. The north magnetic pole is situated about 70° N. latitude and 97° W. longitude, and the south magnetic pole about 74° S. latitude and 147° E. longitude.

4. Poles of the same name repel each other, whilst poles of different names attract each other. If freely suspended in the manner stated, they would assume a position of equilibrium determined by the Earth's directive force acting on the needles, and the mutual attraction of their opposite poles.

5. In the first case, the needle would come to rest in the plane of the magnetic meridian, with its red pole pointing north; in the second case, the needle would still coincide with the magnetic meridian, but its red pole would dip in N. latitude, and its blue pole would dip in S. latitude at an angle from the horizontal depending on the magnetic latitude, the dip being 0 on the magnetic equator, and 90° at the magnetic poles.

6. The variation of the compass is due to the fact that the Earth's magnetic poles do not coincide with the geographical poles.

7. Deviation of the compass is the angle of deflection of its north point from magnetic north, caused by the disturbing influence of iron.

8. "Local attraction" is the name given to the magnetic influences (causing a disturbance of the compass) of the locality in which a ship may be found; for example, when in dock near to iron ships, iron pillars, cranes, etc., also in the neighbourhood of some volcanic islands. It is said that the following places have a disturbing effect on ships' compasses: Ascension,

St. Helena, Cape St. Frances (Labrador), Cossack (North Australia), New Ireland, Solomon Islands, Tumbora, Sumbawa, coast of Madagascar, Iceland, and Ile de Los (west coast of Africa).

9. Malleable iron and cast iron are termed "soft iron." Soft iron has the property of being instantaneously magnetized by induction from the magnetism of the Earth, or of a natural or artificial magnet; also by passing an electric current round it. This induced magnetism is only transient in its character, and is lost in great part or entirely when the inducing cause is removed.

10. "Hard iron" is iron combined with a certain percentage of carbon, and is usually termed steel. Tempered hard iron has the property of acquiring slowly and retaining permanently, or nearly so, the magnetic polarity communicated to it.

11. "Horizontal force," is the horizontal component of the Earth's total magnetic force in the line of dip, and varies as the cosine of the magnetic latitude. It is greatest at the magnetic equator, and 0 at the magnetic poles, and therefore gives the greatest directive force to the compass needle in the former case, whilst there is no directive tendency in the latter.

12. The magnetic equator does not coincide with the geographical equator, but intersects it in two nearly opposite points situated in west longitude and east longitude. From the former point in the Pacific Ocean it runs nearly parallel to the Equator to about 100° W. long., then bends to the south on approaching South America, and in crossing Brazil it recedes from the Equator as far as 16° S. lat. ; then it bends northwards and crosses the Equator in about 6° W. long., whence it passes across the Gulf of Guinea and the African continent, receding to about 10° N. lat., and then very gradually approaches the Equator to the other point of intersection.

13. The values of the dip, etc., may be found on charts prepared for the purpose.

14. The variation changes rapidly in the English Channel, the western portion of the North Atlantic, the South Atlantic near the coast of Brazil, the South Indian Ocean, the neighbourhood of Japan, and especially in the Gulf of St. Lawrence.

In navigating these localities, if the ship's course is across the magnetic meridians, the precaution should be taken of altering the courses steered by compass in accordance with the change of variation. A variation chart is useful to show what magnetic meridians are crossed and the amount of variation.

15. Because the vertical induced magnetism which causes "heeling error" depends on the dip, and the directive force of the needle depends on the Earth's horizontal force.

16. The "vertical force" is the vertical component of the Earth's total magnetic force in the direction of the dip. It is greatest at the magnetic poles, and least at the magnetic equator.

17. The needle would be least affected near the magnetic equator, because the Earth's horizontal force, and therefore the directive power of the needle, are there at their greatest.

18. (a) The card should be accurately centred and graduated, the magnetic axis of the needle (or the common axis, if more than one) coinciding with the north and south points of the card. There should be the least possible friction between the centre cap and the supporting pivot, the needles should have good directive power, the "lubber-point" should be exactly in the fore and aft midship line, and the deviation should be reduced by compensation to a small amount, say under 3°. The gymbals should have little friction, and should not allow any lateral play. When in position in the binnacle, the card and lubber-point should be in full view of the man at the wheel.

(b) It is generally agreed that two or more short needles parallel to each other in the same plane is the best arrangement for securing good compensation.

19. Care should be taken that the magnets are not stowed away with similar poles in contact, because in that position they would tend to weaken or demagnetize each other, or even reverse the poles.

20. A position should be chosen for the steering compass where it would not be near large elongated masses of iron, e.g. stern post, rudder post, tiller, stanchions, funnel, etc.

For a standard compass the same precautions are necessary, and the position chosen should be if possible symmetrical with regard to the various magnetic forces in the ship.

21. Transient induced magnetism is acquired by soft iron by induction from permanent magnets, from electric currents, or from the Earth. This polarity is retained only so long as the inducing cause is present.

22. The blue pole is uppermost in north magnetic latitudes when the cause of induction is the Earth's magnetism; therefore in the northern hemisphere the north end of the needle would be attracted by the upper end of a soft iron bar, and repelled by the lower end.

23. In south magnetic latitude the upper pole of soft iron by induction is red, and the lower end blue; therefore the north point of a compass needle would be repelled by the upper end, and attracted by the lower.

24. No effect.

25. The sub-permanent magnetism of an iron ship is the magnetic polarity acquired in the course of construction by induction from the Earth and retained after launching. It is liable, however, to a considerable reduction after the ship leaves the stocks, and is therefore termed "sub-permanent" to distinguish it from the fixed magnetism of tempered steel.

The red pole is that which was towards the north when the ship was being built, and the blue towards the south.

26. Composition of forces is the method of finding a single force which is equivalent to two or more forces of given amount and direction.

In the "Parallelogram of Forces," two adjacent sides represent in direction and units of amount two given forces acting at the point of meeting, and a diagonal from this point represents in direction and units of amount the equivalent force, or "resultant" of the composition of the two given forces.

A knowledge of this principle enables a compass adjuster to resolve the sub-permanent magnetic force in a ship into fore and aft, athwartship, and vertical components, and to compensate each independently of the others.

27. Coefficients B and C express the maximum deviations caused by magnetic forces acting horizontally in fore and aft and athwartship directions respectively; +B represents a blue pole at the bow, and red at the stern; −B the reverse; +C represents a blue pole at the starboard side, and red at the port side; −C the reverse. They are said to produce semicircular deviation, because the deviation due to each is easterly on one half of the compass, and westerly on the other half.

28. Yes; it may be caused by vertical soft iron, because the induced poles remain the same whatever the direction of ship's head.

29. +B gives E. deviation on all courses from N. through E. to S., and W. deviation from N. through W. to S.; −B gives the opposite.

30. +C gives E. deviation on all courses from W. through N. to E., and W. deviation from W. through S. to E.; −C gives the opposite.

31. See Appendix.

32. The greatest disturbance from sub-permanent magnetism would be found when the ship lay at right angles to the direction in which she was when being built, and least when in the same direction.

33. Quadrantal deviation is the effect of transient induced magnetism in horizontal iron; it is represented by coefficients D and E. D gives the

greatest deviation on N.E., S.E., S.W. and N.W. courses, and E gives the greatest on N., E., S., and W. courses. It is termed quadrantal because it changes its name in successive quadrants.

34. +D gives E. deviation from N. to E. and S. to W. courses, and W. deviation from E. to S. and W. to N. ; −D gives the opposite. +E gives E. deviation from N.W. to N.E. and S.E. to S.W. courses, and W. deviation from N.E. to S.E. and S.W. to N.W. ; −E gives the opposite.

35. Transverse iron produces +D, and fore-and-aft iron −D ; but if end on to the compass, or when a beam is divided for a skylight and a compass placed between the parts, then the signs would be reversed.

36. The following table shows the results :—

Ship's head.	Athwartship iron.	Fore-and-aft iron.	Polarity acquired.
N.E.	End to port.	End towards bow.	Red.
	,, starboard.	,, stern.	Blue.
S.E.	,, port.	,, ,,	Red.
	,, starboard.	,, bow.	Blue.
S.W.	,, ,,	,, stern.	Red.
	,, port.	,, bow.	Blue.
N.W.	,, starboard.	,, ,,	Red.
	,, port.	,, stern.	Blue.

37. Coefficient A is a constant deviation of the same amount and sign on all courses, +A being easterly and −A westerly. It may be due to several causes :

(a) The magnetic axes of the needles not coinciding in direction with the N. and S. points of the card.
(b) The lubber-point not being correctly placed.
(c) Unsymmetrical arrangement of iron with respect to the compass.
(d) An error in the correct magnetic bearing used for finding the deviation.
(e) Retained magnetism when swinging the ship.

38. The object of compensating compasses is to reduce the deviations to the smallest possible amounts. An uncompensated compass having large deviations would be too "sensitive" on some courses, and too "sluggish" on others, so that its indications would not correspond to the actual changes of course ; whilst a compass having its deviations reduced to a minimum by compensation would behave equally well on all courses.

39. When being equipped for sea, it is desirable that the ship's head should point the opposite way to the direction when building, the compensations being made just before sailing.

40. A common practice is to draw two lines on the deck where the binnacle stands—one fore and aft amidships, and the other at right angles, both passing through a point vertically under the centre of the compass ; then, with the ship's head at correct magnetic N. or S., a bar magnet is laid on the deck at right angles to the fore-and-aft line, with its middle point on it and approached to the compass until the needle points correct magnetic N.

Next, the ship's head is steadied on correct magnetic E. or W., and another bar is laid on the deck at right angles to the athwartship line, with its middle point on that line, and approached to the compass until the needle again points correct magnetic N. Both magnets are then fixed in position.

Next, the ship's head is steadied on N.E., S.E., S.W., or N.W. correct magnetic for the purpose of correcting the quadrantal deviation. This is

SYLLABUS. 273

done by placing two soft iron correctors on the supports attached to the binnacle, and moving them towards or from the compass until the deviation is compensated.

Lastly, the ship's head is brought to N. or S. by compass, and she is heeled over, say, 10°, and the heeling error observed. Then a small bar magnet is placed vertically under the centre of the compass, and moved up or down in its groove until the error disappears.

N.B.—The red pole of the magnet must be uppermost when the deviation is towards the higher side, and *vice versâ*.

In Thompson's compass the magnets are inside the binnacle.

41. The magnets should not be nearer to the centre of the card than twice their own length, and the quadrantal correctors not nearer the needles than once and a quarter the length of the needles.

If the magnets are too near, they cause a new deviation called sextantal error, and if the correctors are too near they are liable to receive induced magnetic polarity from the needle, and attract either end indifferently.

42. No. They may be placed either above or below the compass, on a bulkhead, or in the binnacle itself, provided the condition is observed that they are bisected by transverse and fore-and-aft planes passing through the centre of the compass.

43. Coefficient B is partly due to sub-permanent magnetism and partly to induced magnetism in vertical iron before or abaft the compass.

44. The compass would be found to be over-compensated with respect to B on going to a lower latitude, and on crossing the magnetic equator the part due to vertical induction would be reversed, which would necessitate the shifting or removal, or perhaps reversal, of the magnet.

45. One part should be compensated by a vertical soft iron bar amidships, either before or aft the compass, as required, and the other by a bar magnet.

46. The ship's head being at E. or W., the deviation should be observed ; then the Flinders bar should be moved into position before the compass, to correct, say, one-fourth of this deviation as a first approximation, the rest being compensated by a magnet. When on the magnetic equator the whole of coefficient B is due to sub-permanent magnetism, and should be exactly compensated by the fore and aft magnet. Then, as the ship proceeds N. or S., the change observed in the deviation due to coefficient B would be the amount to be compensated by the Flinders bar.

A more correct method would be, when direction of ship's head whilst building is known, to enter the Traverse Table with direction of ship's head when building as a course, and coefficient C in the dep. column, then B will be found in the latitude column. This would be the part of B to be compensated by the magnet, and the remainder by the Flinders bar.

47. Standard compasses *are* liable to be affected by the induced magnetism of vertical iron in the form of stanchions, pillars, funnels, or bulkheads near them. The attraction is not always towards the stern, because the disturbing masses of iron might be forward of the compass, in which case the Flinders bar should be abaft the compass.

48. Vertical iron could not cause any part of coefficient C, unless it was unsymmetrically placed with regard to the compass, therefore the whole of C, when the ship is upright, is usually due to sub-permanent magnetism.

49. Yes ; because the horizontal force of the correctors and the directive force of the needle would vary in the same ratio.

50. When the joint effect is small, and when the constant deviation is W. when the ship is swung to the right, and E. when swung to the left, and consequently the effect of temporarily retained magnetism, they may be disregarded when compensating the compass.

51. (1) (*a*) It would be found to have decreased.
        (*b*) It would increase again and tend to return to its original amount.

T

(2) (a) It would become 0.
(b) It would have a different sign.

52. No deviation would occur from the sub-permanent magnetism so long as it remained constant, because it is exactly counterbalanced by the correcting magnet; nor from the vertical induced magnetism, because it varies in the same ratio as that of the Flinders bar.

53. The deviation from this cause would not change its name and amount, because (a) the red induced pole is towards the north in all latitudes, and (b) the horizontal induced force changes in the same ratio as the directive force of the needle.

54. (1) A compass is set up on shore or in a boat at some distance from the ship, where it is not exposed to local attraction, and an observer stationed by it. The ship is then swung, and as her head comes to the desired points simultaneous bearings are taken by signal of each compass from the other. The bearings by distant compass reversed is the correct magnetic, with which the bearings observed on board are compared, and the deviation found.

(2) Figures on dock walls show either the correct magnetic or the true bearings of a prominent distant object from the observer's position. When, therefore, the ship swings with the tide or is specially swung, the difference is noted between the figure on the wall in line with the object and the bearing of the same object by the compass on board, as the ship's head comes to each point on which the deviation is required. These differences are either "compass errors" or deviation, according as the figures on the wall are true or magnetic bearings.

(3) The correct magnetic bearings of the distant object may be found by means of a compass placed in a direct line with the ship and the object in a position free from local attraction. The deviation for the actual direction of the ship's head is the difference between the compass bearing of the object and the correct magnetic as before.

55. (a) By the sun : Take its bearing by compass and compute for the same time, or take from the Azimuth Tables its true azimuth. The difference is the compass error, from which the deviation may be found when the variation is known.

(b) By a star or planet : Find the hour angle of the body and compute, or take from the Azimuth Tables (when within their limits) its true azimuth, the difference between which and the observed bearing by compass at the time for which the calculation is made will be the compass error, whence the deviation as before.

56. Uses of Napier's diagram :—

(a) From known deviations on a few points, either correct magnetic or by compass, a curve showing the deviation on all points may be drawn.

(b) The curve so drawn may be used as a complete deviation table, from which may be found mechanically the compass course from the correct magnetic, and vice versâ. Its advantages are, that a complete curve of deviations may be drawn from the deviations observed on a few points, whether by compass or correct magnetic, and whether equidistant or not. It also gives a graphic record of the deviations which appeals to the eye, and gives a much better appreciation of the results of compensation than does an ordinary deviation table.

Again, if observations have been obtained of the deviations on two or three points of the quadrant in which the ship's course may lie, a curve will show the deviations on the other neighbouring points.

Further, the deviations obtained by drawing a curve are free from possible errors of observation and calculation.

57. The Napier's diagram consists of a vertical line graduated to points and degrees, with two sets of parallel lines drawn across it at angles of 60°

through each of the thirty-two marks representing the "points" of the compass. The N. is at the top and bottom, the S. in the middle, and the E. and W. midway between the N. and S. marks, the other points following in order.

58. Eight equidistant points, viz. N., N.E., E., S.E., S., S.W., W., and N.W.

59. Two would be the least number, but three would be better, because a circle could be drawn through the three projected points.

Assuming that deviations had been observed on three compass courses in the quadrant in which my course would lie, I would project them on the dotted lines, and then draw a fair curve through the projected points, extending it to comprise the whole quadrant. The parts intercepted by the curve of the dotted lines through the other courses would give the required deviations, measured on the vertical line.

If only two observations were available, a straight line passing through the points of projection would show approximately the deviations on the intermediate courses.

N.B.—If the observations for deviation had been made on correct magnetic courses, the plain lines would be used instead of the dotted.

60. The mean of the equidistant bearings would be the correct magnetic bearing, nearly.

The object should be so far distant that its bearing by the compass on board would not be sensibly affected by the change of position in swinging the ship.

61. See Appendix.
62. See Appendix.

63. An observer being stationed at each compass, bearings are taken simultaneously of a distant object as the ship's head is steadied on seven or eight points correct magnetic (by Pelorus) round the compass. Comparing these bearings with the known correct magnetic, the deviations for each compass are found, and are then marked on the plain lines of a Napier's diagram. A curve is then drawn through each set of projected marks, and from these curves a complete table of deviations on compass courses may be constructed by measuring the parts of the dotted lines intercepted by the curve.

64. If the correct magnetic bearing reads to the *right* of the compass bearing, the deviation is easterly ; if to the *left*, westerly.

65. A knowledge of the coefficients is of advantage, because—

(1) From the coefficients a complete table of deviations can be constructed.

(2) It shows how much of the deviation is due to each of the disturbing forces in a ship.

(3) It enables a compass adjuster to employ the proper means for compensating.

66. (a) "Retentive" or "retained" magnetism is the name given to the temporary magnetic polarity acquired by a ship whilst heading for a considerable time in one direction, and retained for a longer or shorter period after that direction is changed. It is induced by the Earth's magnetism whilst the ship is in dock or on a long-continued course at sea.

(b) It has no effect on the compass needle whilst the ship's head continues on the same course.

(c) A deviation will appear tending to make the ship follow the last course steered.

(d) Allowance should be made for this effect, and an observation for deviation should be made as soon as possible after the course is changed.

67. The "Pelorus" has the general appearance of an azimuth compass, but it has no magnets. The "card" is usually a metal disc, which can be turned about its centre and clamped in any position.

(a) Its use in compensating: First clamp the sight vanes to the known correct magnetic bearing of a distant object; then turn the card round until the desired direction of ship's head is at the lubber-point, and clamp. Next swing the ship until the object is seen through the sight vanes, and place the magnet or corrector in position to effect the compensation.

(b) Its use to find the deviation: Clamp the sight vane to the known correct magnetic bearing of a distant object on land, or a celestial object, and turn the card round until the object is seen through the sight vanes. The course by "Pelorus" is the correct magnetic, the difference between which and the course by compass is the deviation for that course.

68. Deviation is the *difference* between the correction and variation when they are of the same name, but the *sum* if of different names. It has the same name or sign as the correction in all cases except when the correction is less than variation of the same name.

69. See Appendix.

70. See Appendix.

71. When on the magnetic equator, sea smooth, and ship upright, set a watch to show apparent time at ship. Bring the ship's head to near north (or south), and clamp the sight vanes of the Pelorus to the correct magnetic bearing of the Sun—found by taking out the true bearing from the tables for some minutes in advance, and applying the variation; also bring the (N. or S.) point of the Pelorus to the lubber-point. Then steer the ship so that the centre of the Sun's image is on the sight vanes at the moment for which the bearing was computed, and steady her on that course whilst the athwartship magnet is moved into position to make the compass indicate correct magnetic.

In a similar manner place and keep ship's head on correct magnetic east or west, and adjust the position of the fore-and-aft magnet.

72. See Appendix.

73. Yes; a deviation appears which is known as the heeling error.

74. (1) When the ship heels, the vertical component of sub-permanent magnetism takes effect in producing deviation, the upper pole not being vertically under the centre of the compass.

(2) Iron which is horizontal when the ship is upright acquires a vertical component of induced magnetism.

(3) The upper ends of rudder post, stern post, or other vertical iron before or abaft the binnacle, are brought out to one side of a vertical plane passing through the centre of the compass.

75. (a) Towards the higher side.
(b) Towards the lower side.

76. (a) Towards the lower side.
(b) Towards the higher side.

77. Such iron would cause a heeling error.
(a) Towards the higher side.
(b) Towards the lower side.

78. Coefficient C.

79. In the northern hemisphere, the N. point of the compass is generally drawn to windward in ships built head north, also in ships whose compasses have +D, and in others which have large masses of vertical iron below the compass.

80. In some ships built head south, and ships having a large −D.

81. In the majority of cases the N. point of the compass is drawn to windward. If not allowed for, the effect would be to draw the ship to windward on northerly courses, and to leeward on southerly.

82. (a) Keep away in the northern hemisphere; keep closer in the southern hemisphere.
(b) Keep closer in the northern hemisphere; keep away in the southern hemisphere.

83. If in tacking the ship's head is northerly on one tack, and southerly on the other, the heeling error retains the same name; but if on both tacks her head is either northerly or southerly, the error changes its name.

84. Yes.

85. The heeling error has its maximum value when heading north or south magnetic, and nearly vanishes at east and west.

86. It diminishes when the ship goes to a lower latitude, and increases when to a higher, because the directive force of the needle increases in the first case, and decreases in the second, whilst the vertical sub-permanent force is assumed to be constant.

87. It decreases on approaching the equator, and *vice versâ*, becomes 0 on the magnetic equator, and changes its sign after crossing; because the vertical force of induced magnetism decreases to 0 at the equator, and increases with increase of latitude, the uppermost poles being blue in north magnetic latitude, and red in south. The directive force of the needle also varies with change of latitude.

88.

(1) High N. latitude.	(2) Magnetic equator.	(3) High S. latitude.
(a) Nil.	To low side due to sub-permanent only.	Sum to low side.
(b) Difference to low side.	,, ,,	,, ,,
(c) ,, high side.	,, ,,	,, ,,
(d) Sum to ,,	To high side, due to sub-permanent only.	Nil.
(e) ,, ,,	,, ,,	Difference to high side.
(f) ,, ,,	,, ,,	,, low ,,

89. Yes; for a given magnetic latitude, by means of a vertical magnet which can be moved up or down in a groove or tube. There are two ways of proceeding—

(1) Bring the ship's head to N. or S. by compass (assumed to be correct) and give her a list of say 10°; then slide the magnet up or down until the error disappears, and fix in position.

(2) Bring the ship's head to E. or W. magnetic, ship upright; remove the card and place on the supporting pillar a small dipping needle, which has been balanced by a sliding weight so as to show no dip when on shore : then place the magnet in its groove and move up or down until the needle is again perfectly level.

90. No; because the heeling error is partly due to vertical induced magnetism, which varies with change of latitude.

91. The soft iron correctors are found to correct the healing error due to transverse iron for ordinary angles of heeling. (a) In the N. hemisphere they deflect the N. point of the needle to the low side; (b) in the S. hemisphere, towards the high side. The reason is that they produce a co-efficient C due to an induced vertical force in the correctors, which has a different sign from that of the C caused by transverse iron when the ship heels.

92. See Appendix.

93. A clinometer consists of a graduated semicircle, which is fixed on a bulkhead with its diameter horizontal and arc downwards. The zero of the arc should coincide with a plumb-line from the middle of the diameter, so that when the ship heels the plumb-line or weighted index would show on the arc the exact degrees of inclination.

94. Yes; otherwise the deviations would be complicated with the heeling errors.

95. The deviations might be expected to change from several causes—
(1) The place where the ship is swung may have local attraction.
(2) The force of sub-permanent magnetism is liable to change from long continuance on one course, and from concussion and vibration.
(3) The compensation of coefficient B can only be regarded as provisional, there being no certainty that the just proportions were assigned to the magnet and the Flinders bar.
(4) Allowance must be made for a want of scientific knowledge on the part of the adjuster, or a careless performance of his work.
(5) Loss of power in needles or magnets.
96. It is most necessary that such a record should be kept. The following form would serve the purpose :—

Date.	Lat.	Long.	Var.	Ship's head.		Deviation.		Remarks.
				Standard compass.	Steering compass.	Standard compass.	Steering compass.	

97. No, because considerable changes in the deviation may occur. Therefore it is necessary that frequent observations for deviation may be made, especially when the course is altered, or when rapidly changing latitude.

PRACTICAL QUESTIONS IN SYLLABUS OF DEVIATION FOR EXTRA MASTERS.

Explanation.—
31. In the Traverse Table, if the direction of ship's head whilst being built be taken as a *course*, the latitude and departure columns will give the relative values of B and C. Therefore—
(1) With ship's head as a course and B in latitude column, C is found in departure column.
(2) Ship's head as course and C as departure give B in latitude column.
(3) B in latitude and C in departure columns give the direction of ship's head from N. or S.

Examples.—
1. Given the direction of ship's head whilst building N.E. by N., and coefficient B = $-16°·5$. Find coefficient C.
Here the course 3 points and lat. 16·5, give dep. 11·1. Coefficient C is therefore $+11°·1$, because the ship's head was easterly.
2. Given ship's head S.S.W., and coefficient C = $-3°·6$. Required coefficient B.
Course 2 points and dep. 3·6 give lat. 8·7 ; therefore coefficient B = $+8°·7$, because the ship's head was southerly.
3. Given coefficient B = $+7°·3'$, and coefficient C = $-10°·7$. Required the direction of the ship's head when building.
Lat. 7°·3 and dep. 10·7 give the course 56°, which is S. 56° W., because B is + and C is −.

Exercises.—
1. Given coefficient B = $-11°$ and C = $-8°·5$. Find direction of ship's head whilst being built.
2. Given coefficient +C = $12°·5$, and −B = $6°·8$. Required the direction of ship's head.

SYLLABUS. 279

3. Given $+B = 14°·7$ and $+ C = 8°·7$. Required ship's head.
4. Given ship's head N. 65° W., and $-B = 5°·4$. Required coefficient C.
5. Given ship's head S. 14° W., and $-C = 3°·5$. Required coefficient B.
6. Given ship's head S.E. ¾ E., and coefficient $C = +7°·5$. Required coefficient B.

61. **Example.**—(*a*) Magnetic bearing required: S. 53° 45′ W.

Ship's head by compass.	Bearing by compass.	Deviation.	Ship's head by compass.	Bearing by compass.	Deviation.
N.	S. 75° W.	21° 15′ W.	S.	S. 34° W.	19° 45′ E.
N.E.	S. 64° W.	16° 15′ W.	S.W.	S. 31° W.	22° 45′ E.
E.	S. 56° W.	2° 15′ W.	W.	S. 49° W.	4° 45′ E.
S.E.	S. 50° W.	3° 45′ E.	N.W.	S. 71° W.	17° 15′ W.

Sum 8)430°
53° 45′

NOTE.—The magnetic bearing required is the mean of all the bearings by compass, found by adding them and dividing by eight. The deviations are then marked on the dotted lines of a Napier's diagram, and a curve drawn.

(*b*) Magnetic courses: N.N.W. S.S.E. W.N.W. E.S.E.
Compass courses required: N. 1° W. S. 31° E. N. 45° W. S. 68° E.

Explanation.—Place one leg of the compasses on the given course, and extend the other along the *plain* line to the curve, and turn upwards if the line tends upwards, but downwards if it tends downwards, and read off the required course.

(*c*) Compass courses given: N.N.E. E.N.E. S.S.W. W.S.W.
Magnetic courses required: N. 6° E. N. 60° E. S.W. S. 83° W.

Explanation.—Place one leg of the compasses on the given course, and extend the other along the *dotted* lines to the curve, and turn up or down, as the line tends.

(*d*) Ship's head by compass: N.E. ½ E.
Compass bearings: S.E. by S. N.N.W.
Magnetic bearings required: S. 44° E. N. 32½° W.

Explanation.—Find the deviation on the given compass course from the dotted line, and apply it to both bearings (east to right, west to left).

Exercises.—
1. (*a*) Magnetic bearing required—

Ship's head by compass.	Bearing by compass.	Deviation.	Ship's head by compass.	Bearing by compass.	Deviation.
N.	N. 56° E.		S.	N. 55° E.	
N.E.	N. 48° E.		S.W.	N. 64° E.	
E.	N. 45° E.		W.	N. 68° E.	
S.E.	N. 49° E.		N.W.	N. 63° E.	

(*b*) Magnetic courses: N ¼ E. S 70° E. S. 55° W. N. 8° W.
Compass courses required.
(*c*) Compass courses: S.S.W. N.W. ½ W. N. S. 82° E.
Magnetic courses required.

280   TEXT-BOOK ON NAVIGATION.

(d) Ship's head by compass : S.W.
Compass bearings : E. ¼ N.   N. by E.
Magnetic bearings required.

2. (a) Magnetic bearing required—

Ship's head by compass.	Bearing by compass.	Deviation.	Ship's head by compass.	Bearing by compass.	Deviation.
N.	N. 44° W.		S.	S. 79° W.	
N.E.	N. 47° W.		S.W.	S. 89° W.	
E.	N. 80° W.		W.	N. 69° W.	
S.E.	S. 76° W.		N.W.	N. 52° W.	

(b) Magnetic courses : S ¼ E.   S.W. ¼ S.   N. 80° W.   N. 75° E.
Compass courses required.
(c) Compass courses : N.N.W. ¾ W.   W. by S. ¼ S.   S.   N. 80° E.
Magnetic courses required.
(d) Ship's head by compass : W.
Bearings by compass : E. 11° N.   N. 6° W.
Magnetic bearings required.

3. (a) Magnetic bearing required—

Ship's head by compass.	Bearing by compass.	Deviation.	Ship's head by compass.	Bearing by compass.	Deviation.
N.	N. 32° W.		S.	N. 20° E.	
N.E.	N. 20° W.		S.W.	N. 9° E.	
E.	N. 5° E.		W.	N. 15° W.	
S.E.	N. 17° E.		N.W.	N. 31° W.	

(b) Magnetic courses : S. 65° W.   N.W.   N. 32° E.   E.
Compass courses required.
(c) Compass courses : N.   N.E. by N.   E. by S.   S.W. ½ S.
Magnetic courses required.
(d) Ship's head by compass.
Bearings by compass : W. ¼ N.   S. by W. ¾ W.
Magnetic bearings required.

4. (a) Magnetic bearing required—

Ship's head by compass.	Bearing by compass.	Deviation.	Ship's head by compass.	Bearing by compass.	Deviation.
N.	S. 43° W.		S.	W.	
N.E.	S. 50° W.		S.W.	S. 83° W.	
E.	S. 65° W.		W.	S. 65° W.	
S.E.	S. 80° W.		N.W.	S. 45° W.	

(b) Correct magnetic courses : S.E. by E.   N.N.E.   N.W. ½ W.
S. 10° W.
Compass courses required.
(c) Compass courses : N.W. by W.   N. by W.   N.E. ½ E.   S. 50° E.
Magnetic courses required.

(d) Ship's head : E. ¼ N.
Bearings by compass : S.E. ¾ E.   N.E. by N.
Magnetic bearings.

5. (a) Magnetic bearing required—

Ship's head by compass.	Bearing by compass.	Deviation.	Ship's head by compass.	Bearing by compass.	Deviation.
N.	S. 2° W.		S.	S. 29° E.	
N.E.	S. 7° W.		S.W.	S. 26° E.	
E.	S. 6° E.		W.	S. 15° E.	
S.E.	S. 16° E.		N.W.	S. 3° E.	

(b) Magnetic courses : W.   N. 30° E.   S. 56° E.   S. 3° W.
Compass courses.
(c) Compass courses : E.   N.E. by N. ¼ N.   N.W. by W.   S. 3° E.
Magnetic courses.
(d) Ship's head by compass : S.E.
Bearings by compass : N.   E. by N. ¾ N.
Magnetic bearings.

6. (a) Magnetic bearing required—

Ship's head by compass.	Bearing by compass.	Deviation.	Ship's head by compass.	Bearing by compass.	Deviation.
N.	N.		S.	N. 21° W.	
N.E.	N. 7° E.		S.W.	N. 17° W.	
E.	N. 10° W.		W.	N. 18° E.	
S.E.	N. 13° W.		N.W.	N. 16° E.	

(b) Magnetic courses : E.S.E.   S.S.E.   W. by N.   N.E. ¼ N.
Compass courses.
(c) Compass courses : N. ½ W.   W. ¾ N.   N.E. ¾ N.   E.
Magnetic courses.
(d) Ship's head by compass : N.E.
Bearings by compass : S.W. ¼ S.   S.E. ¾ S.
Magnetic bearings.

62. Example.—
Deviation at N. (21° W.)   S.   (20° E.)
N.E. (10 W.)   S.W. (23 E.)
E.   (2 W.)   W.   (5 E.)
S.E. (4 E.)   N.W. (17 W.)

NOTE.—In computing the coefficients, E. deviation is marked +, and W. deviation −.
To find coefficient A.
Add (algebraically) the deviations on N., E., S., and W., and divide by 4.

Deviation on N. −21°
E. − 2
S. +20
W. + 5

4)+ 2

+ ½° = A, i.e. +A = ½°

To find coefficient B.
Add (algebraically) the deviation on E. and W., after reversing the sign of the latter, and divide by 2.

$$\begin{array}{r} \text{Deviation on E.} \quad -2° \\ \text{W.} \quad -5 \text{ (reversed)} \\ \hline 2)-7 \\ \hline -3\tfrac{1}{2}° \text{ i.e. } -B = 3\tfrac{1}{2}° \end{array}$$

To find coefficient C.
Add (algebraically) the deviations on N. and S., after reversing the sign of the latter.

$$\begin{array}{r} \text{Deviation on N.} \quad -21° \\ \text{S.} \quad -20 \text{ (reversed)} \\ \hline 2)-41 \\ \hline -20\tfrac{1}{2}° \text{ i.e. } -C = 20\tfrac{1}{2}° \end{array}$$

To find coefficient D.
Add (algebraically) the deviations on N.E., S.W., S.E., and N.W., after reversing the signs of the two latter, and divide by 4.

$$\begin{array}{r} \text{Deviation on N.E.} \quad -10° \\ \text{S.W.} \quad +23 \\ \text{S.E.} \quad -4 \\ \text{N.W.} \quad +17 \end{array} \Big\} \text{ (reversed)}$$

$$\begin{array}{r} \hline 4)+26 \\ \hline +6° \ 30' \text{ i.e. } +D = 6\tfrac{1}{2}° \end{array}$$

To find coefficient E.
Add (algebraically) the deviations on N., S., E., and W., after reversing the signs of the two latter, and divide by 4.

$$\begin{array}{r} \text{Deviation on N.} \quad -21° \\ \text{S.} \quad +20 \\ \text{E.} \quad +2 \\ \text{W.} \quad -5 \end{array} \Big\} \text{ (reversed)}$$

$$\begin{array}{r} \hline 4)-4 \\ \hline -1° \text{ i.e. } -E = 1° \end{array}$$

SYLLABUS.

Ship's head by compass.	Coefficient A = +0° 30'.	Coefficient B = -3° 30'.	Coefficient C = -20° 30'.	Coefficient D = +6° 30'.	Coefficient E = -1°.	Total deviation.
N.	+0 30	0 0	−20 30	0 0	−1 0	−20 0
N. by E.	+0 30	−0 41	−20 7	+2 29	−0 54	−18 43
N.N.E.	+0 30	−1 20	−18 56	+4 36	−0 42	−15 52
N.E. by N.	+0 30	−1 56	−17 2	+6 0	−0 24	−12 52
N.E.	+0 30	−2 28	−14 30	+6 30	0 0	− 9 58
N.E. by E.	+0 30	−2 55	−11 23	+6 0	+0 24	− 7 24
E.N.E.	+0 30	−3 14	− 7 51	+4 36	+0 42	− 5 17
E. by N.	+0 30	−3 26	− 4 0	+2 29	+0 54	− 3 33
E.	+0 30	−3 30	0 0	0 0	+1 0	− 2 0
E. by S.	+0 30	−3 26	+ 4 0	−2 29	+0 54	− 0 31
E.S.E.	+0 30	−3 14	+ 7 51	−4 36	+0 42	+ 1 13
S.E. by E.	+0 30	−2 55	+11 23	−6 0	+0 24	+ 3 22
S.E.	+0 30	−2 28	+14 30	−6 30	0 0	+ 6 2
S.E. by S.	+0 30	−1 56	+17 2	−6 0	−0 24	+ 9 12
S.S.E.	+0 30	−1 20	+18 56	−4 36	−0 42	+12 48
S. by E.	+0 30	−0 41	+20 7	−2 29	−0 54	+16 33
S.	+0 30	0 0	+20 30	0 0	−1 0	+20 0
S. by W.	+0 30	+0 41	+20 7	+2 29	−0 54	+22 53
S.S.W.	+0 30	+1 20	+18 56	+4 36	−0 42	+24 40
S.W. by S.	+0 30	+1 56	+17 2	+6 0	−0 24	+25 4
S.W.	+0 30	+2 28	+14 30	+6 30	0 0	+23 58
S.W. by W.	+0 30	+2 55	+11 23	+6 0	+0 24	+21 12
W.S.W.	+0 30	+3 14	+ 7 51	+4 36	+0 42	+16 53
W. by S.	+0 30	+3 26	+ 4 0	+2 29	+0 54	+11 19
W.	+0 30	+3 30	0 0	0 0	+1 0	+ 5 0
W. by N.	+0 30	+3 26	− 4 0	−2 29	+0 54	− 1 39
W.N.W.	+0 30	+3 14	− 7 51	−4 36	+0 42	− 8 1
N.W. by W.	+0 30	+2 55	−11 23	−6 0	+0 24	−13 34
N.W.	+0 30	+2 28	−14 30	−6 30	0 0	−18 2
N.W. by N.	+0 30	+1 56	−17 2	−6 0	−0 24	−21 0
N.N.W.	+0 30	+1 20	−18 56	−4 36	−0 42	−22 24
N. by W.	+0 30	+0 41	−20 7	−2 29	−0 54	−22 19
N.	+0 30	0 0	−20 30	0 0	−1 0	−19 0

Exercises.—
1. From the following deviations on ship's head by compass, find the coefficients A, B, C, D, E, and construct a table of deviations from W. by N. to N.E. by E. :—

Deviation at N. (29½° W.)   S. (27¼° E.)
N.E. (26¾ W.)   S.W. (17¾ E.)
E. (6½ E.)   W. (4½ W.)
S.E. (30½ E.)   N.W. (21½ W.)

2. From deviations found on ship's head by compass as follows, find the coefficients A, B, C, D, E, and construct a table of deviations from E. to W. through S. :—

Deviation at N. (22° E.)   S. (25° W.)
N.E. (15 E.)   S.W. (18 W.)
E. (0)   W. (0)
S.E. (15 W.)   N.W. (20 E.)

3. From the following deviations on ship's head by compass, find the coefficients A, B, C, D, E, and make a table of deviations for all points from E.N.E. to S. :—

Deviation at N. (26° E.)   S.   (26° W.)
N.E. (14 E.)   S.W. (15 W.)
E.   (16 W.)   W.   (9 E.)
S.E. (23 W.)   N.W. (25 E.)

4. Given the following deviations on ship's head by compass, required the coefficients A, B, C, D, E, and make a table of deviations for all points from N. through W. to S. :—

Deviation at N. (13° W.)   S.   (18° E.)
N.E. (18 W.)   S.W. (15 E.)
E.   (5 W.)   W.   (4 E.)
S.E. (5 E.)   N.W. (8 W.)

5. Given the following deviations on ship's head correct magnetic, construct a Napier's curve, find the coefficients B, C, D, and make a table of deviations for all points from N.N.W. to S.W. by S. :—

Deviation at N. by E. (mag.) (6° E.)   S.W. by S. (15° W.)
E.N.E.   (15 E.)   W. by S.   (22 W.)
S.E. by E.   (10 E.)   N.W. by W. (16 W.)
S.S.E.   (14 E.)   N.N.W.   (5 W.)

69. **Exercises.—**
1. Given the Sun's declination, 22° 20′ S. ; lat. 52° 10′ N. ; apparent time at ship, 10^h 16^m A.M. ; and the Sun's bearing by compass, S.S.E. Required the Sun's true azimuth ; also the deviation of the compass, supposing the variation to be 20° W.

NOTE.—The azimuth is found in the well-known Burdwood's Tables, and no explanation is required.

2. 1899, January 3^d, at 3^h 30^m P.M. apparent time at ship, in lat. 43° 40′ N., long. 55° 45′ W. Required the Sun's true azimuth.

3. 1899, March 20th, in lat. 22° 15′ S., long. 32° 30′ W., when the correct mean time at Greenwich was 19^d 21^m 50^s. Required the true bearing of the Sun.

4. 1899, June 21st P.M. at ship, in lat. 56° 42′ N., long. 140° W., when a chronometer showed 21^d 15^h 40^m 10^s correct M.T.G. Required the Sun's true azimuth, and the deviation of the compass, supposing the variation to be 33° E., and the bearing by compass, W. 12° 30′ S.

5. 1899, August 1st P.M. at ship, in lat. 38° 35′ S., long. 129° 50′ E., when a chronometer showed 31^d 10^h 32^m 40^s correct M.T.G. Required the Sun's true azimuth ; and, supposing the Sun's bearing by compass to be E. by N., and the variation 0, required the deviation.

6. 1899, October 15th A.M. at ship, on the equator in long. 125° W., when a chronometer showed 15^d 3^h 30^m correct M.T.G. Required the Sun's true azimuth, and the deviation, supposing the variation to be 6° 40′ E., and the bearing by compass E. by S. ¾ S.

70. **Remark.—**In the short time available at an examination, it would be troublesome to find, by direct calculation, what stars would satisfy the conditions as to altitude and hour angle to render them suitable for azimuths. Therefore it is better to make use of tables, where the work is already done.

Table IV. in " Towson's Deviation " will answer the purpose, the sidereal time and latitude being known.

Example.—
1899, October 10th, at 11^h 30^m P.M. mean time at ship, in lat. 36° 50′ N., long. 62° W., what stars would be in a good position for determining the correction of the compass ?

SYLLABUS.

To find the sidereal time—

M.T.S. Oct. $10^d\ 11^h\ 30^m$ Long. $62°$ W. = $4^h\ 8^m$
                4   8
—————
M.T.G.   10 15 38

Sid. time at noon $13^h\ 15^m\ 29^s$ (p. II., N.A.)
Acceleration (15^h) 2 28⎫
             (38^m) 6⎭ (p. 586, N.A.)
M.T.S. 11 30   0
—————
Sid. time of observation 0 48   3

Entering Table IV. (Towson) with the nearest hour of sidereal time and the nearest latitude, the suitable stars are found to be—

East of meridian 3 4 14
West   ,,   10 11

Referring to map (p. 44, Towson), these numbers correspond to—

α Orionis, Castor, Rigel (east of meridian).
Vega, Altair (west of meridian).

Exercises.—

1. 1899, February 15th, $2^h\ 30^m$ A.M. at ship, in lat. $38°$ S., long. $25°$ E. What stars would be suitable for azimuths?
2. 1899, April 20th, $1^h\ 16^m$ A.M. apparent time at ship, in lat. $42°\ 30'$ N., long. $130°$ W.
3. 1899, July 1st, $2^h\ 15^m$ A.M. mean time at ship, in lat. 21 S., long. $93°\ 30'$ E. Give the names of suitable stars for determining the correction of the compass.
4. 1899, September 20th, $11^h\ 48^m$ P.M. mean time at ship, in lat. $49°$ S., long. $168°\ 15'$ W. Name some stars which would be suitable for finding the correction of the compass.
5. 1899, June 23rd, $11^h\ 58^m$ P.M. mean time at ship, in lat. $35°$ N.; long. $44°\ 36'$ W. Give the names of stars suitable for finding the correction of the compass.
6. 1899, October 10th, $1^h\ 45^m$ P.M. apparent time at ship, in lat. $55°$ S., long. $135°$ W. Find what stars are suitable for azimuths.

72. 1899, February 15th, $2^h\ 26^m$ A.M. mean time at ship, in lat. $37°\ 42'$ S., long. $26°$ E.; the bearing by compass of the Star Arcturus was N. $56°\ 15'$ E. Supposing the variation to be $32°$ W., required the deviation.

Example.—

M.T.S. February $14^d\ 14^h\ 26^m$ S.T. at noon $21^h\ 37^m\ 9^s$
                1 44 E. long.   Acceleration   1 58
              —————   —————
M.T.G.   14 12 42   M.T.S. 14 26 0

                                          S.T. of obs. 12 5 14
Dec. of Arcturus $19°\ 42'\ 7''$ N.   R.A. of Arcturus 14 11 5
                                          —————
                                          H.A.   ,,   2 5 51 E.

The dec. and H.A. being within the limit of Burwood's Tables, the true azimuth may be taken out as in the case of the Sun, thus—

286   TEXT-BOOK ON NAVIGATION.

Corrections.   Data.
2)56   H.A. 2ʰ 6ᵐ   By tables azimuth is 146° 55'
 28   Dec. 19° 42' N.   Cor. for 2ᵐ   − 28
 31   Lat. 37 42 S.   Cor. for 42' dec. + 22
 42   ,,   42' lat.   + 11
 ─
 62   True azimuth S. 147 0 E.
124
─
6,0)130,2   or N. 33 0 E.
 22   Compass bearing N. 56 15 E.
 16
 42   Error of compass 23 15 W.
 ─   Variation   32 0 W.
 32
 64
                                          Deviation   8 45 E.
6,0)67,2
 11   By calculation—

             H.A. 2ʰ 6ᵐ

             ½ H.A. 1   3   ... cotan 0·5497   cotan 0·5497
             P. dist. 109° 42'
             Co. lat. 52 18

             Sum 162   0   ½ sum 81° 0'   sec 0·8057   cosec 0·0054
             Diff. 57 24   ½ diff. 28 42   cosine 9·9431   sine 9·6814

                            1st part 87   7   tan 11·2985
                            2nd ,, 59 53   ...   .   tan 10·2365

                  True azimuth S. 147 0 E.¹
                           or N. 33 0 E.
                  Compass bearing N. 56 15 E.

                       Correction 23 15 W.
                       Variation 32 0 W.

                       Deviation 8 45 E.

Exercises.—
   1. 1899, October 9th, 11ʰ 36ᵐ P.M. mean time at ship, in lat. 37° N., long. 63° 30' W., the Star α Orionis bore by compass E. by S. ½ S. Required the compass correction and the deviation, supposing the variation to be 9° W.
   2. 1899, April 21st, 1ʰ 20ᵐ A.M. apparent time at ship, in lat. 41° 50' N., long. 131° W., the Star Regulus bore by compass W. ¾ S. Required the correction and the deviation of the compass when the variation is 23½° E.
   3. 1899, June 30th, 2ʰ 20ᵐ A.M. M.T.S. in lat. 21° 30' S., long. 92° 45' E., the bearing by compass of the Star Antares was S.W. by W. ¼ W.; variation by chart, 9° W. Required the compass correction and the deviation.
   4. 1899, September 21st, when the M.T.G. was 21ᵈ 22ʰ 58ᵐ, lat. 48° 40' S., long. 167° 30' W., the bearing by compass of the Star α Centauri was S. ½ E. Required the correction and deviation, the variation being 19° E.
   5. 1899, June 25th, at ship when the M.T.G. was 25ᵈ 14ʰ 54ᵐ, lat. 35° 30' N., long. 45° 30' W., the bearing of the Star Marcab was E. ¼ S. Required the correction and deviation of the compass, the variation being 16¼° W.

─────────────────────
¹ The sum, because the P. dist. exceeds the Co. latitude.

6. 1899, October 11th, $8^h 41^m$ P.M. apparent time at ship, in lat. 55° 20' S., long. 130° W.; the Star Altair bore by compass W. by N. ¾ N. Required the compass error and the deviation, the variation being 15° 40' E.

7. 1899, February 22nd, at $2^h 5^m$ A.M. apparent time at ship, in lat. 5° 12' S., long. 35° W.; the observed bearing by compass of α Ursæ Majoris was N. 8° 26' E. Required the error and deviation of the compass, the variation being 11° W.

92. **Example.**—Given the angle of heeling 15°, the course N.N.E. on the port tack, and the heeling error — 12°, required the error when steering W.S.W. on the starboard tack, and heeling error 12°.

In the Traverse Table, under 2 points, 12 in latitude column gives dist. 13, which is coefficient C due to heeling. Then under 6 points course and 13 in the dist. column, the latitude is 5°, which would be the heeling error required (in degrees) if the angle of heeling were the same as before; but as the heel has changed from 15 to 12, a calculation is necessary, viz.—

$$15 : 12 :: 5° : \text{ans.}$$
$$12$$
$$15)\overline{60}$$
$$- 4° \text{ (heeling error required)}$$

N.B.—The error retains the same name (see 83, p. 277).

EXERCISES.

1. Given 20° heel on the starboard tack, steering N.W. by N., and observed heeling error +15°, required the error when steering W. by S. on the same tack, and heeling 17°.

2. Given 12° heel on the port tack, steering S.W. ½ W., and observed heeling error +9° 30', required the error when steering S.W. by W. ¾ W. on the starboard tack, and heeling 14°.

3. Given 25° heel on the starboard tack, steering E. by N. ¼ N., and observed heeling error +5° 30', required the error when steering S. by W. ¼ W. on the port tack, and heeling 16°.

4. A ship steering S.E. ½ E., with the wind at S.S.W., and heeling 9°, the observed heeling error was +4°; the wind changes to W.N.W. in a squall, and the ship now lays S.W. ½ S., heeling 19°. Required the new heeling error.

5. Course N., wind E.N.E., heel 13°, and error observed +18°. Required the error when the ship is put on the other tack by change of wind, laying N.N.E., with a list of 20°.

6. Steering S.W. ¾ W., wind S.S.E., heel 12°, and observed heeling error —6°. Required the error when the wind changes to S.W., and the ship lays N.W. by W. ¼ W., heeling 15°.

7. Steering N.W. ¾ N., heeling 12°, wind W.S.W., and error observed —7° 12'. Required the error when the course is changed to S., and heel 15°, with no change of wind.

8. Steering S. by E. ½ E., heeling 20° to starboard, heeling error observed —13° 30'. Required the error when steering E. on the same tack and heeling 18°.

9. Course N., wind W., the error observed when heeling 14° was 12° 30'. Required the error when on the other tack steering S. and heeling 16°.

10. Given heel 8½°, course N.W. by W. on the starboard tack, heeling error observed —2°, required the error when the wind veers so as to allow the ship to lay N. by W. on the same tack, heeling 8½° by clinometer.

288   TEXT-BOOK ON NAVIGATION.

11. Steering E. by S. ⅜ S. on the port tack, and heeling 19°, the observed heeling error was +4° 40'. Required the error when steering S.S.W. on the same tack, the clinometer showing 13½°.

12. Heeling 23° when steering E. by N. ¼ N., and wind at N. by E., the heeling error observed was +4½°. Required the error when steering N.W. by W. ¾ W. on the starboard tack and heeling 17½°.

ANSWERS TO PRACTICAL EXERCISES.

31. 1. N. 37° W. 2. N. 61° E. 3. S. 31° E.
    4. −11°·6. 5. +14°·1. 6. +5°·5.

61. 1. (a) Magnetic bearing required, N. 56° E.
       (b) N. ¼ E., S. 70° E., S. 66° W., N. 7° W.
       (c) S. 19° W., N. 59° W., N., S. 71° E.
       (d) E. 11° N., N. 3° E.
    2. (a) N. 73½° W.
       (b) S. 37° E., S. 18° W., N. 65° W., N. 78° E.
       (c) N.W. by W., W. by S., S. 28° W., S. 80° E.
       (d) N. 74½° E., N. 10¼° W.
    3. (a) N. by W.
       (b) W.S.W., N. 65° W., N. 7° E., S. 75° E.
       (c) N. 26° E., N. 51° E., N. 87° E., S. 21° W.
       (d) S. 78° W., S. 5° W.
    4. (a) S. 65° W.
       (b) S. 40° E., N., N. 63° W., S. 21° W.
       (c) N. 40° W., N. 10° E., N. 64° E., S. 63° E.
       (d) S. 51° E., N. 36° E.
    5. (a) S. 11° E.
       (b) S. 84° W., N. 48° E., S. 58° E., S. 11° E.
       (c) N. 85° E., N. 13° E., N. 61° W., S. 14° W.
       (d) N. 5° E., N. 75° E.
    6. (a) N. 2½° W.
       (b) S. 76° E., S. 35° E., N. 58° W., N. 50° E.
       (c) N. by W., S. 76° W., N. 27° E., S. 82½° E.
       (d) S. 32½° W., S. 46° E.

62. 1. A, 0 ; B, +5½ ; C, −28½ ; D, −4½ ; E, −1.
       −8 20, −12 7, −15 50, −19 32, −23, −25 58, −28 12, −29 30,
       −29 20, −28 6, −25 12, −20 46, −15 2.
    2. A, −0 45 ; B, 0 ; C, +23 30 ; D, −3 15 ; E, −0 45.
       0, −3 51, −7 49, −11 38, −15 22, −18 40, −23, −24 30, −25,
       −25 18, −24 24, −22 28, −19 22, −15 26, −10 37,
       −5 27, 0.
    3. A, −0 30 ; B, −10 ; C, +26 ; D, −0 45 ; E, +0 30.
       −0 35, −5 57, −11, −15 32, −19 21, −22 41, −25 15, −26 49,
       −27 25, −27 15, −26.
    4. A, +1 ; B, −4½ ; C, −15½ ; D, 0 ; E, +1½.
       −13, −11 56, −10 32, −8 49, −6 37, −4 27, −1 50, +1 1, +4,
       +7 2, +10 2, +12 47, +15 9, +16 57, +18 6, +18 28,
       +18.
    5. B, +21 ; C, +3 ; D, +3.
       −8, −14, −18, −20, −21, −21½, −21, −20, −18, −16, −14
       −11.

69. 1. N. 155° 30' E., dev. 18° E. 2. S. 47° 37' W. 3. N. 80° 56' E.
    4. N. 72° 56' W., dev. 13° 7' W. 5. 13° 5' W. 6. 17° 9' W.

70. 1. E. Arcturus, α Pavonis.
       W. Sirius, Procyon, Regulus.
    2. E. Altair, Antares.

W. Regulus.
3. E. Achernar.
   W. Vega, α Centauri, Antares.
4. E. Rigel.
   W. α Crucis, α Centauri, Marcab.
5. E. Marcab.
   W. Arcturus, Spica, Antares.
6. E. Marcab, Canopus.
   Antares, Altair.
72. 1. Corr. 13° 47' W., dev. 4° 47' W.
    2. ,, 10° 7' E., ,, 13° 23' W.
    3. ,, 12° 5' E., ,, 21° 5' E.
    4. ,, 14° 43' W., ,, 33° 43' W.
    5. ,,   5° 25' W., ,, 10° 50' E.
    6. ,, 37° 15' E., ,, 21° 35' E.
    7. ,, 19° 13' W., ,,   8° 13' W.
92. 1. −3°. 2. −7° 15'. 3. +11° 48'. 4. +10° 18'. 5. −25° 30'. 6. +5° 54'. 7. −11° 15'. 8. 0. 9. −14° 18'. 10. −3° 30'. 11. +9° 12'. 12. −5° 56'.

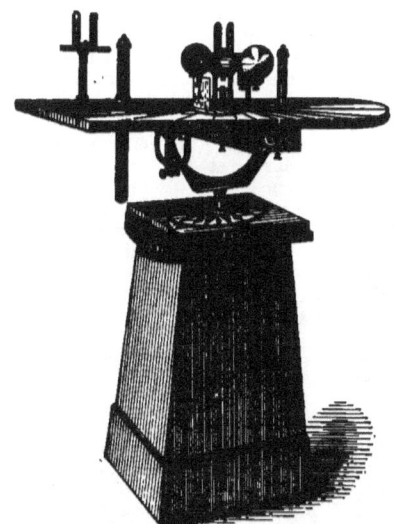

BEALL'S DEVIASCOPE.

This is a model representing the after-deck of a ship, with its binnacle and steering compass. A soft-iron bar at right angles to the deck represents the stern-post, and two soft iron bars under the deck, the athwartship beams. These bars are to give the vertical and horizontal iron effects. To imitate the effects of sub-permanent magnetism in a ship, thin steel magnets are placed in grooves in the deck, which run fore and aft, athwartship, and at intermediate angles. By this means a counterpart of a ship built with her head in any given direction may be obtained. A "Pelorus" stands on the deck. The model has horizontal and heeling motions, with clamping screws, and a clinometer on the binnacle for showing the angle of heeling. On the sides of the binnacle are jointed brass supports for the spherical soft iron

U

correctors. Under the binnacle is a tube for holding a small brass case, having holes for the reception of small round magnets for the compensation of heeling error. Other flat magnets are provided for compensating coefficients B and C, and a "Flinders" bar and dipping needle complete the outfit.

To explain, by the aid of the Deviascope, the ordinary "tentative" method of compensating a ship's compass—

The deck being horizontal by clinometer, the iron bars representing the stern-post and beams in position, and the magnets in one of the grooves, the sight vanes of the Pelorus should be clamped to the assumed correct magnetic bearing of an object as far removed as possible.

1. To compensate coefficient C.

Turn the "Pelorus" card round until the N. or S. point is at the lubber line, then turn the model until the object is seen through the sight vanes. This represents ship's head N. or S. correct magnetic. Now take up one of the compensating magnets and place it on the deck before or abaft the binnacle, at right angles to the midship groove, and move it towards or from the compass until the deviation disappears. Of course, the red end of the magnet must be towards the side which attracts the N. point of the compass.

2. To compensate coefficient B.

Bring the E. or W. point of the Pelorus card to the lubber line, and turn the model until the sight vanes are again directed to the object. Now place the "Flinders" bar on the fore side of the compass (because the vertical iron is abaft), and move it nearer until a part of the deviation is corrected (say one-half for the purpose of illustration); then place a magnet on the deck on either side of the binnacle at right angles to the athwartship groove, and move it towards or from the compass until the balance of the deviation from coefficient B is corrected.

3. To compensate coefficient D.

Turn the Pelorus card until the N.E. point is at the lubber line, and then turn the model until the object is seen through the sight vanes. If the deviation is E. on this course, it shows a coefficient $+D$; but if W., a $-D$. The soft-iron correctors are then placed in position on their supports, and brought near enough to the compass to correct the deviation. For $+D$ the correctors must be at the sides, and for $-D$ they are moved round to the fore-and-aft line.

The joints in the brass supporting arms are to allow the correctors to be moved a little forward on one side, and aft on the other, for compensating coefficient E.

4. To compensate the vertical force producing heeling error.

(1) By heeling the model and observing the heeling error—

Bring the model's bow to N. or S. by compass (which will now show correct magnetic), and incline the deck (say 10° by clinometer), and observe the deviation. If the N. point is drawn to the higher side (which is usually the case), place the small vertical magnet underneath, with the red pole up, and move it up or down until the deviation is corrected; but if the N. point goes towards the lower side, the blue pole of the magnet must be uppermost.

(2) By dipping needle—

Bring the model's head to E. or W. magnetic, and, after removing the compass card and pivot, place the dipping needle [1] on the pillar in the bowl, with its red pole pointing N. If the needle dips, it shows a blue or $+$ force below the compass; but if it rises, it shows the force is red or $-$. In the former case the red pole of the magnet must be uppermost, and in the latter the blue pole. It should then be moved up or down until the needle is horizontal.

---

[1] The dipping needle is weighted to counterbalance the natural dip, so that it is horizontal when not on the model.

APPENDIX

TABLE I.

TIME OF HIGH WATER AT STANDARD PORTS.

Date.	Queenstown.		Brest.		Devonport.		Portsmouth.		Dover.		Greenock.	
	A.M.	P.M.	A.M.	P.M.	A.M.	P.M.	A.M.	P.M.	A.M.	P.M.	A.M.	P.M.
May 2	h. m. 5 5	h. m. 5 22										
June 8			h. m. 9 1	h. m. 9 31	h. m.	h. m.						
Jan. 13					—	0 21						
„ 14					0 57	1 33	h. m.	h. m.				
Feb. 4							1 0	1 18				
„ 5							1 35	1 51	h. m.	h. m.		
May 17									9 54	10 16		
„ 18									10 37	10 58	h. m.	h. m.
Dec. 2											5 53	6 24
„ 3											6 56	7 28

TABLE II.

TIME OF HIGH WATER AND HEIGHTS.

Date.	Holyhead.		Liverpool.		Portishead.	
1897.	Time of H.W.	Height.	Time.	Height.	Time.	Height.
May 15	h. m. 8 36 A.M.	ft. in. 15 10	h. m. 3 56 P.M.	ft. in. 21 11	h. m. 7 18 A.M.	ft. in. 45 3
June 21						
March 19						
Dec. 3	5 39 A.M.	13 2				
Half mean spring range	... 8 ft. 0 in.		13 ft. 9 in.		21 ft. 0 in.	

TABLE III.
TIDAL CONSTANTS.

Port.	Time constant.	Springs.	Standard port for reference.
	h. m.	ft. in.	
Ferrol	−0 47		Brest.
Santander	−0 17	15 0	,,
Arcachon	+0 23		,,
Tour de Cordouan	+0 8		,,
Bordeaux	+3 3	15 6	,,
Ile d'Aix	−0 12		,,
Ile d'Yeu	−0 19	14 9	,,
Ile de Noirmoutier	−0 30		,,
Port Navalo	+0 8		,,
St. Nazaire	0 0		,,
Belle ile	−0 9		,,
Port Louis	−0 23		,,
Port Concarneau	−0 35		,,
Ouessant (Ushant)	0 0	19 3	,,
Gibraltar	−2 0	3 3	,,
Cadiz	−1 51	12 0	,,
(Lisbon bar)	−1 17		,,
Oporto	−1 17		,,
Abervrach	+0 27		,,
Morlaix	+1 6		,,
Plougrescan	+1 30		,,
Bréhat	+2 4		,,
St. Malo	+2 18	36 3	,,
Granville	+2 22	37 0	,,
Ile de Chausey	+2 27		,,
Jersey (St. Helier)	+2 42	31 6	,,
Guernsey (St. Peter port)	+2 50		,,
Ecrehous	+2 45		,,
Alderney	+3 0		,,
Cherbourg	+4 13		,,
Barfleur	+5 12		,,
La Hougue	+5 6		,,
Honfleur	+5 42		,,
Quillebœuf	+6 19		,,
Havre	+5 31	22 0	,,
Fécamp	+7 0	23 6	,,
Dieppe	+7 21		,,
Cayeux	+7 27		,,
Littlehampton	−0 21		Portsmouth.
Selsea bill	+0 4	16 6	,,
Bembridge point	−0 41	14 0	,,
Southampton	−1 11		,,
West Cowes	−1 26	12 6	,,
Hurst camber	−1 41	7 6	,,
Needles point	−1 55	7 6	,,
Christchurch	−2 41		,,
Poole	−2 31	6 6	,,
Portland breakwater	−4 40	6 9	,,
Bridport	+0 22	11 3	Devonport.
Lyme Regis	+0 38		,,
Exmouth	+0 44	11 0	,,
Torbay	+0 17	13 6	,,
Dartmouth	+0 33		,,
Plymouth breakwater	−0 6		,,
East Looe	−0 17		,,
Fowey	−0 29	15 0	,,
Falmouth	−0 46	16 0	,,
Penzance	−1 13		,,
Scilly isles (St. Mary)	−1 16	16 0	,,

APPENDIX. 293

TABLE IV.

TIDE TABLE.

JANUARY, 1897.

WEEK DAY.	MONTH DAY.	MOON'S TRANSIT.	BREST (Entr. of Dockyard basin). Approximate {Rise 6 10, Fall 6 20} h. m.				DEVONPORT (H.M. Dockyard). Approximate {Rise 6 0, Fall 6 10} h. m.				PORTSMOUTH (H.M. Dockyard). Approximate {Rise 7 20, Fall 5 10} h. m.			
			Morning.		Afternoon.		Morning.		Afternoon.		Morning.		Afternoon.	
			Time.	Height	Time.	Height	Time.	Height	Time.	Height	Time.	Height	Time.	Height
		h. m.	h. m.	ft. in.	h. m.	ft. in.	h. m.	ft. in.	h. m.	ft. in.	h. m.	ft. in.	h. m.	ft. in.
F.	1	10 17	1 42	16 6	2 11	17 1	3 18	14 0	3 52	14 2	9 34	12 3	10 5	12 6
S.	2	11 20	2 38	17 8	3 4	18 2	4 24	14 9	4 53	14 7	10 34	12 9	11 1	12 11
☉.	3	0a21	3 29	18 6	3 54	18 8	5 20	15 4	5 46	14 10	11 27	13 1	11 52	13 3
M.	4	1 17	4 18	18 9	4 41	18 10	6 11	15 9	6 34	14 11	—	—	0 16	13 4
Tu.	5	2 8	5 2	18 9	5 22	18 8	6 55	15 11	7 15	14 8	0 39	13 4	1 1	13 3
W.	6	2 55	5 42	18 6	6 1	18 3	7 34	15 9	7 53	14 6	1 22	13 2	1 42	13 1
Th.	7	3 39	6 19	17 11	6 38	17 6	8 11	15 3	8 28	14 0	2 1	13 0	2 20	12 10
F.	8	4 20	6 57	17 1	7 16	16 6	8 44	14 6	9 0	13 5	2 39	12 8	2 58	12 5
S.	9	5 0	7 35	16 0	7 53	15 5	9 16	13 9	9 34	12 10	3 17	12 2	3 35	11 11
☉.	10	5 41	8 11	14 10	8 31	14 4	9 52	13 0	10 10	12 3	3 52	11 8	4 9	11 5
M.	11	6 23	8 54	13 10	9 20	13 5	10 29	12 3	10 51	11 10	4 28	11 1	4 50	10 10
Tu.	12	7 7	9 50	13 2	10 25	12 11	11 17	11 7	11 48	11 6	5 17	10 7	5 45	10 4
W.	13	7 54	11 5	12 10	11 44	12 10	—	—	0 21	11 5	6 17	10 2	6 54	10 1
Th.	14	8 45	—	—	0 22	13 1	0 57	11 7	1 33	11 6	7 32	10 2	8 10	10 4
F.	15	9 38	0 58	13 5	1 31	13 11	2 9	12 0	2 46	11 11	8 47	10 7	9 21	10 10
S.	16	10 33	1 58	14 6	2 21	15 2	3 22	12 10	3 56	12 8	9 50	11 1	10 15	11 6
☉.	17	11 27	2 42	15 10	3 3	16 6	4 25	13 8	4 50	13 4	10 38	11 10	10 59	12 1
M.	18	m.	3 23	17 2	3 43	17 8	5 13	14 6	5 34	13 11	11 19	12 5	11 30	12 8
Tu.	19	0 21	4 3	18 1	4 23	18 5	5 55	15 3	6 16	14 4	11 59	12 11	—	—
W.	20	1 12	4 42	18 9	5 1	19 0	6 36	15 8	6 56	14 9	0 11	13 2	0 39	13 4
Th.	21	2 1	5 20	19 2	5 39	19 3	7 15	15 11	7 34	14 11	0 59	13 5	1 19	13 5
F.	22	2 49	5 58	19 2	6 17	19 1	7 53	15 11	8 12	14 11	1 39	13 6	1 59	13 5
S.	23	3 36	6 37	18 10	6 57	18 5	8 32	15 6	8 52	14 9	2 19	13 5	2 39	13 4
☉.	24	4 25	7 18	17 11	7 40	17 4	9 12	15 1	9 32	14 4	2 59	13 2	3 19	12 11
M.	25	5 15	8 4	16 7	8 29	15 10	9 53	14 4	10 16	13 9	3 40	12 8	4 2	12 4
Tu.	26	6 9	8 56	15 2	9 26	14 8	10 40	13 6	11 7	13 2	4 25	11 11	4 51	11 6
W.	27	7 6	10 3	14	3 10	48 14	0 11	37 12	10	—	—	5 21	11 3	5 56	10 11
Th.	28	8 6	11 36	14	0	—	—	0 12	12 8	0 52	12 4	6 38	10 9	7 24	10 9
F.	29	9 7	0 24	14 9	1 6	14 9	1 35	12 9	2 18	12 6	8 11	11 0	8 56	11 3
S.	30	10 8	1 42	15 5	2 12	16 2	3 2	13 5	3 44	13 3	9 34	11 7	10 6	12 0
☉.	31	11 5	2 38	16 10	3 3	17 6	4 20	14 5	4 50	13 11	10 35	12 5	10 59	12 8

Half-mean spring range ... 9 ft. 6 in. | 7 ft. 9 in. | 6 ft. 9 in.

TABLE B.

FOR FINDING THE HEIGHT OF THE TIDE AT ANY INTERMEDIATE HOUR BETWEEN HIGH AND LOW WATER.

Height above half tide or mean level of the sea.	Time from high water.											
	h. m. 0 0	h. m. 0 30	h. m. 1 0	h. m. 1 30	h. m. 2 0	h. m. 2 30	h. m. 3 0	h. m. 3 30	h. m. 4 0	h. m. 4 30	h. m. 5 0	h. m. 5 30	h. m. 6 0
	Add.						Subtract.					
ft.	ft. in.	ft. in.	ft. in.	ft. in.	ft. in.	ft. in.	ft. in.	ft. in.	ft. in.	ft. in.	ft. in.	ft. in.	ft. in.
3	3 0	2 11	2 7	2 1	1 6	0 9	0 0	0 9	1 6	2 1	2 7	2 11	3 0
4	4 0	3 10	3 6	2 10	2 0	1 0	0 0	1 0	2 0	2 10	3 6	3 10	4 0
5	5 0	4 10	4 4	3 6	2 6	1 3	0 0	1 3	2 6	3 6	4 4	4 10	5 0
6	6 0	5 10	5 2	4 3	3 0	1 7	0 0	1 7	3 0	4 3	5 2	5 10	6 0
7	7 0	6 9	6 1	4 11	3 6	1 10	0 0	1 10	3 6	4 11	6 1	6 9	7 0
8	8 0	7 9	6 11	5 8	4 0	2 1	0 0	2 1	4 0	5 8	6 11	7 9	8 0
9	9 0	8 8	7 9	6 4	4 6	2 4	0 0	2 4	4 6	6 4	7 9	8 8	9 0
10	10 0	9 8	8 8	7 1	5 0	2 7	0 0	2 7	5 0	7 1	8 8	9 8	10 0
11	11 0	10 8	9 6	7 9	5 6	2 10	0 0	2 10	5 6	7 9	9 6	10 8	11 0
12	12 0	11 7	10 5	8 6	6 0	3 1	0 0	3 1	6 0	8 6	10 5	11 7	12 0
13	13 0	12 7	11 3	9 2	6 6	3 4	0 0	3 4	6 6	9 2	11 3	12 7	13 0
14	14 0	13 6	12 1	9 11	7 0	3 7	0 0	3 7	7 0	9 11	12 1	13 6	14 0
15	15 0	14 6	13 0	10 7	7 6	3 11	0 0	3 11	7 6	10 7	13 0	14 6	15 0
16	16 0	15 5	13 10	11 4	8 0	4 2	0 0	4 2	8 0	11 4	13 10	15 5	16 0
17	17 0	16 5	14 9	12 0	8 6	4 5	0 0	4 5	8 6	12 0	14 9	16 5	17 0
18	18 0	17 5	15 7	12 9	9 0	4 8	0 0	4 8	9 0	12 9	15 7	17 5	18 0
19	19 0	18 4	16 5	13 5	9 6	4 11	0 0	4 11	9 6	13 5	16 5	18 4	19 0
20	20 0	19 4	17 4	14 2	10 0	5 2	0 0	5 2	10 0	14 2	17 4	19 4	20 0
21	21 0	20 3	18 2	14 10	10 6	5 5	0 0	5 5	10 6	14 10	18 2	20 3	21 0
22	22 0	21 3	19 1	15 7	11 0	5 8	0 0	5 8	11 0	15 7	19 1	21 3	22 0
23	23 0	22 3	19 11	16 3	11 6	5 11	0 0	5 11	11 6	16 3	19 11	22 3	23 0
24	24 0	23 2	20 9	17 0	12 0	6 2	0 0	6 2	12 0	17 0	20 9	23 2	24 0

Rule.—To find the height of the tide above the zero of the tables at any intermediate hour between *high and low water.*

The zero of the tables is the mean height of the low water of ordinary spring tides.

From the height in the tables subtract the half mean spring range, the remainder will be the height above the half tide or mean level of the sea, with which enter Table B, and, under the time from high water, take out the corresponding correction, and, as directed, add it to, or subtract it from, the half mean spring range; the result will be the height of the tide at that time above zero or the low-water standard of the tables.

APPENDIX.

SUMMARY OF USEFUL FORMULÆ.

I. *Plane Trigonometry.*

1. $\sin A = \dfrac{1}{\operatorname{cosec} A}$, $\cos A = \dfrac{1}{\sec A}$, $\tan A = \dfrac{1}{\cot A}$

   $\operatorname{cosec} A = \dfrac{1}{\sin A}$, $\sec A = \dfrac{1}{\cos A}$, $\cot A = \dfrac{1}{\tan A}$

2. $\tan A = \dfrac{\sin A}{\cos A}$, $\cot A = \dfrac{\cos A}{\sin A}$

3. $\sin^2 A + \cos^2 A = 1$
   whence, $\sin^2 A = 1 - \cos^2 A$, and $\cos^2 A = 1 - \sin^2 A$

4. $\sec^2 A = 1 + \tan^2 A$, and therefore $\tan^2 A = \sec^2 A - 1$
   $\operatorname{cosec}^2 A = 1 + \cot^2 A$, and therefore $\cot^2 A = \operatorname{cosec}^2 A - 1$

5. $\sin A = \cos(90° - A) = \sin(180° - A)$
   $\cos A = \sin(90° - A) = -\cos(180° - A)$

6. $\sin(A + B) = \sin A \cdot \cos B + \cos A \cdot \sin B$
   $\sin(A - B) = \sin A \cdot \cos B - \cos A \cdot \sin B$
   $\cos(A + B) = \cos A \cdot \cos B - \sin A \cdot \sin B$
   $\cos(A - B) = \cos A \cdot \cos B + \sin A \cdot \sin B$

7. $\tan(A + B) = \dfrac{\tan A + \tan B}{1 - \tan A \cdot \tan B}$
   $\tan(A - B) = \dfrac{\tan A - \tan B}{1 + \tan A \cdot \tan B}$
   $\cot(A + B) = \dfrac{\cot A \cdot \cot B - 1}{\cot A + \cot B}$
   $\cot(A - B) = \dfrac{\cot A \cdot \cot B + 1}{\cot B - \cot A}$

8. $\sin A + \sin B = 2 \sin \tfrac{1}{2}(A + B) \cdot \cos \tfrac{1}{2}(A - B)$
   $\sin A - \sin B = 2 \cos \tfrac{1}{2}(A + B) \cdot \sin \tfrac{1}{2}(A - B)$
   $\cos A + \cos B = 2 \cos \tfrac{1}{2}(A + B) \cdot \cos \tfrac{1}{2}(A - B)$
   $\cos A - \cos B = 2 \sin \tfrac{1}{2}(A + B) \cdot \sin \tfrac{1}{2}(B - A)$

9. $\sin 2A = 2 \sin A \cdot \cos A$
10. $\cos 2A = \cos^2 A - \sin^2 A$
    $\quad\;\; = 1 - 2 \sin^2 A$
    $\quad\;\; = 2 \cos^2 A - 1$
    $\cos A = 2 \cos^2 \tfrac{1}{2}A - 1$
    $\quad\;\; = 1 - 2 \sin^2 \tfrac{1}{2}A$
    $\cos^2 \tfrac{1}{2}A = \tfrac{1}{2}(1 + \cos A)$
    $\sin^2 \tfrac{1}{2}A = \tfrac{1}{2}(1 - \cos A)$

NOTE.—These formulæ are equally true of the angles of spherical triangles.

Relations of sides and angles of a plane triangle whose angles are A, B, C, and opposite sides a, b, c.

11. $A + B + C = 180°$.

12. $\dfrac{\sin A}{a} = \dfrac{\sin B}{b} = \dfrac{\sin C}{c}.$

compass ; and, supposing the variation to be 13° 0' E., required the deviation of the compass for the direction of the ship's head.

9. 1899, January 1st A.M. at ship, the observed Alt. of the Sun's lower limb was 6° 55' 32"; index error, −1' 30" ; height of eye, 20 ft. ; the sun's bearing by compass, S. 33° 45' E. ; time by chron., 31^d 12^h 17^m 29^s, which was slow for mean noon at Greenwich, October 12th, 1^m 3^s, and on December 21st was fast 3^m 40^s·5 ; the latitude by Mer. Alt. at noon was 54° 22' N., the ship having run between observations and noon E. ½ S. 40 miles. Required the longitude by chronometer at the time of taking the observations ; also brought up to noon. Required also the true azimuth, and error of the compass ; and, supposing the variation to be 0° 0', required the deviation of the compass for the direction of the ship's head.

10. 1899, January 1st, 4^h 30^m M.T.S. P.M. at ship, in lat. 38° 30' S., long. 15° 45' E., the Sun bore by compass N. 50° 21' W. Required the time azimuth and error of the compass by the "Time Azimuth" Tables ; and, supposing the variation to be 30° 0' W., required the deviation of the compass for the direction of the ship's head.

11. 1899, January 1st A.M. at ship, lat. by account 33° 25' N., long. 52° 10' W., the observed Alt. of Sun's lower limb near the meridian was 32° 59' 15" south of the observer ; index error, −5' 34" ; height of eye, 15 ft. ; time by watch, 1^d 3^h 13^m 31^s, which was fast of A.T.S. 3^h 46^m 15^s ; the difference of longitude made to the east since the error of watch on A.T.S. was determined was 16'. Required the latitude by reduction to the meridian.

12. 1899, January 1st A.M. at ship, at sea, and uncertain of my ship's position, when the chronometer indicated 31^d 23^h 18^m 14^s M.T.G., the observed Alt. of the Sun's lower limb was 10° 17' 0", and again, being P.M. at ship, when the chronometer indicated 1^d 3^h 16^m 44^s M.T.G., the observed Alt. of the Sun's lower limb was 10° 24' 10" ; index error, +3' 30" ; height of eye 20 ft. ; the ship having made 40 miles on a N. 60° E. course (true) during the interval between the observations. Required the line of position when the first altitude was observed, and the true bearing of the Sun and the position of the ship when the second altitude was observed, by Sumner's method of projection on the chart, assuming latitudes 52° 0' and 52° 30' N.

13. 1899, January 2nd P.M. at ship in lat. by account 50° 10' N., long. 30° 15' W. Required approximately the apparent time at ship when the Star Aldebaran will be on the meridian ; also by inspection.

State where tables giving the approximate times of most of the principal stars passing the meridian are to be found, the numbers of the tables, and whether given in mean or apparent time.

Can the approximate meridian passage of a planet and its approximate altitude be found in the same way as in the case of a star ?

State where the times of the meridian passage of the principal planets may be found, and state whether the time there given is mean or apparent.

14. 1899. Supposing you are not familiar with the stars, and wish to take an observation, find the names (from Nautical Almanac, pages 294–304) of the stars of the 1st, 2nd, and 3rd magnitudes; that will be within three-quarters of an hour east or west of your meridian above the pole and above the horizon, at about 10^h 15^m P.M. A.T.S. on January 10th, in lat. 53° 24' N., long. 3° 12' W. Required also the meridian distance or hour angle of each of the stars, and state also whether they will be N. or S. of you.

Are you acquainted with any maps, diagrams, or any other means for facilitating the recognition of the fixed stars and planets ? If so, give a short description of the one you prefer, and how you use it.

15. Having found by preceding question the principal stars about to cross the meridian on January 2nd, 1899, lat. 50° 10' N., long. 30° 15' W., compute the approximate Meridian Alt. of one of them, "Aldebaran," as a guide to setting the sextant for observation, the height of eye being 20 ft., and index error 1' 10" to add.

APPENDIX. 299

16. 1899, the observed Alt. of the Star Alpheratz on the meridian, January 10th, was found to be 70° 41' 10", bearing south of the observer; height of eye, 14 ft.; index error, $-2'$ 10". Required the latitude.

17. 1899, January 23^d A.M. at ship, in lat. by account 49° 57' N.; about five hours after the Star Regulus had passed the meridian, its altitude was observed to be 16° 49' 0"; time by chron., 22^d 22^h 27^m 41^s, which was fast for mean noon at Greenwich on January 8th 15^m 33^s, and losing 4"·4 daily; height of eye, 23 ft.; index error, $+1'$ 10". Required longitude by chronometer.

18. 1899, January 11th, at 6^h 10^m P.M. apparent time at ship, in lat. 40° 45' N., long. 30° 30' W., the Star Aldebaran bore by compass S.E. by E. Required the true azimuth and error of the compass by the "Time Azimuth Tables;" and, supposing the variation to be 27° 30' W., required the deviation of the compass for the direction of the ship's head.

19. 1899, January 20th, 8^h 15^m P.M. mean time at ship, in long. 15° 30' W., the observed Alt. of the Pole Star out of the meridian was 49° 20' 10"; height of eye, 20 ft.; index error, $+2'$ 10". Required the latitude.

20. 1899, January 5th A.M. at ship, in long. 29° 30' W., the observed meridian Alt. of the Moon's lower limb was 30° 14' 30" bearing south of the observer; height of eye, 20 ft.; index error, $+1'$ 10". Required the latitude.

Chart.

21. (a) Using Card A., find the course to steer by compass from Mull of Galloway to the Calf Rock, I.O.M.; also the distance, variation, and deviation.

(b) With the ship's head on the above-named compass course, Ayre Point bore by compass S. 62½° E., and Peel bore S. 6° W. by the same compass. Find the ship's position.

(c) With ship's head as above, Bradda Head bore by compass S. 17° W., and after continuing six miles on the above-named course it bore S. 51¼° E. Find the ship's position and distance from Bradda Head.

(d) Find the course to steer by compass from Mull of Galloway towards Chicken's Rocks, to counteract the effect of a current which set west (cor. mag.) three miles per hour, the vessel sailing eight miles per hour, also the distance made good towards the Chicken's Rocks in four hours.

22. 1897, January 13th, 2^h 13^m A.M. mean time, being off Limerick, took cast of the lead. Required the reduction to be applied to the cast before comparing with the chart.

Napier's Curve.

23. Correct magnetic bearing—

Ship's head by standard compass.	Bearing of distant object by standard compass.	Deviation required.	Ship's head by standard compass.	Bearing of distant object by standard compass.	Deviation required.
N.	N. 7° W.		S.	N. 10° E.	
N.E.	N. 27° W.		S.W.	N. 25° E.	
E.	N. 21° W.		W.	N. 30° E.	
S.E.	N.		N.W.	N. 20° E.	

From the above table construct a Napier curve, and give the courses you would steer by standard compass to make the following courses correct magnetic:—

Correct magnetic courses: E. W.S.W. S.E. by S. N.N.W.
Compass courses.

APPENDIX.

Suppose you steer the following courses by standard compass, find the correct magnetic courses from the curve drawn :—
Compass course : N.W.   N.E. by N. ¼ N.   S.S.W.   E.S.E.
Correct magnetic courses.

Having taken the following compass bearings with ship's head W.S.W., find the correct magnetic bearings W. by S.   North.

24. 1899, January 20th P.M. at ship, in lat. 45° 40′ N., approximate M.T.G. by chron., 20^d 11^h 50^m 30^s. With the following observations, find the true and apparent Altitudes of the Moon, the error of the chronometer for Greenwich mean time, and the longitude of the place : height of eye, 20 ft. ; index error for distance, +1′ 5″ for the altitude 0 ; observed alt. Star Regulus, 24° 21′ 50″ E. ; observed distance between Regulus and the Moon's remote limb, 91° 35′ 30″.

25. January 10th, in lat. by account 50° 0′ N., long. 15° 0′ W. With the following observations of the Sun's lower limb, find the latitude by double Altitudes at the time of taking the second observation :—

A.T.S. 9^h 36^m 00^s A.M.   Obs. Alts. 11° 1′ 50″   Index error, +1′ 20″
A.T.S. 1 21 00 P.M.   Obs. Alts. 15 41 30   Height of eye, 20 ft.

The Sun's true bearing at first observation was S.E. by S. ; run in interval, E.N.E. 10 knots per hour.

26. 1899, January 20th, at St. Paul's, Liverpool, in lat. 53° 25′ N., long. 2° 59′ 30″ W., the following times were noted by ship's chronometer when the Sun had equal Altitudes. Required the error of the chronometer for M.T.S. and M.T.G. :—

A.M. Obs.   P.M. Obs.
19^d 22^h 23^m 14^s 20^d 2^h 43^m 14^s

27. 1899, February 17th, at about 3^h 53^m P.M., the ship's correct position being lat. 30° 25′ S., long. 95° 27′ 15″ W., the observed Alt. of the Sun's lower limb was 36° 8′ 0″ ; index cor. −1′ 17″ ; height of eye, 24 ft. Required the error of a chronometer which showed 10^h 30^m 19^s on M.T.G. and M.T.S.

28. 1899, January 1st P.M. at ship, in lat. by account 35° 36′ N., long. 35° 45′ W., when a chronometer indicated 1^d 8^h 30^m 40^s, which was slow 7^m 20^s on M.T.G., an observed Alt. of the Star Arietis east of the meridian was 71° 44′ 40″, and the observed Altitude of Marcab taken at the same time was 55° 59′ 50″ west of the meridian ; index error, −1′ 30″ ; height of eye, 20 ft. Required the position of the ship and the true bearing of the stars by Sumner's method by projection on the chart, assuming lats. 35° 30′ and 35° 40′ N.

29. Required the first course and distance on the arc of a Great Circle from Wolf Rock in lat. 49° 57′ N., long. 5° 48′ W., to Charleston, lat. 32° 45′ N., long. 79° 52′ W. ; also the position of Vertex, the latitude of points 5°, 10°, 15°, 20° from Wolf Rock, and the latitude at which the 61st W. meridian cuts the arc of the Great Circle.

30. (a) A point in lat. 49° 52′ N. bears N.E. ¼ E. from a ship distant 12 miles. Required the latitude of the ship.

(b) What course would a steamer steaming 9 knots per hour have to steer in order to reach a point 27 miles S.W. by W. of her in a current running S.E. 3 miles an hour ? How long would she be in reaching it ?

(c) In a spherical triangle, $C = 90°$, $A = 36° 38′$, $B = 101° 42′$. Find a, b, and c.

(d) Two places in lat. 32° N. have a difference of longitude of 65°. Find latitude of Vertex.

Paper II.

1 and 2. Multiply 47290 by ·0045, and divide 4765·07 by 32406·5 by logarithms.

3.

Hour.	Courses.	Knts.	10ths.	Winds.	Leeway.	Deviation.	Remarks.
					pts.		
1	S.E. by E. ¼ E.	10	7	S.S.W.	¾	25° W.	A point in lat. 25° 15′ S., long. 47° 20′ W., bearing by compass N.N.E. ½ E., ship's head, S.E. by E. ¼ E; deviation as per log. Distant 15 miles. Variation, 20° W.
2		11	3				
3		9	9				
4		11	1				
5		11	—				
6	E. by S. ¾ S.	10	4	S.	1¼	24° W.	
7		9	7				
8		10	5				
9		9	8				
10		9	6				
11	S.E. by S.	9	7	S.W.	1	18° W.	
12		10	2				
1		10	4				
2		11	7				
3	S.W. by S.	9	6	S.S.E.	1½	25° E.	
4		10	2				
5		11	2				
6	S.	9	5	E.S.E.	1¼	5° E.	
7		10	7				A current set correct magnetic N. ½ E. at 2¼ miles per hour for all the day.
8		9	8				
9	S. by E. ¾ E.	10	4	E.	1	9 W.	
10		11	3				
11		10	9				
12		9	4				

Find the course and distance made good, and latitude and longitude of the ship.

4. 1899, April 20th, in long. 135° E., the observed Mer. Alt. of the Sun's upper limb north of the observer was 65° 14′ 20″ ; index error, −3′ 20″ ; height of eye, 16 ft. Required the latitude.

5. Required the compass course and distance by Mercator's sailing—

From A, lat. 52° 50′ S.		Long. 81° 10′ W.
to B, ,, 41 12 N.		,, 127 55 W.
Variation 18 25 E.		Deviation 22 15 E.

6. A ship from lat. 40° 30′ N., long. 39° 43′ W., sailed due west 89 miles. What is her longitude in ?

7. 1897. Find the times of high water A.M. and P.M. at Oporto on January 15th and 29th.

8. 1899, January 15th, at 6ʰ 49ᵐ P.M. apparent time at ship, in lat. 29° 0′ S., long 48° 10′ W., the Sun's compass bearing was W. ¾ S. Required the true amplitude, and error of the compass ; and, supposing the variation to be 0°, required the deviation for the direction of the ship's head.

302   APPENDIX.

9. 1899, January 1st A.M. at ship, lat. 0° 20' N., the observed Alt. of the Sun's lower limb was 32° 2' 20"; index correction, +3' 20"; height of eye, 18 ft.; when a chronometer showed 31ᵈ 20ʰ 27ᵐ 57ˢ, which was slow on October 9th 1ᵐ 9ˢ, and on November 26th was fast 29ˢ. Required the longitude; and if the compass bearing be S.E. by E., and variation 10° W., find the deviation for the direction of the ship's head.

10. 1899, February 10th, mean time at ship, 8ʰ 10ᵐ A.M., in lat. 51° 25' N., long. 42° 30' W.; the Sun's compass bearing was S. ¼ W. Required the Sun's true azimuth by the "Tables," error of the compass, and deviation, if the variation be 40° W.

11. Using Deviation Card A., find the compass course and distance from the Tuskar to the Skerries.
With the ship's head on the above course, Lucifer L.V. bore by compass N. 80° W., and Blackwater L.V. bore N. 18½° W. by the same compass. Find the ship's position.
Steering as above, Carnarvon L.V. bore by compass N. 84° E., and after continuing 12 miles on the above course it bore S. 27° E. by the same compass. Find the ship's position and distance from Carnarvon L.V.

12. 1899. At what time will Fomalhaut pass the meridian of a place, in long. 125° 30' W., on March 23rd?

13. 1899. What bright stars are within about 2 hours of the meridian of a place in lat. 32° 40' N., long. 130° E., on May 17th at midnight? What is the approximate Mer. Alt. of Arcturus?

14. 1899, May 17th, the Mer. Alt. of Arcturus was 77° 9' 30" bearing south; index correction, −2' 11"; height of eye, 22 ft. Required the latitude.

15. 1899, July 1ˢᵗ P.M. at ship, the mean time being about 10ʰ 30ᵐ in lat. 45° 15' S., long. by account 123° 40' E., the observed Alt. of Spica west of the meridian was 30° 31' 10"; index correction, +1' 40"; height of eye, 14 ft.; time by chron., 2ʰ 14ᵐ 37ˢ, which was fast 8ˢ on March 13th, and slow 28ˢ on April 22nd of G.M.T. Required the longitude.

16. 1899, May 18ᵈ, when the mean time at ship was 4ʰ 15ᵐ A.M., in lat. 32° 40' N., long. 130° E., the compass bearing of Arcturus was W. ¼ S. Required the true azimuth by the "Tables;" and if the variation be 1° 30' W., find the deviation.

17. 1899, August 25th A.M. at ship, in lat. by account 25° 16' S , long. 35° 15' W.; the observed Alt. of α Eridani south of the observer was 57° 27' 50"; index correction, +2' 11"; height of eye, 26 ft.; time by a chron., 24ᵈ 17ʰ 47ᵐ 32ˢ, which had been found to be 21ᵐ 7ˢ fast of G.M.T. Required the latitude by ex-meridian.

18. 1899, July 26th A.M.; at sea and uncertain of my position, when the chronometer showed 25ᵈ 7ʰ 39ᵐ 17ˢ G.M.T., the observed Alt. of Sun's lower limb was 20° 59' 40"; and again A.M. on the same day, when the chronometer showed 25ᵈ 12ʰ 7ᵐ 57ˢ G.M.T., the observed Alt. of Sun's lower limb was 57° 6' 10"; index correction, −3' 7"; height of eye, 36 ft.; the ship having made a true N.W. ½ W. course 40 miles in the interval. Required the line of position when first observation was taken; also the Sun's true azimuth and position of the ship at the second observation, assuming lats. 49° 50' and 50° 20' N.

19. 1899, September 7th, when the G.M.T. was 7th noon in long. 125° 40' E., the observed Alt. of Polaris out of the meridian in an artificial horizon was 58° 37' 20"; index correction, − 5' 40". Required the latitude.

20. 1899, September 18th P.M., in long. 9° 33' W., the observed Mer. Alt. of Moon's lower limb, 38° 57' 50" bearing north; index correction, +6' 24"; height of eye, 29 ft. Required the latitude.

APPENDIX. 303

21.

Ship's head by compass.	Bearing of distant object.	Deviation.	Ship's head by compass.	Bearing of distant object.	Deviation.
N.	S. 43° W.		S.	W.	
N.E.	S. 50° W.		S.W.	S. 83° W.	
E.	S. 65° W.		W.	S. 65° W.	
S.E.	S. 80° W.		N.W.	S. 45° W.	

Required correct magnetic bearing and deviations, and the following by Napier's curve :—

Correct magnetic : S.E. by E.   N.N.E.   N.W. ½ W.   S. 10° W.
Courses to steer.

Courses steered : N.W. by W.   N. by W.   N.E. ½ E.   S. 50° E.
Correct magnetic.

Ship's head : E ¼ N. by compass.
Compass bearings : S.E. ¾ E.   N.E. by N.
Correct magnetic.

Required the course to steer from Tuskar to the Skerries to counteract the effect of a current which set N. 55° E. 3 miles per hour, the ship making by log 10 knots per hour, and the distance she would make good towards the Skerries in 4 hours.

22. 1897, April 10th, 4 P.M., off Llanelly took a cast of the lead. Required the correction to be applied to the cast before comparing it with the depth marked on the chart.

23. 1899, March 7ᵈ A.M. at ship in lat. 31° 25′ N.; time by chron., 6ᵈ 16ʰ 10ᵐ 15ˢ, which was estimated to be correct for G.M.T. With following observations, find true and apparent Alts. of the Moon, error of chronometer for G.M.T., and long. of ship ; height of eye, 27 ft. :—

Observed Alt. of Jupiter's centre.   Observed distance of Moon's F.L. and Jupiter's centre.
44° 19′ 50″ W.   63° 25′ 45″
Index cor.   − 1 20   Index cor.   − 2 15

24. 1899, July 27th in lat. N., long. 156° 27′ E. With following observations, find latitude by "Ivory" at time of taking second observation :—

First M.T.S. 27ᵈ 10ʰ 30ᵐ 15ˢ   Obs. Alt. Altair 49° 20′ 50″ bearing S. 22° E.
Second M.T.S. 27 12 54 24   ,,   ,,   47 0 0

The ship's true course and distance East 32 miles during interval; index correction, − 2′ 13″ ; height of eye, 25 ft.

25. 1899, February 23ᵈ P.M. at ship, when a chronometer showed 23ᵈ 14ʰ 16ᵐ 58ˢ correct G.M.T., the observed Alt. of Spica east of meridian 26° 1′ 0″, and observed Alt. of Capella taken at the same time was 31° 9′ 50″ west of the meridian ; height of eye, 26 ft. Required the position of the ship, and the true bearings of the stars by Sumner's method, by projection on the chart, assuming latitudes 19° 30′ and 20° N.

26. 1899, August 22nd, in lat. 27° 10′ N., long. 56° 17′ E., the following times were observed when the Sun had equal altitudes. Find the error of the chronometer for A.T.S., A.T.G., and M.T.G. :—

A.M. 4ʰ 19ᵐ 27ˢ   P.M. 0ʰ 7ᵐ 30ˢ
      4 25 32   0 12 15

27. Find the initial course and distance, and latitude and longitude of Vertex, by Great Circle sailing :—

From lat. 34° 2' S., long. 25° 42' E.
To ,, 5 35 N., ,, 95 19 E.

28. Construct a Mercator's chart for the above simultaneous Alts. problem, scale 10' long. = 1 in. By dead reckoning, latitude of ship is 19° 32' N., long. 45° 20' W. ; required set and drift of the current.

29. A stick 2 ft. 10 in. long placed vertically casts a shadow 3 ft. 9 in. long. Required Alt. of Sun.

30. AB is 1000 yards long, B is due N. of A. At B a distant point P bears 70° E. of N., at A it bears 41° 22' E. of N. Find AP.

31. In a right-angled spherical triangle, given C = 90°, B = 103° 20', $a = 75°$, find the other sides and angle.

32. At a place in lat. 45° 20' N., when the Sun's declination was 15° 25' N., find his hour angle, and altitude, when he bears due east.

33. A ship making 10 miles per hour through the water, has to make an E.N.E. course through a current setting S.E. by E. 3 miles per hour. Find the course to steer, and distance made good per hour in the required direction.

PAPER III.

1. Multiply ·0765 by ·000006201 by common logarithms.
2. Divide ·00652 by ·9876 by common logarithms.
3. In the following day's work, correct the courses for leeway, variation, and deviation. Find the course and distance made good, also the latitude and longitude of the ship :—

Hour.	Courses.	Knts.	10ths.	Winds.	Lee-way.	Devia-tion.	Remarks.
1	W.	7		N.N.W.	pts. 0	10° W.	Departure was
2		7					taken from a
3		7					point in lat.
4		7					60° 45' N.; long.
5	N.W.	7	5	N.N.E.	¼	20° W.	4° 43' E.; bear-
6		7	5				ing by compass,
7		7	5				S. W.; dist., 15
8		7	5				miles, ship's
9	W. by N. ¾ N.	9		S.W. by S.	¾	13° W.	head, W.; de-
10		9					viation, 10° W.
11		9					Variation, 17° W.
12		9					
1	W.S.W.	10		S.	½	7° W.	
2		10					
3		10					
4		10					
5	S.W.	10	5	W.N.W.	¼	0°.	A current set
6		10	5				N.N.E. correct
7		10	5				magnetic all the
8		10	5				day at the rate
9	W. by S.	8		S. by W.	1¼	8° W.	of ¾ mile per
10		8					hour.
11		8					
12		8					

APPENDIX.

4. 1899, January 21st, in long. 45° 0' W., the observed Mer. Alt. of the Sun's lower limb was 34° 25' 30" bearing south of the observer; index cor. 6' 40" to add; height of eye, 15 ft. Required the latitude.

5. How many miles of departure equal 5° 47' of longitude in the parallel of 54° 37' N. ?

6. Required the true course and distance from A to B by Mercator sailing, also the compass course, assuming the variation to be 25° W., and the deviation of the compass 20° 30' E.

Lat. A, 50° 10' N., long. 13° 25' W.
„ B, 47 53 N., „ 62 30 W.

7. 1897, January 13th. Find the time of high water at Black Ball Harbour A.M. and P.M.

8. 1899, January 31^d 16^h 22^m 37^s A.T.S. in lat. 53° 2' S., long. 115° 30' W., the Sun rose bearing by compass E. 36° 33' 45" S. Required the true amplitude and error of the compass; and, supposing the variation to be 21° W., required the deviation of the compass for the direction of the ship's head.

9. 1899, February 1st A.M. at ship, the observed Alt. of the Sun's lower limb was 43° 40' 50"; index error, +2' 10"; height of eye, 25 ft.; the Sun's bearing by compass, E ¼ S.; time by chron., 31^d 21^h 3^m 3^s, which was correct M.T.G. December 31st, and on January 28th was slow 21ˢ.; the latitude at noon was 36° 4' S.; course and distance since observation, N.N.W. 26 miles. Required longitude at noon and sights; required also the true azimuth, and if the variation is 25° 30' W., what is the deviation of the compass for the direction of the ship's head ?

10. 1899, March 21st A.M. at ship, time by watch 6^h 48^m M.T.S., lat. 49° 30' S., long. 155° 30' E.; the Sun bore by compass S. 75° 56' 15" E. Required the true azimuth and error of the compass by the "Time Azimuth Tables;" and, supposing the variation to be 15° 0' E., required the deviation of the compass for the direction of the ship's head.

11. 1899, January 10th P.M. at ship, in lat. by account 53° 23' N., long. 3° 0' W., the M.T. by watch was 12^h 0^m 0^s, when an observed Alt. of Castor east of the meridian was 68° 46' 10" bearing S.; height of eye, 32 ft. Required the latitude by reduction to the meridian.

12. 1899, February 2nd A.M. at ship, at sea and uncertain of ship's position, when the chronometer showed 2^d 1^h 1^m 6^s M.T.G., the observed Alt. of the Sun's lower limb was 12° 54' 30", and again being P.M. at ship, when the chronometer showed 2^d 5^h 10^m 57^s M.T.G., the observed Alt. of the Sun's lower limb was 19° 0' 15", the ship having made 42 miles on an E. (true) course during the interval; height of eye, 21 ft.; index error, +2' 40". Required both lines of position and azimuths, also the position of the ship by Sumner's method when the second Alt. was taken, assuming the lats. 51° 0' and 51° 30' N.

13. 1899, February 11th at ship, in lat. 33° 17' N., long. 30° 30' E. Find the mean time when Capella will be on the meridian.

14. 1899. Find from pages 294-304 of the Nautical Almanac, what stars of the 1st, 2nd, and 3rd magnitudes will be within three-quarters of an hour east or west of your meridian above the pole and above the horizon at about 6 A.M. A.T.S. on the 26th of February, in lat. 10° 15' N., long. 55° 30' W. Required also the hour angle of each of the stars, and state if they will be north or south of you.

15. Find the Alt. of Capella to place on the sextant, using data from question 13; height of eye, 20 ft.; index error, +1' 10".

16. 1899, the observed Alt. of the Star Algenib on the meridian above the pole on February 20th was 54° 14' 30" S. of the observer; height of eye, 15 ft.; index error, 3' 45" to add. Required the latitude.

x

306   APPENDIX.

17. 1899, February 25th, at about 8^h 30^m P.M. M.T.S. in lat. 52° 17' N., the observed Alt. of Star Dubhe, 57° 15' 41" E. of the meridian; index error, $-2'$ 10"; height of eye, 10 ft.; time by chron., 25^d 9^h 43^m $12^s \cdot 75$, which was found to be slow on M.T.G. noon January 21st, 5^m 10^s, and on January 31st was 5^m $9^s \cdot 5$ slow. Required the longitude by chronometer.

18. 1899, February 14th, at 4^h 17^m A.M. M.T.S. in lat. 37° 25' N., long. 52° 15' W., the Star Regulus bore by compass West. Required the true azimuth by the "Time Azimuth Tables;" and, supposing the variation to be 21° 30' W., required the deviation for the direction of the ship's head.

19. 1899, February 14th A.M. at ship at 4^h 47^m A.T.S., long. 72° 29' W., the observed Alt. of Star Polaris out of the meridian was 42° 21' 30"; index error, $-1'$ 10"; height of eye, 15 ft. Required the latitude.

20. 1899, February 14th P.M. at ship, in long. 70° 15' E., the observed Mer. Alt. of the Moon's lower limb was 75° 49' 30" bearing north of the observer; height of eye, 21 ft.; index error, $-2'$ 10". Required the latitude.

21.

(a) Using deviation Card B, find the course to steer by compass from St. John's Point to Great Ross; also the deviation, variation, and distance. ·

(b) With ship's head on the above compass course, a point, South Rock L.V., bore by compass N. 67° W., and Skulimartin L.V. bore N. 33° W. by the same compass. Find the ship's position.

(c) With the ship's head as above, a point, Mull of Galloway, bore by compass N. 34° E., and after continuing 10 miles on the above-named course it bore N. 30° W. Find the ship's position and distance from Mull of Galloway.

(d) Find the course from St. John's Point to Great Ross in a current which set S.E. corr. mag. 3 miles per hour, the vessel sailing 10 miles per hour; and the distance she makes good towards Great Ross in 3 hours.

22. 1897, January 11th, 4^h 18^m P.M. M.T.S., being off Ballycastle Bay, took cast of the lead. Required the reduction to be applied to the cast before comparing with the soundings on the chart.

23. Correct magnetic bearing—

Ship's head by standard compass.	Bearing of distant object by standard compass.	Deviation required.	Ship's head by standard compass.	Bearing of distant object by standard compass.	Deviation required.
N.	N. 41° E.		S.	N. 33° W.	
N.E.	N. 30° E.		S.W.	N. 27° W.	
E.	N. 11° E.		W.	N.	
S.E.	N. 12° W.		N.W.	N. 28° E.	

From the above table construct a Napier's curve, and give the courses you would steer by compass to make the following courses correct magnetic :—

Cor. mag. courses : N. by E. ½ E.   E. ¾ S.   S.S.W.   N.W. by N. ¼ N.
Compass courses.

Steering the following compass courses, find the correct magnetic courses from the curve :—

Compass courses : N. ½ E.   E.N.E.   S.E. by E. ¼ E.   S.W. ¾ W.
Correct magnetic courses.

Having taken the following compass bearings with the ship's head at E. ¾ S., find the correct magnetic bearings :—

E.S.E. and W. by N. ¾ N.

24. 1899, February 4th A.M. at ship, in lat. 20° 28′ N., and approximate long. 27° 20′ W. ; time by chronometer, which was supposed to be correct for M.T G., was $3^d 22^h 48^m 48^s$; observed Alt. of Sun's lower limb, 29° 34′ 30″ ; index correction, +1′ 30″ ; height of eye, 21 ft.; observed distance between the Sun's and Moon's near limbs, 80° 49′ 30″ ; index error, 0. Required Moon's true and apparent Altitudes and error of chronometer.

25. 1899, February 25th, in lat. by account 34° 2′ S., long. 28° 30′ E. With the following observations, find the latitude by double Altitude of the Sun at the second observation :—

A.T.S. $8^h 20^m$ A.M.   Alt. of Sun's L. L. 33° 30′ 30″   Index error, +2′ 15″
„ $11^h 10^m$ A.M.   „   „   62 20 50   Height of eye, 21 ft.

The Sun's bearing at the first observation was N..77° 30′ E. ; the run during the interval, S.E. (true) 12 knots per hour.

26. 1899, March 29th, at Alcatraz Island, San Francisco, in lat. 37° 49′ N., long. 122° 24′ 19″ W., the following times were observed : $8^h 44^m 13^s$ A.M., $3^h 25^m 16^s$ P.M., when the Sun had equal Altitudes. Required error on A.T.S., M.T.S., and M.T.G.

27. 1899, February 15th A.M. On shore in lat. 13° 38′ 20″ S., long. 76° 24′ 15″ W. ; observed Alt. of the Sun's lower limb by artificial horizon, 72° 35′ 20″ ; index cor. +2′ 14″ ; a chronometer, which was supposed to be $16^m 40^s$ fast of M.T.G., showed $1^h 42^m 20^s$. Required the true error of chronometer on M.T.G. and A.T.S.

28. 1899, February 12th A.M. at ship, supposed position lat. 47° 48′ 30″ N., long. 130° 30′ W. ; time by chron., $12^d 2^h 12^m 45^s$ M.T.G. ; observed Alt. of Spica W. of meridian, 27° 34′ 30″, and Antares 13° 50′ 20″ E. of meridian ; index error, −2′ 10″ ; height of eye, 16 ft. Required ship's position and azimuth of the two stars by Sumner's method, using lats. 47° 35′ and 47° 55′ N.

29. Find the first and last courses and distance on the arc of a Great Circle from Cape Gardafui, lat. 11° 51′ N., long. 51° 16′ E., to Campbell Island, in lat. 52° 33′ S., long. 169° 9′ E. ; also the latitude and longitude of Vertex, and the latitudes of the intersection of the following meridians with the arc of the Great Circle :—

Long. 60° E.
„ 70° „
„ 80° „
„ 90° „
„ 160° „

30. (a) From a reef awash at low water, the angle of elevation to the summit of a hill distant 1950 yards in a horizontal direction is 1° 3′ ; the height of eye is 5 ft. ; tidal rise, 18 ft. Required the height of the hill above high water.

(b) Two ships, P and Q, leave a port at the same time, P steering E. by S., and Q steering S.W. ; after 2½ hours P bears E.N.E. from Q, distant 25 miles. Find their rates of sailing.

(c) In a spherical triangle, R = 90°, $p = 119° 8′$, $q = 168° 20′$. Find P, Q, and r.

(d) In lat. 51° 24′ N. the Sun's amplitude at rising was E. by N. ¾ N. Find his declination and hour angle.

Paper IV.

1. Multiply 300·07 by ·0246; divide 5623·87 by 48960·02 by logs.
2.

Hour.	Courses.	Knts.	10ths.	Winds.	Lee-way.	Devia-tion.	Remarks.
1	W.S.W.	12	8	N.W. ¼ N.	pts. 1¾	26° E.	A point in lat. 37° 3' N., long. 4° 25' E., bearing by compass E. by N.; dist., 20 miles, ship's head, S.W.; deviation, 29° E. Variation, 21° W.
2		13	1				
3		12	5				
4		11	6				
5		11	—				
6	W.	12	9	N.N.W.	2	19½° E.	
7		13	—				
8		12	8				
9		12	3				
10	W. ½ N.	12	2	N.	1½	17° E.	
11		12	7				
12		12	1				
1	W. by S.	11	9	N.N.W.	¾	23° E.	
2		11	8				
3		11	9				
4		11	8				
5		11	6				
6	W. ½ S.	11	5	N.N.W.	1¼	20° E.	A current set correct magnetic South 1¾ miles per hour for all the day.
7		12	9				
8		12	8				
9		12	8				
10	W.N.W.	13	1	N.	1	11° E.	
11		13	6				
12		13	3				

Find the course and distance made good; and latitude and longitude of the ship.

3. 1899, January 1st, in long. 64° 5' E., the observed Mer. Alt. of Sun's lower limb, 56° 36' 20" bearing S.; index correction, − 2' 40"; height of eye, 14 ft. Required the latitude.

4. Find the compass course and distance by Mercator's sailing—

From A, lat. 42° 35' N.   Long. 71° 18' W.
To B,   ,, 53° 22' N.   ,,   12° 30' W.
Variation, 4° 30' W.   Deviation, 19° 20' E.

5. How far due east must a vessel sail on the parallel of 39° 27' S. to change her longitude 1° 29' ?

6. 1897. Find times of high water A.M. and P.M. at Exmouth on February 12th and 25th.

7. 1899, March 27^d 19^h 41^m 40^s A.T.G., in lat. 3° 22' S., long. 25° 10' W.,

the Sun's compass bearing was E. by S. ¼ S. Find the true amplitude, and compass error; and if the variation be 18° W., what is the deviation?

8. 1899, February 21st A.M., in lat. at noon, 40° 14' S.; the observed Alt. of the Sun's lower limb, 45° 50' 20"; index correction, —2' 22"; height of eye, 29 ft.; when a chronometer which was 50' slow on January 9ᵈ, and losing daily 1'·66, showed 21ᵈ 0ʰ 46ᵐ 1ˢ; run since observation, N. by W. ¾ W. 17 miles; the Sun's compass bearing was E.N.E.; variation, 14° E. Required the longitude at sights and noon, compass error, and deviation.

9. 1899, March 20th M.T. at ship, 5ʰ 32ᵐ P.M., in lat. 24° 40' N., long. 179° 30' W.; the Sun's compass bearing, W. by S. ¼ S. Required the Sun's true azimuth by the "Tables," compass error, and deviation, if variation be 12° E.

10. Using card B, find compass course and distance from Bardsey Island to Rockabill.

11. With ship's head on the above course, Codling L.V. bore by compass N. 87° 30' W., and Kish L.V. bore N. 20° 30' W. by the same compass. Find the ship's position.

12. Steering as above, Howth Head bore by compass N. 26° 30' W., and after continuing 13 miles on the above-named course it bore N. 77° W. by the same compass. Find the ship's position, and distance from Howth Head.

13. 1899. At what time will α Eridani pass meridian of a place long. 90° W. on January 18th civil list?

14. 1899. What bright stars are within 2 hours of the meridian of a place in lat. 24° S., long. 160° W., on April 13th, at 4ʰ 30ᵐ A.M.? What is the approximate Mer. Alt. of Vega?

15. 1899. April 13th, the Mer. Alt. of Vega bearing north was 27° 23' 40"; index correction, +3' 13"; height of eye, 24 ft. Required the latitude.

16. 1899, April 13th, when the M.T. ship was 2ʰ 45ᵐ A.M., in lat. 23° 45' S., long. 160° W., the compass bearing of Altair was E.N.E. Required the true azimuth by the "Tables," and deviation, if variation be 8° 30' E.

17. 1899, March 23ᵈ A.M. at ship, approx. M.T.S. 0ʰ 15ᵐ A.M., in long. 100° E., lat. 10° 19' N.; the observed Alt. of Antares east of meridian, 19° 49'; index correction, —3' 1"; height of eye, 26 ft.; time by chron., 4ʰ 33ᵐ 36ˢ·5, which was 59ᵐ 58ˢ slow of G.M.T. on January 23ᵈ, and on February 12ᵈ was slow 1ʰ 0ᵐ 48ˢ. Required the longitude.

18. 1899, February 19th P.M. at ship, in lat. by account 35° 17' S., long. 77° 30' E., the observed Alt. of Sun's lower limb bearing north was 64° 37' 10"; index correction, +2' 32"; height of eye, 16 ft.; time by chron., 18ᵈ 17ʰ 5ᵐ 11ˢ, which was slow of A.T.S. 7ʰ 29ᵐ 25ˢ, the diff. longitude made to the west after the error on A.T.S. was determined being 19'. Required the latitude by reduction to the meridian.

19. 1899, October 20th P.M. at sea and uncertain of my position, when the chronometer showed 20ᵈ 8ʰ 59ᵐ 52ˢ G.M.T., the observed Alt. of the Sun's lower limb was 29° 26' 10"; and again P.M. on the same day, when the chronometer showed 12ʰ 12ᵐ 46ˢ G.M.T., the observed Alt. of the Sun's lower limb was 9° 2' 40"; index correction, —4' 19"; height of eye, 31 ft.; the ship having made a true N. by W. ½ W. course, 33 miles in the interval. Required the line of position and the Sun's true azimuth at the first observation, and ship's position at the second observation by Sumner's method, assuming lats. 49° and 49° 40' N.

20. 1899, August 26th, when the G.M.T. was 26ᵈ 10ʰ 14ᵐ 16ˢ in long. 96° 46' E. the observed Alt. of Polaris out of the meridian was 28° 14' 50"; index correction, +4' 32"; height of eye, 27 ft. Required the latitude.

21. 1899, August 5th in long. 156° 36' W., the observed Mer. Alt. of the Moon's lower limb in an artificial horizon, below the pole was 16° 15' 40"; index correction, —6' 16". Required the latitude.

APPENDIX.

22.

Ship's head by compass.	Bearing of distant object.	Deviation.	Ship's head by compass.	Bearing of distant object.	Deviation.
N.	N. 32° W.		S.	N. 20° E.	
N.E.	N. 20° W.		S.W.	N. 9° E.	
E.	N. 5° E.		W.	N. 15° W.	
S.E.	N. 17° E.		N.W.	N. 31° W.	

Required correct magnetic bearing and deviations, also courses and bearings as follows by Napier's curve:—

Corr. mag.: S.E. ½ E.   N.E. by N.   W. by N. ¼ N.   S. 72° W.
Courses to steer.

Courses steered: E. by N. ¼ N.   S.E. ¾ S.   S.S.W.   W. 15° N.
Corr. mag.

Ship's head E.S.E. by compass.
Bearings by compass: S. 50° W.   S. 35° E.
Corr. mag.

23. Find the course to steer from Bardsey Island to Rockabill to counteract the effect of a current which set (C.M.) N. 26° E. 4 knots per hour, the ship making by log 10 knots per hour, and the distance she would then make good towards Rockabill in 4 hours.

1897, April 25th, 9ʰ 58ᵐ A.M., off the Smalls took a cast of the lead. Required the correction to be applied to the cast before comparing it with the depth marked on the chart.

24. 1899, January 18ᵈ P.M. at ship, in lat. 24° 30′ N.; time by a chron., 18ᵈ 14ʰ 30ᵐ 20ˢ, which was estimated to be 5ᵐ 22ˢ slow of G.M.T. With following observations, find true and apparent Altitudes of Aldebaran, error of chronometer for G.M.T., and longitude of place; height of eye, 32 ft.

Observed Alt. Moon's lower limb west of mer.
44° 51′ 10″
Index correction, +1 12

Observed dist. between a Tauri and Moon's remote limb.
36° 56′ 0″
Index correction, −1 11

25. 1899, January 7ᵈ A.M. at ship, in lat. 40° S., long. 100° 40′ E., with following observations, find the latitude by "Ivory" at the time of taking the second observation:—

Correct A.T.G.
First observation   6ᵈ 14ʰ 31ᵐ 33ˢ
Second observation 6 21 1 45

Observed Alt. of the Sun's lower limb.
50° 19′ 0″ bearing E. by N. ¼ N.
39 34 30

The ship's true course and distance during the interval, S.E. ½ S. 45 miles; index correction, +1′ 21″; height of eye, 29 ft.

26. 1899, February 4ᵈ A.M. at ship, when a chronometer showed 3ᵈ 9ʰ 28ᵐ 20ˢ; the observed Alt. of the Sun's lower limb was 16° 7′ 20″, and the observed Alt. of the Moon's upper limb was 6° 37′ 15″, taken at the same instant west of meridian; index correction, −1′ 15″; height of eye, 29 ft. Required the position of the ship, and true bearings of Sun and Moon by Sumner's method by projection, assuming lats. 47° 15′ and 47° 45′ N.

27. 1899, March 25ᵈ, in lat. 53° 23′ N., long. 3° W., the following times were observed when Regulus had equal altitudes:—

Star E. 7ʰ 59ᵐ 46ˢ   Star W. 0ʰ 10ᵐ 19ˢ

Required error of chron. for M.T.S., M.T.G., and A.T.G.

APPENDIX.

28. Find the initial course and distance, and latitude and longitude of Vertex by Great Circle sailing.

From Buenos Ayres, lat. 34° 36′ S., long. 58° 22′ W.
To Cape Town, ,,   33° 56′ S., ,, 18° 29′ E.

29. Construct a Mercator's chart, scale 1° long. = 2·75 in., to extend from long. 120° to 124° E., and from lat. 38° to 42° S. A ship from lat. 38° 17′ S., long. 120° 27′ E., made the following mag. courses and distances: S. 34° E. 25 miles, S. 5° W. 56 miles. Find her position, variation 6° W.

30. AB is a horizontal line 1300 ft. long; a vertical line is drawn from B upwards, and in it two points, P and Q, are taken, such that BQ = 3 BP, BAP = 10° 30′ 20″. Find BP and BAQ.

31. XY is 2000 ft. long; Y is due east of X; at Y a point Z bears N. 46° 20′ W., at X it bears N. 8° 45′ E. Find XZ.

32. Two places, A and B, are on the same parallel 35° N., and their difference of longitude is 85°. Find the distance between them on the Great Circle; also latitude of Vertex.

33. From a rock 7 ft. above high water, the angle of elevation to the summit of a cliff distant 5320 ft. in a horizontal direction is 1° 30′; height of eye, 5 ft. Required height of cliff above high water.

PAPER V.

1. Multiply 1·0009 by 62·7014 by common logs.
2. Divide 7·5 by 1000 by common logs.
3.

Hour.	Courses.	Knts.	10ths.	Winds.	Lee-way.	Devia-tion.	Remarks.
					pts.		
1	N.E. by E.	10		N. by W.	¼	16° E.	A point of land in
2		10					lat. 62° 30′ N.,
3		10					long. 178° 54′ E.,
4		10					bore by compass
5	E. by N.	10	5	S.S.E.	1¼	18° E.	W.N.W. with
6		10	5				the ship head-
7		10	5				ing N.N.E.; de-
8		10	5				viation on that
9	N.E.	11		E.S.E.	½	10° E.	point, 5° E.; the
10		11					distance off was
11		11					20 miles.
12		11					Variation, 15° E.
1	E.N.E.	11	5	S.E.	1	17° E.	
2		11	5				
3		11	5				
4		11	5				
5	E.	12		S. by E.	2¼	20° E.	A current set
6		12					E.N.E. correct
7		12					magnetic, 2¼
8		12					knots per hour
9	N.E. by E. ½ E.	11	5	S.E. by E.	¼	16° E.	for the 24 hours.
10		11	5				
11		11	5				
12		11	5				

Find the course and distance made good, also latitude and longitude of ship.

312   APPENDIX.

4. 1899, February 20th, in long. 162° 0' E., the observed Mer. Alt. of the Sun's upper limb was 78° 55' 55" bearing north; index error, +7' 30"; height of eye, 26 ft. Find the latitude.

5. How many miles must a ship sail east in lat. 45° 50' N. to make a change of longitude equal to 800'?

6. Find the true course and distance from A to B by Mercator sailing; and if the variation is 30° 0' W., and the deviation 10° 30' E., find the course to steer by compass.

A, lat. 33° 24' S., long. 20° 15' E.
B, ,, 53° 40' S., ,, 63° 37' W.

7. 1897. Find the time of high water at Mellon, February 9th, A.M. and P.M.

8. 1899, March 1st A.T.S. 6ᵈ 39ᵐ 29ˢ P.M., in lat. 52° 0' N., long. 160° 35' E., the observed bearing of the Sun by compass was W. ¾ N. Find the true amplitude and the error of the compass ; and if the variation was 9° 0' E., find the deviation of the compass for the point the ship's head is on.

9. 1899, March 21st A.M. at ship, the observed Alt. of the Sun's lower limb was 29° 21' 30"; index error, +1' 10"; height of eye, 21 ft.; the Sun's bearing by compass, E.S.E.; time by chron., 20ᵈ 18ʰ 10ᵐ 20ˢ·5, which was slow of mean noon Greenwich 5ᵐ 24ˢ on January 1st, and on March 2nd was correct at mean noon Greenwich; the latitude at noon was 38° 10' N.; course and distance since observation, S. 30° E. 37 miles. Required longitude at noon and at sights; required also the true azimuth; and if the variation was 9° 30' W., what is the deviation of the compass?

10. 1899, in lat. 25° 42' S., long. 120° 30' W.; time by chron., April 4ᵈ 12ʰ 12ᵐ 0ˢ M.T.G.; the Sun bore by standard compass, N. 75° 56' 15" W. Find the true azimuth by the "Time Azimuth Tables" and compass error. If the variation for place should be 9° 30' E., find also the deviation of the compass.

11. 1899, March 3rd A.M. at ship, lat. by D.R. 55° 23' N., long. 47° 50' W.; time by watch, 3ᵈ 1ʰ 37ᵐ 52ˢ, which was fast of A.T.S. 2ʰ 11ᵐ 40ˢ; the difference of longitude made to the east after the error on A.T.S. was determined was 14 miles; the observed Alt. of the Sun's lower limb bearing south from the observer was 27° 22' 30"; index error, 3' 10" to subtract; height of eye, 17 ft. Required the latitude at observation by reduction to the meridian ; and if the ship sailed W. 2° S. 3 miles between then and noon, required the latitude at noon.

12. 1899, April 15th A.M. at ship, when a chronometer showed 15ᵈ 4ʰ 43ᵐ 17ˢ M.T.G., the observed Alt. of the Sun's lower limb was 26° 30' 20"; and again being A.M. at ship, when the chronometer showed 15ᵈ 7ʰ 43ᵐ 15ˢ M.T.G., the observed Alt. of Sun's lower limb was 47° 34' 0", the ship having made 30 miles on a S. 36° ½ E. course during the interval ; height of eye, 22 ft. Required the line of position when the first altitude was taken, and the Sun's true azimuth ; also the position of the ship by Sumner's method when the second altitude was observed, assuming lats. 50° 0' and 51° 0' N.

13. 1899, March 6th, at sea, lat. 42° 20' N., long. 72° 30' W. Find the apparent time when Sirius will be on the meridian.

14. 1899. Find from pages 294–304 of the Nautical Almanac, what stars of the 1st, 2nd, and 3rd magnitudes will be within ¾ of an hour east and west of your meridian above the pole and horizon at about 10ʰ 30ᵐ P.M. March 18th, lat. 45° 40' S., long. 54° 45' W. ; required also the meridian distance of the stars obtained, and state if they are north or south of you.

15. Find the altitude of Sirius to place on the sextant, using lat. 42° 20' N.; index error, −3' 15"; height of eye, 22 ft.

16. 1899. The observed Mer. Alt. of the Star Schedir below the pole

on March 21st was 15° 17' 15''; index error, −2' 15''; height of eye, 16 ft. Required the latitude.

17. 1899, March 18th P.M. at ship, in lat. 60° 44' 0'' S., the observed Alt. of Star a' Crucis was 47° 42' 40'', about 6ʰ 18ᵐ east of the meridian; index error, +1' 10''; height of eye, 30 ft.; the chronometer showed 18ᵈ 10ʰ 11ᵐ 11ˢ, which was fast on Greenwich mean noon 3ᵐ 16ˢ on January 5th, and on February 10th was slow 1ᵐ 5ˢ. Find the longitude by chronometer.

18. 1899, March 21st, 1ʰ 20ᵐ A.M. A.T.S., in lat. 34° 30' S., long. 30° 10' E., the Star β Herculis bore N.E. by N. Required the true azimuth and compass error; and if the variation is 29° 30' W., required the deviation of the compass.

19. 1899, March 20th, 7ʰ 15ᵐ P.M. M.T.S., long. 165° 20' E., the observed Alt. of Star Polaris out of the meridian was 37° 15' 30''; index error, 0; height of eye, 23 ft. Required the latitude.

20. 1899, March 25th, long. 37° 30' E.; observed Mer. Alt. of the Moon's upper limb was 36° 40' 0'', bearing N. of the observer; index error, +2' 10''; height of eye, 24 ft. Required the latitude.

21. (a) Using deviation card A, find the course to steer by compass from Inishtrahull Island, to Craigani Point; also the deviation, variation, and distance.

(b) With ship's head on the above course, a point, Inishowen Head, bore by compass S. 60° W., and Bengore river bore S. 39° E. by the same compass. Find the ship's position.

(c) With the ship's head as above, Sanda Island bore by compass N. 52° E., and after continuing 11 miles on the above-named course, it bore N. 11° W. Find the ship's position and distance from Sanda Island.

(d) Find the course from Inishtrahull to Craigani Point in a current which set South correct magnetic 3 miles per hour, the vessel making 8 miles per hour, and the distance she makes good towards Inishtrahull in 2 hours.

22. 1897, February 8th, at noon, off Wexford, obtained cast of lead. Required reduction to be applied to cast.

Napier's Curve.

23. Correct magnetic bearing—

Ship's head by standard compass.	Bearing of distant object by standard compass.	Deviation.	Ship's head by standard compass.	Bearing of distant object by standard compass.	Deviation.
N.	N 56° E.		S.	N. 56° E.	
N.E.	N. 48° E.		S.W.	N. 63° E.	
E.	N. 45° E.		W.	N. 68° E.	
S.E.	N. 49° E.		N.W.	N. 64° E.	

With the deviations as above, give the courses to steer in order to make the following correct magnetic courses:—

Correct magnetic : E. by N.   S.E. by E.   W.S.W.   N.N.W.
Course to steer.

Having steered the following courses by compass, find correct magnetic courses :—

Compass courses : W. by S.   S.S.E.   N.E. by N.   N.W.
Correct magnetic courses.

APPENDIX.

You have taken the following bearings by compass as above, with ship's head on E.N.E. Find the correct magnetic bearings :—

Bearings by compass: S. by W. ½ W.   S.E. by E. ¼ E.
Correct magnetic.

24. 1899, March 22nd P.M. at ship, lat. 42° 5′ N., long. by D.R. 5° 0′ E. ; time by chron., 22ᵈ 10ʰ 20ᵐ 30ˢ, supposed to be 10ᵐ 12ˢ slow of M.T.G. Observed the following altitude and distance : Observed Alt. Mars' centre, 46° 52′ 30″ W. of the meridian ; index error, +2′ 10″ ; height of eye, 20 ft. Observed distance of Moon's near limb and Mars' centre, 21° 21′ 44″ ; index error, +2′ 0″. Required error of chronometer on M.T.G., and the longitude.

25. 1899, January 20th, in N. lat., long. 70° 39′ W. With the following observation of Venus' lower limb, find the latitude by double altitude at second observation.

M.T.G. 20ᵈ 2ʰ 42ᵐ 36ˢ   Obs. Alt. 56° 37′ 10″   Index error, −2′ 10″
       20 6 28 6   13 37 30   Height of eye, 20 ft.

Bearing of Venus at the first observation, S. 26° W., run in the interval, E. 10 knots per hour.

26. 1899, June 7th P.M. at Perim, in lat. 12° 39′ 40″ N., long. 43° 24′ E., the following times were noted by watch when Star Vega had equal altitudes. Required the error on A.T.S., M.T.S., and M.T.G.

27. 1899, October 16th, about 11ʰ 50ᵐ P.M., in lat. 51° 38′ N., long. 55° 53′ 40″ W. ; observed alt. of Aldebaran by artificial horizon, 80° 17′ 30″ ; index cor. −5′ 30″. A chronometer showed 3ʰ 23ᵐ 5ˢ, supposed to be 10ᵐ 30ˢ slow of M.T.G. Required the true error of chronometer on M.T.G. and M.T.S.

28. 1899, March 31st P.M. at ship, in lat. by account 53° 24′ N., long. 3° 0′ W. ; time by chron., 31ᵈ 7ʰ 48ᵐ 49ˢ M.T.G. ; observed simultaneous altitudes of two stars, Sirius 17° 23′ 20″ W. of the meridian, and Regulus 43° 38′ 45″ E. of the meridian ; index error, 0 ; height of eye, 15 ft. Required ship's position and the azimuth of both stars by Sumner's method of projection on the chart, assuming lats. 53° 10′ and 53° 40′ N.

29. Find the initial and final courses from Cape of Good Hope, lat. 34° 21′ S., long. 18° 30′ E., to Port Philip Head in lat. 38° 18′ S., long. 144° 39′ E., by composite sailing. Easting to be run on the parallel of 46° S. lat. Also required longitudes of the meridians that cut the parallel, and the three distances on the composite track.

30. (a) From a boat the angle of elevation of the truck, known to be 165 ft. above the water-level, was found to be 9° 17′. Find the distance of the observer from the ship.

(b) A vessel sails by log S.S.E. ½ E. 137 miles, but by observation she is known to be 101 miles S. ¼ W. of her former position. Find the set and drift of the current.

(c) In a right-angled spherical triangle, hypotenuse = 105° 20′, and one angle is 35° 4′. Find the side opposite the given angle and the remaining angle.

(d) Two points are in lat. 20° S., one in long. 80° W., the other in long. 150° E. Find the arc of a Great Circle joining them.

APPENDIX. 315

PAPER VI.

1 and 2. Multiply ·00483506 by 100·0706, and divide 6789 by 78900 by logs.

3.

Hour.	Courses.	Knts.	10ths.	Winds.	Lee-way.	Devia-tion.	Remarks.
1	N.E. ¼ E.	14	6	N.W. ¼ W.	pts. 1	20° E.	A point in lat. 37° 45′ S., long. 178° 32′ E., bearing by compass E. ¼ N.; ship's head,
2		14	8				
3		13	9				
4		14	3				
5		14	4				
6	E.N.E.	14	9	W.N.W.	¼	21° E.	N.E. ¼ E.; deviation as per log; distant 17 miles.
7		14	5				
8		13	6				
9	E. by N. ¼ N.	13	9	W.	—	22° E.	Variation, 14° E.
10		13	8				
11		14	3				
12		14	—				
1	E. by S. ½ S.	14	7	S.W.	1¼	19° E.	
2		13	6				
3		14	3				
4		14	4				
5	E.S.E.	14	2	S.S.W.	1	18° E.	A current set N. ¾ E. correct magnetic, 2½ miles per hour for 24 hours.
6		13	8				
7		14	1				
8		13	9				
9	E.N.E.	14	2	S.S.W.	½	14° E.	
10		14	1				
11		13	9				
12		13	8				

Find the course and distance made good; also latitude and longitude of the ship.

4. 1899, February 28th, in long. 45° 45′ W., the observed Mer. Alt. of the Sun's upper limb south of the observer was 45° 23′ 10″; index correction, −5′ 45″; height of eye, 22 ft. Required the latitude.

5. Find the compass course and distance by Mercator's sailing—

From A., lat. 55° 13′ S. Long. 43° 30′ E.
To B., „ 10 19 S.   „ 120 12 E.
Variation, 36 40 W. Deviation, 14 10 W.

6. Along what parallel is a vessel sailing when for a distance sailed due west of 76 miles, she alters her longitude 1° 30′ ?

7. 1897, Find the times of high water at Wexford on March 11th and 26th.

8. 1899, March 20th, when the G.A.T. was $7^h 37^m 48^s$ in lat. 13° N., long. 155° 33′ E., the Sun's compass bearing was E.N.E. ¾ E. Find the true amplitude and deviation, if the variation is 7° E.

9. 1899, May 26th A.M. at ship, in lat. at noon 48° 11′ N.; observed Alt. of the Sun's lower limb, 44° 23′ 20″; index correction, +3′ 13″; height of

APPENDIX.

eye, 42 ft.; time by chron., $26^d\ 0^h\ 53^m\ 57^s$, which was correct for G.M.T. on April 19th, and losing daily 4^s·06; run since observation, S.E. by E. 27 miles; the Sun's compass bearing was S.E. ¼ E.; variation, 26° W. Required the longitude at observation and noon, compass error, and deviation.

10. 1899, April 15th, M.T.G. $14^d\ 21^h\ 11^m$ in lat. 56° 55′ N., long. 11° 45′ W., the Sun's compass bearing was S.E. by E. ¼ E. Required the Sun's true azimuth by the "Tables," error of the compass, and deviation, variation being 32¼° W.

11. Using deviation card A, find the compass course and distance from Dinas Head to Wicklow Head.

With the ship's head on the above course, Kenmaes Head bore by compass S. 64½° E., and Strumble Head bore S. 25° W. by the same compass. Find the ship's position.

Steering as above, South Arklow L.V. bore by compass N. 71° W., and after continuing 7 miles on the above course it bore S. 55° W. by the same compass. Find the ship's position and distance from South Arklow L.V.

12. 1899. At what time will Vega cross the meridian of a place in long. 120° 25′ W. on August 14th?

13. 1899. What bright stars are within 2^h E. and W. of the meridian of a place in lat. 10° 15′ N., long. 160° E., on November 22nd, at $3^h\ 35^m$ A.M. M.T.S. ? What is the approximate Mer. Alt. of Alphard?

14. 1899, November 22nd, the Mer. Alt. of Alphard bearing south was 71° 32′ 30″; index correction, +3′ 13″; height of eye, 25 ft. Required the latitude.

15. 1899, November 6th A.M. at ship, in lat. 39° 18′ S., about 3^h before Canopus came to the meridian of a place whose estimated longitude was 20° W., the chronometer showed $1^h\ 46^m\ 11^s$, which was slow $4^m\ 35^s$·8 on August 4^d, and slow $5^m\ 14^s$·8 on September 25^d. The observed Alt. of Canopus was 58° 12′ 20″; index correction, −1′ 13″; height of eye, 27 ft. Required the longitude.

16. 1899, November 22nd, when the M.T.S. was $1^h\ 35^m$ A.M. in lat. 10° 15′ N., long. 160° E., the compass bearing of Alphard was E. ¼ N. Required the true azimuth by the "Tables;" and if the variation be 7° 45′ E., find the deviation.

17. 1899, July 20th A.M. at ship, in lat. by account 10° 15′ N., long. 150° 30′ E., the observed Alt. of a Cygni bearing north was 55° 6′ 40″; index correction, −1′ 12″; height of eye, 24 ft.; time by chron., $19^d\ 11^h\ 59^m\ 24^s$, which had been found to be $29^m\ 28^s$ slow of M.T. ship; run since error was determined, S.W. ½ S. 39 miles. Required the latitude by reduction to the meridian.

18. 1899, April 26th A.M. at ship, at sea and uncertain of my position, when the chronometer showed $25^d\ 10^h\ 20^m\ 56^s$ G.M.T., the observed Alt. of Sun's lower limb was 59° 16′ 50″; and again P.M. on the same day, when the chronometer showed $25^d\ 15^h\ 41^m\ 13^s$ G.M.T., the observed Alt. of the Sun's lower limb was 36° 45′ 40″; index correction, −2′ 9″; height of eye, 28 ft.; the ship having made a true N.N.E. course 40 miles in the interval. Required the line of position and Sun's true azimuth at second observation; also the ship's position at second observation by Sumner's method by projection, assuming lats. 33° and 34° N.

19. 1899, December 31st, when the M.T.S. was $7^h\ 15^m\ 24^s$ A.M. in long. 74° 36′ W., the observed Alt. of Polaris in artificial horizon, out of the meridian, was 97° 56′ 20″; index correction, −8′ 30″. Required the latitude.

20. 1899, December 19th A.M., in long. 89° 42′ E., the observed Mer. Alt. of the Moon's lower limb bearing north was 84° 5′ 20″; index correction, −6′ 54″; height of eye, 35 ft. Find the latitude.

21. Napier's curve.

APPENDIX. 317

Ship's head by compass.	Bearing of distant object.	Deviation.	Ship's head by compass.	Bearing of distant object.	Deviation.
N.	S. 5° W.		S.	S. 10° E.	
N.E.	S. 6° E.		S.W.	S. 7° W.	
E.	S. 18° E.		W.	S. 18° W.	
S.E.	S. 17° E.		N.W.	S. 21° W.	

Find correct magnetic bearing :—

Correct magnetic : N. by W.　　W. ¾ S.　　S.E. ¼ E.　　N. 32° E.
Courses to steer.

Courses steered : E. by S. ½ S.　S.W. ¾ S.　W. ½ N.　　N. 8° W.
Correct Magnetic.

Ship's head by compass : E. by S.
Bearings by compass : N.W. ¼ W.　　S.W. ¾ W.
Correct Magnetic.

22. Find the course to steer from Dinas Head to Wicklow Head to counteract the effect of a current which set (C.M.) S. by W. ¾ W. 3 miles per hour, the ship making by log 8 miles per hour, and the distance she would make good towards Wicklow Head in 3 hours.

23. 1897, May 18th, at noon off Pwllheli, took a cast of the lead. Find the correction to be applied to the cast before comparing it with the depth marked on the chart.

24. 1899, August 30th A.M. at ship, in lat. 40° 20′ N. ; time by chron., $29^d\ 22^h\ 30^m\ 5^s$, which was estimated to be slow $15^m\ 14^s$ of G.M.T.

With following observations, find true and apparent Alts. of the Moon, error of chronometer for G.M.T., and longitude of place ; height of eye, 33 ft.

　　　　Obs. Alt. of Sun's lower limb.　　　　Obs. Dist. between Sun and Moon.
　　　　　　　39° 42′ 20″　　　　　　　　　　　　　　61° 48′ 56″
　Index correction,　−3 12　　　Index correction,　+2 8·5

25. 1899, February 21st A.M. at ship, in lat. by account 50° S., long. 55° 20′ W. With following observations, find latitude by "Ivory" at second observation :—

1st A.T.S. $6^h\ 19^m\ 25^s$ A.M.　　Obs. Alt. Sun's L.L. 11° 22′ 30″ bearing E. ¼ S.
2nd A.T.S. 10 49 25 A.M.　　　　　　　　　　　　　47 17 20

The ship's course and distance during the interval, N.N.W. ¾ W. 36 miles ; index correction, +2′ 30″ ; height of eye, 27 ft.

26. 1899, June 29th P.M. at ship, when a chronometer showed $29^d\ 20^h\ 29^m\ 23^s$ correct G.M.T., the observed Alt. of Vega E. of meridian was 78° 29′ 45″ ; and the observed Alt. of Arcturus W. of meridian at the same time was 38° 45′ 20″ ; height of eye, 27 ft. ; index correction, +4′ 14″. Find the position of the ship and true bearings of the stars by Sumner's method by projection, assuming lats. 46° and 46° 40′ N.

27. 1899, July 29th, in lat. 17° 29′ S., long. 150° 44′ W., the following times were observed when the Sun had equal altitudes :—

　　　E. of mer. $8^h\ 25^m\ 19^s$　　　　W. of mer. $0^h\ 18^m\ 17^s$

Find error of chronometer for A.T.S., A.T.G., and M.T.G.

318   APPENDIX.

28. Find initial course and the distance on a composite track, highest lat. 50° S.; also longitude of points of reaching and leaving the parallel—
From lat. 33° 3' S., long. 71° 38' W.
To   ,,   41 16 S.,   ,,   174 53 E.

29. Construct a Mercator's chart on scale 1·8 in. to 1° long., extending from 59° to 61° S. and 28° 30' W. to 31° 30' W. A ship sails by compass from lat. 59° 40' S., long. 30° 30' W., as follows : S.E. by S., 57' (deviation, 3° W.); W., 25' (deviation, 8° 30' E.); variation, 10° E. Find latitude and longitude in, and course and distance made good.

30. In the triangle ABC, C = 90°, AC = 15,866 ft., BC = 13,000 ft. Find the angles, and length of line drawn from B to AC bisecting angle ABC.

31. The elevation of a tower at A is 28° 17' ; at B, in the same horizontal plane and 155 ft. nearer the tower, the elevation is 62° 37'. Find height of the tower.

32. The Sun's declination being 18° 30' N., and his true amplitude N.E. by E. ¾ E., find the latitude of the place and hour angle at rising.

33. A vessel sails for 18 hours S.E. by S. 95 miles by the log, in a current setting S. by W. ½ W. 2 knots per hour. What is the course and distance made good ?

PAPER VII.

1. Multiply 13·5426 by 51400 by common logs.
2. Divide 7560009 by 387·003 by common logs.
3.

Hour.	Courses.	Knts.	10ths.	Winds.	Lee-way.	Devia-tion.	Remarks.
					pts.		
1	N.W.	10		W.S.W.	¼	10° E.	The departure was
2		10					taken from point
3	N.N.W.	11		W.	¼	15° E.	in lat. 50° 10' S.,
4		11					long. 110° 15'
5	W.S.W.	12		N.W.	¾	8° W.	W.
6		12					
7	S.W.	11	5	W.N.W.	1	15° W.	
8		11	5				
9		11	5				
10		11	5				
11	N.N.W. ½ W.	12		W.	0	15° E.	Variation, 20° E.
12		12					
1		12					
2		12					
3	N.	12		W.N.W.	½	25° E.	
4		12					
5	S.W.	12				¼	15° W.	A current set N.
6		12					20° W. correct
7		12					magnetic, 24
8		12					miles for the
9	W. S.	12		N.W. by N.	¾	5° W.	day.
10		12					
11		12					
12		12					

Find the course and distance made good ; also the latitude and longitude of the ship.

APPENDIX. 319

4. 1899, March 20th, in long. 15° 32' W., the observed Mer. Alt. of the Sun's lower limb was 62° 23' 35" S. of the observer; index error, +4' 12"; height of eye, 19 ft. Required the latitude.

5. A ship sails west 250 miles, in lat. 51° 23' N., long. 9° 36' W. Required present position.

6. Find the course and distance from A to B by Mercator sailing; and if the variation was 13° E., and the deviation 5° 30' W., find the course to steer by compass.

A, lat. 38° 13' S., long. 178° 43' W.
B, ,, 52 10 N., ,, 165 15 E.

7. 1897, July 24th. Find the time of high water at Bridlington, A.M. and P.M.

8. 1899, April 30^d 1^h 0^m 0^s A.T.G., on the Equator, long. 105° 0' W., the amplitude by compass was E. Required the true amplitude and error of the compass; and if the variation was 7° 0' E., what is the deviation of the compass?

9. 1899, April 16th, at about 2^h 52^m P.M. the observed Alt. of the Sun's upper limb was 21° 7' 30"; index error, 3' 32" to add; height of eye, 26 ft.; Sun's bearing by compass, N. 46° W.; time by chron., 15^d 15^h 6^m 53^s, which was fast for mean noon at Greenwich 10^m 24' on the 26th January, and on the 23rd March was found to be 12^m 2^s fast; the lat. at noon, 47° 45' S.; course and distance since noon, W. 30 miles. Required the longitude at sights and noon; required also the true azimuth and error of the compass; and if the variation is 16° 30' E., find the deviation.

10. 1899, May 1st P.M. at ship, in lat. 42° 15' N., long. 107° 40' E.; time by chron., 30^d 18^h 24^m 20^s M.T.G.; Sun's bearing, N. 87° 11' 15" W. by compass. Required the true azimuth and error of the compass by the "Time Azimuth Tables;" and if the variation is 8° 30' E., required the deviation of the compass.

11. 1899, July 2nd, in lat. by account 41° 2' S., long. 65° 20' W.; time by chron., 2^d 11^h 5^m 10^s, which was slow 10^m 20^s on M.T.G. when observation was made of β Centauri east of the meridian 71° 0' bearing south; index error, 2' 21" on the arc; height of eye, 31 ft. Required latitude by reduction to the meridian.

12. 1899, June 30th P.M. at ship, and uncertain of the position, when the chronometer showed 29^d 15^h 40^m 17^s M.T.G., the observed Alt. of Sun's lower limb was 54° 33' 50"; and again P.M. at ship, the chronometer showed 29^d 19^h 10^m 19^s M.T.G.; the observed Alt. of Sun's lower limb was 21° 51' 40", the ship having made 35 miles on a S. 54° W. course during the interval; height of eye, 19 ft.; index error, -1' 10'. Required the lines of position and both azimuths; also the position of the ship by Sumner's method when the second altitude was taken, assuming lats. 47° 50' and 48° 20' N.

13. 1899, April 1st M.T.S., lat. 62° 19' S., long. 120° 15' E. Find the M.T.S. when Canopus will be on the meridian of the observer.

14. 1899. Find from pages 294–304 Nautical Almanac what 1st, 2nd, 3rd magnitude stars will be within three-quarters of an hour of the meridian east and west above the pole and above the horizon at about 4.15 A.M. April 27th A.T.S., in lat. 27° 48' N., long. 33° 45' E. Required also the meridian distance, and state if they will be north or south of you.

15. 1899, April 1st M.T.S. Find the approximate altitude of Canopus, lat. 62° 19' S., long. 120° 15' E.; index error, 2' 10" on the arc; height of eye, 23 ft.

16. 1899, April 1st, the observed Alt. of Achernar on the meridian below the pole was 12° 37' 20"; index error, -1' 10"; height of eye, 17 ft. Required the latitude.

17. 1899, April 27th, at about 9^h 15^m P.M. the observed Alt. of the Star Spica was 47° 58' 10" E. of meridian; index error, -2' 10"; height of eye,

14 ft.; time by chron., 0^h 51^m 18', which was fast 10^m 31' on the 17th of January for Greenwich mean noon, and on February 6th it was 10^m 30' fast; lat. by ex-meridian of Star Deneb, 45° 50' S.; long. by D.R., 127° 15' E. Required the longitude at time of observation, by chronometer.

18. 1899, April 27th, 14^h 10^m 20' M.T.G., lat. 50° 40' N.; long., 53° 20' W.; Star Procyon bore by compass N. 80° W. Find the true azimuth and error of the compass by the "Time Azimuth Tables;" and, supposing the variation to be 40° 0' W., required the deviation of the compass for the point the ship's head is on.

19. 1899, April 12^d 16^h 55^m 5^s M.T.G., long. 15° 20' E., the observed Alt. of Star Polaris out of the meridian was 36° 24' 40''; index error, $-3'$ 15''; height of eye, 26 ft. Required the latitude.

20. 1899, April 21st, in long. 119° 15' W., the observed Alt. of the Moon's lower limb was 60° 10' 20'', zenith north of object; index error, $+1'$ 10''; height of eye, 20 ft. Required the latitude.

21. (a) Using card B, find the course to steer by compass from Mull of Cantyre to New Island; also the deviation, variation, and distance.

(b) With the ship's head on the above course, a point, Garron Point tower, bore by compass S. 84° ¼ W., and another point, Maiden's, bore by compass S. 12° ¼ W. Find the ship's position.

(c) With the ship's head as above, a point, Gobbins Head, bore by compass S. 19° W., and after continuing 9 miles on the same course it bore N. 71° W. Find the ship's position and distance from Gobbins Head at the second bearing.

(d) Find the course from Mull of Cantyre to New Island in a current which set N.N.W. ¼ W. cor. mag. 2 miles per hour, the vessel making 8 miles per hour by log; also the distance made good towards New Island in 4 hours.

22. 1897, February 11th, 11^h 46^m A.M. M.T.S., when off Tuskar took a cast. What reduction must be applied to the cast before comparing it with the chart?

23. Correct magnetic bearing—

Ship's head by standard compass.	Bearing of distant object by standard compass.	Deviation.	Ship's head by standard compass.	Bearing of distant object by standard compass.	Deviation.
N.	N. 44° W.		S.	S. 79° W.	
N.E.	N. 47° W.		S.W.	S. 89° W.	
E.	N. 80° W.		W.	N. 69° W.	
S.E.	S. 76° W.		N.W.	N. 52° W.	

Cor. mag. course: N. 30° W.   S. 55° E.   S. 66° W.   N. 35° E.
Courses to steer.

Courses steered: E. ¾ S.   S.S.W. ¼ W.   W. ¼ N.   N.W. by W.
Correct mag. course.

Ship's head N. ½ E. by compass.
Bearings by compass: S. 55° W.   N. 32° W.
Corrected bearings.

24. 1899, May 30th A.M. at ship, lat. 30° 35' S.; time by chron., 29^d 16^h 30^m 12^s, supposed to be correct for M.T.G. Altitude of Moon's upper limb, 57° 31' 10'' W.; altitude of Star Fomalhaut, 87° 49' 10'' W.; index error, $-2'$ 30''; height of eye, 23 ft. Observed distance between Star and Moon's N.L., 30° 27' 20''; index error, $-3'$ 0''. Required error of chronometer and the longitude.

APPENDIX.

25. 1899, October 15th, lat. N., long. 135° 17' 30" W. With the following altitude of Markab, find the latitude by double altitudes at second observation :—

M.T.S. 6ʰ 30ᵐ 0ˢ Observed Alt. Star, 40° 39' 0"   Index error, −1' 30"
„   10 47 0   „   „   50 52 20   Height of eye, 24 ft.

Star bearing at first observation, S. 60° E. ; run during interval, S. 30° W. 7 knots per hour.

26. 1899, April 15th, at Melbourne, lat. 37° 49' 53" S., long. 144° 58' 42" E. Equal altitudes of the sun were taken with the artificial horizon, and the following times noted:—

Chron. 11ʰ 9ᵐ 58ˢ A.M. ; 5ʰ 30ᵐ 28ˢ P.M.

27. 1899, July 26th A.M., the true latitude and longitude being 50° 26' N., 165° 40' E. ; the altitude of the Sun's lower limb, 57° 6' 10" ; index cor. −3' 7" ; height of eye, 36 ft. ; time by chronometer, July 25ᵈ 12ʰ 12ᵐ 25ˢ. Required the error of chronometer on apparent and mean time at place and mean time at Greenwich.

28. 1899, April 16th A.M. at ship, in latitude by D.R. 50° 49' N., long. 30° 48' W., M.T.S. about 4ʰ 30ᵐ A.M. When the chronometer showed 6ʰ 33ᵐ 12ˢ M.T.G., simultaneous altitudes were observed of Markab and Arcturus : Marcab 21° 55' 50" E., Arcturus 34° 41' 40" W. of the meridian ; index error, 2' 25" off the arc ; height of eye, 16 ft. Required the position of the ship by Sumner's method ; also the two azimuths of the stars, assuming lats. 50° 30' and 51° 0' N.

Great Circle.

29. Find the first and last courses on the arc of a Great Circle from Tarifa, lat. 36° 0' N., long. 5° 37' W., to Monte Video, in lat. 34° 57' S., long. 55° 55' W. ; also the latitude and longitude of vertex, the distance, the course to steer crossing the Equator, and the longitude to cross the Equator.

30 (a) The summit of a spire is vertically over the middle point of a horizontal square enclosure, whose side is 1000 feet long ; the shadow of the spire reaches a corner of the square when the Sun has an altitude of 27° 29' 48". Required the height of the summit of the spire above the level of the square.

(b) An island bears from a cape S.E. ½ E. 35 miles ; a ship bound to the island meets a tide running to the eastward, and finds herself after some time 28 miles from the island and 40 miles from the cape. Required the bearing of the island and the set of the tide.

(c) Calculate angles A and B,.and side AC, of a spherical triangle having given C = 90°, AB = 74° 49', and BC = 30°.

(d) Given the Sun's true amplitude W.N.W., dec. 20° 20' N., find the latitude.

Y

APPENDIX.

PAPER VIII.

1. Multiply ·0043 by ·00025, and divide 9999·99 by 100,000 by logs.

2.

Hour.	Courses.	Knts.	10ths.	Winds.	Lee-way.	Devia-tion.	Remarks.
1	N.N.W. ¼ W.	12	7	E.N.E.	1¼ pts.	3° W.	A point in lat.
2		12	9				52° 10′ N., long.
3		13	1				178° 15′ W.;
4		13	3				bearing by compass N.N.E. ¼ E.
5	S.W. ½ S.	13	2	N.E. by E.	—	27° E.	(ship's head,
6		12	6				N.N.W. ¼ W.;
7		12	5				deviation as per
8		11	7				log), distant 19 miles.
9	W.	11	9	N.E.	¼	18° E.	Variation, 14° E.
10		12	4				
11		12	6				
12		12	1				
1	S.	12	2	N.E.	¼	5° E.	
2		12	4				
3		11	8				
4		11	6				
5	S. by E. ¼ E.	11	7	N.E. ½ E.	½	7° W.	
6		11	9				A current set correct magnetic
7		12	3				South 28 miles
8		12	1				during the day.
9	S. by W.	12	4	E.N.E.	¾	15° E.	
10		12	5				
11		11	9				
12		12	2				

Find the course and distance made good, and latitude and longitude of the ship.

3. 1899, March 20th, in long. 151° 15′ W., the observed Meridian Alt. of the Sun's lower limb, 35° 32′ 10″, zenith north of the Sun; index correction, −10′ 20″; height of eye, 15 ft. Find the latitude.

4. Find the compass course and distance by Mercator's sailing—

From A, lat. 35° 7′ N.   Long. 141° 6′ E.
To B, ,, 54 53 S.   ,, 75 35 W.
Variation, 1 20 E.   Deviation, 18 40 W.

5. At noon the latitude is 51° 20′ S.; by morning sights the longitude was found to be 85° 40′ W.; run since sights, due East 76 miles. What is the noon longitude?

6. 1897. Find the times of high water at Crinan on April 1st and 12th.

7. 1899, May 31st, when the A.T.S. was $3^h\ 23^m$ P.M. in lat. 57° 15′ S., long. 160° 45′ E., the Sun's compass bearing was N.W. by W. ¾ W. Find the compass error and deviation, if the variation is 20° 30′ E.

8. 1899, June 14th A.M. at ship; latitude at noon, Equator; the observed

Alt. of the Sun's upper limb, 53° 22' 10"; index error, +1' 41"; height of eye, 31 ft.; time by chron. 14ᵈ 10ʰ 1ᵐ 18ˢ·5, which was fast 30ˢ on March 27th, and on May 6th was correct for G.M.T.; run since sights, West 16'; the Sun's compass bearing was E. by N. ¾ N.; variation 10° E. Required the longitude at observation and noon, also compass error, and deviation.

9. 1899, May 26th, when the M.T.S. was 9ʰ 24ᵐ A.M. in lat. 24° 37' S., long. 70° 48' E.; the Sun's compass bearing was N.E. ¾ E. Required the Sun's true azimuth by the "Tables," also compass error, and deviation, the variation being 11¾° W.

10. Using deviation card B, find the course and distance from N.W. light-ship to Point Lynus.

With the ship's head on the above course, Orme's Head bore by compass S. 2° E., and Puffin Island light bore S. 41° W. by the same compass. Find ship's position.

Steering as above, Puffin Island light bore by compass S. 18° W., and after continuing five miles on the above course it bore S. 20 E. by the same compass. Find the ship's position and distance from Puffin Island light.

11. 1899. At what time will Regulus pass the meridian of a place in long. 150° 17' E. on November 5th?

12. 1899. What bright stars are within two hours of the meridian of a place lat. 50° 15' S., long 52° W., on October 7th, at 10ʰ 20ᵐ P.M. ? What is the approximate Mer. Alt. of Fomalhaut?

13. 1899, October 7th, the Mer. Alt. of Fomalhaut bearing North was 70° 1' 10"; index correction, −3' 22"; height of eye, 23 ft. Required the latitude.

14. 1899, September 7th, at about 7ʰ P.M. at ship, in lat. 35° 20' N., long. by account, 165° 15' W., when a chronometer showed 7ʰ 54ᵐ 50ˢ, which was fast 1ʰ 58ᵐ 5ˢ·4 on June 29th, and was fast 2ʰ 3ᵐ 13ˢ·4 on August 3rd on G.M.T., the observed Alt. of Altair east of the meridian was 53° 1' 40"; index correction, +1' 11"; height of eye, 29 ft. Required the longitude.

15. 1899, October 7th, when the M.T.G. was 7ᵈ 17ʰ 19ᵐ in lat. 50° 15' S., long. 52° W.; the compass bearing of Markab was N.W. ¾ W. Required the true azimuth by the "Tables;" and if the variation be 14° 20' E., find the deviation.

16. 1899, April 15th P.M. at ship, in lat. by account 14° 50' S., long. 15' 45" W., the observed Alt. of Sun's lower limb north of the observer, 63° 48' 40"; index correction, +7' 9"; height of eye, 20 ft.; time by watch, 14ᵈ 22ʰ 55ᵐ 8ˢ, which was found to be slow of A.T.S. 1ʰ 38ᵐ 20ˢ; the diff. longitude made to the west being 21' since the error was determined. Required the latitude by reduction to the meridian.

17. 1899, November 18th, A.M. at ship, at sea and uncertain of my position, when the chronometer showed 18ᵈ 5ʰ 41ᵐ 39ˢ G.M.T., the observed Alt. of Sun's lower limb was 15° 47' 20"; and again P.M. on the same day, when the chronometer showed 18ᵈ 9ʰ 20ᵐ 31ˢ the observed Alt. of Sun's lower limb was 21° 7' 10"; index correction, − 4' 11"; height of eye, 30 ft. ; the ship having made a true S.W. by W. ¼ W. course 32 miles in the interval. Required the line of position, Sun's true azimuth, and position of the ship at second observation by Sumner's method by projection, assuming lats. 47° and 47° 30' N.

18. 1899, February 12th, the M.T.G. 12ᵈ 14ʰ 15ᵐ 26ˢ, in long. 89° 57' W., the observed Alt. of Polaris out of the meridian was 44° 23' 40"; index correction, +10'; height of eye, 26 ft. Required the latitude.

19. 1899, February 28th A.M., in long. 130° 27' E.; the observed Mer. Alt. of the Moon's lower limb, 34° 27' 40" bearing South; index correction, +4' 17"; height of eye, 31 ft. Required the latitude.

324   APPENDIX.

20.

Ship's head by compass.	Bearing of distant object.	Deviation.	Ship's head by compass.	Bearing of distant object.	Deviation.
N.	N.		S.	N. 21° W.	
N.E.	N. 7° E.		S.W.	N. 17° W.	
E.	N. 10° W.		W.	N. 18° E.	
S.E.	N. 13° W.		N.W.	N. 16° E.	

Find correct magnetic bearing, courses to steer, etc., as follows :—
Correct magnetic courses : N. ¼ E. S.E. by E. ½ E. W.S.W. W. 12° N.
Courses to steer.

Courses steered : E. ¾ N.   N.W. ½ W.   S. 65° W.   S. 32° E.
Correct magnetic courses.

Ship's head at E. ½ S. by compass.
Bearings by compass : S. by W. ¼ W.   S.E. by E.
Correct magnetic bearings.

21. Find the course to steer from N.W. lightship to Point Lynus to counteract the effect of a current which set N. 19° E. 2 miles per hour ; the ship making by log 8 miles per hour, and the distance she would then make good towards Point Lynus in three hours.

22. 1897, November 13th, 8ʰ 34ᵐ A.M., off Flushing, took a cast of the lead. Find the correction to be applied to the cast before comparing it with the depth marked on chart.

23. 1899, July 25th P.M. at ship, in lat. 40° 20′ S. ; time by chron., 25ᵈ 14ʰ 22ᵐ 14ˢ, which was supposed to be correct for G.M.T. With following observations, find the error of chronometer for G.M.T. and the longitude ; height of eye, 40 ft. :—

Obs. Alt. of Moon's L.L. east
of Meridian.   Obs. Alt. of Antares.   Obs. dist. between Moon's
                                                  R.L. and star.
27° 22′ 40″   46° 51′ 40″   101 6′ 0″
I.C. − 1 15   I.C. − 1 23   I.C. − 2 11

24. 1899, March 20th P.M. at ship, lat. N., long. 45° 30′. W. With following observations, find latitude by " Ivory " at time of taking the second observation :—

M.T.G. 20ᵈ 4ʰ 24ᵐ 54ˢ   Obs. Alt. Sun's L.L. 59° 30′ 40″   Bearing S.W. ½ S.
,,   20 8 30 6   ,,   ,,   8 56 20

The ship's true course and distance during the interval, S.E. ½ E. 40 miles ; index correction, − 2′ 12″ ; height of eye, 32 ft.

25. 1899, May 19th P.M. at ship, when a chronometer showed 19ᵈ 18ʰ 47ᵐ 35ˢ correct G.M.T., the observed Alt. of Jupiter's centre East of meridian was 30° 9′ 55″ ; and at the same time observed Alt. of Pollux West of meridian, 23° 42′ 0″ ; index correction, − 1′ 33″ ; height of eye, 29 ft. Required the position of the ship, and true bearings of Jupiter and star by Sumner's method by projection, assuming lats. 48° and 48° 40′ N.

26. 1899, July 28th A.M., at a place whose correct latitude and longitude are 47° 25′ N. and 150° 27′ E. ; the observed alt. of the Star Altair was 47° 0′ 0″ ; index cor. − 2′ 13″ ; height of eye, 25 ft. ; the M.T.G. by chron. 27ᵈ 2ʰ 28ᵐ 46ˢ. Required the error of chronometer on mean time at place and M.T.G.

27. Find the initial course and distance on a composite track from Cape Town to Albany, the latitude of highest parallel being 42° S.; also longitudes of points of arrival at and departure from parallel.

Cape Town, lat. 33° 56′ S.; long. 18° 29′ E.
Albany, lat. 35 2 S.; long. 117 54 E.

28. Construct a Mercator chart, scale 1″·2 = 1° long., extending from 67° to 70° N., and 2° W. to 2° E. A ship sails from 69° 30′ N.; 0° 30′ E. the following true courses: S.W. by S., 30′; E. by S., 40′; S.S.W., 56′; and W. by N., 25′. What is her latitude and longitude in?

29. If the diff. lat. be 224·6 N., and departure 157·9 miles E., find the course and distance by calculation.

30. The angular elevation of a tower at A is 30°; at B in the same horizontal plane, and 100 ft. nearer the tower, the elevation is 60°. Find the height of the tower.

31. In a right-angled spherical triangle PQR, P = 90°, Q = 68° 30′, PQ = 101° 20′. Find R, PR, and RQ.

32. In a right-angled spherical triangle ABC, C = 90°, AC = 43° 45′, BC = 81° 27′. Find the angles and hypotenuse.

Paper IX.

1. Multiply 30,000,000 by ·00765 by common logs.
2. Divide ·00854006 by ·000002 by common logs.
3.

Hour.	Courses.	Knts.	10ths.	Winds.	Lee-way.	Deviation.	Remarks.
					pts.		
1	S.	12	5	E.S.E.	¼	0°	A point in lat.
2		12	5				51° 53′ N., long.
3		12	5				55° 22′ W.; bear-
4		12	5				ing by compass,
5	S.S.E. ¼ E.	13		E. ½ N.	¼	11° W.	N.N.W.; ship's
6		13					head, S.E. ½ S.;
7		13					deviation as per
8		13					log; distance off,
9	S.E. ½ S.	12	5	E.N.E.	¾	18° W.	48 miles.
10		12	5				Variation, 36° W.
11		12	5				
12		12	5				
1	S.E. by E. ½ E.	12		N.E.	¼	27° W.	
2		12					
3		12					
4		12					
5	E.	11	5	N.N.E.	½	35° W.	
6		11	5				A current set cor-
7		11	5				rect magnetic
8		11	5				E. 36 miles for
9	S. by E. ½ E.	11		E.	¾	8° W.	the day.
10		11					
11		11					
12		11					

Find the course and distance made good; also the ship's position.

4. In what latitude does 520 miles of departure correspond with 640 minutes of longitude?

5. Find the compass course and distance from A to B by Mercator sailing—

A, lat. 51° 9′ S. Long. 23° 30′ W. Dev. 10° 15′ W.
B, ,, 38 17 S. ,, 118 47 E. Var. 7 30 W.

6. 1897, March 5th. Find the time of high water at Dublin bar A.M. and P.M.; also March 12th at Padstow.

7. 1899, March 20th, in long. 13° 30′ E., the observed Mer. Alt. of the Sun's upper limb was 55° 38′ 40″, zenith South of the Sun; index error, 6′ 40″ off the arc; height of eye, 12 ft. Required the latitude.

8. 1899, June 21st, at $9^h\ 57^m$ P.M. A.T.S., in lat. by account 63° 12′ N., long. 5° 30′ W., the Sun set bearing N. by E. ½ E. by compass. Required the true amplitude and error of the compass; and, supposing the variation to be 27° 20′ W., required the deviation for the direction of the ship's head.

9. 1899, May 21st A.M. at ship, the observed Alt. of the Sun's lower limb was 14° 24′ 45″; height of eye, 12 ft.; the Sun's bearing by compass, E.; time by chron., $20^d\ 18^h\ 9^m\ 6^s$, which was fast for M.N.G. $9^m\ 39^s$ on April 10th, and losing 5^s·5 daily; the lat. at noon was 54° 26′ N.; run between observation and noon, N. 61 miles. Required the longitude by chronometer at observation and at noon. Required also the true azimuth and error of the compass; and, supposing the variation to be 19° 30′ W., what is the deviation?

10. 1899, June $13^d\ 5^h\ 17^m\ 40^s$ M.T.G., lat. 13° 52′ S., long. 5° 40′ W.; the Sun bore by compass N.W. ¼ W. Find the true azimuth and error of the compass by the "Time Azimuth Tables;" and, supposing the variation to be 25° W., required the deviation of the compass.

11. 1899, June 30th P.M. at ship, in lat. by account 14° 35′ S., long. 100° 44′ E., the observed Alt. of the Sun's lower limb was 50° 42′ 45″ N. of the observer; index error, +5′ 20″; height of eye, 11 ft.; time by chron., $29^d\ 17^h\ 53^m\ 21^s$, which was found to be slow of A.T.S. $6^h\ 45^m\ 10^s$; the diff. of long. made to the east being 15½′ since the error was determined; and if the course and distance since noon was S.S.W. 3 miles, find the latitude at noon.

12. 1899, September 23rd, being A.M. at ship and uncertain of the position, when the chronometer showed $23^d\ 9^h\ 51^m\ 13^s$ M.T.G., the observed Alt. of the Sun's lower limb was 32° 50′ 30″; and again, being P.M. on the same day, when the chronometer showed $23^d\ 13^h\ 51^m\ 9^s$ M.T.G., the observed Alt. of the Sun's lower limb was 32° 54′ 10″, the ship being becalmed heading East during the interval; height of eye, 17 ft.; index error, −1′ 20″. Required the lines of bearing and the azimuths from both observations, and the ship's position at the last observation by Sumner's method of projection on the chart, assuming lats. 51° 15′ and 51° 45′ S.

13. 1899. What is the M.T.S. of the meridian passage of Regulus in lat. 40° 10′ N., long. 74° 15 W. on May 21st? Also by inspection.

14. 1899, on May 10th, at about $9^h\ 35^m$ P.M. M.T.S., what stars of the first, second, and third magnitudes are within three-quarters of an hour of the meridian above the horizon and the pole in lat. 34° 39′ S., long. 22° 15′ E. Give the meridian distances, and state if they are N. or S. of observer.

15. 1899, May 21st, in lat. 40° 10′ N., long. 74° 15′ W., if the index error is 0, and the height of eye 24 ft., find the approximate Altitude of the Star Regulus to place on the sextant.

16. 1899, May 9th, the observed Mer. Alt. of the Star Menkar was 15° 20′ 30″ N. of the observer; index error, +2′ 30″; height of eye, 18 ft. Required the latitude.

17. 1899, May 20th A.M. at ship on the Equator, observed Alt. of the star Antares W. of the meridian was 42° 38′; index error, +3′ 15″;

APPENDIX.

height of eye, 25 ft. ; time by chron., $19^d\ 22^h\ 4^m\ 43^s$, which was slow M.T.G. $5^m\ 2^s\cdot 4$ on February 28th, and on March 30th was correct. Required longitude by chronometer.

18. 1899, May $4^d\ 2^h\ 10^m\ 30^s$ M.T.G., in lat. 30° 42′ S., long. 79° 30′ W., Altair bore by compass N. 67° 30′ W. Find the true azimuth and error of the compass ; and if the variation was 10° E., required the deviation of the compass for the direction of the ship's head.

19. 1899, May 21st, $4^h\ 30^m$ A.M. M.T.S., long. 30° 17′ E., the observed Alt. of Polaris out of the meridian was 31° 15′ 10″ ; height of eye, 20 ft. Required the latitude.

20. May 4th, in long. 30° 30′ E., the observed Mer. Alt. of the Moon's lower limb south of the observer was 57° 2′ 30″ ; index error, −2′ 30″ ; height of eye, 29 ft. Required the latitude.

21. (a) Using card A, find the course to steer by compass from Orme's Head to Point Lynus ; also the deviation, variation, and distance.

(b) With the ship's head on the above course, a point, Orme's Head, bore by compass S. 41° E., and another point, Puffin Island, bore by compass S. 57° W. Find the ship's position.

(c) With the ship's head as above, a point, Moelfre Island, bore by compass S. 84° W., and after continuing 3 miles on the same course it bore S. 46° W. Find the ship's position and distance from Moelfre Island at the second bearing.

(d) Find the course from Orme's Head to Point Lynus in a current which set East (cor. mag.) 3 miles per hour, the vessel making 12 miles per hour by log. ; also the distance made good towards Point Lynus in 1 hour.

22. 1897, March 4th, 11.30 P.M. mean time, being off St. Ives, took cast of the lead. Required the reduction to be applied to the cast before comparing with the chart.

23. Correct magnetic bearing—

Ship's head by compass.	Bearing of distant object by compass.	Deviation.	Ship's head by compass.	Bearing of distant object by compass.	Deviation.
N.	S. 2° W.		S.	S. 28° E.	
N.E.	S. 7° W.		S.W.	S. 26° E.	
E.	S. 6° W.		W.	S. 5° E.	
S.E.	S. 16° E.		N.W.	S. 3° E.	

(a) Magnetic courses : S.E. by E.   W.   N. ¼ W.   S. 70° W.
Required courses to steer by compass.

(b) Compass courses : W. ¾ N.   S.W. ½ W.   S.E. by S.   E. by N. ½ N.
Required the correct magnetic courses.

(c) Ship's head N.N.W. by compass.
Bearings by compass : S. 72° W.   N. 20° W.
Required the correct magnetic bearings.

24. 1899, June 2nd A.M. at ship, in lat. 37° 30′ S., when the approximate M.T.G. by chronometer was $1^d\ 8^h\ 54^m\ 42^s$, observed the following : Alt. of Saturn, 12° 16′ 35″ W. ; the observed distance between Saturn and the Moon's remote limb was 92° 12′ 0″ ; index error for altitude, +1′ 10″, for the distance, +1′ 30″ ; height of eye, 24 ft. Required the Moon's altitude, the true distance, error of chronometer, and the longitude.

25. 1899, April 16th, in N. lat., long. 70° 45′ W., with the following observations find the latitude by double altitudes at the first observation :—

A.T.S. $10^h\ 14^m\ 50^s$ A.M.   Obs. alt. Sun's U.L. 52° 0′20″ Index error, −3′ 15″
   ,,   1 45 30 P.M.   ,,   ,,   51 21 40 Height of eye, 23 ft.

The Sun's bearing at the second observation was S. 45° W. ; the run during the interval, N.E. 11 knots ~~per hour~~.

26. 1099, July 15th P.M., at St. Elmo's, Malta, lat. 35° 54′ N., long. 14° 31′ 30″ E., the following times were noted when Altair had equal altitudes: M.T.G. $7^h\ 58^m\ 10^s$, $14^h\ 28^m\ 48^s$. Required the index error of chronometer on A.T.S., M.T.S., and M.T.G.

27. 1899, May 20th P.M. at ship, when chronometer showed $20^d\ 14^h\ 31^m\ 40^s$ M.T.G., simultaneous Altitudes were taken of Antares, 13° 15′ 40″ E. of meridian, and Vega, 44° 36′ 30″, also E. of the meridian; index error, −3′ 15″; height of eye, 19 ft. Required the two azimuths and the ship's position by Sumner's method, assuming lats. 45° 30′ and 46° 0′ N.

Great Circle Sailing.

28. Find the initial and final courses, the distance, latitude and longitude of Vertex, and the latitudes of a succession of points for every 5° from A to B on the arc of a Great Circle.

A, Cape Clear, lat. 51° 26′ N., long. 9° 29′ W.
B, Cape Race, ,, 46 39 N., ,, 53 4 W.

29. (*a*) If the difference of lat. is 225·3 N., and dep. 265·7 E., find the course and distance by calculation.

(*b*) In a plane triangle ABC, $a = 17$, $b = 20$, $c = 27$. Find A.

(*c*) In a right-angled spherical triangle, C = 90°, B = 45°, $c = 120°$. Find A, a, and b.

(*d*) Two places in lat. 44° S. have a difference of longitude of 104°. Find the distance gained by following the great circle instead of the parallel.

Paper X.

1. Multiply 2000 by ·05; and divide ·00467 by ·000467 by logs.
2.

Hour.	Courses.	Knts.	10ths.	Winds.	Lee-way.	Devia-tion.	Remarks.
					pts.		
1	E.	7	6	S.S.E.	¾	10° E.	A point in lat.
2		6	5				37° 18′ N., long.
3		7	4				18° 24′ W.; bear-
4		6	5				ing by compass,
							distant 16 miles,
5	S.W. ½ S.	6	7		1½	6° W.	W. ½ N.; ship's
6		6	8				head, N.N.E.;
7		7	3				deviation, 7° E.,
8		7	5				Variation 2½ pts. W.
9		6	9				
10		6	8				
11	S. by W. ½ W.	6	4	S.E. by E.	1	5° W.	
12		6	8				
1		6	8				
2	S.	6	9	E. by S.		3° W.	
3		7	5				
4		8	7				
5		7	9				
6	N.E. ¾ E.	7	8	E. by S. ½ S.	½	10° E.	A current set cor-
7		6	9				rect magnetic
8		7	7				S.S.W. ½ W., 20
9		7	6				miles during the
							day.
10	S.S.W.	8	2	E.S.E.	¾	5° W.	
11		7	7				
12		8	1				

APPENDIX.   329

Find the course and distance made good; also latitude and longitude of the ship.

3. 1899, March 21st, in long. 11° 45' W.; the observed Mer. Alt. of Sun's lower limb, 57° 52' 50"; index correction, −8' 20"; height of eye, 29 ft.; zenith north of the Sun. Find the latitude.

4. Find the compass course and distance by Mercator's sailing—

From A, lat. 52° 27' S.   Long. 37° 7' W.
To B, ,,   39 58 S.   ,, 142 53 E.
Variation, 10 45 E.   Deviation, 19 25 E.

5. A ship sailing due west on the parallel of 45° N. has changed her longitude 1° 50'. How far has she sailed?

6. 1897, May 5th and 21st. Find the times of high water at Lerwick.

7. 1899, in lat. 60° 0' N., long. 179° 30' E., when the correct A.T.G. was July $4^d\ 2^h\ 55^m$, the Sun's compass bearing was N.N.W. ¼ W. Find the compass error, and deviation if the variation be 14° 50' E.

8. 1899, July 5th, P.M. at ship, about $2^h\ 10^m$ after the Sun had passed the meridian in lat. 40° 15' S.; approximate long. 94° E.; the observed Alt. of Sun's lower limb, 19° 37' 0"; index correction, −1' 19"; height of eye, 18 ft.; chronometer shows $9^h\ 21^m\ 45^s$, which was fast $1^h\ 15^m\ 30^s$ on March 13th, and on May 19th was fast $1^h\ 17^m\ 46^s$ of G.M.T.; the Sun's compass bearing was N. by W. ¾ W.; variation, 18° 30' W. Find the longitude, compass error, and deviation.

9. 1899, June 27th, mean time at ship, $7^h\ 11^m\ 8^s$ A.M., in lat. 34° 17' N., long. 130° 58' W.; the Sun's compass bearing was E. ¾ N. Required the Sun's true azimuth by the "Tables," compass error, and deviation, the variation being 15° E.

10. Using deviation card A, find the compass course and distance from Great Orme's Head to Blackpool.

With the ship's head on the above course, N.W. lightship bore by compass S. 21° W., and Bar L.V. bore S. 55° E. by same compass. Find ship's position.

Steering as above, Stanner Point bore by compass N. 61½° E., and after continuing 8 miles on the above course it bore S. 70½° E. by the same compass. Find the ship's position and distance from Stanner Point.

11. 1899. At what time will Canopus pass the meridian of a place long. 73° 24' W. on October 19th.

12. 1899. What bright stars are 1½ hour from the meridian of a place lat. 45° S., long. 30° E., on February 26th at $10^h\ 40^m$ P.M.? What is the approximate Mer. Alt. of Procyon?

13. 1899, February 26th, the Mer. Alt. of Procyon bearing N. was 39° 36' 30"; index correction, +1' 11"; height of eye, 32 ft. Required the latitude.

14. 1899, August 31st, about 6^h P.M. at ship, in lat. 24° 19' S., long. by account 180° E., a chronometer showed $6^h\ 4^m\ 3^s$, which was slow $1^m\ 53^s\cdot 5$ on June 16th, and was fast $1^m\ 29^s\cdot 5$ on August 5th, for G.M.T.; the observed Alt. of Vega E. of meridian, 21° 29' 10"; index correction, −2' 15"; height of eye, 28 ft. Required the longitude.

15. 1899, February 26th, M.T.S. $10^h\ 30^m$ P.M., in lat. 45° S., long. 30° E., the compass bearing of Procyon was N.N.W. Required the true azimuth by the "Tables;" and if the variation be 33° W., find the deviation.

16. 1899, May 1st P.M. at ship, in lat. by account 13° 36' S., long. 110° 32' W., the observed Alt. of Sun's upper limb was 61° 24' 40" bearing north; index correction, +1' 30"; height of eye, 23 ft.; time by chron., $1^d\ 10^h\ 0^m\ 8^s$, which was fast $2^h\ 30^m\ 12^s$ of M.T.G. Required the latitude by reduction to the meridian.

17. 1899, May 15th A.M. at ship, at sea and uncertain of my position,

when the chronometer showed 15ᵈ 2ʰ 52ᵐ 38ˢ G.M.T., the observed Alt. of Sun's lower limb, 19° 27' 45"; and again A.M. on the same day, when the chronometer showed 15ᵈ 7ʰ 12ᵐ 52ˢ; the observed Alt. of Sun's lower limb, 56° 18' 10"; index correction, +2' 4"; height of eye, 31 ft.; the ship having made a true S.W. course, 36 miles in the interval. Required the line of position at the first observation; the Sun's true azimuth and position of the ship at second observation by Sumner's method by projection on the chart, assuming lats. 49° and 49° 30' N.

18. 1899, November 25th, M.T.S. 5ʰ 49ᵐ 18ˢ A.M. in long. 20° 13' E., the observed Alt. of Polaris out of the meridian was 36° 19' 20"; index correction, −4' 27"; height of eye, 19 ft. Find the latitude.

19. 1899, November 12th P.M., in long. 45° 21' E., the observed Mer. Alt. of the Moon's upper limb, 36° 29' 50" bearing north; index correction, +4' 57"; height of eye, 28 ft. Find the latitude.

20.

Ship's head by compass.	Bearings of distant object.	Deviation.	Ship's head by compass.	Bearings of distant object.	Deviation.
N.	S. 32° W.		S.	S.	
N.E.	S. 28° W.		S.W.	S. 3° W.	
E.	S. 7° W.		W.	S. 10° W.	
S.E.	S.		N.W.	S. 16° W.	

Find correct magnetic bearing.

(a) Correct magnetic courses : W. ¾ N.   N. 6° W.   S. 15° E.   S. 35° W.
Courses to steer.

(b) Courses steered : E. by N.   N. ¾ E.   S. 25° W.   S. 80° E.
Correct magnetic courses.

(c) Ship's head at S.E. ¼ S. by compass.
Bearings by compass : N.E. ¼ E.   S.E. by S. ¼ S.
Correct magnetic bearings.

21. Find the course to steer from Great Orme's Head to Blackpool to counteract the effect of a current which set (C.M.) W.N.W. 3 miles per hour, the ship making by log 6 miles per hour, and the distance she would then make good towards Blackpool in 4 hours.

22. 1897, November 20th, 9ʰ A.M., off Noel Bay took a cast of the lead. Find the correction to be applied to the cast before comparing it with the depth marked on the chart.

23. 1899, April 21st P.M. at ship, in lat. 25° 10' S.; time by chron., 21ᵈ 16ʰ 46ᵐ, which was estimated 32ˢ slow of G.M.T. With following observations, find the error of chronometer for G.M.T.; also longitude, and true and apparent Alts. of the Moon; height of eye, 19 ft. :—

Obs. Alt. of Pollux W. of meridian.   Obs. distance between Star and Moon's N.L.
23° 16' 20"   57° 35' 10"
Index cor. + 1 20   Index cor. + 1 10

24. 1899, September 17th, lat. N., long. 153° 27' E. With following observations, find latitude by "Ivory" at time of taking the second observation.

M.T.S. 17ᵈ 8ʰ 45ᵐ 13ˢ   Obs. Alt. Moon's L.L. 28° 56' 50" bearing S. 21° E.
 ,,   17 13 15 22   ,,   ,,   19 4 20

The ship's true course and distance during the interval, S. 34° W. 30 miles; index correction, −1' 9"; height of eye, 25 ft.

APPENDIX.

25. 1899, October 23rd A.M. at ship, when a chronometer showed 22nd, 4ʰ 52ᵐ 49ˢ correct G.M.T., the observed Alt. of Rigel E. of meridian was 26° 14′ 40″, and the observed Alt. of Markab W. of meridian was 27° 11′ 20″ at the same time; index correction, +4′ 11″; height of eye, 22 ft. Required the position of the ship and true bearings of the stars by Sumner's method by projection on the chart, assuming lats. 50° and 50° 45′ N.

26. 1899, May 19th, in lat. 33° 20′ S., long. 119° 47′ E.; the following times were observed when Antares had equal altitudes:—

    Star E. 1ʰ 30ᵐ 52ˢ·   Star W. 7ʰ 53ᵐ 21ˢ·

Find the error of the chronometer for M.T.S., M.T.G., and A.T.G.

27. Find the initial course and the distance on a Great Circle, latitude and longitude of Vertex, and succession of points at 10° long. apart—

    From lat. 6° 2′ N., long. 80° 13′ E.
    To ,, 35 2 S., ,, 117 54 E.

28. Construct a Mercator's chart, scale 1″·4 = 1° long., extending from lat. 53° to 54° 30′ N., and long. 3° W. to 6° W., and insert the following positions: Chicken Rock, lat. 54° 2′ N., long. 4° 50′ W.; Langness Point, lat. 54° 3′ N., long. 4° 37′ W.; N.W. Light, lat. 53° 31′ N., long. 3° 31′ W. At 2ʰ P.M., Chicken Rock bore N. 49° W., Langness Point N. 22° E.; shaped course S. 76° E., steaming 15 knots. At 4ʰ 40ᵐ P.M., altered course to S. 8° E. At 5ʰ 50ᵐ P.M., N.W. Light bore S. 54° E. At 6ʰ 10ᵐ P.M., N.W. Light bore N. 47° E. Find true position, and set and drift of current.

29. From a cliff 250 feet above the sea's surface, the angle of depression of a boat was observed to be 15° 32′. How far was the boat from the base of the cliff.

30. Wishing to know the height of a flagstaff, I measured the length of its shadow when the sun had an altitude of 43° 27′, and found it to be 59 ft. Find its height.

31. How long will a ship be reaching a port that bears N.N.W. 30 miles ? The wind comes from north, and the ship can lie within 6½ points of the wind, sailing 8 knots per hour.

32. The declination of the sun is 19° 43′ N., and his true altitude when on the prime vertical is 20° 30′. Find the latitude.

33. In a spherical triangle XYZ, Z = 90°, XZ = 64° 23′, Y = 70° 19′. Find the other parts.

Paper XI.

1. Multiply ·999 by ·88801 by common logs.
2. Divide ·0004629 by ·00000542 by common logs.

332   APPENDIX.

3.

Hour.	Courses.	Knts.	10ths.	Winds.	Lee-way.	Deviation.	Remarks.
					pts.		
1	N.W. ½ N.	12		N.E. by N.	¼	20° W.	Departure taken
2		12					from Cape Fare-
3		12					well in lat. 59°
4		12					49' N., long. 43°
5	W.N.W.	13		N. ¼ E.	¼	25° W.	54' W.; bearing
6		13					by compass, E.
7		13					by N., distant
8		13					30 miles; ship's
9	W. by N.	14		N.	0	27° W.	head, N.W.½N.;
10		14					deviation as per
11		14					log.
12		14					Variation, 47° W.
1	W.	14		N.N.W.	¾	30° W.	
2		14					
3		12					
4		12					
5	N.W. by W. ½ W.	15		S.W. ½ S.	½	21° W.	A current set
6		15					S. 60° E., cor-
7		15					rect magnetic
8		15					10 miles for the
9	N.W. by N.	15		W. by S. ½ S.	¾	18° W.	24 hours.
10		15					
11		15					
12		15					

Find the true course and the distance made good, also the ship's position.

4. Longitude left, 62° 30' W.; long. in 72° 30' W., lat. 50° 9' S.; course, west true. Required the distance run.

5. Find the compass course and distance from A to B by Mercator sailing; variation, 6° 30' E. ; deviation, 19° 44' W.—

Lat. A, 1° 15' S. ; long. A, 153° 13' E.
,, B, 20 49 N.;   ,, B, 110 57 W.

6. 1897, May 9th. Find the time of high water at Cardigan A.M. and P.M.

7. 1899, March 21st, long. 150° 27' E.; observed Mer. Alt. of the Sun's lower limb, 56° 21' 10", bearing north of the observer; index error, +4' 15" ; height of eye, 32 ft. Required the latitude.

8. 1899, August 1st, in lat. 59° 30' N., long. 60° 30' W.; at 8ʰ 12ᵐ 57ˢ A.T.S. the Sun set, bearing by compass N. by E. ¼ E. Required the true amplitude and error of the compass; and, supposing the variation to be 54° 30' W., required the deviation for the direction of the ship's head.

9. 1899, June 14th P.M. at ship, the observed Alt. of the Sun's lower limb was 46° 25' 30"; index correction, +2' 30"; height of eye, 21 ft.; the Sun's bearing by compass, N. 56° 15' W.; time by chron., 14ᵈ 5ʰ 37ᵐ 36ˢ, which was fast 11ᵐ 43ˢ·5 on February 4th, and on April 15th was correct for M.T.G.; the lat. at noon, 0°; ship becalmed all afternoon. Required ship's position by chronometer. Required also the Sun's true azimuth and error of the compass ; and if the variation was 4° 45' 30" W., required the deviation of the compass.

10. 1899, July 3rd, 8ʰ 11ᵐ 30ˢ A.M. mean time at ship, in lat. 40° 17' 30" S.,

long. 122° 7' 30" E. ; the Sun bore by compass, N. 42° 11' 15" E. Find the true azimuth and error of the compass by the "Time Azimuth Tables ;" and, supposing the variation to be 14° 30' E., required the deviation for the direction of the ship's head.

11. 1899, June 23rd A.M. at ship, in lat. by D.R. 45° 30' N., long. 45° 15' W.; time by watch, 22^d 20^h 40^m 10^s, which was 10^m 5^s fast of M.T.G. ; observed Alt. of the Star α Andromedæ E. of the meridian, 72° 8'0" bearing south ; index error, + 2' 30" ; height of eye, 20 ft. Required the latitude by reduction to the meridian.

12. 1899, June 23rd, being A.M. at ship, when a chronometer showed 22^d 22^h 40^m 10^s M.T.G., the observed Alt. of the Sun's lower limb was 59° 4' 4"; and again, being P.M. at ship on the same day, when a chronometer showed 23^d 2^h 59^m 55^s M.T.G., the observed Alt. of the Sun's lower limb was 46° 19' 40"; the ship having made 39 miles on a S. 83° W. course during the interval ; height of eye, 26 ft. ; index error, -3' 15". Required both lines of position and the azimuths ; also the ship's position by Sumner's method when the second altitude was taken, assuming the lats. 50° 0' and 51° 0' N.

13. 1899, June 6th, in lat. 10° 9' N., long. 74° 34' E. Find the M.T. when the Star Antares will be on the meridian.

14. 1899. Find from pages 294-304 Nautical Almanac, what first, second, and third magnitude stars will be within three-quarters of an hour of the meridian above the horizon and pole on June 23rd, at about 2.30 A.M. A.T.S., in lat. 55° 20' N., long. 27° 20" W. State how far east or west they are, and if they are north or south of you.

15. 1899, June 6th, lat. 10° 9' N., long. 74° 34' E. Find the approximate Mer. Alt. of Antares; index error, -3' 5" ; height of eye, 25 ft.

16. 1899, June 19th, observed Mer. Alt. of the Star Aldebaran was 65° 27' 10", zenith south; height of eye, 19 ft. Required the latitude.

17. 1899, June 23rd, at about 5^h 45^m A.M., latitude by ex-meridian of Star β Gruis was 40° 15' 30" S., long. by D.R. 179° 58' E. ; observed Alt. of Star Altair, 15° 54' 0" W. of meridian ; height of eye, 26 ft. ; time by chron., 5^h 46^m 18^s, which was fast 45^s·5 for mean noon Greenwich on May 31st, and was losing 2^s·05 daily. Required longitude by chronometer.

18. 1899, June 21st, 9^h 4^m 50^s P.M. mean time at ship, in lat. 30° 42' S., long. 79° 30' W. ; Star β Leonis bore by compass W. by S. Required the true azimuth and compass error by the "Time Azimuth Tables ;" and, supposing the variation to be 5° E., required the deviation for the direction of the ship's head.

19. 1899, June 29th, at 10^h 34^m P.M. A.T.S., in long. 86° 24' E., the observed Alt. of Polaris out of the meridian was 15° 24' 0" ; index error, -1' 10" ; height of eye, 19 ft. Required the latitude.

20. 1899, June 22nd, in long. 179° 30' W., the observed Mer. Alt. of the Moon's upper limb was 15° 20' 30" bearing south of the observer ; height of eye, 26 ft. Required the latitude.

21. (a) Using card B, find the course to steer by compass from Piel Harbour to Rhyl ; also the deviation, variation, and distance.

(b) With the ship's head on the above course, a point, Fleetwood, bore by compass N. 53° E., and another point, Blackpool, bore by compass S. 82½° E. Find the ship's position.

(c) With the ship's head as above, the N.W. Lightship bore by compass S. 40½° W., and after continuing 7 miles on the same course it bore N. 48½° W. Find the ship's position and distance from the N.W. Lightship at the second bearing.

(d) Find the course from Piel Harbour to Rhyl in a current which set W. ¼ S. (cor. mag.) 3 miles per hour, the vessel making 10 miles per hour by log ; also the distance made good towards Rhyl in 3 hours.

22. 1897, March 20th, 0^h 28^m P.M. M.T.S., being off Ilfracombe, took cast

of the lead. Required the reduction to be applied to the cast before comparing with the chart.
23. Correct magnetic bearing—

Ship's head by compass.	Bearing of distant object by compass.	Deviation.	Ship's head by compass.	Bearings of distant object by compass.	Deviation.
N.	N. 6° W.		S.	N. 8° E.	
N.E.	N. 7° E.		S.W.	N. 6° W.	
E.	N. 17° E.		W.	N. 18° W.	
S.E.	N. 16° E.		N.W.	N. 20° W.	

(a) Magnetic courses : N. 5° W.   W.S.W.   S. 20° E.   E. by N.
Required compass courses.

(b) Compass courses : W. ¾ S.   N. by W.   E. by N. ½ N.   N. 45° W.
Required correct magnetic courses.

(c) Ship's head by compass : N.N.W.
Bearings by compass : N.E. ¾ N.   S.E. ¼ S.
Required the correct magnetic bearings.

24. 1899, July 26th A.M., lat. 0° 10′ 0″ N. ; time by chron., 26ᵈ 0ʰ 20ᵐ 30ˢ, supposed to be slow 10ᵐ of M.T.G. ; observed Alt. Moon's lower limb, 30° 55′ 57″ west of meridian ; index error, +2′ 10″ ; height of eye, 20 ft. ; observed distance between the Sun's and Moon's near limbs, 128° 55′ 40″ ; index correction, 3′ 10″ to add. Required the Sun's altitude, the error of the chronometer, and the longitude.

25. 1899, March 20th, in N.′ lat., long. 116° 15′ W. With the following observations, find the latitude by double altitudes at the second observation:—

A.T.S. 9ʰ 5ᵐ A.M.   Obs. Alt. Sun's L.L. 46° 32′ 7″   Index error, 0°
„   2 41 P.M.   „   „   49 16 10   Height of eye, 24 ft.

Sun's observed bearing at first altitude, S. 84° W. ; run during interval, East, rate 8 knots per hour. Required the latitude.

26. 1899, September 23rd, in lat. 15° 37′ N., long. 80° 42′ E., with the following times equal altitudes of the Sun was observed : M.T.S. 7ʰ 7ᵐ 21ˢ A.M., 4ʰ 37ᵐ 26ˢ P.M. Required the error on A.T.S., M.T.S., and M.T.G.

27. 1899, September 23rd, 2ʰ 49ᵐ P.M., at a place whose correct latitude and longitude are known to be 23° 28′ S. and 158° 35′ E., the observed Alt. of Sun's upper limb, 41° 32′ 15″ ; index cor. −2′ 30″ ; height of eye, 34 ft. ; the chronometer shows 4ʰ 17ᵐ 7ˢ. Required its error on M.T.G.

28. 1899, June 23rd A.M. at ship, lat. by D.R. 49° 40′ N., long. 178° 14′ W. ; time by chron., 23ᵈ 3ʰ 3ᵐ 13ˢ M.T.G. ; the observed Alt. of a Cygni, 82° 1′ 50″ W. of the meridian, and at the same time Altair, 44° 54′ 10″, also west of the meridian; height of eye, 28 ft. Required the lines of position, the azimuths, and the ship's position by Sumner's method, assuming lat. 49° and 50° N.

29. Find initial and final course from A to B., easting to be run down on parallel of 55° S. Required also total distance and the longitudes of the points where the Great Circles cut the parallel.

A, Hobarton, lat. 42° 53′ S., long. 147° 21′ E.
B, Valparaiso, „ 33 2 S., „ 71 38 W.

30. (a) A ship from lat. 50° 29′ S. sails E.N.E. until her departure is 69 miles. Find the distance sailed and latitude in.

(*b*) In a plane triangle PQR, PQ = 1046 ft. ; QR = 792 ft. ; Q = 65° 40′. Find the other angles and side.

(*c*) In a spherical triangle ABC, C = 90°; a = 135°; c = 120°. Find A, B, and b.

(*d*) A ship sails on a Great Circle from a place in north latitude to a place in south latitude; the latitude of Vertex is 55° 30′ N. What are her courses on crossing the Equator?

Paper XII.

1. Multiply ·5829 by ·377707, and divide 1256 by 78943 by logs.
2.

Hour.	Courses.	Knts.	10ths.	Winds.	Lee-way.	Devia-tion.	Remarks.
					pts.		
1	W. by S. ½ S.	6	8	S.	¼	8° W.	A point in lat. 50° 44′ N., long. 0° 13′ E.; bearing by compass, N.E. ¾ E., distant 12 miles; ship's head, W. by S. ½ S.; deviation as per log. Variation, 18° W.
2		7	3				
3		8	4				
4		6	5				
5	S.	7	2	W.S.W.	¾	3° W.	
6		7	2				
7		6	9				
8		6	7				
9	S.W. by W. ½ W.	8	–	N.W.	¼	7° W.	
10		7	2				
11		7	7				
12		7	1				
1	N.E. ½ N.	6	2	N.N.W.	1¾	2° W.	
2		6	5				
3		7	3				
4	W. by N. ½ N.	7	8	N.	¼	7° W.	
5		7	6				A current set correct magnetic S. 1¾ mile per hour for all the day.
6		6	9				
7		6	4				
8		6	3				
9	N.E. ½ E.	7	4	N. ½ W.	1¼	—	
10		6	5				
11		6	3				
12		6	8				

Find the course and distance made good, and latitude and longitude of ship.

3. 1899, April 30th, in long. 12° 41′ E.; the observed Mer. Alt. of Sun's lower limb, 35° 28′ 40″ bearing North; index correction, −3′ 22″; height of eye, 20 ft. Required the latitude.

4. Find the compass course and distance by Mercator's sailing—

From A, lat. 33° 10′ S.	Long. 171° 17′ E.
To B, ,, 41 27 S.	,, 7 39 W.
Variation, 11 15 E.	Deviation, 14 45 W.

5. To change my long. 2° 30′, I actually sail a distance of 100 miles due East. What is my latitude?

336   APPENDIX.

6. 1898. Find the times of high water at St. Ives on June 10th and 22nd.
7. 1899, September 23rd, 6ʰ P.M. app. T. ship, in lat. 30° N., long. 170° 42′ E., the Sun's compass bearing was S.W. by W. ¼ W. Required the true amplitude, error of the compass, and deviation, if the variation is 13° 45′ E.
8. 1899, November 15th, at about 5ʰ 15ᵐ P.M. at ship, in lat. 34° 36′ S., long. by account 59° W., when a chronometer shows 10ʰ 9ᵐ 10ˢ, which was fast 1ʰ 0ᵐ 54ˢ on May 29th, and on August 7th was fast 1ʰ 0ᵐ 5ˢ, the observed Alt. of Sun's lower limb, 16° 15′ 30″ ; index correction, −5′ 15″ ; height of eye, 25 ft. Find the longitude ; and if the compass bearing is W. by S. ¼ S., and variation 12° E., find the compass error and deviation.
9. 1899, July 22ᵈ 6ʰ 20ᵐ mean time Greenwich, in lat. 60° 12′ N., long. 3° 45′ E., the Sun's compass bearing W. by N. ¼ N. Required the Sun's true azimuth by the "Tables," error of the compass, and deviation, the variation being 19° 15′ W.
10. Using deviation card B, find the compass course and distance from Howth Head to N.W. lightship.

With the ship's head on the above course, S. Stack bore by compass S. 13° ½ E., and Skerries bore S. 55° ½ E. by the same compass. Find the ship's position.

Steering as above, Point Lynus bore by compass S. 49° E., and after continuing 10 miles on the above course it bore S. 49° W. by the same compass. Find the ship's position and distance from Point Lynus.

11. 1899. At what time will Capella pass the meridian of a place in long. 74° 30′ W. on June 20th ?
12. 1899. What bright stars are within 2 hours of the meridian of a place in lat. 0°, long. 0°, on August 24th at 6 A.M. ? What is the approximate Mer. Alt. of Capella ?
13. 1899, August 24th, the observed Mer. Alt. of Capella bearing North was 44° 7′ 50″ ; index correction, +4′ 47″ ; height of eye, 20 ft. Required the latitude.
14. 1899, January 15th P.M. at ship, about 1ʰ 45ᵐ before Aldebaran was on the meridian of a place, lat. 40° 20′ N., long. by account 43° W. ; its observed Alt. was 57° 33′ 20″ ; index correction, +1′ 11″ ; height of eye, 32 ft. ; time by chron., 9ʰ 53ᵐ 55ˢ, which was slow 12ᵐ 13ˢ on October 14th, and on November 28th was slow 10ᵐ 25ˢ. Find the longitude.
15. 1899, August 24th, when the mean time at ship was 2ʰ 15ᵐ A.M., lat. equator, long. Greenwich, the compass bearing of Rigel was E. ½ S. Required the true azimuth by the "Tables;" and the deviation, the variation being 20° 15′ W.
16. 1899, February 3rd A.M. at ship, in lat. by account 20° 30′ S., long. 80° 40′ W., the observed Alt. of Regulus near the meridian bearing North was 56° 13′ 20″ ; index correction, −2′ 9″ ; height of eye, 28 ft. ; chronometer showed 8ʰ 3ᵐ 32ˢ, which had been found to be 59ᵐ 24ˢ fast of G.M.T. ; the difference of longitude made to the eastward since determining the error was 16′. Required the latitude by reduction to the meridian.
17. 1899, September 29th P.M. at sea and uncertain of my position, when the chronometer showed 29ᵈ 9ʰ 26ᵐ 12ˢ G.M.T., the observed Alt. of Sun's lower limb was 36° 33′ 40″ ; and again P.M. on the same day, when the chronometer showed 29ᵈ 12ʰ 2ᵐ 10ˢ G.M.T., the observed Alt. of Sun's lower limb was 18° 39′ 15″ ; index correction, −2′ 3″ ; height of eye, 34 ft. ; the ship having made a true E. ¼ N. course 43 miles in the interval. Required the line of position when the first observation was taken ; the Sun's true azimuth and position of the ship at the second observation by Sumner's method by projection on the chart, assuming lats. 48° and 48° 30′ N.
18. 1899, May 19th A.M. at ship, when the M.T.G. was 18ᵈ 5ʰ 26ᵐ 26ˢ in long. 150° 42′ E., the observed Alt. of Polaris out of the meridian in artificial horizon was 102° 37′ 40″ ; index correction, −6′ 46″ ; height of eye, 50 feet. Required the latitude.

APPENDIX. 337

19. 1899, May 6th A.M. at ship, in long. 165° 48' E., the observed Mer. Alt. of the Moon's lower limb was 69° 52' 50" bearing North ; index correction, +3' 12" ; height of eye, 27 ft. Find the latitude.
20.

Ship's head by compass.	Bearing of distant object.	Deviation.	Ship's head by compass.	Bearing of distant object.	Deviation.
N.	E.		S.	E. 20° S.	
N.E.	E.		S.W.	E. 8° S.	
E.	E. 7° S.		W.	E. 8° N.	
S.E.	E. 18° S.		N.W.	E. 3° N.	

Required the correct magnetic bearing and deviations.
(a) Correct magnetic courses : S.E.   S. 50° W.   N.   E.N.E. Courses to steer by compass.
(b) Courses steered by compass : W. by N. ½ N. N.E. ¼ E. S. ¼ W. N. ½ E. Correct magnetic courses.
(c) Ship's head at S.W. ¾ W. by compass.
Bearings by compass : E. ¼ S.   S. by E. ¼ E.
Correct magnetic bearings.

21. Required the course to steer from Howth Head to N.W. lightship to counteract the effect of a current which set (C.M.) E. by S. ¼ S. 3 miles per hour, the ship making by log 10 miles per hour, and the distance she would make good towards N.W. lightship in 4 hours.

22. 1897, December 1st, 4ʰ 6ᵐ P.M., off Port Gallegos took a cast of the lead. Find the correction to be applied to the cast before comparing it with the depth marked on the chart.

23. 1899, August 20th A.M. at ship, in lat. 10° 14' N. ; time by chron., 19ᵈ 13ʰ 24ᵐ 39ˢ, which was estimated to be correct for G.M.T.
With the following observations, find the true and apparent Altitudes of the Moon, error of chronometer for G.M.T., and longitude of ship ; height of eye, 23 ft. :—

Obs. Alt. Fomalhaut, E.   Obs. distance between Star and Moon's R.L.
48° 21' 10"   32° 56' 50"
I.C. −2 11   I.C. +3 19

24. 1899, September 29th, in long. 125° 35' W., lat. N. With following observations, find the latitude at the time of taking the second observation :—

A.T.S. 29ᵈ 1ʰ 13ᵐ 40ˢ   Alt. Sun's L.L. 36° 33' 40" bearing S. 23° W.
A.T.S. 29 3 49 40   ,,   ,,   18 39 15

The ship's true course and distance, E. 3° N., 16·5 knots per hour during the interval ; index correction, −2' 3" ; height of eye, 34 ft.

25. 1899, March 19th P.M. at ship, when a chronometer showed 19ᵈ 11ʰ 6ᵐ 55ˢ G.M.T., the observed Alt. of Regulus East of meridian was 46° 26' 30", and the observed Alt. of Aldebaran taken at the same time West of meridian was 34° 10' 10" ; index correction, −2' 24" ; height of eye, 36 ft. Required the position of the ship, and true bearings of the stars by Sumner's method by projection, assuming lats. 50° and 50° 30' N.

26. 1899, December 2nd, in lat. 49° 20' N., long. 18° 3' 30" E., the following times were observed when Pollux had equal altitudes ; find the error of the chronometer for M.T.S., M.T.G., and A.T.G.:—

E. of Mer. 11ʰ 6ᵐ 25ˢ   W. of Mer. 4ʰ 26ᵐ 37ˢ

Z

338 APPENDIX.

27. Find the initial course, and distance on the Great Circle, latitude and longitude of Vertex, and succession of points 10° long. apart—

From lat. 32° 43' N., long. 129° 46' E.
To ,, 37 47 N., ,, 122 30 W.

28. Construct a Mercator's chart, scale $1'' = 1°$ long., extending from lat. 68° to 71° N., and 6° to 10° E. long. A ship leaves 69° 38' N., 8° 20' E., and sails as follows by compass : N.E. ¾ E., 30' (Dev. 12° E.) ; N.W. by W. ¼ W., 55' (Dev. 8° W.) ; South, 80' (Dev. 4° W.) ; variation, 20° W. Find her latitude and longitude in, course and distance made good.

29. From the summit of a cliff 225 ft. above the water-level, I observed the angle of depression of a boat bearing South to be 24° 35', and of another boat bearing East to be 18° 55'. Required distance of the boats from base of cliff, and from each other ; also the bearing of second boat from the first.

30. With the wind at N.E. ¼ E., a ship plying to windward, after running 38 miles on the port tack and 38 miles on the starboard tack, finds she has made 27 miles directly to windward. What were her courses ?

31. In a right-angled spherical triangle ABC, $C = 90°$, $B = 104° 25'$, $b = 107° 39'$. Find a, c, and A.

32. A ship sailing on a Great Circle from lat. 34° 36' S. to lat. 33° 56' S., finds the latitude of Vertex to be 41° 1' S. Find the difference of longitude between the places, supposing the Vertex falls between them.

ANSWERS.

LOGARITHMS.

1.	0·477121	0·532627
	1·672098	0·875061
	1·698970	2·213385
	2·000000	2·847579
	2·439333	4·513697
	3·922206	9·698970
	1·758079	8·690196
	2·563173	7·556302
	0·000389	7.080626
	4·876391	7·827886
	1·797278	6·447794
	2·243058	5·876795
	4·361779	8·539087
	2·215643	8·878545
	4·333447	9·564667
	4·370328	

NUMBERS.

2.	1598	149899·7
	139·9	3400015·6
	50·15	10010023·1
	5·199	230100
	1·010	11000
	48026·9	1121000
	490000	1·001
	1408961	4679·01
	10969874	18880043·5
	63980	239901·1

APPENDIX.

MULTIPLICATION BY LOGS.

1. 19639
2. 23360
3. 278966·6
4. 358900
5. 709·3934
6. 47.808
7. 2110·41
8. 288·651
9. 3540683
10. 5·9585 ·
11. ·0000081005
12. 6·4831
13. ·0000129
14. ·000259
15. ·5
16. 24951·6
17. 100000
18. 20·42
19. ·2303
20. ·001795
21. 1267·618
22. 229500
23. 18·16
24. 143·21

DIVISION BY LOGS.

1. 46·3572
2. 1608·302
3. 9040
4. 176016·22
5. 4·613
6. 24·7069
7. 1328856·7
8. 1026·903
9. 1·424
10. 142199·67
11. 1·02217
12. ·00001029
13. 1490996·5
14. 44826082·5
15. ·000001
16. 20
17. 1·2187
18. 5·8916
19. 57·454
20. 20967·1
21. 20·399
22. ·032725
23. 7·9876
24. 957·94

INVOLUTION BY LOGS.

1. 547·56
2. 6.602
3. ·000005797
4. ·000000001411
5. ·000002191

EVOLUTION BY LOGS.

1. 17·5
2. 3·6
3. ·30012
4. 230
5. ·08

ADDITIONAL EXERCISES.

1. 143·8, 27·74, 777·4
2. 891·9 and 9754·5
3. 172974·6, ·03528
4. 1·439

ANSWERS TO DAYS' WORKS.

1. N. 31 W., 49. N. 34 W., 61. N. 17 E., 63. N. 60 W., 55. N. 15 W., 57. S. 27 W., 62. S. 25 E., 57. S. 30 E., 4. Lat. 125·0. Dep. 105·3. Lat. in 1° 31′ N. Long. 2° 2′ W. Course N. 40° W. Dist. 163.

340   APPENDIX.

2. S. 80 E., 10.   S. 55 E., 36.   N. 73 W., 28.   S. 9 W., 25.   N. 9 E., 19.
S. 49 E., 29.   N. 86 W., 26.   S. 40 W., 11.   Lat. 45·6.   Dep. 0·5.   Lat.
64° 43' N.   Long. 179° 46' E.   Course S. 1° E.   Dist. 46.
    3. N. 41 W., 19.   N. 38 E., 37.   S. 87 E., 25.   S. 78 E., 23.   S. 88 W.,
21.   S. 9 W., 34.   N. 62 E., 40.   S. 22 E., 15.   Lat. 8.   Dep. 72·4.   Lat.
18° 51' S.   Long. 178° 36' W.   Course N. 83¼ E., 73.
    4. S. 14 E., 60.   S. 31 W., 53.   N. 82 W., 51.   S. 64 W., 58.   N. 44 E.,
49.   N. 26 E., 65.   S. 84 W., 21.   N. 75 W., 39.   Lat. 20·4.   Dep. 111·5.
Lat. 58° 50' S.   Long. 68° 28' W.   S. 80 W., 113.
    5. S. 23 W., 15.   N. 49 W., 12.   S. 37 E., 13.   S. 16 W., 25.   N. 34 E.,
20.   N. 78 W., 15.   N. 15 E., 27.   N. 46 E., 30.   Lat. 26·3.   Dep. 11·0.
Lat. 30° 25' N.   Long. 133° 7' E.   N. 23 E., 28.
    6. S. 48 E., 36.   S. 6 W., 49.   N. 24 E., 53.   S. 65 E., 49.   S. 30 E., 44.
S. 25 W., 54.   S. 66 E., 52.   S. 12 W., 44.   Lat. 196·3.   Dep. 125·3.   Lat.
51° 6' N.   Long. 178° 7' W.   S. 32½ E., 233.

ANSWERS TO PARALLEL SAILING.

1. 174'·8.
2. 116'·7.
3. 373'·1.
4. 70° 31' 44".
5. 12'·01.
6. 57° 52' N.   41° 52' S.   Diff. Lat. 5984 miles.
7. 34° 44' W.
8. 45° 20' N.   Diff. long. 210·9.
9. 13·09.
10. 38° 8' S.
11. 1266.
12. 60°.
13. 136° 49' 24" E.
14. 44° 21' 18" E.
15. 96·24.
16. 50° 44'.
17. 92·99.
18. 41° 24½'.

ANSWERS TO MERCATOR SAILINGS.

No	True courses.	Compass course.	Distance.
1	S. 85 55 41 W.	N. 54 19·19 W.	2141 miles.
2	N. 3 8 15 E.	N. 16 21 15 E.	672 ,,
3	S. 12 5 34 W.	S. 2 5 34 W.	5673 ,,
4	N. 19 27 0 W.	N. 33 57 0 W.	900·4 ,,
5	S. 71 5 33 E.	S. 48 5 33 E.	3009 ,,
6	S. 60 27 26 E. or W.	{ S. 49 27 26 E. or S. 71 27 26 W. }	10951·7 ,,
7	S. 54 6 30 W.	S. 31 6 30 W.	6261 ,,
8	N. 53 43 0 W.	N. 50 23 0 W.	5428 ,,
9	N. 89 41 19 W.	N. 44 0 19 W.	1292 ,,
10	N. 2 5 E.	N. 1 5 0 W.	3810 ,,
11	N. 71 56 W.	N. 53 52 0 W.	909·3 ,,
12	N. 89 16 43 E. or W.	{ S. 83 33 17 W. or N. 82 6 43 W. }	10801 ,,

APPENDIX.

Answers to Tides.

No.	A.M.	P.M.	A.M.	P.M.
	h. m.	h. m.	h. m.	h. m.
1	11 57	—	1 51	2 39
2	—	0 22	1 42	2 12
3	11 40	—	—	0 21
4	0 55	1 24	11 35	—
5	—	0 14	0 41	2 4
6	11 39	—	11 41	—
7	11 49	—	9 28	9 48
8	—	0 20	—	0 3
9	0 2	0 23	0 22	0 41
10	11 5	11 28	—	0 8
11	—	0 1	11 24	11 51
12	—	0 20	10 24	10 59

Answers to Tidal Soundings.

No.	Interval.	Half daily range.	Table B.	Reduction.
	h. m.	ft. in.	ft. in.	ft. in.
1	1 14	3 5	+2 8	− 9 9
2	5 59	9 9	−9 9	+ 0 5
3	4 36	6 0	−4 5	− 1 4
4	0 3	4 4	+4 4	−11 3
5	1 0	9 6	+8 2	−13 8
6	0 16	3 4	+3 4	−12 10
7	6 0	9 8	−9 8	+ 0 4
8	3 0	8 5	—	−11 0
9	2 30	5 11	+1 7	− 4 0
10	0 0	5 6	+5 6	−13 2
11	5 59	8 2	−8 2	+ 0 6
12	3 30	3 7	−0 11	− 7 6
13	6 0	3 4	−3 4	− 1 11
14	4 30	5 4	−3 9	− 1 6
15	2 20	5 8	+2 0	− 2 0
16	5 55	9 8	−9 8	+ 0 2
17	0 10	8 2	+8 1	−13 9
18	1 15	6 7	+5 1	−13 0
19	4 0	5 8	−2 10	− 3 7
20	6 1	9 9	−9 9	+ 0 5
21	0 57	6 3	+5 5	− 9 4
22	0 30	5 0	+4 10	−11 4
23	4 40	8 2	−6 2	− 2 1
24	6 0	8 2	−8 2	+ 0 4

Answers to Time Equivalents.

$11^m 58^s$; $11^h 56^m 3^s$; $1^h 0^m 1^s$; $4^h 31^m 31.6^s$; $1^m 9.3^s$; $41^m 18.6^s$; $10^h 20^m 46.4^s$; $6^h 12^m 3^s$; $7^h 50^m 34^s$; $9^h 39^m 36^s$; $11^h 46^m 38.3^s$; 13.53^s ; $3^m 59^s$; 1^s ; 1^m ; $3^m 12^s$; $11^h 12^m 36^s$; $4^h 17^m 15^s$.

342   APPENDIX.

Answers to Arc Equivalents.

18° 7′ 30″ ; 154° 58′ 45″ ; 120° 2′ 30″ ; 104° 49′ 30″ ; 179° 59′ 30″ ; 45° 4′ 45″ ; 89° 50′ 30″ ; 135° 24′ 15″ ; 3° 48′ 45″ ; 36° 15′ ; 2° 38′ 45″ ; 11′ 15″ ; 36° 15′ ; 165° 4′ 45″ ; 168° 22′ 30″ ; 75° 1′ 15″ ; 127° 13′ ; 150° 3′ ; 3′ ; 25′.

Answers to Difference of Times.

41^m 20^s less ; 1^h 11^m more ; 7^h 21^m 8^s more ; 11^h 30^m 24^s more ; 6^h 41^m 6^s more ; 6^h 0^m 19^s less ; 29^s more ; 11^h 59^m 47^s less ; 4^m 9^s less ; 7^h 40^m 2^s more ; 11^s more ; 11^h 0^m 1^s less.

Answers to Longitudes.

1. 96° 25′ 45″ E.
2. 106 50 15 W.
3. 110   5   0 W.
4. 122 27 30 E.
5. 78 48 45 W.
6. 117 58 30 E.
7.   0 34 45 E.

8.   9° 7′ 30″ E.
9.   0 35   0 W.
10. 174   1 45 W.
11. 22   9 45 W.
12. 167   0   0 E.
13. 80 17 30 W.
14. 179   2   0 W.

Answers to Finding Greenwich Times.

1. 16^d 7^h 9^m 36^s.
2. 1   2 13 41·3.
3. 12 20 50 36.
4. 30   6   4 56.
5. 31 21 36 41.
6. 17 12 42 26.

7. 28^d 5^h 48^m 17^s.
8. 20 21   9 34.
9. 31 15 50 34.
10. 10   0 16 34.
11. 22   7 58 15.
12. 29 16 25 20.

Answers to Finding Ship Times.

1. October 15th, 5^h 29^m 14^s A.M.
2. March 19th,   10 41 18 P.M.
3. June 2nd,   4 58 27 A.M.
4. May 1st,   4 52   4 P.M.
5. May 6th,   11 43 46 A.M.

6. January 13th, 1^h 3^m 52^s A.M.
7. October 19th, 11 52 46 P.M.
8. August 31st,   7   3 33 P.M.
9. July 1st,   4 14 27 P.M.

Answers to Sun's Declination.

1. 12° 33′ 33″ S.
    0   8 37·5 S.
   10 30 16 N.
   20   9   3 N.
    0   6 41 N.

   23° 14′ 36″ N.
   20 49 37 N.
    0   1 16 S.
   23 27   9 S.
   17 25 10 S.

APPENDIX.

Answers to Sun's Declination and Equation of Time.

1. (a) Dec. 21° 3' 40" S. ; Eq. T. 9m 45·2s.
   (b) ,,   8 40 22 S. ; ,, 13 4·7.
   (c) ,,   0 2 30 N. ; ,,   7 29·5.
   (d) ,,   9 55 22 N. ; ,,   0 2.
   (e) ,, 23 27 2 N. ; ,,   1 39·8.
   (f) ,,   0 2 21 S. ; ,,   7 35·8.
2. (a) Dec. 0 10 26 N. ; Eq. T. 7 23·5.
   (b) ,,   9 52 46 N. ; ,,   0 0.
   (c) ,, 23 8 52 N. ; ,,   3 28·7.
   (d) ,,   8 21 1 N. ; ,,   0 1.
   (e) ,,   0 9 28 S. ; ,,   7 42·1.
   (f) ,, 23 24 57 S. ; ,,   0 2·5.

Answers to Sun's R.A.

1. A.R.A. 19h 49m 40·9s ; M.R.A. 19h 39m 55·7s.
2.   ,,   3h 39m 7.6s.
3.   ,,   22h 37m 39·6s ; ,, 22h 24m 34·8s.
4.   ,,   0h 0m 23·1s ; ,, 23h 52m 53·6s.
5.   ,,   8h 26m 51·7s.
6.   ,,   12h 0m 21·5s ; ,, 12h 7m 57·4s.
7.   ,,   13h 39m 22·6s.

Answers to Moon's S.D. and H.P.

1. S.D. 15' 47·9" ; H.P. 57' 8·4". | 6. S.D. 16' 15" ; H.P. 58' 38'.
2. ,, 16 26·0 ; ,, 59 36. | 7. ,, 16 52·1 ; ,, 61 2·4.
3. ,, 15 10·6 ; ,, 55 5·7. | 8. ,, 16 17·5 ; ,, 59 26·9.
4. ,, 16 26·7 ; ,, 59 26·2. | 9. ,, 15 20·6 ; ,, 55 20·7.
5. ,, 16 16·3 ; ,, 58 50.

Answers to Moon's Mer. Pass.

	M.T.S.	M.T.G.
1.	18d 7h 10·9m.	18d 18h 40·9s.
2.	4 20 32·4.	5 2 58.
3.	27 15 55·0.	27 7 33·3.
4.	12 3 55·2.	12 21 56·2.
5.	30 18 21·2.	30 8 50.
6.	20 12 10·6.	20 23 2·2.
7.	19 11 44·3.	18 23 47.
8.	6 19 57·1.	7 6 27·2.
9.	31 15 38·0.	31 9 11·2.
10.	12 0 22·5.	11 17 39·3.
11.	18 6 55·1.	18 9 38·7.
12.	27 16 40·9.	27 18 53·9.
13.	3 22 53·6.	4 10 12·3.
14.	18 11 38·7.	18 10 0·7.
15.	22 16 24·6.	22 12 42·3.

APPENDIX.

Answers to Moon's R.A. and Dec.

	R.A.				Dec.			
1.	R.A.	9ʰ 16ᵐ 27·8ˢ			Dec.	12° 22′ 38″ N.
2.	,,	15 27 14·5.			,,	22 15 37 S.
3.	,,	1 15 15·2.			,,	13 3 51 N.
4.	,,	23 8 0·4.			,,	0 1 7 N.
5.	,,	11 17 58·7.			,,	1 9 41 S.
6.	,,	7 38 10·7.			,,	19 23 40 N.
7.	,,	1 8 15·3.			,,	12 49 52 N.
8.	,,	17 10 19·7.			,,	24 41 13 S.
9.	,,	11 11 19·8.			,,	0 1 3 S.
10.	,,	23 11 14·4.			,,	0 3 58 N.
11.	,,	23 8 49·7.			,,	0 2 9 S.
12.	,,	0 0 22·1.			,,	5 47 43 N.

Answers to Planet's R.A. and Dec.

1.	R.A.	2ʰ 45ᵐ 18·13ˢ.		Dec.	14° 15′ 45″ N.
2.	,,	9 13 55·6.		,,	17 45 52 N.
3.	,,	14 19 34.		,,	12 40 40 S.
4.	,,	17 11 41·1.		,,	21 32 16 S.
5.	,,	16 19 37·9.		,,	21 20 15 S.
6.	,,	12 17 20·5.		,,	0 18 12 S.
7.	,,	14 8 23·8.		,,	11 53 33 S.

Answers to Planet's S.D. and Par.

1.	S.D. 10·7″		Par. in Alt. 6·2″.
2.	,, 7·1.		,, 5·5.
3.	,, 20·6.		,, 2.
4.	,, 8·4.		,, 0.
5.	,, 6·4.		,, 5.
6.	,, 16·4.		,, 12.

Answers to Planet's Mer. Pass.

1. M.T.G. 25ᵈ 6ʰ 40·6ᵐ.
2. M.T.S. 11ᵈ 12ʰ 59·2ᵐ ; M.T.G. 11ᵈ 21ʰ 22·2ᵐ.
3. W. of Mer.
4. M.T.G. Dec. 30ᵈ 10ʰ 15·4ᵐ
5. M.T.S. 13ᵈ 15ʰ 4·5ᵐ ; M.T.G. 13ᵈ 17ʰ 32·5ᵐ.
6. On the Mer.

Answers to Star's R.A. and Dec.

1.	R.A.	4ʰ 30ᵐ 13·1ˢ.		Dec.	16° 18′ 33″ N.
2.	,,	7 34 3·6.		,,	5 28 52 N.
3.	,,	1 34 1·6.		,,	57 44 28 S.
4.	,,	14 29 5·5.		,,	41 42 50 S.
5.	,,	16 23 16.		,,	26 12 35 S.
6.	,,	17 54 18·6.		,,	51 30 15 N.
7.	,,	20 17 44·1.		,,	57 3 10 S.
8.	,,	9 12 8·7.		,,	69 18 24 S.
9.	,,	16 23 16·7.		,,	26 12 37 S.
10.	,,	22 1 56·9.		,,	47 26 50 S.

Answers to Sidereal Times.

1. 12ʰ 27ᵐ 15ˢ·63
2. 10 22 17·50
3. 7 34 29·73
4. 5ʰ 37ᵐ 19ˢ·41
5. 8 2 52·34
6. 21 17 50·23

Answers to Solar Times.

1. M.T.S. 6th 0ʰ 26ᵐ 23·7ˢ P.M.
2. M.T.G. 14ᵈ 12 51 37·6
3. M.T.G. 29 12 59 20·7
4. M.T.S. 1st 8ʰ 50ᵐ 21ˢ·5 A.M.
5. A.T.S. 16th 3 39 6·8 A.M.
6. M.T.S. 18th 11 33 35·2 A.M.

APPENDIX. 345

Answers to Star's Meridian Passage.

1. 6th 6ʰ 24ᵐ 37ˢ P.M.
2. 24th 4 47 51 P.M.
3. 31st 3 48 45 A.M.
4. 15th 6 14 40 A.M.
5. 25th 3 11 55 P.M.
6. 26th 8 7 0 P.M.
7. 10th 3ʰ 53ᵐ 59ˢ A.M.
8. 9th 2 40 0 A.M.
9. 17th 4 17 39 P.M.
10. 1st 0 25 18 P.M.
11. 25th 0 19 50 A.M.
12. 31st 8 35 30 A.M.

Answers to Bright Stars near Meridian.

	Name.	Hour angle.	Bearing from observer.
		h. m.	
1.	Pollux	2 58 W.	S.
	α Hydræ	1 14 W.	S.
	Regulus	0 34 W.	S.
	α Ursæ Majoris	0 20 E.	N.
	β Leonis	1 7 E.	S.
	ε Ursæ Majoris	2 13 E.	N.
	Spica	2 37 E.	S.
2.	α Coronæ	2 39 W.	N.
	Antares	1 46 W.	N.
	α Triang. Aust.	1 31 W.	S.
	λ Scorpii	0 43 W.	N.
	Vega	0 34 E.	N.
	Altair	1 37 E.	N.
	α Pavonis	2 8 E.	S.
	α Cygni	2 28 E.	N.
3.	λ Scorpii	2 41 W.	S.
	θ „	2 38 W.	S.
	α Ophiuchi	2 38 W.	S.
	Vega	1 35 W.	N.
	Altair	0 23 W.	S.
	α Cygni	0 29 E.	N.
	α Gruis	1 54 E.	S.
	Fomalhaut	2 44 E.	S.
	β Pegasi	2 50 E.	S.
	Markab	2 51 E.	S.
4.	β Gruis	2 36 W.	S.
	Fomalhaut	2 20 W.	S.
	β Pegasi	2 13 W.	S.
	Markab	2 12 W.	S.
	α Andromedæ	1 9 W.	N.
	β Cassiopeiæ	1 8 W.	N.
	γ Pegasi	1 4 W.	S.
	β Ceti	0 33 W.	S.
	γ Cassiopeiæ	0 21 W.	N.
	α Ursæ Minoris	0 10 E.	N.
	α Eridani	0 22 E.	S.
	α Arietis	0 49 E.	N.
	α Persei	2 5 E.	N.

APPENDIX.

ANSWERS TO LATITUDES BY SUN'S MERIDIAN ALTITUDE.

1. Dec. 0° 0' 0"; true alt. 90° 0' 0"; lat. 0° 0' 0"
2. ,, 18 56 56 N.; ,, 11 20 35; ,, 59 42 29 S.
3. ,, 23 27 6 N.; ,, 16 7 20; ,, 50 25 34 S.
4. ,, 14 1 24 N.; ,, 76 0 53; ,, 0 2 19 N.
5. ,, 0 4 33 N.; ,, 37 10 34; ,, 52 53 59 N.
6. ,, 0 6 1 S.; ,, 33 49 23; ,, 56 4 36 N.
7. ,, 21 44 52 S.; ,, 68 15 8; ,, 0 0 0
8. ,, 23 16 3 N.; ,, 61 51 11; ,, 51 24 52 N.
9. ,, 0 0 37 N.; ,, 53 1 59; ,, 36 57 24 S.
10. ,, 23 27 7 S.; ,, 90 0 0; ,, 23 27 7 S.
11. ,, 23 7 22 N.; ,, 19 7 58; ,, 47 44 40 S.
12. ,, 0 6 32 N.; ,, 40 52 53; ,, 49 13 39 N.
13. ,, 3 3 20 S.; ,, 31 32 9; ,, 61 31 11 S.

ANSWERS TO LATITUDES BY STAR'S MERIDIAN ALTITUDE.

1. Dec. 45° 53' 39" N.; true alt. 54° 21' 1"; lat. 10° 14' 40" N.
2. ,, 8 18 52 S.; ,, 84 50 55; ,, 13 27 57 S.
3. ,, 16 34 32 S.; ,, 45 26 22; ,, 27 59 6 N.
4. ,, 10 38 18 S.; ,, 36 16 5; ,, 43 5 37 N.
5. ,, 30 9 11 S.; ,, 10 9 29; ,, 49 41 20 N.
6. ,, 16 18 21 N.; ,, 56 18 30; ,, 49 59 51 N.
7. ,, 10 38 21 S.; ,, 43 21 30; ,, 36 0 9 N.
8. ,, 14 39 56 N.; ,, 30 56 0; ,, 44 24 4 S.
9. ,, 22 59 17 N.; ,, 75 10 0; ,, 37 49 17 N.
10. ,, 14 37 48 N.; ,, 56 2 0; ,, 19 20 12 S.
11. ,, 12 27 12 N.; ,, 57 7 0; ,, 45 20 12 N.

ANSWERS TO LATITUDES BY MOON'S MERIDIAN ALTITUDE.

No.	M.T.G.	S.D.	H.P.	Declination.	True alt.	Latitude.
	d. h. m.	' "	' "	° ' "	° ' "	° ' "
1	22 11 4·1	17 1	61 12	17 21 26 S.	73 22 11	33 59 15 S.
2	19 15 43·1	16 54	61 13	13 48 48 S.	34 33 6	41 38 6 S.
3	19 0 22·4	16 47	60 40	3 5 40 N.	40 38 32	46 15 48 S.
4	18 20 27·8	16 26	59 11	15 40 23 N.	67 39 32	38 0 51 N.
5	24 17 54·7	14 58	54 7	5 55 1 N.	45 49 46	50 5 15 N.
6	23 8 24·5	14 56	54 6	0 5 24 N.	36 22 38	53 81 58 S.
7	4 3 52·5	15 31	55 59	8 14 34 S.	61 21 24	36 53 10 S.
8	18 20 34·3	15 3	54 35	23 46 43 N.	33 29 20	32 43 57 S.
9	19 9 55·1	15 2	54 18	10 50 38 N.	49 4 6	51 46 32 N.
10	29 5 9·0	16 15	58 39	5 25 13 N.	54 58 24	29 36 23 S.
11	2 10 25·0	15 1	54 20	0 3 31 N.	46 58 4	42 58 25 N.

ANSWERS TO LATITUDES BY PLANET'S MERIDIAN ALTITUDE.

1. Dec. 19° 0' 49" S.; true alt. 35° 52' 28"; lat. 35° 6' 25" N.
2. ,, 25 53 27 N.; ,, 33 54 18; ,, 30 12 15 S.
3. ,, 13 16 49 S.; ,, 52 36 24; ,, 50 40 25 S.
4. ,, 24 15 9 N.; ,, 65 44 51; ,, 0 0 0

APPENDIX.

Answers to Meridian Altitudes below Pole.

1. Dec. 52° 38' 8" S. ; true alt. 29° 4' 12" ; lat. 66° 26' 4" S.
2. M.T.G. July 22ᵈ 2ʰ 53ᵐ·2 ; S.D. 16' 45" ; H.P. 60' 58".
   Dec. 18° 46' 47" S. ; true alt. 10° 35' 12" ; lat. 81° 48' 25" S.

Answers to Star's calculated Altitudes.

1. 39° 48'
2. 42 6
3. 19 38
4. 57 0
5. 67 50
6. 32° 15'
7. 61 44
8. 37 57
9. 51 4
10. 25 1

Answers to Latitudes by Ex-Meridian Altitude of Sun.

No.	Hour angle.	Declination.	True altitude.	Latitude at obs.	Latitude at noon.
	m. s.	° ' "	° ' "	° ' "	° ' "
1	32 47	0 3 19 S.	37 1 32	52 34 44 S.	—
2	48 57	21 15 15 N.	67 52 50	2 37 0 N.	—
3	33 52	18 8 37 N.	70 47 43	0 50 0 N.	—
4	19 3	16 25 14 N.	72 35 56	33 17 0 N.	33 13 6 N.
5	47 57	12 55 25 N.	57 27 48	17 23 0 S.	17 21 12 S.
6	11 33	0 4 51 S.	52 26 27	37 23 0 N.	37 23 42 N.
7	47 28	6 7 38 S.	48 6 44	46 45 0 S.	—
8	35 3	16 0 8 S.	81 33 4	15 29 0 S.	—
9	11 4	23 4 8 S.	53 9 54	13 40 0 N.	13 38 30 N.
10	10 18	21 22 6 N.	37 1 6	31 32 0 S.	—
11	10 49	23 18 3 N.	53 24 52	59 50 0 N.	59 47 12 N.
12	32 4	17 55 21 S.	56 47 22	14 20 0 N.	14 12 18 N.
13	32 33	23 25 25 S.	54 24 28	11 16 0 N.	11 11 48 N.
14	19 27	10 49 33 S.	78 8 40	0 0 0	0 5 42 S.

Answers to Latitudes by Ex-Meridian Altitude of Stars, etc.

No.	Hour angle.	Declination.	True altitude.	Latitude.
	m. s.	° ' "	° ' "	° ' "
1	8 58	26 12 31 S.	29 50 16	33 55 0 N.
2	24 23	10 38 23 S.	27 46 7	51 22 0 N.
3	15 38	60 25 24 S.	29 30 4	0 0 0
4	78 34	44 55 30 N.	11 30 38	31 30 0 S.
5	38 34	30 9 3 S.	8 39 7	50 44 15 N.
6	46 35	38 41 8 N.	49 52 26	0 0 0
7	39 11	10 38 23 S.	25 28 57	53 20 0 N.
8	26 12	16 34 53 S.	30 46 35	42 20 0 N.
9	30 2	12 27 18 N.	42 30 23	34 30 0 S.
10	28 26	8 18 51 S.	38 49 0	42 27 0 N.
11	21 15	5 28 52 N.	44 56 35	50 19 0 N.
12	29 44	28 15 55 N.	50 52 6	10 12 0 S.
13	22 42	21 41 43 S.	40 30 1	27 30 0 N.
14	41 14	14 26 11 S.	58 3 22	45 10 0 S.
15	24 52	10 25 58 S.	60 40 14	39 13 30 S.
16	24 39	8 36 25 N.	63 9 21	34 51 0 N.

APPENDIX.

Answers to Latitudes by Pole Star.

No.	Sid. time.	First cor.	Second cor.	Third cor.	True altitude.	Latitude.
	h. m. s.	° ′ ″	′ ″	′ ″	° ′ ″	° ′ ″
1	23 22 20	−1 3 16	+0 3	+0 23	12 30 36	11 27 46 N.
2	19 56 59	− 11 6	1 5	1 10	55 10 51	55 2 0
3	23 1 36	− 59 43	11	45	36 32 28	35 33 41
4	1 55 20	−1 12 12	—	31	19 13 16	18 1 35
5	8 52 41	+ 28 8	54	43	53 28 0	53 57 45
6	0 38 8	−1 11 40	—	55	10 9 23	8 58 45
7	5 36 30	− 32 25	41	45	48 48 12	48 16 13
8	10 24 25	+ 52 9	20	1 33	42 23 38	43 16 40
9	13 47 27	+1 12 32	1	1 45	39 10 0	40 23 18
10	17 18 7	+ 37 34	49	1 34	55 21 26	56 0 23
11	22 12 19	− 49 23	23	31	39 31 53	38 42 24
12	6 29 38	− 16 34	49	25	47 31 3	47 14 43

Answers to Amplitudes.

1. Dec. 12° 10′ 51″ S. ; true amp. E. 24° 53′ 37″ S.; dev. 13° 19′ 52″ E.
2. ,, 23 26 33 S.; ,, W. 57 32 0 S.; ,, 36 9 15 E.
3. ,, 23 19 59 N.; ,, E. 40 15 32 N.; ,, 33 40 43 E.
4. ,, 20 51 15 N.; ,, E. 20 51 15 N.; ,, 6 51 15 W.
5. ,, 20 46 27 S.; ,, W. 38 12 9 S.; ,, 33 30 54 W.
6. ,, 10 16 16 S.; ,, W. 15 44 36 S.; ,, 5 3 21 W.
7. ,, 23 17 37 S.; ,, E. 52 16 6 S.; ,, 46 40 9 W.
8. ,, 23 0 30 S.; ,, W. 72 6 45 S.; ,, 5 57 0 E.
9. ,, 17 40 10 S.; ,, E. 30 40 30 S.; ,, 11 6 30 E.
10. ,, 23 5 52 S.; ,, E. 51 30 30 S.; ,, 59 28 30 W.
11. ,, 15 21 16 S.; ,, E. 15 36 30 S.; ,, 37 56 30 E.
12. ,, 16 12 49 S.; ,, E. 29 18 0 S.; ,, 14 33 0 E.
13. ,, 7 31 31 S.; ,, E. 9 59 30 S.; ,, 9 33 30 E.
14. ,, 23 26 55 S.; ,, W. 47 30 0 S.; ,, 27 45 0 E.
15. ,, 23 2 43 S.; ,, E. 51 32 0 S.; ,, 17 21 0 E.

Answers to Altazimuths.

No.	M.T.G.	True alt.	P.D.	True azimuth.	Compass error.	Deviation.
	d. b. m. s.	° ′ ″	° ′ ″	° ′ ″	° ′ ″	° ′ ″
1	31 10 17 47	25 40 2	104 16 41	S. 37 16 4 W.	0 42 19 E.	17 47 41 W.
2	26 13 45 45	37 57 23	111 4 32	S. 33 19 54 E.	11 40 6 E.	13 55 6 E.
3	22 18 29 38	21 56 21	90 0 0	N. 61 40 46 W.	0 11 44 E.	0 11 44 E.
4	20 8 18 49	16 31 20	90 0 32	N. 68 32 0 W.	29 17 0 E.	7 17 0 E.
5	31 21 22 37	34 3 52	81 41 8	S. 65 25 30 E.	14 48 0 W.	10 12 0 E.
6	10 12 57 36	35 0 27	112 58 49	S. 39 58 0 W.	0 35 14 E.	9 24 46 W.

APPENDIX.

ANSWERS TO SUN TIME AZIMUTHS.

No.	A.T.S.	Declination.	Sum of corrections.	True azimuth.	Compass error.	Deviation.
	h. m. s.	° ′ ″	′	° ′	° ′	° ′
1	7 20 6 A.M.	15 1 10 N.	− 1	N. 74 4 E.	7 30 W.	13 0 E.
2	7 58 0 A.M.	7 40 6 N.	−35	N. 88 57 E.	18 38 E.	19 48 E.
3	5 8 25 P.M.	8 59 17 S.	−49	S. 90 32 W.	19 9 W.	0 32 E.
4	4 50 54 P.M.	17 21 40 S.	−47	S. 92 33 W.	10 59 E.	16 30 E.
5	7 32 51 A.M.	23 24 35 S.	−18	S. 74 41 E.	0 2 W.	26 58 E.
6	6 55 41 A.M.	21 27 4 S.	−14	S. 77 51 E.	14 58 E.	20 2 W.
7	4 5 0 P.M.	8 21 35 N.	+ 3	S. 116 38 W.	21 0 E.	10 45 E.
8	4 38 0 P.M.	—	—	W.	—	9 0 W.
9	7 52 0 A.M.	5 24 22 S.	−27	S. 83 53 E.	24 49 W.	4 49 W.
10	9 37 0 A.M.	19 39 1 S.	+35	N. 146 34 E.	36 15 W.	64 15 W.
11	4 17 0 P.M.	22 8 44 S.	− 1	S. 84 55 W.	5 5 W.	22 55 E.
12	3 32 0 P.M.	8 23 51 N.	+13	S. 125 4 W.	26 38 E.	3 22 W.

ANSWERS TO STAR TIME AZIMUTHS.

No.	Hour angle.	True azimuth.	Compass error.	Deviation.
	h. m. s.	° ′	° ′	° ′
1	2 49 52 E.	N. 126 56 E.	3 41 E.	11 24 W.
2	0 57 42 E.	S. 165 47 E.	19 32 W.	19 32 W.
3	2 50 53 W.	N. 115 36 W.	2 31½ E.	12 48½ W.
4	2 51 51 E.	N. 138 12 E.	3 12 E.	36 27 E.
5	2 4 2 E.	S. 147 36 E.	6 58 W.	9 28 W.
6	1 10 1 E.	N. 151 22 E.	28 38 W.	9 8 W.
7	2 31 30 E.	N. 124 11 E.	1 25 E.	33 25 E.
8	1 7 51 E.	N. 157 44 E.	13 50 W.	7 20 W.
9	—	N.	19 41 W.	10 4 E.
10	5 0 46 E.	N. 79 31 E.	10 29 W.	4 16 E.
11	0 27 32 E.	S. 168 15 E.	0 30 E.	9 45 W.
12	1 22 58 W.	N. 145 12 W.	41 8 W.	7 58 W.

ANSWERS TO CHRONOMETER RATES.

1. 2·7 losing.
2. ·75 losing.
3. 5·3 gaining.
4. ·75 losing.
5. ·3 gaining.
6. 6·06 losing.
7. 10·05 gaining.
8. ·9 gaining.
9. ·7 gaining.
10. ·5 gaining.
11. ·5 losing.
12. 5·4 gaining.
13. ·7 gaining.
14. 2·05 losing.
15. ·8 gaining.
16. ·9 gaining.
17. 1·6 gaining.
18. 10·05 losing.

APPENDIX.

ANSWERS TO LONGITUDES BY SUN CHRONOMETER.

No.	Rate.	M.T.G.	True altitude.	P. dist.	Eq. time.	Hour angle.	Longitude at sights.	Longitude at noon.
	s.	d. h. m. s.	° ′ ″	° ′ ″	m. s.	h. m. s.	° ′ ″	° ′ ″
1	1·01 l.	31 10 17 47	25 40 2	104 16 41	−16 19	2 17 16	124 12 30 W.	124 1 1 W.
2	8·5 g.	26 13 45 45	37 57 23	111 4 32	−12 24	1 50 40	122 47 45 E.	123 17 58 E.
3	—	22 18 29 38	21 56 21	90 0 0	− 7 34	3 38 58	135 24 0 E.	135 24 0 E.
4	3·67 l.	20 8 18 49	16 31 20	90 0 32	+ 7 31	3 56 29	63 42 15 W.	—
5	5 04 g.	31 21 22 37	34 3 52	81 41 8	− 0 3	3 18 20	10 15 0 W.	10 42 37 W.
6	2·25 g.	10 12 57 36	35 0 27	112 58 49	− 6 48	2 19 24	161 15 0 W.	161 47 0 W.

ANSWERS TO LONGITUDES BY STAR CHRONOMETER.

No.	Rate.	M.T.G.	True altitude.	R.A.M.S.	Hour angle.	Longitude.
	s.	d. h. m. s.	° ′ ″	h. m. s.	h. m. s.	° ′ ″
1	·75 g.	3 23 58 22	54 2 5	6 49 7	4 15 34 E.	70 20 15 W.
2	10·05 g.	28 6 58 18	34 50 51	10 27 6	2 51 21 W.	5 44 30 W.
3	·8 g.	27 19 24 11	68 24 57	12 27 25	2 5 7 E.	160 19 45 E.
4	1·03 g.	9 10 21 52	35 29 3	13 13 15	5 14 13 W.	108 17 0 E.
5	·65 g.	26 19 23 52	42 39 12	16 23 58	4 0 54 W.	4 40 15 W.
6	9·3 l.	14 5 23 24	49 31 17	17 32 39	3 44 6 W.	9 36 0 W.
7	2·5 g.	2 15 11 20	26 21 10	20 52 20	4 51 20 W.	30 45 45 W.
8	3·5 l.	5 7 7 53	33 16 1	0 55 27	2 46 37 W.	34 20 0 E.
9	·4 g.	1 0 24 10	25 43 38	2 36 51	3 57 20 W.	91 30 15 E.
10	3·0 g.	9 0 36 53	35 36 1	5 10 39	4 22 13 E.	169 24 30 W.
11	3 4 l.	17 11 3 59	58 50 6	13 44 54	2 30 32 E.	50 20 0 E.
12	10·5 g.	14 11 12 49	41 35 45	17 33 36	3 51 13 E.	14 36 0 W.

APPENDIX.

ANSWERS TO CHRONOMETER AZIMUTHS.

No.	Daily rate.	M.T.G.	True altitude.	P.D.	Eq. time.	Hour angle.	Sight longitude.	Noon longitude.	True azimuth.	Deviation.
	s.	d. h. m. s.	° ′ ″	° ′ ″	m. s.	h. m. s.	° ′ ″	° ′ ″	° ′	° ′ ″
1	1·75 l.	1 3 11 54	9 3 2	113 6 40	+ 3 35	1 20 45	67 16 0 W.	67 37 29 W.	N. 18 45 E.	1 11 19 W.
2	·09 g.	7 0 0 0	54 6 0	73 34 25	+ 5 35	2 53 35	30 12 45 W.	30 45 10 W.	S. 59 1¼ E.	0 46 30 W.
3	3·7 l.	22 15 45 53	43 8 8	90 2 49	− 7 31	1 45 44	148 5 0 E.	148 30 25 E.	N. 37 36 W.	3 20 48 W.
4	4·04 g.	22 1 28 26	30 38 38	89 43 25	− 7 19	3 7 15	70 45 0 W.	—	S. 57 56 E.	1 22 37 E.
5	·5 l.	24 21 9 44	19 38 53	113 24 29	+ 0 11	1 10 19	60 11 30 E.	—	S. 17 7 W.	15 37 6 E.
6	1·0 l.	1 0 6 0	21 8 59	113 1 22	+ 3 40	1 2 30	15 0 0 E.	15 11 52 E.	S. 15 22 W.	28 55 59 E.
7	10·06 g.	15 19 17 16	30 14 47	79 55 19	− 0 8	4 2 54	131 22 15 E.	130 59 27 E.	S. 83 51 W.	9 46 30 W.
8	·7 l.	3 16 53 9	22 38 46	72 40 30	+ 5 55	4 42 18	37 37 0 E.	36 42 30 E.	S. 102 42¼ E.	4 54 0 W.
9	2·07 g.	22 16 14 51	41 7 5	90 2 11	− 7 32	2 56 43	158 35 0 E.	—	N. 67 40¼ W.	6 40 30 W.
10	10·2 g.	16 23 4 1	34 47 13	80 44 3	−14 34	3 38 44	44 19 45 W.	44 35 45 W.	N. 101 18½ E.	2 45 30 W.
11	6·02 l.	24 2 3 36	12 18 16	111 20 50	−11 53	2 15 42	0 3 15 E.	—	S. 32 8¼ W.	9 20 0 E.
12	9·02 g.	24 15 4 12	33 51 3	113 24 53	+ 0 4	3 20 24	175 56 0 W.	176 7 32 W.	S. 57 58 W.	22 39 0 E.

351

352

APPENDIX.

ANSWERS TO SUMNER'S METHOD.

1. First dec. 0° 2' 31"·6 N. ; eq. time, −7ᵐ 31'·43; true alt. 32° 20' 25" ; log (a), 9·023228 ; H.A. 2ʰ 31ᵐ 50ˢ; long. 82° 0' 0" E. ; log (b), 9·013801; H.A. 2ʰ 29ᵐ 56ˢ; long. 82° 28' 30" E. ; line $\begin{Bmatrix} \text{N. } 41° \text{ E.} \\ \text{S. } 41° \text{ W.} \end{Bmatrix}$; azi. S. 49° E. Second dec. 0° 2' 20"·7 S ; eq. time, − 7ᵐ 35'·77 ; true alt. 32° 20' 32" ; log (c), 9·020966 ; H.A. 2ʰ 31ᵐ 13ˢ; long. 82° 42' 45" E. ; log (d), 9·010418 ; long. 82° 22' 30" E. ; line $\begin{Bmatrix} \text{N. } 24° \text{ W.} \\ \text{S. } 24° \text{ E.} \end{Bmatrix}$; azi. S. 66° W. ; position, lat. 47° 37' N., long. 82° 30' E.

2. First dec. 23° 3' 4" S. ; eq. time, +3ᵐ 31'·22 ; true alt. 14° 22' 13" ; log (a), 8·696152 ; H.A. 1ʰ 43ᵐ 2ˢ; long. 178° 50' E. ; log (b), 8·635748; H.A. 1° 36' 0" ; long. 179° 24' 30" W. ; line $\begin{Bmatrix} \text{N. } 66\frac{1}{2}° \text{ E.} \\ \text{S. } 66\frac{1}{2}° \text{ W.} \end{Bmatrix}$; azi. S. 23½° E. Second dec. 23° 2' 14"·9 S. ; eq. time, +3ᵐ 35'·83 ; true alt. 11° 23' 40" ; log (c), 8·962769 ; H.A. 2ʰ 21ᵐ 5ˢ; long. 179° 46' 15" E. ; log (d), 8·933306 ; H.A. 2ʰ 16ᵐ 14ˢ; long. 178° 34' 0" E. ; line $\begin{Bmatrix} \text{N. } 15° \text{ W.} \\ \text{S. } 15° \text{ E.} \end{Bmatrix}$; azi. S. 75° W. ; position, lat. 48° 54' N., long. 180° 0'.

3. First dec. 23° 26' 27"·8 N. ; eq. time, +1ᵐ 54'·29; true alt. 59° 11' 0"; log (a), 8·480546 ; H.A. 1ʰ 20ᵐ 7ˢ; long. 0° 24' 15" E. ; log (b), 8·38037 long. 2° 36' 15" E. ; line $\begin{Bmatrix} \text{N. } 54\frac{1}{2}° \text{ E.} \\ \text{S. } 54\frac{1}{2}° \text{ W.} \end{Bmatrix}$; azi. S. 35½° E. Second dec. 23° 26' 19"·6 N. ; eq. time, +1ᵐ 56'·58 ; true alt. 46° 26' 24" ; log (c), 9·158254; H.A. 2° 58' 23"; long. 0° 6' 0" E. ; log (d), 9·146651 ; H.A. 2ʰ 55ᵐ 54ˢ; long. 0° 31' 15" W. ; line, $\begin{Bmatrix} \text{N. } 22° 30' \text{ W.} \\ \text{S. } 22° 30' \text{ E.} \end{Bmatrix}$; azi. S. 67° 30' W. ; position, lat. 50° 10' N., long. 0° 0'.

4. *First observation:* True alt. 11° 31' 42"; dec. 10° 33' 5" S. ; eq. time, +13ᵐ 50ˢ ; E.H.A.'s. 5ʰ 38ᵐ 25ˢ and 5ʰ 38ᵐ 35ˢ ; long. (a), 54° 47' 30" W. ; long. (b), 54° 50' W.
Second observation: True alt. 47° 30' 11" ; dec. 10° 29' 1" S. ; eq. time, +13ᵐ 48ˢ ; E.H.A.'s. 1ʰ 13ᵐ 41ˢ and 1ʰ 6ᵐ 9ˢ ; long. (c), 56° 6' 30" W. ; long. (d), 54° 13' 30" W. ; first line of position, N. 3° E. or S. 3° W. ; second true azi. N. 26° E. ; lat. 50° 25' S., long. 55° 19½° W.

5. *First observation:* True alt. 59° 39' 57" ; dec. 23° 18' 11" N. ; eq. time, +6ˢ ; W.H.A.'s. 1ʰ 28ᵐ 41ˢ and 1ʰ 25ᵐ 23ˢ ; long. (a), 177° 36' 30" W. ; long. (b), 178° 26' W.
Second observation: True alt. 26° 55' 52" ; dec. 23° 18' 34" N. ; eq. time, +8ˢ ; W.H.A.'s. 5ʰ 0ᵐ 4ˢ and 5ʰ 0ᵐ 22ˢ ; long. (c), 179° 59' 15" E. ; long. (d), 179° 56' 15" W. ; second line of position, N. 6° E. or S. 6° W. ; true azi. N. 84° W. ; lat. 48° 34' N., long. 179° 55½° W.

6. *First observation:* True alt. 24° 9' 41" ; dec. 11° 9' 49" N. ; eq. time, +2ᵐ 18ˢ ; E.H.A.'s. 4ʰ 22ᵐ 25ˢ and 4ʰ 21ᵐ 46ˢ ; long. (a), 0° 32' 15" E. ; long. (b), 0° 42' E.
Second observation: True alt. 48° 46' 26" ; dec. 11° 6' 54" N. ; eq. time, +2ᵐ 15ˢ ; E.H.A.'s. 1ʰ 8ᵐ 34ˢ and 1ʰ 1ᵐ 47ˢ ; long. (c), 2° 0' W. ; long. (d), 0° 18' 15" W. ; first line of position, N. 12° E. or S. 12° W. ; first azi. S. 78° E. ; lat. 50° 26' N. ; long. 0° 1' E.

7. *First observation:* True alt. 24° 6' 51" ; dec. 6° 50' 41" S. ; eq. time, −13ᵐ 4ˢ ; E.H.A.'s. 2ʰ 51ᵐ 45ˢ and 2ʰ 46ᵐ 14ˢ ; long. (a), 146° 34' 30" E. ; long. (b), 147° 57' 45" E.
Second observation: True alt. 31° 2' 10" ; dec. 6° 54' 54" S.; eq. time, −13ᵐ 7ˢ ; W.H.A.'s. 1ʰ 47ᵐ 43ˢ and 1ʰ 37ᵐ 30ˢ ; long. (c), 150° 25' E. ; long. (d), 147° 51' 45" E. ; first line of position, N. 43½° E. or S. 43½° W. ; first azi. S. 46½° E. ; second line of position, N. 60° W. or S. 60° E. ; second azi. S. 30° W. ; lat. 40° 47½' N., long. 148° 22½' E.

APPENDIX. 353

Answers to Chart.

1. (a) C.C. N. 15° W.; dev. 1° E.; dist. 103 miles.
   (b) Lat. 52° 24' N.; long. 4° 34' W.
   (c) Lat. 53° 14' N.; long. 5° 34' W.; dist. 12½ miles.
   (d) N. 6° W.; 69½ miles.
2. (a) N. 26° W.; dev. 6° W.; dist. 33 miles.
   (b) Lat. 52° 3' N.; long. 4° 16' W.
   (c) Lat. 52° 41' N.; long. 4° 38' W.; dist. 9½ miles.
   (d) N. 12° W.; 19 miles.
3. (a) S. 69½° W.; dev. 27½° W.; dist. 147 miles.
   (b) Lat. 50° 39½' N.; long. 179° 39' W.
   (c) Lat. 50° 50' N.; long. 179° 0' W.; dist. 39 miles.
   (d) S. 52° W.; 47 miles.
4. (a) S. 5° E.; dev. 34° E.; dist. 46 miles.
   (b) Lat. 52° 31' N.; long. 4° 52' W.
   (c) Lat. 52° 10' N.; long. 4° 52½' W.; dist. 7 miles.
   (d) S. 26° E.; 26 miles.
5. (a) S. 72° W.; dev. 28° W.; dist. 25¼ miles.
   (b) Lat. 55° 27½' N.; long. 4° 58' W.
   (c) Lat. 55° 19' N.; long. 5° 2' W.; dist. 7 miles.
   (d) N. 89° W.; 20 miles.
6. (a) N. 89° E.; dev. 7° W.; dist. 30¾ miles.
   (b) Lat. 54° 23½' N.; long. 4° 8½' W.
   (c) Lat. 54° 32' N.; long. 3° 38' W.; dist. 1¾ mile.
   (d) S. 82° E.; 24 miles.
7. (a) N. 29° E.; dev. 29° W.; dist. 78 miles.
   (b) Lat. 54° 55¼' N.; long. 5° 22' W.
   (c) Lat. 55° 9' N.; long. 5° 30' W.; dist. 14½ miles.
   (d) N. 18° E.; 43½ miles.
8. (a) N. 71° E.; dev. 15° W.; dist. 36 miles.
   (b) Lat. 52° 16' N.; long. 6° 0' W.
   (c) Lat. 52° 29' N.; long. 5° 46' W.; dist. 11 miles.
   (d) N. 70° E.; dist. 23 miles.
9. (a) N. 55° E.; dev. 21° W.; dist. 108 miles.
   (b) Lat. 50° 15' N.; long. 6° 13¼' W.
   (c) Lat. 51° 9' N.; long. 5° 56' W.; dist. 23½ miles.
   (d) N. 76° E.; dist. 23 miles.
10. (a) N. 80° E.; dev. 12° W.; dist. 64 miles.
    (b) Lat. 53° 29' N.; long. 5° 37½' W.
    (c) Lat. 54° 0' N.; long. 4° 43' W.; dist. 5 miles.
    (d) N. 66½° E.; dist. 23 miles.

Answers to Napier's Curve.

1. Correct magnetic bearing: N. 66° E.
   Deviations: 17° W. 28° W. 24° W. 9° W. 16° E. 32° E. 26° E. 5° E.
   Compass courses: N. 76° E. S. 80° E. S. 28° E. N. 16° E.
   Magnetic courses: N. 44° W. S. 16° W. N. 79½° E. N. 9° W.
   Deviation: 31° E.
   Bearings: S. 53° W. N. 76° E.
2. Correct magnetic bearing: N. 82½° W.
   Deviations: 12½° E. 18½° E. 9½° E. 6½° W. 9½° W. 14½° W. 8½° W. 3½° W.
   Compass courses: N. 71° E. S. 69° E. S. 20° W. N. 42° W.
   Correct magnetic bearings: N. 12½° E. S. 44° E. S. 37° W. N. 35° W.
   Deviation: 0.
   Correct bearings: S. 80° W. N. 40° E.

2 A

354   APPENDIX.

3. Correct magnetic bearing: S. 14° W.
Deviations: 10° E.   15° E.   14° E.   8° W.   20° W.   18° W.   3° E.
5° E.
Compass courses: N. 16° W.   S. 75° W.   S. 75½° E.   N. 51° E.
Correct magnetic courses: N. 56° W.   N. 56° E.   S. 27° E.   N. 33° W.
Deviation: 4° E.
Correct bearings: S. 57° W.   N. 58° W.
4. Correct magnetic bearing: N. 8½° W.
Deviations: 23½° E.   18½° E.   14½° W.   10½° W.   12½° W.   12½° W.
1½° E.   6½° E.
Compass courses: S. 57° W.   N. 18½° W.   S. 73° E.   S. 88° E.
Correct magnetic courses: S. 80° W.   N. 40° E.   S. 45° E.   S. 1° E.
Deviation: 1° E.
Bearings: N. 52° E.   S. 39° W.
5. Correct magnetic bearing: N. 83¾° E.
Deviations: 21° 45' E.   17° 45' E.   0° 45' E.   19° 15' W.   21° 15' W.
15° 15' W.   1° 15' W.   16° 45' E.
Compass courses: N. 29° E.   S. 86° E.   S. 25° W.   N. 26° W.
Correct magnetic courses: N. 58½° E.   S. 61° E.   N. 87½° W.   N. 21° W.
Deviation: 22° E.   Bearings: S. 3½° E.   S. 73½° E.
6. Correct magnetic bearing: S. 13¼° E.
Deviations: 18¾° E.   8½° W.   29¼° W.   34¾° W.   20° 5' W.   12¾° E.
29¾° E.   31° 20' E.
Compass courses: N. 2½° E.   S. 60½° E.   S. 26° W.   N. 81½° W.
Correct magnetic courses: N. 32° E.   N. 50° E.   S. 56½° E.   N. 11° W.
Deviation: 35° W.
Correct bearings: N. 15° 38' E.   N. 49° 4' W.

ANSWERS TO GREAT CIRCLE SAILING.

1. First course, S. 84° 44' 50" W.; last course, N. 63° 10' 40" W.; lat. of vertex, 34° 43' S., long. 9° 15' E.; distance, 3269 miles.

Points.	Longitude.	Latitude.
Long. from V. ° '	° '	° '
8 25	18 30 E.	34 22 S.
5	14 15 E.	34 37 S.
0	9 15 E.	34 43 S.
5	4 15 E.	34 37 S.
10	0 45 W.	34 19 S.
15	5 45 W.	33 47½ S.
20	10 45 W.	33 14 S.
25	15 45 W.	32 7½ S.
30	20 45 W.	30 57½ S.
35	25 45 W.	29 34½ S.
40	30 45 W.	27 57½ S.
45	35 45 W.	26 6½ S.
50	40 45 W.	24 0½ S.
52 44	43 9 W.	22 55 S.

2. Initial course, N. 56° 54' W.; final course, S. 54° 18' W.; distance, 4469 miles; lat. and long. of vertex, 48° 34' N., 169° 14' W.

Points 1. 40° 57½' N., 129° 14' W.
,, 2. 46° 47¾' N., 149° 14' W.
,, 3. 48° 34' N., 169° 14' W.
,, 4. 46° 47½' N., 170° 46' E.
,, 5. 40° 57½' N., 150° 46' E.

APPENDIX. 355

3. Courses, N. 55° 43' E., and N. 60° 15' E.; dist. 6441 miles; vertex, lat. 46° 40' S., long. 100° 31' E.
4. Courses, N. 48° 47½' E., and S. 45° 23½' E.; dist. 9904 miles; vertex, lat. 45° 52½' N., long. 179° 14' W.
5. Courses, S. 67° 35¼' W., and S. 35° 34½' W.; dist. 3452 miles; vertex, lat. 55° 29' N., long. 21° 21½' E.

ANSWERS TO COMPOSITE TRACKS.

No.	First course.	Last course.	Distance.	Longitude arrival.	Longitude departure.
1	S. 65 27½ E.	N. 59 6½ E.	7832·4	25 40½ W.	68 32½ E.
2	S. 53 5½ W.	N. 68 44½ W.	5600·6	19 1½ W.	21 55½ W.
3	N. 52 5 E.	S. 75 26 E.	4096·1	167 1 W.	143 54 W.
4	S. 64 17¼ W.	N. 47 20 W.	7785·9	59 0 E.	106 48½ E.
5	S. 63 0 E.	N. 63 16½ E.	4602·3	61 44 E.	76 39 E.
6	S. 52 57 W.	N. 67 31 W.	5108·9	125 37½ W.	156 12 W.

ANSWERS TO CALCULATION OF ALTITUDES.

1. Dec. 23° 18' 11" N.; eq. time, +6*; H.A. 1^h 19^m 23^s; true alt. 60° 16' 56"; appar. alt. 60° 17' 26".
2. Dec. 11° 9' 49" N.; H.A. 4^h 24^m 34^s; true alt. 23° 42' 29"; appar. alt. 23° 44' 30".
3. R.A.M.S. 19^h 41^m $38^s\cdot3$; E.H.A. 3^h 43^m 47^s; true alt. 26° 38' 54"; appar. alt. 26° 40' 47".
4. R.A.M.S. 22^h 14^m $59^s\cdot1$; W.H.A. 4^h 21^m $8^s\cdot3$; true alt. 31° 3' 16"; appar. alt. 31° 4' 50".
5. R.A.M.S. 20^h 55^m $20^s\cdot5$; R.A. 14^h 53^m $8^s\cdot9$; dec. 20° 47' 34" S.; H.P. 57' 35"; W.H.A. 3^h 30^m $29^s\cdot3$; true alt. 7° 4' 2"; appar. alt. 6° 14' 57".
6. R.A.M.S. 11^h 45^m $18^s\cdot5$; R.A. 21^h 53^m 20^s; dec. 8° 5' 50" S.; H.P. 60' 50"; W.H.A. 3^h 7^m $20^s\cdot5$; true alt. 20° 9' 44"; appar. alt. 19° 15' 0".
7. R.A.M.S. 19^h 46^m $21^s\cdot6$; R.A. 8^h 10^m $2^s\cdot3$; dec. 24° 32' 35" N.; H.P. 10"; W.H.A. 51^m $46^s\cdot3$; true alt. 62° 33' 18"; appar. alt. 62° 33' 43".
8. R.A.M.S. 3^h 50^m $50^s\cdot6$; R.A. 14^h 2^m $50^s\cdot6$; dec. 11° 0' 24" S.; E.H.A. 49^m 45^s; true alt. 30° 1' 30"; appar. alt. 30° 3' 6".

ANSWERS TO LUNARS.

1. Sun's elements: R.A. 22^h 4^m 51^s; dec. 11° 48' 2" S.; R.A.M.S. 21^h 50^m 40^s; H.A. 3^h 38^m 39^s; appar. alt. 36° 18' 7"; true alt. 36° 16' 57".
Moon's elements: S.D. 15' 11"; H.P. 55' 12"·5; R.A. 4^h 11^m 47^s; dec. 23° 53' 42" N.; H.A. 2^h 28^m 17^s; appar. alt. 24° 17' 14"; true alt. 25° 5' 27".
App. dist. 96° 52' 0"; B = 25° 26' 45"; true dist. 96° 18' 32". Second cor. −4"; M.T.G. 17^d 10^h 14^m 40^s; error, fast 5^m 30^s; long. 95° 27' 18" W.
2. Sun's elements: R.A. 2^h 48^m 39^s; dec. 16° 14' 2" N.; R.A.M.S. 2^h 52^m 2^s; H.A. 1^h 10^m 31^s; appar. alt. 65° 15' 36"; true alt. 65° 15' 13".
Moon's elements: S.D. 16' 14"·5; H.P. 58' 44"; R.A. 23^h 12^m $14^s\cdot5$; Dec. 0° 20' 41" N.; W.H.A. 2^h 25^m $52^s\cdot5$; appar. alt. 40° 32' 16"; true alt. 41° 15' 48". App. dist. 56° 14' 32"; B = 21° 59' 23·5"; true dist. 55° 37' 4". Second cor. 0; M.T.G. 4^d 20^h 48^m $22^s\cdot5$; error, fast 2^m $7^s\cdot5$; long. 29° 25' 36" E.
3. Saturn's elements: R.A. 17^h 12^m 58^s; dec. 21° 33' 9" S. R.A.M.S. 6^h 31^m $42^s\cdot6$; W.H.A. 6^h 26^m 8^s; appar. alt. 9° 51' 21"; true alt. 9° 46' 2".

Moon's elements: S.D. 16' 8"; H.P. 58' 20"; R.A. 0ʰ 15ᵐ 49ˢ; dec. 7° 25' 46" N.; E.H.A. 0ʰ 36ᵐ 42ˢ; appar. alt. 38° 43' 0"; true alt. 39° 27' 20". App. dist. 107° 35' 18"; B = 24° 56' 56"; true dist. 107° 17' 39". Second cor. 3"·5; M.T.G. 20ᵈ 14ʰ 2ᵐ 25ˢ·5; error, slow ·5ˢ; long. 46° 15' 6" E.

4. Mars' elements: R.A. 13ʰ 39ᵐ 1ˢ; dec. 10° 16' 8" S.; R.A.M.S. 11ʰ 39ᵐ 56ˢ; W.H.A. 5ʰ 18ᵐ 3ˢ; appar. alt. 14° 36' 7"; true alt. 14° 32' 35". Moon's elements: S.D. 16' 44"; H.P. 60' 19"; R.A. 20ʰ 35ᵐ 28ˢ·6; dec. 15° 10' 45" S.; E.H.A. 1ʰ 38ᵐ 24ˢ; appar. alt. 48° 10' 5"; true alt. 48° 49' 29". App. dist. 101° 0' 4"; B = 21° 16' 39"·5; true dist. 100°·39' 17". Second cor. +1ˢ; M.T.G. 15ᵈ 18ʰ 20ᵐ 40ˢ; error, slow 10ˢ; long. 165° 53' W.

5. Moon's semidiameter, 16' 50"·31; H.P. 61' 13"·26; R.A. 20ʰ 44ᵐ 54ˢ·95; dec. 14° 32' 48"·5 S.; R.A.M.S. 9ʰ 52ᵐ 29ˢ. α Arietis R.A. 2ʰ 1ᵐ 32ˢ·59; dec. 22° 59' 22" N.; moon's H.A. 3ʰ 30ᵐ 40ˢ W.; star's H.A. 1ʰ 45ᵐ 58ˢ; moon's appar. alt. 21° 38' 2", true alt. 22° 32' 33"; star's appar. alt. 65° 21' 40", true alt. 65° 21' 14"; arc B = 13° 19' 29"; true dist. 86° 0' 47"; interval, 12ᵐ 52ˢ; cor. 0; chron. 2ˢ fast; M.T.S. 19ᵈ 14ʰ 23ᵐ 6ˢ; long. 32° 33' 30" E.

6. Moon's semidiameter, 16' 36"·85; H.P. 60' 27"·21; R.A. 20ʰ 47ᵐ 56ˢ·52; dec. 14° 9' 51"·2 S.; R.A.M.S. 11ʰ 40ᵐ 47ˢ·21. Sun's R.A. 11ʰ 35ᵐ 37ˢ·5; dec. 2° 38' 12·5" N.; appar. alt. 19° 38' 39"; true alt. 19° 36' 9"; Sun's H.A. 3ʰ 58ᵐ 28ˢ W.; R.A.M. 15ʰ 34ᵐ 6ˢ. Moon's H.A. 5ʰ 13ᵐ 51ˢ, true alt. 18° 0' 2"; appar. alt. 17° 5' 19"; arc B = 9° 21' 54"; dist. 137° 40' 17"; intervals, 2ʰ 30ᵐ 16ˢ·5; cor. +2ˢ; chron. 9ᵐ 59ˢ fast; M.T.S. 16ᵈ 3ʰ 53ᵐ 19ˢ; long. 65° 45' 7"·5 E.

7. Moon's semidiameter, 14' 58"·42; H.P. 54' 24"·92; R.A. 8ʰ 59ᵐ 5ˢ·07; dec. 12° 42' 18"·7 N.; R.A.M.S. 17ʰ 57ᵐ 22ˢ·27. Jupiter's R.A. 15ʰ 47ᵐ 32ˢ·03; dec. 19° 8' 7"·6 S.; H.A. 2ʰ 27ᵐ 15ˢ; true alt. 34° 17' 18"; appar. alt. 34° 18' 40". Moon's H.A. 4ʰ 21ᵐ 12ˢ W.; R.A.M. 13ʰ 20ᵐ 17ˢ; appar. alt. 26° 42' 18"; true alt. 27° 29' 1"; arc B = 18° 47' 20"; true dist. 105° 23' 48"; interval, 2ʰ 54ᵐ 39ˢ; cor. 0; chron. fast 15ˢ; M.T.S. 20ᵈ 19ʰ 22ᵐ 55ˢ; long. 112° 4' 0" E.

ANSWERS TO DOUBLE ALTITUDES.

1. A.T.G. May 10ᵈ 8ʰ 51ᵐ; H.E.T. 2ʰ 9ᵐ 30ˢ; Dec. 17° 44' 6"·2 N.; angle, 90°; dist. 21·6 miles; first true alt. 75° 11' 35"; second true alt. 43° 27' 25"; arc 1, 30° 39' 52"; arc 2, 15° 52' 23"; arc 3, 0° 45' 0"; arc 4, 69° 15' 35"; arc 5, 68° 30' 35"; approx. lat. 20° 37' 57"; true lat. 20° 37' 11" N.

2. A.T.G. June 23ᵈ 0ʰ 50ᵐ 1ˢ·35; H.E.T. 2ʰ 9ᵐ 51ˢ·35; dec. 23° 26' 23"·7 N.; angle, 62° 45'; dist. 39; cor. −17' 52" first alt.; first true alt. 58° 53' 8"; second true alt. 46° 26' 24"; arc 1, 29° 30' 7"; arc 2, 7° 40' 20"; arc 3, 23° 36' 4"; arc 4, 62° 48' 11"; arc 5, 39° 12' 7"; approx. lat. 50° 10' 41" N.; cor. +2"; lat. 50° 10' 43" N.

3. A.T.G. Dec. 31ᵈ 12ʰ 22ᵐ 20ˢ·5; H.E.T. 1ʰ 59ᵐ 44ˢ·5; dec. 23° 3' 37"·6 S., angle 45°; dist. 40'; cor. +28' 18" first alt.; first true alt. 14° 50' 31"; second true alt. 11° 23' 40"; arc 1, 27° 19' 56"; arc 2, 3° 39' 30"; arc 3, 75° 10' 25"; arc 4, 116° 9' 47"; arc 5, 50° 59' 22"; approx. lat. 48° 52' 35" N.; cor. −5"; lat. 48° 52' 30" N.

4. A.T.G. Sept. 22ᵈ 18ʰ 30ᵐ 15ˢ; H.E.T. 2ʰ 30ᵐ 5ˢ; dec. 0° 0' 7" N.; angle, +74°; dist 29'; cor. +7' 59" to first alt.; first true alt. 32° 28' 24"; second true alt. 32° 20' 32"; arc 1, 37° 31' 15"; arc 2, 0° 5' 27"·5; arc 3, 47° 30' 39"·5; arc 4, 89° 59' 51"·5; arc 5, 42° 29' 12"; lat. = 47° 30' 48" N.

5. M.T.G. Jan. 20ᵈ 4ʰ 35ᵐ 16ˢ; H.E.T. 1ʰ 52ᵐ 47ˢ·50; dec. 18° 7' 26"·9 S., angle, 64°; dist. 37'·5; cor. first alt. −16' 27"; first true alt. 56° 14' 1"; second true alt. 13° 27' 37"; arc 1, 26° 41' 10"; arc 2, 41° 47' 15"; arc 3, 37° 0' 6"; arc 4, 110° 22' 30"; arc 5, 73° 22' 24"; approx. lat. 12° 19' 7"; cor. +46"; lat. 12° 19' 53" N.

APPENDIX. 357

6. M.T.G. 23ᵈ 5ʰ 47ᵐ ; H.E.T. 2ʰ 0ᵐ 20ˢ ; dec. 10° 25' 3"·1 S. ; angle, 16° ; dist. 25 ; cor. for first alt. —24' ; first true alt. 36° 3' 52" ; second true alt. 36° 19' 4" ; arc 1, 29° 32' 16" ; arc 2, 0° 12' 24" ; arc 3, 47° 15' 33" ; arc 4, 101° 59' 43" ; arc 5, 54° 44' 10" ; lat. 35° 15' 49" N.
7. M.T.G. Nov. 26ᵈ 15ʰ 25ᵐ 40ˢ ; H.E.T. 1ʰ 37ᵐ 36ˢ ; dec. 16° 34' 40' S. ; angle, 180° ; dist. 26 miles ; cor. —26' to first alt. ; first true alt. 40° 56' 54" ; second true alt. 8° 31' 19" ; arc 1, 23° 19' 27" ; arc 2, 39° 49' 46" ; arc 3, 55° 16' 2" ; arc 4, 108° 6' 7" ; arc 5, 52° 50' 5" ; lat. 27° 38' 28" N.
8. H.E.T. 2ʰ 40ᵐ 10ˢ ; mid. dec. 13° 23' 19" N. ; first true alt. 59° 17' 10" ; cor. for run, —7' 48" ; second true alt. 36° 53' 6" ; arc. 1, 38° 44' 43" ; arc 2, 11° 57' 57" ; arc 3, 16° 55' 5" ; arc 4, 72° 43' 45" ; arc 5, 55° 48' 40" ; lat. 33° 20' 50" N. ; cor. for lat. —50" ; true lat. 33° 20' N.
9. H.E.T. 2ʰ 19ᵐ 40ˢ ; mid. dec. 23° 2' 20" N. ; cor. for run, +42' 0" ; first true alt. 10° 21' 57" ; second true alt. 10° 21' 57" ; arc 1, 31° 47' 6" ; arc 2, 0.; arc 3, 77° 46' 46" ; arc 4, 117° 24' 47" ; arc 5, 39° 38' 1" ; lat. 50° 21' 59" S. ; cor. for lat. 0.
10. H.E.T. 2ʰ 12ᵐ 3ˢ ; mid. dec. 6° 52' 46" S. ; cor. for run, +20' 6" ; first true alt. 24° 26' 56" ; second true alt. 31° 2' 10" ; arc 1, 32° 44' 41" ; arc 2, 5° 23' 37" ; arc 3, 56° 17' 24" ; arc 4, 98° 11' 14" ; arc 5, 41° 54' 20" ; lat. 47° 49' 15" N. ; cor. for lat. —32" ; true lat. 47° 48' 43" N.
11. Sid. int. 3ʰ 20ᵐ 42ˢ ; H.E.T. 1ʰ 40ᵐ 21ˢ ; dec. 16° 34' 44" S. ; cor. for run, —13' 18" ; first true alt. 26° 38' 54" ; second true alt. 58° 26' 56" ; arc 1, 23° 58' 38" ; arc 2, 29° 46' 46" ; arc 3, 34° 54' 40" ; arc 4, 108° 11' 49" ; arc 5, 73° 17' 9" ; true lat. 14° 27' 12" N.
12. Sid. int. 2ʰ 25ᵐ 35ˢ·8 ; cor. for R.A. +10·4ˢ ; H.E.T. 1ʰ 12ᵐ 53ˢ·1 ; mid. dec. 24° 32' 53" N. ; cor. for run, 0 ; first true alt. 62° 33' 18" ; second true alt. 44° 19' 4" ; arc 1, 16° 31' 28" ; arc. 2, 19° 23' 9" ; arc 3, 28° 43' 34" ; arc 4, 64° 19' 10" ; arc 5, 35° 35' 36" ; lat. 50° 5' 30" N. ; cor. for lat. —30" ; true lat. 50° 5' N.

ANSWERS TO SIMULTANEOUS ALTITUDES.

1. *Venus*, R.A. 19ʰ 5ᵐ 25ˢ·94 ; dec. 24° 0' 7"·45 ; R.A.M.S. 17ʰ 32ᵐ49ˢ·36 ; alt. 9° 3' 41" ; first H.A. = 2ʰ 55ᵐ 54ˢ W. ; second H.A. = 2ʰ 52ᵐ 22ˢ W. ; first long. 30° 7' 15" W. ; second long. 31° 0' 15" W. ; first line $\left\{ \begin{array}{l} \text{N. 5. E.} \\ \text{S. 5 W.} \end{array} \right\}$; azi. S. 85° E.
Star α *Arietis*, R.A = 2ʰ 1ᵐ 33ˢ·9 ; dec. 22° 59' 34" N. ; alt. 37° 6' 33" ; third H.A. = 3ʰ 58ᵐ 41ˢ E. ; fourth H.A. = 3ʰ 58ᵐ 25ˢ E. ; third long. 29° 44' W. ; fourth long. 29° 40' W. ; second line $\left\{ \begin{array}{l} \text{N. 50° W.} \\ \text{S. 50° E.} \end{array} \right\}$; azi. S. 40° W. ; ship's position, lat. 46° 18' N., long. 29° 45' W.
2. Moon's semid. 15' 3"·6 ; H.P. 54' 49" ; R.A. 12ʰ 37ᵐ 54ˢ·54 ; dec. 9° 22' 55"·6 S. ; R.A.M.S. 18ʰ 16ᵐ 47ˢ·87 ; alt. 14° 17' 1" ; first H.A. = 3ʰ 51ᵐ 41ˢ W. ; second H.A. = 3ʰ 49ᵐ 58ˢ W. ; first long. 179° 26' 30" W. ; second long. 179° 52' 15" W.; line $\left\{ \begin{array}{l} \text{N. 31° W.} \\ \text{S. 31° E.} \end{array} \right\}$; azi. S. 59° W.
Sun's R.A. = 18ʰ 17ᵐ 15ˢ·59 ; dec. 23° 23' 34"·6 S. ; alt. 15° 40' 1" ; first H.A. = 1ʰ 57ᵐ 27ˢ E. ; second, 1ʰ 51ᵐ 50ˢ E. ; first long. 178° 9' 15" E. ; second long. 179° 33' 15" E. ; line $\left\{ \begin{array}{l} \text{N. 62° E.} \\ \text{S. 62° W.} \end{array} \right\}$; azi. S. 28° E. ; ship's position, lat. 46° 40' N., long. 179° 58' 30" E.
3. *Mars*, R.A. = 8ʰ 11ᵐ 27ˢ·31 ; dec. 22° 24' 51"·6 N. ; R.A.M.S. 1ʰ 34ᵐ 54ˢ·11 ; alt. 55° 37' 52" ; first H.A. = 2ʰ 23ᵐ 18ˢ W. ; second H. A. 2ʰ 22ᵐ 47ˢ W. ; first long. 26° 4' 15" E. ; second long. 25° 56' 30" E. ; first line, $\left\{ \begin{array}{l} \text{N. 18° W.} \\ \text{S. 18° E.} \end{array} \right\}$; azi. S. 72° W.

Jupiter, R.A. = 14ʰ 19ᵐ 0ˢ·82 ; dec. 12° 21′ 26″·9 S. ; alt. 17° 16′ 46″ ; third H.A. 3ʰ 44ᵐ 35ˢ E. ; fourth H.A. = 3ʰ 43ᵐ 32ˢ E. ; third long. 25° 54′ 30″ E. ; fourth, 26° 10′ 15″ E. ; line $\begin{Bmatrix} \text{N. } 52° \text{ E.} \\ \text{S. } 52° \text{ W.} \end{Bmatrix}$; azi. S. 38° E. ; ship's position, lat. 38° 9′ N., long. 26° 1′ 30″ E.

4. *Venus's elements:* R.A. 22ʰ 26ᵐ 38ˢ ; dec. 10° 22′ 44″ S. ; R.A.M.S. 0ʰ 53ᵐ 51ˢ ; true alt. 16° 17′ 20″ ; E.H.A.'s. 3ʰ 32ᵐ 28ˢ and 3ʰ 29ᵐ 44ˢ ; long. (*a*), 50° 49½′ W. ; long (*b*), 50° 8′ W.

Saturn's elements: R.A. 17ʰ 32ᵐ 45ˢ ; dec. 21° 50′ 4″ S. ; true alt. 19° 4′ 46″ ; W.H.A.'s. 1ʰ 31ᵐ 13ˢ and 1ʰ 21ᵐ 14ˢ ; long. (*c*), 48° 22′ W. ; long. (*d*), 50° 51′ 45″ W.

Bearing of Venus, S. 54½° E. ; bearing of Saturn, S. 21° W. ; lat. 46° 31′ N. ; long. 50° 20′ W.

5. *Altair's elements:* R.A. 19ʰ 45ᵐ 55ˢ ; dec. 8° 36′ 17″ N. ; R.A.M.S. 7ʰ 47ᵐ 46ˢ ; true alt. 36° 27′ 16″ ; E.H.A.'s. 2ʰ 53ᵐ 42ˢ and 2ʰ 51ᵐ 8ˢ ; long. (*a*), 179° 50¼′ W. ; long. (*b*), 179° 11′ 45″ W.

Spica's elements: R.A. 13ʰ 19ᵐ 55ˢ ; dec. 10° 38′ 12″ S. ; true alt. 15° 1′ 53″ ; W.H.A.'s. 3ʰ 33ᵐ 5ˢ and 3ʰ 30ᵐ 14ˢ ; long. (*c*), 179° 38½′ W. ; long. (*d*), 179° 38′ 45″ E.

Altair's bearing, S. 56½ E. ; Spica's bearing, S. 54° W ; lat. 47° 46½′ N., long. 179° 44′ W.

6. *Capella's elements:* R.A. 5ʰ 9ᵐ 21ˢ ; dec. 45° 53′ 49″ N. ; R.A.M.S. 17ʰ 13ᵐ 40ˢ ; true alt. 37° 19′ 55″ ; E.H.A.'s. 5ʰ 24ᵐ 2ˢ and 5ʰ 25ᵐ 46ˢ ; long (*a*), 50° 25′ 15″ W. ; long. (*b*), 50° 51′ 15″ W.

Altair's elements: R.A. 19ʰ 45ᵐ 54ˢ ; dec. 8° 36′ 22″ N. ; true alt. 26° 23′ 43″ ; W.H.A.'s. 3ʰ 58ᵐ 59ˢ and 3ʰ 58ᵐ 1ˢ ; long. (*c*), 50° 31′ 45″ W. ; long. (*d*), 50° 46′ 15″ W.

Capella's bearing, N. 61° E. ; Altair's bearing, S. 73° W. ; lat. 48° 18½′ N. ; long. 50° 40′ W.

ANSWERS TO CHRONOMETER ERROR BY SINGLE ALTITUDES.

1. Dec. 16° 14′ 2″ N. ; eq. time, −3ᵐ 25ˢ ; true alt. 65° 15′ 13″ ; H.A. 1ʰ 10ᵐ 31ˢ ; slow M.T.S. 1ʰ 55ᵐ 34ˢ ; fast M.T.G. 2ᵐ 6ˢ.

2. Dec. 9° 0′ 57″ N. ; eq. time, +33ˢ ; true alt. 39° 48′ 25″ ; H.A. 2ʰ 57ᵐ 14ˢ ; slow M.T.G. 4ˢ ; fast M.T.S. 1ʰ 41ᵐ 56ˢ.

3. Dec. 22° 22′ 58″ S. ; eq. time, +6ᵐ 27ˢ ; true alt. 39° 45′ 50″ ; H.A. 3ʰ 44ᵐ 25ˢ ; slow A.T.S. 6ʰ 34ᵐ 13ˢ ; fast M.T.G. 2ᵐ.

4. Dec. 10° 29′ 1″ S. ; eq. time, +13ᵐ 48ˢ ; true alt. 47° 30′ 11″ ; H.A. 1ʰ 10ᵐ 35ˢ ; fast A.T.S. 4ʰ 16ᵐ 1ˢ ; fast M.T.G. 20ᵐ 53ˢ.

5. Dec. 0° 0′ 44″ N. ; eq. time, +7ᵐ 31ˢ ; true alt. 8° 59′ 5″ ; H.A. 5ʰ 20ᵐ 35ˢ ; fast A.T.S. 3ʰ 9ᵐ 31ˢ ; correct for M.T.G.

6. Dec. 13° 25′ 28″ N. ; eq. time, −2ᵐ 12ˢ ; true alt. 36° 53′ 6″ ; H.A. 3ʰ 37ᵐ 45ˢ ; fast M.T.G. 22ˢ.

7. Dec. 18° 57′ 16″ N. ; eq. time, −3ᵐ 49ˢ ; true alt. 56° 30′ 5″ ; H.A. 1ʰ 9ᵐ 19ˢ ; slow M.T.G. 2ᵐ 32ˢ ; fast A.T.S. 8ʰ 19ᵐ 33ˢ.

8. Dec. 23° 18′ 34″ N. ; eq. time, +8ˢ ; true alt. 26° 55′ 53″ ; H.A. 5ʰ 0ᵐ 23ˢ ; slow M.T.G. 11ᵐ 36ˢ.

9. R.A.M.S. 2ʰ 0ᵐ 7ˢ ; true alt. 23° 11′ 11″ ; W.H.A. 2ʰ 46ᵐ 49ˢ ; slow M.T.G. 1ᵐ 32ˢ.

10. R.A.M.S. 8ʰ 14ᵐ 21ˢ·6 ; true alt. 46° 43′ 11″ ; W.H.A. 3ʰ 18ᵐ 53ˢ·3 ; fast M.T.S. 2ʰ 57ᵐ 11ˢ·3 ; fast M.T.G. 2ᵐ 51ˢ·3.

11. R.A.M.S. 22ʰ 14ᵐ 3ˢ·59; true alt. 25° 54′ 4″; E.H.A. 4ʰ 24ᵐ 28ˢ·7; fast M.T.G. 32ᵐ 37ˢ ; fast M.T.S. 3ʰ 34ᵐ 9ˢ·2.

12. R.A.M.S. 17ʰ 13ᵐ 40ˢ·2 ; true alt. 37° 19′ 55″; E.H.A. 5ʰ 25ᵐ 1ˢ ; fast M.T.G. 6ᵐ 40ˢ·1

APPENDIX.

ANSWERS TO EQUAL ALTITUDES.

1. Dec. 21° 36′ 0″ S.; eq. time, $+8^m 37^{s} \cdot 95$; int. $2^h 57^m 53^s$; Part I. $+3^{s} \cdot 78$; Part II. $-2^{s} \cdot 36$; slow A.T.S. $4^m 43^{s} \cdot 08$; slow M.T.G. $5^h 6^m 57^{s} \cdot 03$.
2. Dec. 0° 6′ 58″ N.; eq. time, $+7^m 26^{s} \cdot 11$; int. $6^h 7^m 56^s$; Part I. $-12^{s} \cdot 07$; Part II. $-\cdot 02^s$; slow A.T.S. $2^h 38^m 29^{s} \cdot 09$; fast M.T.G. $6^h 32^m 58^{s} \cdot 8$.
3. Dec. 5° 2′ 14″ N.; eq. time, $+3^m 35^{s} \cdot 1$; int. $5^h 3^m 46^s$; Part I. $-5^{s} \cdot 12$; Part II. $+1^{s} \cdot 1$; fast A.T.S. $5^h 47^m 15^{s} \cdot 98$; fast M.T.S. $5^h 43^m 40^{s} \cdot 88$; fast M.T.G. $36^m 16^{s} \cdot 88$.
4. Dec. 21° 5′ 7″ N.; eq. time, $-3^m 13^{s} \cdot 4$; int. $4^h 17^m 17^s$; Part I. $+4^{s} \cdot 37$; Part II. $+2^{s} \cdot 27$; slow A.T.S. $7^h 12^m 17^{s} \cdot 86$; fast A.T.G. $30^m 44^{s} \cdot 14$; fast M.T.G. $33^m 57^{s} \cdot 54$.
5. Dec. 23° 16′ 43″ N.; eq. time, 0; int. $4^h 45^m 36^s$; Part I. $-\cdot 8^s$; Part II. $+\cdot 7^s$; fast A.T.S. $1^h 14^m 6^{s} \cdot 9$; slow A.T.G. and M.T.G. $11^m 5^{s} \cdot 1$.
6. R.A. $13^h 19^m 55^{s} \cdot 5$; M.T. transit, $9^h 30^m 57^{s} \cdot 33$; eq. time, $+3^m 41^{s} \cdot 73$; slow A.T.S. $2^h 9^m 0^{s} \cdot 06$; slow M.T.S. $2^h 5^m 18^{s} \cdot 33$; fast M.T.G. $2^m 41^{s} \cdot 67$.
7. R.A. $6^h 21^m 44^{s} \cdot 7$; M.T. transit $9^h 0^m 28^{s} \cdot 08$; eq. time, $-14^m 26^{s} \cdot 82$; slow M.T.S. $9^h 32^m 26^{s} \cdot 08$; fast M.T.G. $7^m 27^{s} \cdot 92$; fast A.T.G. $21^m 54^{s} \cdot 74$.
8. R.A. $14^h 11^m 4^{s} \cdot 7$; M.T. transit $17^h 41^m 10^{s} \cdot 62$; eq. time, $-13^m 9^{s} \cdot 96$; fast A.T.S. $5^h 12^m 29^{s} \cdot 34$; fast M.T.G. $5^m 0^{s} \cdot 38$.
9. R.A. $14^h 11^m 6^{s} \cdot 4$; M.T. transit $11^h 35^m 0^{s} \cdot 81$; eq. time, $+2^m 58^{s} \cdot 3$; fast M.T.S. $8^h 14^m 50^{s} \cdot 19$; fast M.T.G. $4^m 10^{s} \cdot 19$; fast A.T.G. $1^m 11^{s} \cdot 89$.

ANSWERS TO MERCATOR'S CHARTS.

1. Lat. 57° 31½′ N., long. 27° 5¼′ W.
2. Lat. 64° 10¼′ S., long. 121° 22′ E., magnetic bearing S. 22° W.
3. Lat. 55° 51½′ N., long. 30° 24′ W.
   True course, N. 47° E.; distance, 110 miles.
4. Obs. lat. 54° 29′ N., long. 5° 21½′ W.
   D.R. lat. 54° 22½′ N., long. 5° 15′ W.
   Current, N. 28° W. 7¾ miles.

ANSWERS TO RIGHT-ANGLED PLANE TRIANGLES.

1. $A = 44° 47′ 10″$; $B = 45° 12′ 50″$; $BC = 354 \cdot 3$.
2. $P = 28° 48′ 30″$; $R = 61° 11′ 30″$; $PR = 423 \cdot 4$.
3. $B = 10° 25′ 30″$; $P = 79° 34′ 30″$; $OB = 683 \cdot 5$.
4. $AC = 443 \cdot 4$; $BC = 937 \cdot 4$.
5. $AB = 1095$; $AC = 617$.
6. $A = 52° 6′ 11″$; $B = 37° 53′ 49″$; $Ac = 90 \cdot 3$.
7. $PR = 5227$; $QR = 2864$.
8. $A = 54° 15′ 30″$; $B = 35° 44′ 30″$; $AB = 1208 \cdot 5$.
9. Height, 91·55 ft.
10. Course, S. 42° 22′ W.; dist. 320·1 miles.
11. Height, 15 ft. 7 in.
12. First dist. 9·08 miles; second dist. 5·75 miles.
13. 6° 15½′.
14. 113·8 ft.

ANSWERS TO OBLIQUE-ANGLED PLANE TRIANGLES.

1. $A = 47° 7′ 8″$; $B = 14° 25′ 32″$; $C = 118° 27′ 20″$.
2. $A = 42° 16′ 41″$; $B = 23° 38′ 38″$; $C = 114° 4′ 41″$.
3. $A = 83° 25′ 54″$; $B = 55° 52′ 48″$; $C = 40° 41′ 18″$.
4. $C = 67° 41′ 0″$; $b = 164$; $c = 157 \cdot 5$.
5. $A = 64° 27′ 0″$; $a = 1078 \cdot 7$; $c = 1172 \cdot 8$.
6. $B = 32° 10′ 0″$; $a = 20 \cdot 62$; $b = 11 \cdot 08$.
7. $A = 96° 16′ 25″$; $B = 44° 3′ 35″$; $c = 4466$.
8. $A = 45° 31′ 53″$; $C = 65° 42′ 7″$; $b = 348 \cdot 7$.

APPENDIX.

9. B = 46° 50′ 2″ ; C = 35° 51′ 58″ ; a = 82·05.
10. B = 42° 50′ 18″ ; C = 84° 43′ 42″ ; c = 1086.
11. A = 19° 4′ 29″ ; B = 121° 51′ 31″ ; a = 180·5 ;
    or 82° 47′ 31″ ; or 58° 8′ 29″ ; or 547·8.
12. B = 90° 30′ 11″ ; C = 39° 53′ 49″ ; b = 1398·4.
13. B = 35° 4′ 5″ ; C = 123° 32′ 55″ ; c = 1342 ;
    or 144° 55′ 55″ ; or 13° 41′ 5″ ; or 380·9.
14. A = 48° 11′ 22″ ; B = 58° 24′ 42″ ; C = 73° 23′ 56″.
15. A = 27° 7′ 36″ ; B = 22° 19′ 54″ ; C = 130° 32′ 30″.
16. A = 44° 24′ 54″ ; B = 57° 7′ 20″ ; C = 78° 27′ 46″.
17. C = 11° 40′ 0″ ; a = 6291 ; c = 4044.
18. C = 61° 25′ 0″ ; a = 206·3 ; b = 184·7.
19. B = 88° 30′ 1″ ; C = 33° 30′ 59″ ; a = 201·2.
20. A = 48° 41′ 9″ ; C = 73° 1′ 51″ ; b = 2015.
21. A = 46° 35′ 31″ ; B = 16° 13′ 31″ ; c = 1591·8.
22. B = 62° 31′ 25″ ; C = 102° 17′ 35″ ; c = 6056·8 ;
    or 117° 28′ 35″ ; or 47° 20′ 25″ ; or 4455.
23. B = 59° 57′ 53″ ; C = 84° 47′ 7″ ; c = 345·1 ;
    or 120° 2′ 7″ ; or 24° 42′ 53″ ; or 144·9.
24. 212 ft.
25. 134·7 ft.

ANSWERS.—RIGHT-ANGLED SPHERICAL TRIANGLES.

1. A = 54° 15′ 52″ ; B = 70° 5′ 36″ ; c = 74° 53′ 52″.
2. a = 71° 20′ 56″ ; b = 38° 18′ 16″ ; c = 75° 27′ 57″.
3. A = 62° 17′ 0″ ; B = 79° 8′ 34″ ; b = 77° 42′ 53″.
4. B = 77° 2′ 15″ ; a = 50° 17′ 53″ ; b = 73° 20′ 39″.
5. A = 45° 40′ 14″ ; b = 39° 6′ 20″ ; c = 49° 18′ 54″.
6. B = 139° 26′ 36″, or 40° 33′ 24″ ; b = 145° 5′ 45″, or 34° 54′ 15″ ;
   c = 118° 21′ 6″, or 61° 38′ 54″.
7. A = 70° 55′ 17″ ; B = 101° 55′ 44″ ; c = 94° 11′ 25″.
8. A = 110° 3′ 55″ ; B = 98° 13′ 51″ ; c = 86° 58′ 17″.
9. a = 117° 2′ 58″ ; b = 65° 5′ 37″ ; c = 101° 2′ 29″.
10. a = 92° 52′ 13″ ; b = 118° 45′ 49″ ; c = 88° 37′ 9″.
11. A = 78° 10′ 28″ ; B = 144° 52′ 26″ ; b = 146° 41′ 0″.
12. A = 96° 57′ 4″ ; B = 101° 19′ 54″ ; a = 97° 5′ 24″.
13. A = 55° 2′ 52″ ; B = 96° 0′ 20″ ; a = 54° 49′ 36″.
14. A = 144° 52′ 47″ ; a = 148° 4′ 7″ ; b = 62° 23′ 2″.
15. A = 129° 14′ 9″ ; a = 129° 43′ 36″ ; b = 100° 40′ 35″.
16. B = 95° 41′ 50″ ; a = 76° 8′ 49″ ; c = 91° 24′ 9″.
17. A = 106° 24′ 22″ ; b = 94° 18′ 33″ ; c = 88° 47′ 0″.
18. A = 49° 37′ 0″, or 130° 23′ 0″ ; a = 46° 57′ 45″, or 133° 2′ 15″ ;
    c = 106° 21′ 15″, or 73° 38′ 45″.

ANSWERS.—OBLIQUE-ANGLED SPHERICAL TRIANGLES.

1. A = 82° 42′ 30″ ; B = 90° 9′ 16″ ; C = 104° 19′ 36″.
2. A = 153° 7′ 14″ ; B = 41° 22′ 24″ ; C = 43° 50′ 32″.
3. A = 116° 19′ 7″ ; B = 104° 58′ 43″ ; c = 137° 29′ 6″.
4. B = 59° 39′ 27″ ; C = 82° 37′ 19″ ; a = 91° 29′ 30″.
5. A = 119° 15′ ; C = 73° 59′ ; b = 70° 30′.
6. c = 97° 10′ 50″ ; a = 57° 28′ 20″ ; b = 13° 3′ 24″.
7. c = 62° 38′ 18″ ; a = 131° 7′ 42″ ; b = 74° 23′ 24″.
8. A = 115° 38′ 35″.
9. B = 79° 15′ 40″ ; C = 40° 0′ 16″ ; c = 22° 11′ 12″ ;
   or 100° 44′ 20″ ; or 63° 28′ 32″ ; or 31° 42′ 32″.
10. A = 91° 30′ 10″ ; B = 75° 24′ 27″ ; C = 84° 1′ 31″.
11. C = 45° 55′ 57″ ; a = 97° 27′ 52″ ; c = 46° 24′ 2″.
12. B = 99° 14′ 14″ ; C = 85° 53′ 51″ ; a = 63° 49′ 50″.

APPENDIX. 361

MISCELLANEOUS EXERCISES.

1. 244·7 miles.
2. 39° 16″.
3. Lat. 35° 20′ N., long. 30° 17′·5 W.
4. 36·2 hours.
5. N. 7° 54′ W., 44·8 miles.
6. $6^h 33^m$.
7. Cliff, 131·6 ft.; lighthouse, 107·2 ft.
8. 315·3 ft.
9. 2507·6.
10. 4275·5.
11. 32° 39½′ W.
12. 45° 52½′ N.
13. 15″.
14. 15″.
15. Divide 9 divisions into 20 equal parts.
16. Capella, Vega, Great Bear; Achernar, Canopus, Crucis.
17. $6^h 55^m 22^s$ A.M., or $5^h 4^m 38^s$ P.M.
18. $15^h 14^m 20^s$.
19. $9^h 32^m 4^s$.
20. 49° 2½′.
21. S. 60° 9′ W., 13·36 miles.
22. 116·82 miles.
23. N. 56° 13′ E., 116·9 miles.
24. S. 7° 50′ W., 5·6 miles.
25. $6^h 45^m$.
26. 12·34 miles.

ANSWERS TO PAPER I.

1. Log 2·000000, product 100.
2. Log 1·002554, quotient 10·059.
3. S. 15° W., 9′; S. 89° E., 21′; N. 67° E., 22′; S. 86° E., 25′; N. 75° E., 27′; N. 71° E., 28′; N. 30° E., 26′; E., 54′; diff. lat. 36′·4 N.; dep. 183·5 E.; lat. in, 59° 4′ 54″ N.; long. in, 2° 51′ 6″ E.; course N., 78° 47′ E.; distance, 187·1.
4. A.T.G. $1^d 11^h 35^m$; reduced dec. 22° 57′ 49″ S.; true alt. 78° 40′ 36″ N.; lat. 34° 17′ 13″ S.
5. Log 1·564403; dep. 36′·68.
6. Diff. lat. 2472′; mer. diff. lat. 2569′; diff. long. 876′; true course, S. 18° 49′ 44″ W.; compass course, S. 46° 19′ 44″ W.; distance, 2612 miles.
7. Constant, $-0^h 52^m$, $7^h 56^m$ A.M., $8^h 13^m$ P.M.
8. A.T.G. $27^d 21^h 41^m$; reduced dec. 18° 11′ 47″·9 S.; sin 9·534096; true amp. E. 20° 0′ 8″ S.; error 34° 3′ 53″ E.; dev. 21° 3′ 53″ E.
9. Daily rate, 4^s·05, gaining M.T.G. $31^d 12^h 13^m 6^s$; P. dist. 113° 2′ 42″·8; eq. time, $+3^m 33^s$·30; true alt. 6° 58′ 35″; log 8·951947; H.A. $2^h 19^m 17^s$; M.T.S. $31^d 21^h 44^m 16^s$; lat. at obs. 54° 24′ 0″ N.; long. at obs. 142° 47′ 30″ E.; noon 143° 56′ 11″ E.; azi. log 9·439831; true azi. S. 31° 57′ 42″ E.; error, 1° 47′ 18″ E.; dev. 1° 47′ 18″ E.
10. A.T.S. $4^h 26^m 9^s$ P.M.; dec. 22° 59′ 31″ S.; true azi. S. 84° 46′ W.; error, 44° 53′ W.; dev. 14° 53′ W.
11. A.T.G. $1^d 2^h 57^m$; time from noon, $31^m 40^s$; reduced dec. 22° 59′ 36″·5 S.; true alt. 33° 4′ 52″; lat. 33° 25′ 23″ N.; arc 1, 23° 11½′ S.; arc 2, 56° 37′ N.
12. First true alt. 10° 27′ 31″; dec. 23° 0′ 23″·38 S.; eq. time, $+3^m 46^s$·43;

log (a) 8·833162, (b) 8·787770; hour angle, (a) $2^h 1^m 1^s$, (b) $1^h 54^m 43^s$; long. (a) 18° 52′ W., (b) 17° 18′ 0″ W.; line $\{^{N.\ 63°\ E.}_{S.\ 63°\ W.}\}$; azi. S. 27° E. Second true alt. 10° 34′ 44″; dec. 22° 59′ 33″·97 S.; eq. time $+3^m 51^s·08$; log (c) 8·822672, (d) 8·775984; hour angle (c) $1^h 59^m 32^s$, (d) $1^h 53^m 9^s$; long. (c) 18° 20′ 15″, (d) 19° 56′ 0″; line $\{^{N.\ 62\frac{1}{2}°\ W.}_{S.\ 62\frac{1}{2}°\ E.}\}$; azi. S. $27\frac{1}{2}$° W. Ship's position, lat. 52° 4′ 30″ N., long. 18° 36′ 30″ W.

13. M.P. Aldebaran, $9^h 36^m$ P.M.; by Norie's Tables, $9^h 34^m$.
14. R.A.M. $5^h 43^m 50^s·58$; Capella, S. $0^h 34^m 37^s$ W.; β Orion, S. $0^h 34^m 10^s$ W.; γ Orionis, S. $0^h 24^m 8^s$ W.; β Tauri, S. $0^h 23^m 57^s$ W.; α Leporis, S. $0^h 15^m 35^s$ W.; κ Orionis, S. $0^h 0^m 57^s$ W.; α Orionis, S. $0^h 5^m 51^s$ E.; β Auriga, S. $0^h 8^m 16^s$ E.; β Canis Major, S. $0^h 34^m 24^s$ E.
15. Alt. of Aldebaran, 56° 12′ 19″ S. of observer.
16. True alt. 70° 35′ 0″ S.; dec. 28° 32′ 13″ N.; lat. 47° 57′ 13″ N.
17. Daily rate, 4ˢ·4; M.T.G. $22^d 22^h 13^m 14^s$; true alt. 16° 42′ 20″; P.D. 77° 32′ 34″; log 9·604929; H.A. = $5^h 15^m 6^s$ W.; R.A.M. $15^h 18^m 8^s·13$; R.A.M.S. $20^h 10^m 7^s·53$; M.T.S. $22^d 19^h 8^m 0^s·60$; long. 46° 18′ 21″ W.
18. R.A.M. $1^h 42^m 45^s$; H.A. = $2^h 47^m 25^s$ E.; dec. 16° 18′ 27″ N.; true azi. N. 111° 37′ E.; error 12° 8′ W.; dev. 15° 22′ E.
19. R.A.M. $4^h 15^m 06^s·89$; true alt. 49° 17′ 8″; first cor. −0° 53′ 7″·8; second cor. +24″; third cor. 39″·5; lat. 48° 24′ 3″·7 N.
20. M.T.G. $4^d 19^h 54^m·2$; semidiameter, 15′ 34″·42; H. px. 56′ 29″; dec. 11° 32′ 30″·1 S.; true alt. 31° 13′ 56″; lat. 47° 13′ 34″ N.
21. (a) Mag. course, S. $18\frac{1}{2}$° W.; compass course, S. $37\frac{1}{2}$° W.; var. $19\frac{1}{2}$° W.; dev. 19° W.; distance, 36 miles.
   (b) Lat. 54° 22′ N.; long. 4° 51′ W.
   (c) Lat. 54° $5\frac{3}{4}$′ N.; long. 4° $50\frac{1}{2}$′ W.; distance, $2\frac{1}{2}$ miles from Bradda Head.
   (d) Mag. course, S. 3° E.; compass course, S. 7° W.; dev. 10° W.; distance made good towards Chicken's Rocks, 34 miles.
22. Interval, $1^h 8^m$; table B, −2 ft. 3 in.; cor. for Galway, 9 ft. 7 in.; cor. for Limerick, −12 ft. 2 in.
23. *Napier's Curve.*—Cor. mag. bearing: N. $3\frac{3}{4}$° E.
   Deviations: $10\frac{3}{4}$° E. $30\frac{1}{4}$° E. $24\frac{3}{4}$° E. $3\frac{3}{4}$° E. $6\frac{1}{4}$° W. 21° W. $26\frac{1}{4}$° W. $16\frac{1}{4}$° W.
   Compass courses: N. 59° E. N. 86° W. S. 37° E. N. 21° W.
   Magnetic courses: N. 61° W. N. 52° E. S. 9° W. S. 53° E.
   Dev. on ship's head, 25° W.; mag. bearings, S. 54° W., N. 25° W.
24. Moon's semidiameter, 15′ 15″·13; H.P. 55′ 3″·11; R.A. $3^h 37^m 5^s·23$; dec. 22° 58′ 5″·5 N. Regulus, R.A. $10^h 3^m 2^s·13$; dec. 12° 27′ 26″ N.; R.A.M.S. $20^h 0^m 32^s·10$. Star's true alt. 24° 15′ 25″; appar. alt. 24° 17′ 30″. Moon's true alt. 57° 35′ 9″; appar. 58° 7′ 53″. Star's H.A. $4^h 31^m 40^s$ E.; R.A.M. $5^h 31^m 22^s$; Moon's H.A. $1^h 54^m 17^s$; Z.D. 22° 41′ 54″; log 4·894050; nat. num. 78352; arc. B, 14° 26′ 35″; sin 9·396928; half dist. 45° 30′ 2″; sin 9·853246; N.A. time, $20^d 11^h 50^m 0^s$; corr. −1^s; chron: fast 31^s; long. 34° 47′ 15″ W.
25. Middle time, $10^d 0^h 28^m 30^s$; dec. 21° 56′ 43″·9 S.; first alt. 11° 17′ 44″; second alt. 15° 51′ 31″; angle, 7 pts.; cor. for first alt. +7′ 18″; arc 1, 25° 55′ 40″; arc 2, 5° 4′ 40″; arc 3, 74° 49′ 12″; arc 4, 114° 33′ 14″; arc 5, 39° 44′ 2″; approx. lat. 49° 59′ 47″ N.; second cor. +13″; lat. 50° 0′ 0″ N.
26. M.T.G. January $20^d 0^h 23^m 14^s·55$; interval, $4^h 20^m$; dec. 20° 6′ 1″·7 S.; eq. time, +$11^m 16^s·55$; two-day change, $1565^s·4$; Part I. −$11^s·81$; Part II. −$2^s·71$; eq. time, −$14^s·52$; fast on A.T.S. $32^m 59^s·48$; fast on M.T.S. $21^m 42^s·93$; fast on M.T.G. $9^m 44^s·93$.
27. Dec. 11° 48′ 2″ S.; eq. time, +$14^m 11^s$; true alt. 36° 16′ 57″; H.A. $3^h 38^m 39^s$; error fast, $15^m 40^s$.

APPENDIX. 363

28. Star α Arietis: R.A. 2^h 1^m $30^s\cdot 31$; dec. 22° 59' 18"·1 N.; R.A.M.S. 18^h 45^m $05^s\cdot 9$; alt. 71° 38' 32"; first H.A. $8\cdot 258241 = 1^h$ 1^m $53^s\cdot 5$ E.; second H.A. $8\cdot 248888 = 1^h$ 1^m $13^s\cdot 5$ E.; first M.T.S. 1^d 6^h 14^m 31^s; second M.T.S. 1^d 6^h 15^m 11^s; first long. 35° 52' 15" W.; second long. 35° 42' 15" W. Star Marcab: R.A. 22^h 59^m 44^s; dec. 14° 39' 52"·2 N.; alt. 55° 53' 22"; first H.A. $8\cdot 830619 = 2^h$ 0^m 39' W.; second H.A. $8\cdot 827275 = 2^h$ 0^m 11' W.; first M.T.S. $= 1^d$ 6^h 15^m 17^s; second M.T.S. 1^d 6^h 14^m 49^s; first long. 35° 40' 45" W.; second long. 35° 47' 45" W. Position, lat. 35° 36' N., long. 35° 45' W.

First line: Star α Arietis $\begin{Bmatrix} N. & 36° & E. \\ S. & 36° & W. \end{Bmatrix}$; azi. S. 54° E.

Second line: Star Marcab $\begin{Bmatrix} N. & 26\frac{1}{2}° & W. \\ S. & 26\frac{1}{2}° & E. \end{Bmatrix}$; azi. S. $63\frac{1}{2}°$ W.

29. First course, N. 78°·03 W.; distance, 3344 miles; vertex lat. 51° 0 N.; long. 21° 16' W. at 5° W. of Wolf Rock; lat. 50° 33' N., 10° W., 50° 53' N., 15° W., 50° $59\frac{1}{2}'$ N., 20° W., 50° 54' N.; long. 61° W.; lat. 43° $24\frac{1}{2}'$ W.

30. (a) 49° 44' N.
(b) S. 75° 20' W.; 3^h 25^m.
(c) $a = 34° 57' 55''$, $b = 109° 52' 5''$, $c = 106° 10' 16''$.
(d) 36° 32' N.

ANSWERS TO PAPER II.

1, 2. *Logs*: $212\cdot 8$ log $= 2\cdot 327982$; $\cdot 147$ log $= 9\cdot 167437$.
3. *Day's work*: S. 17° E., 15'; N. 68° E., 54'; N. 52° E., 50'; S. 83° E., 42'; S. 56° W., 31'; S. 1° E., 30'; S. 37° E., 42'; N. 14° W., 54'; d. lat. 3'·2 N.; dep. 122'·6 E.; course, N. 88° E.; dist. 123'; lat. in, 25° 12' S.; long. in, 45° $4\frac{1}{2}'$ W.
4. *Mer. alt.*: Dec. 11° 24' 47" N.; true alt. 64° 50' 46"; lat. 13° 44' 27" S.
5. *Mercator*: Compass course, N. 64° 7' W.; distance, 6150 miles; true course, N. 23° 27' W.
6. *Parallel*: 41° 40' W.
7. *Tides*: No A.M.; 0^h 14^m P.M.; and 11^h 49^m A.M.; no P.M.
8. *Amplitude*: A.T.G. 15^d 10^h 1^m 40^s; dec. 21° 1' 56" S.; true amp. W. 24° $13\frac{1}{2}$ S.; error, 15° $47\frac{1}{2}'$ W.; dev. 15° $47\frac{1}{2}'$ W.
9. *Chron. azimuth*: Rate, 2·04 gaining; M.T.G. 31^d 20^h 20^m 15^s; true alt. 32° 16' 27"; dec. 23° 0' 59" S.; eq. time, 3^m 43^s +; H. angle, 3^h 37^m 28"; true azi. S. 62° 13' E.; long. 0°; error, 5° 58' W.; dev. 4° 2' E.
10. *Time azimuth*: A.T.S. 7^h 55^m 34^s A.M.; dec. 14° 19' 58" S.; cor. for azi. 56'; true azi. N. 121° 30' E.; error, 61° 19' W.; dev. 21° 19' W.
11. *Chart work*: Compass course, N. $31\frac{1}{2}°$ E.; distance, 93 miles; dev. 26° E.; lat. 52° $19\frac{1}{2}'$ N., long. 5° $59\frac{1}{2}'$ W.; lat. 53° $15\frac{1}{2}'$ N., long. 4° 50' W.; distance, 10 miles.
12. *Meridian pass.*: March 23^d 10^h 47^m 52^s A.M.
13. *Stars near meridian*: R.A.M. 15^h 40^m 25^s; *E. meridian*, Antares; *W. meridian*, Arcturus, η Ursæ Majoris; *appar. alt.* 77° 2'.
14. *Meridian altitude of star*: True alt. 77° 2' 30"; dec. 19° 42' 15" N.; lat. 32° 39' 45" N.
15. *Star chron.*: Rate, ·9^s losing; M.T.G. 1^d 2^h 16^m 8^s; true alt. 30° 27' 33"; dec. 10° 38' 22" S.; R.A.M.S. 6^h 37^m 40^s; W.H.A. 3^h 48^m 25^s; long. 123° 38' E.
16. *Star azimuth*: Dec. 19° 42' 15" N.; R.A. 14^h 11^m 6^s; W.H.A. 5^h 45^m 1^s; cor. to azi. —35'; true azi. N. 75° 4' W.; error, 17° 45' E.; dev. 19° 15' E.
17. *Ex-meridian*: Dec. 57° 44' 23" S.; R.A. 1^h 34^m 1^s; R.A.M.S. 10^h 13^m 3^s; E.H.A. 15^m 33^s; true alt. 57° 24' 25"; nat. No. 1111; lat. 25° 16' S.; arc 1, 57° 48' 33" S.; arc 2, 32° 31' 55" N.
18. *Sumner.—First obs.*: Dec. 19° 35' 37" N.; true alt. 21° 4' 8"; H.A.'s,

364   APPENDIX.

$5^h\ 20^m\ 53^s$ and $5^h\ 21^m\ 12^s$; long. (a), 166° 31′ 45″ E. ; long. (b), 166° 27′ E. ;
first line, N. 6° W. or S. 6° E. ; eq. time, $+6^m\ 17^s$.
   Second obs. : Dec. 19° 33′ 12″ N. ; true alt. 57° 12′ 23″ ; H.A.'s, $1^h\ 3^m\ 10^s$
and $0^h\ 57^m\ 1^s$; long. (c), 163° 47′ 30″ E. ; long. (d), 165° 19′ 45″ E. ; second
azi. S. 27° E. ; lat. 50° 26½′ N. ; long. 165° 41′ E. ; eq. time, $+6^m\ 17^s$.
   19. *Pole Star*: R.A.M. $19^h\ 28^m\ 3^s$; true alt. 29° 14′ 9″ ; first cor. - 1′ 56″ ;
second cor. +0′ 27″; third cor. +1′ 23″ ; lat. 29° 13′ 3″ N.
   20. *Moon's meridian altitude*: M.T.G. $18^d\ 11^h\ 57^m\cdot 2$; S.D. 16′ 48″ ; H.P.
60′ 48″ ; dec. 0° 0′ 10″ N. ; appar. alt. 39° 15′ 46″ ; cor. +45′ 55″ ; true alt.
40° 1′ 41″ ; lat. 49° 58′ 9″ S.
   21. *Curve:* Cor. mag. bearings, S. 65° W. ; dev. 22° E., 15° E., 0, 15° W.,
25° W., 18° W., 0, 20° E. ; S. 40° E., N., N. 65° W., S. 32° W. ; N. 40° W.,
N. 12° E., N. 64° E., S. 64° E. ; bearings S. 51° E., N. 36° E.
   Current course: N. 32° E. ; dev. 26° E. ; distance made good, 51½ miles.
   22. *Tidal sounding:* Table B. – 4 ft. 4 in. ; correction, 7 ft. 8 in. to
subtract.
   23. *Lunar.—Jupiter's elements:* Dec. 13° 31′ 11″ S. ; R.A. $14^h\ 32^m\ 11^s$;
R.A.M.S. $22^h\ 58^m\ 40^s$; A.A. 44° 13′ 25″ ; true alt. 44° 12′ 29″ ; W.H.A.
$36^m\ 43^s$.
   Moon's elements: Dec. 22° 16′ 56″ S. ; R.A. $18^h\ 54^m\ 12^s$; S.D. 16′ 22″ ;
H.P. 59′ 40″ ; A.A. 13° 0′ 4″ ; true alt. 13° 54′ 9″ ; E.H.A. $3^h\ 45^m\ 17^s$.
   A. dist. 63° 7′ 8″ ; angle B. 44° 42′ 50″ ; true dist. 62° 30′ 4″ ; second cor.
$+3\cdot 5^s$; M.T.G. $6^d\ 16^h\ 10^m\ 14^s\cdot 5$; error, ·5° fast ; long. 0° 0′ 4″ E.
   24. *Ivory:* H.E.T. $1^h\ 12^m\ 16^s$; dec. 8° 36′ 21″ ; cor. to alt. +12′ 0″ ; first
true alt. 49° 24′ 54″ ; second true alt. 46° 52′ 0″ ; arc. 1, 17° 51′ 28″ ; arc 2,
2° 46′ 25″ ; arc 3, 38° 26′ 41″ ; arc 4, 80° 57′ 18″ ; arc 5, 42° 30′ 37″ ; lat.
47° 25′ N.
   25. *Simultaneous altitudes—Spica's elements:* Dec. 10° 38′ 18″ S. ; R.A.
$13^h\ 19^m\ 55^s$; R.A.M.S. $22^h\ 14^m\ 59^s$; true alt. 25° 54′ 4″ ; first H.A. $3^h\ 49^m\ 48^s$;
second H.A. $3^h\ 48^m\ 54^s$; first long. 45° 27′ 30″ W. ; second long. 45° 14′ W. ;
bearing, S. 67° E.
   Capella's elements: Dec. 45° 53′ 54″ N. ; R.A. $5^h\ 9^m\ 16^s$; true alt. 31° 3′ 16″;
first H.A. $4^h\ 20^m\ 25^s$; second H.A. $4^h\ 22^m\ 22^s$; third long. 45° 34′ W. ; fourth
long. 45° 4′ 45″ W. ; bearing N. 46° W. ; lat. 19° 42′ N. ; long. 45° 22′ W.
   26. *Equal altitudes:* M.T.G. $21^d\ 20^h\ 17^m\ 40^s$; dec. 11° 49′ 55″ N. ; eq. time,
$2^m\ 48^s\cdot 3\ +$; two days' change, $2420″\cdot 7$; Part I. $+7^s\cdot 89$; Part II. $-1^s\cdot 69$;
slow A.T.S. $3^h\ 43^m\ 42^s\cdot 8$; fast A.T.G. $1^m\ 25^s\cdot 2$; slow M.T.G. $1^m\ 23^s\cdot 1$.
   27. *Great circle:* Initial course, N. 73° 36′ E. ; distance, 4592 miles ; lat.
ver. 37° 20½′ S. ; long. ver. 2° 4′ W.
   28. *Mercator's chart:* Set, N. 10° W. ; drift, 10′.
   29. *Trigonometrical problems:* Sun's alt. 37° 4½′.
   30. A.P. = 1961 yards.
   31. A = 75° 24′ 48″ ; c = 93° 32′ 9″ ; b = 103° 47′ 9″.
   32. H.A. $4^h\ 56^m\ 44^s$; alt. 21° 56′ 57″.
   33. Course, N. 53° 3′ 20″ E. ; distance, 11·35 miles.

ANSWERS TO PAPER III.

1. Log 3·676123, product ·000000474376.
2. Log 7·819667, quotient ·00660186363.
3. N. 18° E., 15′; S. 63° W., 28′; N. 85° W., 30′; S. 88° W., 36′;
S. 49° W., 40′; S. 25° W., 42′; S. 68° W., 32′; N. 6′ E., 18′; diff. lat.
55′·5 S. ; dep. 161′·9 W. ; lat. in, 59° 49′ 30″ N. ; long. in, 0° 43′ 36″ W. ;
course, S. 71° W.; distance, 171 miles.
4. A.T.G. Jan. $21^d\ 3^h\ 0^m\ 0^s$; cor. dec. 19° 51′ 13″·6 S.; true alt. 34° 43′ 23″ S.;
lat. 35° 25′ 23″ N.
5. Log 2·303041 ; dep. 201 miles.
6. Diff. lat. 140′; M.D.L. 213′; diff. long. 2945′; true course, S.85° 51′48″ W.;
compass course, N. 89° 37′ 34″ W. ; dist. 1941′.

APPENDIX. 365

7. H.W. Black Ball harbour, $10^h 45^m$ A.M., $11^h 23^m$ P.M.
8. A.T.G. Feb. $1^d 0^h 4^m 37^s$; dec. $17° 4' 15''·7$ S.; true amp. E. $29° 13' 12''$ S.; compass error, $7° 20' 33''$ W.; dev. $13° 39' 27''$ E.
9. Daily rate, $·75^s$ losing M.T.G. Jan. $31^d 21^h 3^m 27^s$; P.D. $72° 53' 25''$; eq. time, $+13^m 47^s·48$; true alt. $43° 53' 28''$; lat. sight, $36° 28' 0''$ S.; log $9·911516 =$ H.A. $3^h 10^m 20^s$ E.; M.T.S. $31^d 21^h 3^m 27^s·5$; long. obs. $0° 0' 0''$; at noon, $0° 12' 24''$ W.; azi., log $19·600227$; true azi. N. $78° 15' 58''$ E.; compass error, $14° 32' 47''$ W.; dev. $10° 57' 13''$ E.
10. A.T.S. $6^d 40^m 29^s$ A.M.; dec. $0° 0' 38''$ N.; eq. time, $-7^m 30^s·95$; true azi. S. $97° 38'$ E.; compass error, $21° 41' 45''$ W.; dev. $36° 41' 45''$ W.
11. M.T.G. Jan. $10^d 12^h 12^m$; R.A. $7^h 28^m 12^s$; dec. $32° 6' 32''$ N.; R.A.M.S. $19^h 21^m 10^s·07$; R.A.M. $12^h 7^m 2^s$; H.A. $7^h 2^s$; number, 238; true alt. $68° 40' 16$; M.Z. dist. $21° 17' 34''$ N.; lat. $53° 24' 6''$ N.; arc 1, $32° 6' 45''$ N.; arc 2, $21° 16' 43''$ N.
12. First true alt. $13° 5' 0''$; dec. $16° 46' 29''·3$ S.; eq. time, $+13^m 56^s·17$; (a) log $9·100472$; H.A. $= 2^h 46^m 21^s$; (b) log $9·081414$; H.A. $= 2^h 42^m 35^s$; (a) long. $53° 22' 45''$ W.; (b) long. $52° 26' 15''$ W.; line $\begin{Bmatrix} \text{N. } 50° \text{ E.} \\ \text{S. } 50° \text{ W.} \end{Bmatrix}$; azi. S. $40°$ E. Second true alt. $19° 12' 4''$; dec. $16° 43' 27''·4$; eq. time, $+13^m 57^s·39$; (a) log $8·619490$; H.A. $= 1^h 34^m 12^s$; (d) log $8·547737$; H.A. $= 1^h 26^m 38^s$; (c) long. $50° 42' 0''$ W.; (d) long. $52° 35' 30''$ W.; line $\begin{Bmatrix} \text{N. } 68° \text{ W.} \\ \text{S. } 68° \text{ E.} \end{Bmatrix}$; azi. S. $22°$ W.; position, lat. $51° 17'$ N., long. $51° 45'$ W.
13. Mer. pass. Capella, $7^h 43^m 1^s$ P.M.
14. R.A.M. $16^h 37^m 6^s·77$; δ Scorpii, S. $0^h 42^m 45^s$ W.; α Ophiuchi, S. $0^h 28^m 4^s$ W.; η Draconis, N. $14^m 30^s$ W.; α Scorpii, S. $0^h 13^m 54^s$ W.; β Herculis, N. $0^h 11^m 14^s$ W.; β Arae, S. $0^h 39^m 47^s$ E.
15. Alt. of Capella, $77° 26' 19''$ N. of observer.
16. True alt. $54° 13' 46''$; dec. $14° 37' 25''$ N.; lat. $50° 23' 39''$ N.
17. Daily rate, $·05^s$ gaining; M.T.G. Feb. $25^d 9^h 48^m 21^s$; true alt. $57° 9' 49''$; P.D. $27° 42' 31''$; log $9·405012$; H.A. $= 4^h 2^m 10^s$ E.; R.A.M.S. $22^h 22^m 8^s·05$; R.A.M. $6^h 55^m 24^s·55$; M.T.S. $25^d 8^h 33^m 16^s·50$; long. $18° 46' 7''·5$ W.
18. R.A.M. $13^h 53^m 27^s·57$; H.A. $= 3^h 50^m 25^s·07$; dec. $12° 27' 24''$ N.; true azi. N. $100° 13'$ W.; compass error, $15°$ W.; dev. $11° 17'$ E.
19. R.A.M. $14^h 38^m 11^s·75$; true alt. $42° 14' 30''$; first cor. $1° 8' 59''·8$; second, $+4''·5$; third, $+1' 24''·5$; lat. $43° 24' 58''·8$ N.
20. M.T.G. February $13^d 22^h 50^m$; semidiameter, $16' 8''·4$; H.P. $58' 9''·83$; dec. $12° 39' 14''·4$ N.; true alt. $76° 12' 51''$; lat. $1° 7' 55''$ S.
21. (a) Magnetic course: N. $80°$ E.; compass course, N. $87\frac{1}{2}°$ E.; dev. $7\frac{1}{2}°$ W.; var. $19°$ W.; distance, $64\frac{1}{2}$ miles.
(b) Lat. $54° 25\frac{1}{4}'$ N.; long. $5° 4'$ W.
(c) Lat. $54° 33\frac{3}{4}'$ N.; long. $4° 38'$ W.; distance, 9 miles from Mull of Galloway.
(d) Mag. N. $67°$ E.; compass, N. $79°$ E.; dev. $12°$ W.; distance, $34\frac{1}{2}$ miles.
22. Interval, $4^h 18^m$; Table B, -2 ft. 0 in.; cor. Belfast, -2 ft. 9 in.; cor. for Ballycastle, -10 in.
23. *Napier's curve.*—Correct magnetic bearing: N. $4° 45'$ E. Deviations: $36\frac{1}{4}$ W. $25\frac{1}{4}$ W. $6\frac{1}{4}$ W. $16\frac{3}{4}$ E. $37\frac{3}{4}$ E. $31\frac{3}{4}$ E. $4\frac{3}{4}$ E. $23\frac{1}{4}$ W. Compass courses: N. $43°$ E. S. $81°$ E. S. by E. N. $7°$ E. Magnetic courses: N. $30°$ W. N. $50°$ E. S. $49°$ E. S. $81°$ W. Dev. on ship's head: $1°$ W. Magnetic bearings: S. $68\frac{1}{2}°$ E. N. $71°$ W.
24. Moon's semidiameter, $16' 1''$; H.P. $58' 8''·84$; R.A. $15^h 24^m 43^s·05$; dec. $22° 22' 35''·7$ S. Sun's R.A. $21^h 11^m 39^s·98$; dec. $16° 12' 44''·4$ S.; R.A.M.S. $20^h 57^m 32^s·3$; semidiameter, $16' 14''·90$. Sun's appar. alt. $29° 47' 45''$; true alt. $29° 46' 14''$; log $9·229877$; H.A. $= 3^h 14^m 40^s$ E.; R.A.M. $17^h 57^m 0^s$. Moon's H.A. $2^h 32^m 16^s$; log $5·265467$; number 184276. Moon's true alt. $33° 17' 40''$; appar. alt. $32° 30' 6''$; arc B,

366   APPENDIX.

33° 31' 3"; sin 9·742090; half distance, 40° 29' 27"; sin 9·812438; interval, 1ʰ 48ᵐ 30ˢ; second cor. +5ˢ; chron., fast 43ˢ; M.T.S. 3ᵈ 20ʰ 59ᵐ 28ˢ; long. 27° 9' 15" W.
    25. H.E.T. 1ʰ 25ᵐ 0ˢ; middle time, 25ᵈ 19ʰ 51ᵐ 0ˢ; dec. 9° 7' 3" S.; first alt. 33° 56' 22"; second alt. 62° 34' 25"; angle, 57° 30'; distance, 34'; cor. for first alt. +18' 15"; arc 1, 20° 58' 8"; arc 2, 27° 23' 30"; arc 3, 29° 18' 20"; arc 4, 80° 13' 47"; arc 5, 50° 55' 27"; approx. lat. 34° 2' 0"; cor. 119"·6; lat. 34° 0' 0" S.
    26. M.T.G. March 29ᵈ 8ʰ 14ᵐ 22·79ˢ; interval, 6ʰ 41ᵐ 3ˢ; dec. 3° 32' 25"·3 N.; eq. time, +4ᵐ 45ˢ·63; two-day change, 2801·6; Part I., 1·1190 = −13"·15; Part II., 9·8277 = +·67 : G.M.T. 29ᵈ 0ʰ 4ᵐ 32ˢ·02; fast on A.T.S. 4ᵐ 32ˢ·02; slow on M.T.S. 13ˢ·61; slow on M.T.G. 8ʰ 9ᵐ 50ˢ·77.
    27. Dec. 12° 37' 27" S.; eq. time, +14ᵐ 20ˢ·7; true alt. 36° 33' 51"; E.H.A. 3ʰ 39ᵐ 55ˢ; fast M.T.G. 2ᵐ 17ˢ·3; fast A.T.S. 5ʰ 22ᵐ 15ˢ.
    28. Star Spica, R.A. = 13ʰ 19ᵐ 54ˢ·38; dec. 10° 38' 16" S.; R.A.M.S. 21ʰ 29ᵐ 37ˢ·99; alt. 27° 26' 35"; first H.A. 1ʰ 42ᵐ 57ˢ W.; second H.A. 1ʰ 39ᵐ 15ˢ W.; first long. 129° 53' W.; second long. 130° 48' 30" W.; lines, $\begin{Bmatrix} N.\ 61°\ W. \\ S.\ 61°\ E. \end{Bmatrix}$; azi. S. 29° W.
    Star Antares, R.A. = 16ʰ 23ᵐ 14ˢ·10; dec. 26° 12' 31" S.; alt. 13° 40' 29"; first H.A. 1ʰ 26ᵐ 38ˢ E.; second H.A. 1ʰ 20ᵐ 57ˢ E.; first long. 131° 26' 45" W.; second long. 130° 1' 30"; line, $\begin{Bmatrix} N.\ 70°\ E. \\ S.\ 70°\ W. \end{Bmatrix}$; azi. S. 20° E. Lat. 47° 47' N., long. 130° 26' W.
    29. First course = S. 36° 48' E.; lat. course, N. 74° 35' E.; distance, 6971 miles; vertex in, lat. 54° 6' 30" S.; long. 150° 0' E.; first intersection, 60° E., lat. 0° 0'; 70° E., lat. 13° 30' S.; 80° E., lat. 25° 18' S.; 90° E., lat. 34° 38' S.; 160° E., lat. 53° 42' S.
    30. (a) 94·2 ft.
    (b) P 4·6 miles per hour; Q 6·7 miles per hour.
    (c) P = 96° 25' 20"; Q = 166° 42'; r = 61° 31' 28".
    (d) Dec. = 12° 7' 58" N.; H. angle = 7ʰ 2ᵐ 30ˢ.

                    ANSWERS TO PAPER IV.

    1. *Logs*: 7·382 log = 1·868158; ·1149 log = 9·060194.
    2. *Days' work*: S. 87° W., 20'; S. 53° W., 61'; S. 66° W., 51'; S. 75° W., 37'; S. 72° W., 59'; S. 69° W., 50'; N. 89° W., 40'; S. 21° E., 42; D. lat. 142'·6 S.; dep. 278'·7 W.; course, S. 63° W.; distance, 313'; lat. in, 34° 40' N.; long. in, 1° 19' W.
    3. *Mer. alt.*: Dec. 23° 1' 8" S.; true alt. 56° 45' 46"; lat. 10° 13' 6" N.
    4. *Mercator*: True course, N. 74° 36½' E.; compass course, N. 59° 46½' E.; distance, 2438 miles.
    5. *Parallel*: 68·73 miles.
    6. *Tides*: 0ʰ 32ᵐ A.M., 1ʰ 14ᵐ P.M.; and 11ʰ 53ᵐ A.M., no P.M.
    7. *Amplitude*: Dec. 2° 56' 55" N.; true amp. E. 2° 57½' N.; error, 17° 1¼' W.; dev. 0° 58½' E.
    8. *Chron. azimuth*: Rate, 1ˢ·66 losing; M.T.G. 21ᵈ 9ʰ 48ᵐ 3ˢ; true alt. 45° 58' 4"; st. lat. 40° 30' S.; dec. 10° 22' 37" S.; H.A. 2ʰ 25ᵐ 38ˢ; eq. time, 13ᵐ 46ˢ +; st. long. 179° 58' 45" W.; noon long. 179° 53' 46" E.; true azi. N. 57° 8' E.; error, 10° 22' W.; dev. 24° 22' W.
    9. *Time azimuth*: A.T.S. 5ʰ 24ᵐ 36ˢ P.M.; dec. 0° 9' 37" N.; cor. for azi. −7'; true azi. N. 93° 34' W.; error, 16° 7' E.; dev. 4° 7' E.
    10, 11, 12. *Chart work*: Compass course, N. 11½° E.; distance, 67 miles; dev. 33° W.; lat. 53° 14½' N., long. 5° 29½' W.; lat. 53° 30½' N., long. 5° 51' W.; distance, 10½ miles.
    13. *Meridian pass. star*: 5ʰ 41ᵐ 21ˢ P.M.

APPENDIX. 367

14, *Stars near meridian:* R.A.M. $17^h 56^m 45^s$; *E. mer.*, Altair, Vega; *W. mer.*, Antares; approx. alt. Vega, $27° 19'$.
15. *Star's mer. alt.:* True alt. $27° 20' 16''$; dec. $38° 41' 6'' $ N.; lat. $23° 58' 38''$ S.
16. *Star azimuth:* Dec. $8° 36'$ N.; R.A. $19^h 45^m 53^s$; E.H.A. $3^h 34^m 50^s$; cor. for azi. $+36'$; true azi. S. $115° 8'$ E.; error, $2° 38'$ W.; dev. $11° 8'$ W.
17. *Star chron.:* Rate, $2^s\cdot5$ losing; M.T.G. $22^d 5^h 36^m 0^s$; R.A. $16^h 23^m 15^s$; dec. $26° 12' 34''$ S.; R.A.M.S. $0^h 0^m 0^s$; true alt. $19° 38' 21''$; E.H.A. $4^h 7^m 46^s$; long. $99° 52' 15''$ E.
18. *Ex-meridian:* Time from noon, $33^m 20^s$; true alt. $64° 51' 36''$; dec. $11° 18' 29''$ S.; arc 1, $11° 25' 35''$ S.; arc 2, $23° 51' 28''$ S.; lat. $35° 17' S$.
19. *Sumner.*—*First observation:* True alt. $29° 31' 0''$; dec. $10° 30' 8''$ S.; eq. time, $15^m 12^s$ —; W.H.A.'s, $0^h 49^m 14^s$ and $0^h 28^m 4^s$; long. (a), $126° 27' 30''$ W.; long. (b), $131° 45'$ W.; line position, N. $79°$ W. or S. $79°$ E.; azi. S. $11°$ W.
Second observation: True alt. $9° 3' 18''$; dec. $10° 33' 1''$ S.; eq. time, $15^m 14^s$ —; W.H.A's. $4^h 10^m 54^s$ and $4^h 8^m 43^s$; long. (c), $124° 16' 30''$; long. (d), $124° 49' 15''$ W.; lat. $49° 14'$ N., long. $124° 27'$ W.
20. *Pole Star:* True alt. $28° 12' 31''$; R.A.M. $3^h 1^m 6^s$; first cor. $-1° 6' 17''$; second cor. $+4''$; third cor. $+12''$; lat. $27° 5' 30''$ N.
21. *Meridian altitude of Moon:* M.T.G. $5^d 22^h 25^m\cdot2$; S.D. $14' 45''$; H.P. $53' 46''$; dec. $13° 27' 9''$ N.; A.A. $8° 19' 27''$; cor. $+56' 51''$; true alt. $9° 16' 18''$; lat. $85° 39' 7''$ N.
22. *Curve:* Cor. mag. bearing, N. $6°$ W.; devs. $26°$ E., $14°$ E., $11°$ W., $23°$ W., $26°$ W., $15°$ W., $9°$ E., $25°$ E.; courses to steer, S. $27\tfrac{1}{2}°$ E., N. $10°$ E., N. $84°$ W., S. $75°$ W.; cor. mag. N. $74°$ E., S. $61°$ E., S., N. $60°$ W.; bearings, S. $30°$ W., S. $54°$ E.
23. *Current:* N. $9° 30'$ W.; dev. $34°$ W.; dist. made good, 51 miles.
Sounding: Table B, *nil*; cor. 10 ft. 6 in. to subtract.
24. *Lunar*—*Moon's elements:* S.D. $15' 34''$; H.P. $56' 23''$; R.A. $1^h 59^m 28^s$; dec. $17° 16' 23''$ N.; R.A.M.S. $19^h 53^m 6^s$; A.A. $45° 2' 31''$; cor. $+38' 54''$; true alt. $45° 41' 25''$; W.H.A. $3^h 8^m 4^s$.
Star's elements: R.A. $4^h 30^m 10^s$; dec. $16° 18' 27''$ N.; A.A. $78° 1' 16''$; true alt. $78° 1' 4''$; W.H.A. $0^h 37^m 22^s$.
App. dist. $36° 39' 16''$; angle B. $20° 52' 11''$; true dist. $36° 1' 18''$; second cor. -4^s; M.T.G. $18^d 14^h 35^m 37^s\cdot5$; error, slow $5^m 17^s\cdot5$; long. $80° 17' 53''$ W.
25. *Ivory:* H.E.T. $3^h 15^m 6^s$; mid. dec. $22° 24' 1''$ S.; first true alt. $50° 49' 53''$; run, $+19' 12''$; second true alt. $39° 45' 50''$; arc 1, $44° 3' 25''$; arc 2, $5° 35' 49''$; arc 3, $8° 27' 2''$; arc 4, $57° 58' 36''$; arc 5, $49° 31' 34''$; lat. cor. $+11''$; true lat. $40° 14' 39''$ S.
26. *Simultaneous altitudes.—Sun's elements:* dec. $16° 22' 41''$ S.; eq. time, $14^m 5^s +$; true alt. $16° 13' 50''$; H.A.'s, $2^h 46^m 38^s$ and $2^h 43^m 15^s$; first long. $179° 46' 45''$ E.; second long. $179° 22' 30''$ W.; bearing, S. $41°$ E.
Moon's elements: S.D. $15' 47''$; H.P. $57' 35''$; R.A. $14^h 53^m 9^s$; dec. $20° 47' 34''$ S.; R.A.M.S. $20^h 55^m 20^s$; true alt. $7° 4' 2''$; W.H.A.'s, $3^h 31^m 11^s$ and $3^h 28^m 33^s$; third long. $179° 50'$ W.; fourth long. $179° 30' 30''$ E.; bearing, S. $47°$ W.; lat. $47° 23'$ N.; long. $180°$.
27. *Equal altitudes:* Error M.T.S. fast $8^h 14^m 50^s\cdot9$; fast A.T.G. $1^m 12^s\cdot6$.
28. *Great circle:* First course, S. $66° 26\tfrac{1}{2}'$ E.; distance, 3708 miles; vertex lat. $41° 1'$ S., long. $20° 51'$ W.
29. *Chart:* Lat. $39° 33'$ S.; long. $120° 48'$ E.
30. *Trigonometrical problems:* BP = 241·1 ft.; BAQ = $29° 5\tfrac{1}{4}'$.
31. XZ = 1684 ft.
32. Distance, 4032 miles; lat. vertex, $43° 31' 22''$ N.
33. Height above H.W. 151·3 ft.

368   APPENDIX.

ANSWERS TO PAPER V.

1. Log 1·797667, product 62·7577.
2. Log 7·875061, quotient ·0075.
3. S. 48° E., 20'; N. 90° E., 40'; S. 82° E., 42'; N. 64° E., 44';
N. 88° E., 46'; S. 80° E., 48'; N. 90° E., 46'; N. 83° E., 54'; D. lat.
0' 0'; dep. 328'·9 E.; course E. dist. 329 miles; lat. in, 62° 30' N.; long.
in, 169° 13' 42" W.
4. A.T.G. Feb. 20^d 10^h 48^m; dec. 11° 2' 36"·4 S.; true alt. 78° 42' 4" N.;
lat. 22° 20' 32" S.
5. Log 2·746166; dep. 557·4.
6. D. lat. 1216; M.D. lat. 1703; D. long. 5032; true course, S. 71° 18' W.;
distance, 3793; comp. course, N. 89° 12' W.
7. Con. +1^h 26^m; H.W. Feb. 9^d 10^h 26^m A.M., 10^h 48^m P.M.
8. A.T.G. Feb. 28^d 19^h 57^m 9^s; dec. 7° 36' 48"·4 S.; sin 9·332830; true
amp. W. 12° 25' 37" S.; compass error, 20° 51' 52" W.; dev. 29° 51' 52" W.
9. Daily rate, 5^s·49 gaining; M.T.G. March 20^d 18^h 8^m 39^s; dec.
0° 10' 15"·6 N.; eq. time, +7^m 23^s·60; true alt. 29° 32' 40"; lat. sights,
38° 42' N.; log 9·267775 H.A. 3^h 23^m 56^s; M.T.S. 20^d 20^h 43^m 28^s; long.
obs. 38° 42' 15" E.; at noon, 39° 6' 15" E.; azi. log 19·439569; true azi.
S. 63° 16' 32" E.; compass error, 4° 13' 28" E.; dev. 13° 43' 28" E.
10. A.T.S. 4^h 7^m 6^s P.M.; dec. 5° 54' 48"·7 N.; eq. time, −2^m 54^s·30;
true azi. N. 71° 14' W.; compass error, 4° 42' 15" E.; dev. 4° 47' 45" W.
11. A.T.G. March 3^d 2^h 38^m 28^s; T.F.N. 32^m 52^s; dec. 6° 44' 40" S.; true
alt. 27° 29' 50"; log 3·762784; number, 5791; zen. dist. 62° 7' 40" N.; lat.
55° 23' N., at obs. 55° 22' 54" N. at noon.
12. First true alt. 26° 39' 57"; dec. 9° 51' 42" N.; eq. time, +0^m 0^s·56;
(a) log 9·396643 H.A. 3^h 59^m 36^s; (b) log 9·389913 H.A. 3^h 57^m 34^s;
(a) long. 130° 43' W.; (b) long. 130° 12' 30" W.; line, $\left\{ \begin{array}{l} \text{N. } 17\frac{1}{2}° \text{ E.} \\ \text{S. } 17\frac{1}{2}° \text{ W.} \end{array} \right\}$; azi.
S. 72½° E. Second true alt. 47° 44' 36"; dec. 9° 54' 22"·4 N.; eq. time,
−0^m 1^s·29; (c) log 8·292757, H.A. 1^h 4^m 25^s; (d) log 8·036750, H.A.
0^h 47^m 54^s; (c) long. 131° 55' 15" W.; (d) long. 127° 47' 30" W.; line,
$\left\{ \begin{array}{l} \text{N. } 69° \text{ E.} \\ \text{S. } 69° \text{ W.} \end{array} \right\}$; azi. S. 21° E.; position, lat. 50° 16' N., long. 130° 50'. W.
13. M.P. of Sirius, 7^h 31^m 16^s·56 P.M.
14. R.A.M. 10^h 15^m 38^s; Regulus, N. 0^h 12^m 38^s W.; θ Argus, S.
0^h 23^m 43^s E.; η Argus, S. 0^h 26^m 47^s E.
15. Alt. of Sirius, 31° 14' 34" S. of observer.
16. True alt. 15° 7' 37"; P.D. 34° 0' 52"; lat. 49° 8' 29" N.
17. Daily rate, 7^s·25 losing; M.T.G. Mar. 18^d 10^h 16^m 40^s. Star's R.A. =
12^h 21^m 3^s; dec. 62° 32' 35" S.; R.A.M.S. 23^h 45^m 0^s·31. True alt.
47° 37' 39"; log 9·762172; H.A. = 6^h 36^m 3^s E.; R.A.M. 5^h 45^m 0^s; M.T.S.
18^d 6^h 0^m 0^s; long. 64° 10' W.
18. R.A.M. = 13^h 20^m 32^s·49; H.A. 3^h 5^m 23^s·27 E.; dec. 21° 42' 15" N.;
true azi. S. 134° 54' E.; compass error, 11° 21' E.; dev. 40° 51' E.
19. R.A.M. = 7^h 5^m 34^s·94; true alt. 37° 8' 33"; first cor. −5' 13"·4;
second cor. +35"; third cor. +1' 7"; lat. 37° 5' 1"·6 N.
20. M.T.G. March .25^d 8^h 28^m·4; semid. 15' 12"·6; H.P. 55' 4"·16; dec.
0° 3' 23"·3 N.; true alt. 37° 5' 12"; lat. 52° 51' 25" N.
21. (a) True course, S. 73° E.; mag. course, S. 53° E.; compass course,
S. 68° E.; dev. 15° E.; var. 20° W.; distance, 79 miles.
(b) Lat. 55° 21' N.; long. 6° 38' W.
(c) 55° 6½' N.; long. 5° 30' W.; dist. 10½'.
(d) True course, N. 89½° E.; mag. N. 109½° E.; compass, S. 45½° E.;
distance, 18¾ miles.

APPENDIX.

22. Interval, $1^h 2^m$; Table B, +4 ft. 4 in.; cor. for Waterford, −10 ft. 6 in.; cor. for Wexford, −4 ft. 3 in.

23. *Napier's curve*: Cor. mag. bearing, N. 56° E.; devs. 0, 8° E., 11° E., 7° E., 0, 7° W., 12° W., 8° W.; compass courses, E.N.E., S. 67° E., W. by S., N. 19° W.; mag. courses, W.S.W., S. 18° E., N. 40° E., N. 53° W.; ship's head, E.N.E.; dev. 10° E.; bearings, S. 27° W., S. 49° E.

24. Moon's semidiameter, $14' 59''\cdot65$; H.P. $54' 10''\cdot25$; R.A. $8^h 59^m 18^s\cdot56$; dec. $13° 48' 56''\cdot5$ N.; R.A.M.S. $0^h 0^m 48^s\cdot84$; H.A. $1^h 52^m 12^s$; log. 4·927654; num. 84655; true alt. 52° 45′ 30″, appar. 52° 13′ 3″; M.T.S. $22^d 10^h 50^m 42^s$.

Mars' R.A. $7^h 39^m 15^s\cdot99$; dec. $24° 29' 49''\cdot8$ N.; H.P. $8''\cdot6$; reduced $7''$; true alt. 46° 49′ 27″; log. 9·219692 = H.A. $3^h 12^m 15^s$ W.; R.A.M. $10^h 51^m 31^s$; arc B, 38° 8′ 28″; log sin 9·790706; half dist. 10° 50′ 8″; sin 9·274095; interval, $1^h 30^m 38^s$; second cor. $+6^s$; chron., slow $10^m 14^s$; long. 4° 59′ 30″ E.

25. M.T. interval, $3^h 45^m 30^s$; sid. int. $3^h 46^m 7^s\cdot07$; planetary int. $3^h 45^m 35^s\cdot40$; dec. 18° 7′ 26″·9 S.; first alt. 56° 14′ 1″; second alt. 13° 27′ 37″; angle, 64°; dist. 37·5; cor. −16′ 27″ to first alt.; arc 1, 26° 41′ 10″; arc 2, 41° 47′ 15″; arc 3, 37° 0′ 6″; arc 4, 110° 22′ 30″; arc 5, 73° 22′ 24″; approx. lat. 12° 19′ 7″; cor. +46″; lat. 12° 19′ 53″ N.

26. A.T.S. meridian passage, $7^d 13^h 30^m 29^s\cdot76$; M.T.S. $13^h 29^m 9^s\cdot98$; M.T.G. $10^h 35^m 33^s\cdot98$; watch fast on A.T.S. $\cdot24^s$; on M.T.S. $1^m 30^s\cdot02$; fast on M.T.G. $2^h 54^m 56^s\cdot02$.

27. R.A.M.S. $13^h 41^m 42^s$; true alt. 40° 4′ 52″; E.H.A. $2^h 58^m 39^s\cdot4$; slow M.T.G. $10^m 20^s 6$; fast M.T.S. $3^h 33^m 14^s\cdot1$.

28. *Sirius*, R.A. = $6^h 40^m 43^s\cdot33$; dec. 16° 34′ 53″ S.; true alt. 17° 16′ 35″; log (a) 8·631315; log (b) 8·557179; H.A. (a) $1^h 35^m 30^s$; H.A. (b) $1^h 27^m 35^s$ W.; (a) M.T.S. $7^h 40^m 22^s$; (b) M.T.S. $7^h 42^m 27^s$; long. (a) 2° 6′ 45″; long. (b) 4° 5′ 30″ W.

Regulus, R.A. $10^h 3^m 2^s\cdot44$; dec. 12° 27′ 24″ N.; R.A.M.S. $0^h 35^m 51^s\cdot20$; true alt. 43° 33′ 57″; log (c) 8·769311; log (d) 8·736734; H.A. (c) $1^h 52^m 15^s$ E.; H.A. (d) $1^h 48^m 3^s$ E.; M.T.S. (c) $7^h 34^m 56^s$; M.T.S. (d) $7^h 39^m \cdot 08^s$; long. (c), 3° 28′ 15″ W.; long. (d), 2° 25′ 15″ W.; first line, $\begin{Bmatrix} N.\ 67°\ W. \\ S.\ 67°\ W. \end{Bmatrix}$; azi. S. 23° W.; second line, $\begin{Bmatrix} N.\ 61°\ E. \\ S.\ 61°\ W. \end{Bmatrix}$; azi. S. 29° E.; position, lat. 53° 23′ N.; long. 3° 0′ W.

29. First course, S. 57° 17′ 30″ E.; final course, N. 62° 17′ 30″ E.; first long. 67° 12′ E.; final long. 104° 21′ E.; first dist. 2300 miles; second, 1549 miles; third, 1830 miles; total dist. 5679 miles.

30. (a) 1009 ft.
(b) N. 70° 7′ 25″ W.; drift, 67·55 miles.
(c) Side = 33° 38′ 50″; angle = 100° 30′ 56″.
(d) 116° 47′.

Answers to Paper VI.

1, 2. *Logs*: ·4839 log = 9·684706; ·08604 log = 8·934729.

3. *Days' work*: N. 62° W., 17′; S. 84° E., 72′; S. 75° E., 43′; S. 68° E., 56′; S. 54° E., 57′; S. 47° E., 56′; E. 56′; N. 22° E., 60′; diff. lat. 47′·7; dep. 315′·6; course, S. $81\tfrac{1}{2}$° E.; dist. 318′; lat. in, 38° 33′ S.; long. in, 174° 47′ W.

4. *Meridian altitude*: Dec. 7° 52′ 49″ S.; true alt. 44° 55′ 48″; lat. 37° 11′ 23″ N.

5. *Mercator*: True course, N. 53° 48′ E.; com. course, S. 75° 22′ E. dist. 4561′.

6. *Parallel*: 32° 23′.

7. *Tides*: $11^h 35^m$ A.M., no P.M.; and $0^h 26^m$ A.M., $1^h 9^m$ P.M.

8. *Amplitude*: Dec. 0; true amp. E.; error, 14° 4′ E.; dev. 7° 4′ E.

9. *Chron. azimuth*: Rate, $4^s\cdot06$ losing; M.T.G. $26^d 0^h 56^m 27^s$; true alt.

APPENDIX.

44° 35' 9"; st. lat. 48° 26'; dec. 21° 8' 50" N.; H.A. 3ʰ 2ᵐ 52ˢ; eq. time, 3ᵐ 11ˢ —; st. long. 60° 37' 30" W.; noon long. 60° 3' 50" W.; true azi. S. 69° 38' E.; error, 19° 0½' W.; dev. 6° 59½' E.
 10. *Time azimuth*: A.T.S. 8ʰ 24ᵐ A.M.; dec. 9° 45' N.; cor. for azi. 0; true azi. N. 116° 14' E.; error, 4° 42' W.; dev. 27° 33' E.
 11. *Chart work*: Dev. 0; compass course, N. 17° W.; dist. 68 miles; lat. 52° 8¼' N., long. 5° 2½' W.; lat. 52° 47' N., long. 5° 50' W.; dist. 7 miles.
 12. *Meridian passage star*: 8ʰ 59ᵐ 59ˢ P.M.
 13. *Stars near meridian.*—*East*: α Hydræ; Pollux; β Argûs. *West*: Procyon; Castor; Sirius; α Argûs. Approx. mer. alt. 71° 31'. R.A.M. 7ʰ 22ᵐ 52ˢ.
 14. *Star's meridian altitude*: True alt. 71° 30' 30"; dec. 8° 13' 27" S.; lat. 10° 16' 3" N.
 15. *Star chron.*: Rate, ·75ˢ losing; M.T.G. 5ᵈ 13ʰ 51ᵐ 57ˢ; R.A. 6ʰ 21ᵐ 46ˢ; dec. 52° 38' 14" S.; R.A.M.S. 15ʰ 0ᵐ 16ˢ; true alt. 58° 5' 36"; E.H.A. 2ʰ 50ᵐ 32ˢ; long. 20° 15' W.
 16. *Star azimuth*: Dec. 8° 13' 27" S.; R.A. 9ʰ 22ᵐ 41ˢ; R.A.M.S. 16ʰ 1ᵐ 32ˢ; E.H.A. 3ʰ 46ᵐ 9ˢ; cor. for azi. +14'; true azi. N. 106° 6' E.; error, 18° 55' E.; dev. 11° 10' E.
 17. *Ex-Meridian*: E.H.A. 22ᵐ 12ˢ; true alt. 55°; dec. 44° 55' 21" N.; lat. 10° 15' N.; arc 1, 45° 3' 26" N.; arc 2, 34° 48' 4" S.
 18. *Sumner.*—*First obs.*: True alt. 59° 24' 55"; dec. 13° 21' 11" N.; eq. time, 2ᵐ 9ˢ —; E.H.A.'s, 1ʰ 42ᵐ 55ˢ and 1ʰ 39ᵐ 33ˢ; long. (*a*), 178° 30' E.; long. (*b*), 179° 20' 30" E.
 Second obs.: True alt. 36° 53' 6"; dec. 13° 25' 28" N.; eq. time, 2ᵐ 12ˢ —; W.H.A.'s, 3ʰ 37ᵐ 59ˢ and 3ʰ 37ᵐ 16ˢ; long. (*c*), 178° 38' 30" E.; long. (*d*), 178° 27' 45" W.; line position, N. 9° W. or S. 9° E.; Sun's azi. S. 81° W.; lat. 33° 19' N.; long. 178° 35' E.
 19. *Polaris*: True alt. 48° 53' 6"; R.A.M. 13ʰ 54ᵐ 13ˢ; first cor. +1° 12' 16"; second cor. +2"; third cor. +1' 11"; lat. 50° 5' 35" N.
 20. *Meridian altitude Moon*: M.T.G. 18ᵈ 7ʰ 24ᵐ·2; S.D. 15' 20"; H.P. 55' 17"; dec. 20° 1' 52" N.; A.A. 84° 7' 58"; cor. +5' 33"; true alt. 84° 13' 31"; lat. 14° 15' 23" N.
 21. *Curve*: Cor. mag. S.; deviations, 5° W., 6° E., 18° E., 17° E., 10° E., 7° W., 18° W., 21° W.; compass courses, N. 3° W., N. 80° W., S. 66° E., N. 31° E.; mag. courses, S. 55° E., S. 30° W., S. 77° W., N. 19° W.; bearings, N. 29½° W., S. 72° W.
 22. *Current*: N. 9½° W.; dev. 5½° E.; dist. made good, 16 miles.
 23. *Sounding*: Table B, 1 ft. 6 in.; cor. 6 ft. 0 in. to subtract.
 24. *Lunar.*—*Moon's elements*: S.D. 15' 9"·5; H.P. 54' 39"·5; R.A. 6ʰ 20ᵐ 10ˢ; dec. 22° 41' 48" N.; R.A.M.S. 10ʰ 33ᵐ 38ˢ; A.A. 65° 40' 46"; cor. −22' 5"; true alt. 66° 2' 51"; W.H.A. 1ʰ 16ᵐ 47ˢ.
 Sun's elements: R.A. 10ʰ 34ᵐ 11ˢ; dec. 9° 0' 57" N.; A.A. 39° 49' 23"; T.A. 39° 48' 21"; E.H.A. 2ʰ 57ᵐ 14ˢ; app. dist. 62° 22' 8"; angle B, 18° 8' 4"; true dist. 62° 9' 53"; second cor. −3ˢ; M.T.G. 29ᵈ 22ʰ 45ᵐ 18ˢ·5; error, slow 15ᵐ 13ˢ·5; long. 25° 30' W.
 25. *Ivory*: H.E.T. 2ʰ 15ᵐ 0ˢ; mid. dec. 10° 31' 3" S.; cor. −20' 0"; first T.A. 11° 11' 42"; second T.A. 47° 30' 10"; arc 1, 33° 6' 33"; arc 2, 29° 48' 51"; arc 3, 50° 8' 47"; arc 4, 77° 24' 50"; arc 5, 27° 16' 3"; lat. cor. −2' 47"; true lat. 50° 25' 4" S.
 26. *Simul. altitudes.*—*Vega's elements*: R.A. 18ʰ 33ᵐ 35ˢ; dec. 38° 41' 27" N.; R.A.M.S. 6ʰ 32ᵐ 46ˢ; true alt. 78° 28' 42"; E.H.A.'s, 0ʰ 48ᵐ 22ˢ and 0ʰ 45ᵐ 24ˢ; long. (*a*) 139° 13' W.; long. (*b*) 138° 29' 30" W.; bearing, S. 52° E.
 Arcturus: R.A. 14ʰ 11ᵐ 6ˢ; dec. 19° 42' 23" N.; true alt. 38° 43' 19"; W.H.A.'s, 3ʰ 36ᵐ 36ˢ and 3ʰ 35ᵐ 46ˢ; long. (*c*), 138° 36' 45" W.; long. (*d*), 138° 49' 15" W.; bearing, S. 76° W.; lat. 46° 26½' N.; long. 138° 44' W.
 27. *Equal altitudes*: Dec. 18° 38' 25" N.; eq. time, 6ᵐ 14ˢ·09; Part I.,

APPENDIX. 371

$-2^s\cdot98$; Part II., $-2^s\cdot 78$; fast A.T.S. $10^h\,21^m\,42^s\cdot24$; fast A.T.G. $18^m\,46^s\cdot24$; fast M.T.G. $12^m\,32^s\cdot14$.
  28. *Great circle*: First course, S. 50° 4½' W.; final, N. 58° 46½' W.; long. arrival, 128° 32½' W.; long. leaving, 142° 32' W.; dist. 5050·1 miles.
  29. *Chart*: Lat. 60° 22' S.; long. 30° 20' W.; course, S. 3° E.; dist. 43'.
  30. *Trig. problems*: B = 50° 40' 12''; A = 39° 19' 48''; BD = 14,383 ft.
  31. Height, 115·6 ft.
  32. Lat. 42° 4'; H.A. $7^h\,10^m\,18^s$.
  33. Course, S. 78½° E.; dist. 92 miles.

Answers to Paper VII.

  1. Log 5·842665; product, 696090·3.
  2. Log 4·290809; quotient, 19534·82.
  3. N. 12° W., 20'; N. 15° E., 22'; S. 71° W., 24'; S. 39° W., 46'; N. 7° E., 48'; N. 51° E., 24'; S. 47° W., 48'; S. 88° W., 48'; N. 24'; diff. lat. 49'·7 N.; dep. 108'·7 W.; course, N. 65° 25' W.; dist. 119'·5; lat. in, 49° 20' S.; long. 113° 3' 12'' W.
  4. A.T.G. March $20^d\,1^h\,2^m\,8^s$; dec. 0° 6' 31'' S.; true alt. 62° 39' 10'' S.; lat. 27° 14' 19'' N.
  5. 16° 16' 30'' W. long.
  6. D. lat. 5423; M.D. lat. 6166·3; long. 962; course by compass, N. 16° 22' W.; distance, 5489.
  7. Hull, con. $-1^h\,50^m$; H.W. no A.M., $0^h\,10^m$ P.M.
  8. A.T.G. April $30^d\,1^h\,0^m\,0^s$; dec. 14° 48' 16''·8 N.; true amp. E. 14° 48' 17'' N.; dev. 21° 48' 17'' W.
  9. Daily rate, $1^s\cdot 75$ gain: M.T.G. April $15^d\,14^h\,54^m\,10^s$; dec. 10° 0' 47''·4 N.; eq. time, $-0^m\,5^s\cdot68$; true alt. 20° 47' 47''; log 8·129361; H.A. $2^h\,52^m\,15^s$; long. obs. 179° 29' 45'' E.; long. noon, 179° 45' 38'' W.; azi. log 9·183642; true azi. N. 45° 59' 36'' W.; dev. 16° 30' W.
  10. A.T.S. $5^h\,38^m$ P.M.; dec. 15° 1' 34''·7 N.; eq. time, $+2^m\,57^s\cdot86$; true azi. N. 82° 18' W.; dev. 3° 37' 0'' W.
  11. M.T.G. July $2^d\,11^h\,15^m\,30^s$; R.A.M. $13^h\,37^m\,14^s\cdot84$; R.A. $13^h\,56^m\,46^s\cdot19$; dec. 59° 53' 40'' S.; log 3·137090; true alt. 70° 51' 52''; Z. dist. 18° 53' 10''; lat. 41° 0' 30'' S.; H.A. $19^m\,31^s$.
  12. First dec. 23° 12' 17''·6 N.; eq. time, $+3^m\,17^s\cdot69$; true alt. 54° 43' 48''; second dec. 23° 11' 44''·7 N.; eq. time, $+3^m\,19^s\cdot37$; true alt. 22° 0' 2''; log (a) 8·875260; (b) 8·861910; (c) 9·636490; (d) 9·637716; H.A. (a) $2^h\,7^m\,11^s$; (b) $2^h\,5^m\,12^s$; (c) $5^h\,29^m\,12^s$; (d) $5^h\,29^m\,45^s$; long. (a), 157° 33' 0'' E.; (b) 157° 3' 15'' E.; (c) 155° 33' 0'' E.; (d) 155° 41' 15'' E.; first line, {N. 34° W. / S. 34° E.}; azi. S. 56° W.; second line, {N. 9° E. / S. 9° W.}; azi. N. 81° W.; position, lat. 48° 37' 30'' N., long. 155° 46' 0'' E.
  13. $5^h\,43^m\,34^s\cdot87$.
  14. R.A.M. $18^h\,32^m\,1^s$; γ Draconis, N. $0^h\,37^m\,45^s$ W.; δ Sagittarii, S. $0^h\,17^m\,29^s$ W.; Vega, N. $0^h\,1^m\,30^s$ E.; σ Sagittarii, S. $0^h\,16^m\,59^s$ E.
  15. Approx. obs. alt. 80° 26' 47'' N.
  16. True alt. 12° 27' 56''; P.D. 32° 15' 5''; lat. 44° 43' 1'' S.
  17. Daily rate, ·05 losing; M.T.G. April $27^d\,0^h\,40^m\,52^s$; R.A. $13^h\,19^m\,55^s\cdot46$; dec. 10° 38' 23'' S.; R.A.M.S. $2^h\,21^m\,7^s\cdot84$; true alt. 47° 51' 32''; log 8·742589; H.A. $1^h\,48^m\,47^s$ E.; M.T.S. $27^d\,9^h\,10^m\,0^s\cdot62$; long. 127° 17' 0'' E.
  18. R.A.M. $12^h\,40^m\,21^s$; R.A. $7^h\,34^m\,2^s\cdot77$; dec. 5° 28' 52''; true azi. N. 96° 58' W.; dev. 23° 2' E.
  19. R.A.M. $19^h\,21^m\,4^s\cdot57$; true alt. 36° 14' 8''; first cor. $+17^s\cdot5$; second cor. $+32^s$; third cor. $+53''\cdot5$; lat. 36° 15' 51'' N.
  20. M.T.G. April $21^d\,17^h\,8^m\cdot5$; semidiameter, 15' 17''·09; H.P. 55' 13''·95; dec. 0° 10' 42''·8 S.; true alt. 60° 46' 52''; lat. 29° 2' 25'' N.

APPENDIX.

21. *Chart work*: (a) True course, S. 15° E.; mag. course, 55° W.; compass course, S. 22½° E.; var. 20° W.; dev. 27½° E.; dist. 38 miles; (b) lat. 55° 2′ N., long. 5° 40′ W.; (c) lat. 54° 45½′ N., long. 5° 33′ W.; distant 6½ miles; (d) true course, S. 22½° E.; mag. course, S. 2½° E.; compass course, S. 27½° E.; D. 25 miles.
22. Interval, 0; half daily range, 3 ft.; cor. for Waterford, 9 ft. 2 in.; for Tuskar, −6 ft. 8 in.
23. *Napier's curve*: C.M.B. N. 73½° W.; devs. 29½° W., 26½° W., 6½° E., 30½° E., 27½° E., 17½ E., 4½° W., 21½° W.
Compass bearings: N. 2° W. S. 74° E. S. 51° W. N. 55° E.
Mag. courses: S. 67° E. S. 48° W. S. 86½° W. N. 74° W.
Correct mag. bearings: S. 25½° W. N. 61½° W.
24. Moon's semidiameter, 16′ 31″·66; H. P. 59′ 37″·08; R.A. 21ʰ 5ᵐ 35ˢ·76; dec. 12° 44′ 9″·8 S.; R.A.M.S. 4ʰ 29ᵐ 53ˢ·6; star's R.A. 22ʰ 52ᵐ 6ˢ·5; star's dec. 30° 9′ 5″ S.; Moon's true alt. 57° 39′ 7″; app. alt. 57° 7′ 23″; Moon's H.A. = 1ʰ 57ᵐ 0ˢ. W.; R.A.M. 23ʰ 2ᵐ 36ˢ; Star's true alt. 87° 42′ 0″; app. alt. 87° 42′ 2″; log arc B, 18·322590 = 8° 20′ 9″; log half dist. 18·830274; ½ dist. 15° 4′ 33″; interval, 1ʰ 32ᵐ 2ˢ; second cor. −110ˢ; M.T.S. = 29ᵈ 18ʰ 32ᵐ 42ˢ; cor. long. 30° 37′ 30″ E., chron. correct.
25. Sid. H.E.T. 2ʰ 8ᵐ 51ˢ·1; dec. 14° 40′ 15″ N.; angle, 90°; first alt. 40° 31′ 35″; second alt. 50° 45′ 19″; arc 1, 31° 2′ 35″; arc 2, 6° 56′ 31″; arc 3, 33° 8′ 40″; arc 4, 72° 48′ 23″; arc 5, 39° 39′ 43″; lat. 49° 50′ 3″ N.
26. A.T.G. April 14ᵈ 14ʰ 20ᵐ 5ˢ·5; dec. 9° 38′ 51″·6 N.; eq. time, +0ᵐ 9ˢ·44 two-day change, 2571·1; first log, 1·0761 = Part I. + 11ˢ·91; second log, 0·2455; Part II. +1ˢ·76; G.M.T. 14ᵈ 14ʰ 20ᵐ 26ˢ·67; slow on A.T.S. 9ʰ 39ᵐ 33ˢ·33; slow on M.T.S. 9ʰ 39ᵐ 42ˢ·77; fast on M.T.G. 11ˢ·73.
27. Dec. 19° 33′ 10″ N.; eq. time, +6ᵐ 17ˢ·4; true alt. 57° 12′ 19″; chr. fast of M.T.G. 4ᵐ 27ˢ·9.
28. Star Markab, R.A. = 22ʰ 59ᵐ 44ˢ·48; dec. 14° 39′ 44″ N.; R.A.M.S. = 1ʰ 36ᵐ 45ˢ·37; true alt. = 21° 51′ 58′; log (a), 9·551592; log (b), 9·551108; H.A. (a) 4ʰ 53ᵐ 6ˢ E.; H.A. (b) 4ʰ 52ᵐ 55ˢ E.; M.T.S. (a) 15ᵈ 16ʰ 29ᵐ 53ˢ; M.T.S. (b), 15ᵈ 16ʰ 30ᵐ 4ˢ; long. (a), 30° 49′ 45″ W.; long. (b), 30° 47′ 0″ W.; Star Arcturus, R.A. = 14ʰ 11ᵐ 6ˢ·26; dec. 19° 42′ 10″ N.; true alt. 34° 38′ 34″; log (c), 9·384847; log (d), 9·382705; H.A. (c), = 3ʰ 56ᵐ 3ˢ; H.A. (d) = 3ʰ 55ᵐ 25ˢ; M.T.S. (c) 15ᵈ 16ʰ 30ᵐ 24ˢ; M.T.S. (d) 15ᵈ 16ʰ 29ᵐ 46ˢ; long. (c), 30° 42′ 0″ W.; long. (d) 30° 51′ 30″ W.; first line, $\begin{Bmatrix} N.\ 3°\ E.\\ S.\ 3°\ W. \end{Bmatrix}$; azi. S. 87° E.; second line, $\begin{Bmatrix} N.\ 11°\ W.\\ S.\ 11°\ E. \end{Bmatrix}$; azi. S. 79° W.; position, lat. 50° 49′ N., long. 30° 48′ W.
29. First course, S. 39° 16½′ W.; last course, S. 38° 40½′ W.; vertex, lat. 59° 11′ 30″ N., long. 58° 43½′ E.; cross equator steering, S. 31° W.; dist. 5034 miles; long. 31° 16½′ W.
30. (a) 368 ft.
(b) Bearing, S. 27° 19′ 49″ W.; set, E. 3° 49′ 42″ N.
(c) AC. = 72° 23′ 48″. A = 31° 12′ 14″; B = 80° 59′ 8″.
(d) Lat. 24° 46′ N.

ANSWERS TO PAPER VIII.

1. *Logs*: ·000001075 log = 4·031408; ·1 log = 9·000000.
2. *Days' work*: S. 39° W., 19′; N. 31° W., 52′; S. 80° W., 50′; N. 61° W., 49′; S. 22° W., 48′; S. 4° E., 48′; S. 49° W., 49′; S. 14° W., 28′; diff. lat. 106′·8; dep. 189′·4; course, S. 61° W.; dist. 217 miles; lat. in, 50° 23′ N.; long. in, 176° 42′ E.
3. *Meridian altitude*: Dec. 0° 2′ 26″ N.; true alt. 35° 32′ 54″; lat. 54° 29′ 32″ N.

APPENDIX. 373

4. *Mercator*: True course, S. 54° 10' E.; Compass course, S. 36° 50' E.; dist. 9224 miles.

5. *Parallel*: Noon long. 83° 38'·4 W.

6. *Tides*: $3^h 58^m$ A.M., $4^h 16^m$ P.M.; and $11^h 50^m$ A.M., no P.M.

7. *Amplitude*: Dec 21° 53' 7" N.; true amplitude, W. 43° 33' N.; error, 18° 14' E.; dev. 2° 16' W.

8. *Chron. azimuth*: Rate, ·75 losing; M.T.G. $14^d 10^h 1^m 48^s$; true alt. 53° 2' 0"; St. lat. Equator; dec. 23° 17' 47" N.; H.A. $1^h 58^m 12^s$·5; eq. time, +4^s·5; St. long. 179° 59' W.; noon long. 179° 45' E.; true azi. N. 48° 53' E.; error, 21° 26' W.; dev. 31° 26' W.

9. *Time azimuth*: A.T.S. $9^h 27^m$ A.M.; dec. 21° 5' N.; cor. for azi. +36 + 4 + 15 = 55'; true azi. S. 137° 34' E.; error, 11° 0' W.; dev. 0° 45' E.

10. *Chart work*: Dev. 1° E.; compass course, N. 84° W.; dist. 28¼ miles; lat. 53° 27¾' N., long. 3° 56' W.; lat. 53° 26' N., long. 4° 10¼' W.; dist. 8¾ miles.

11. *Meridian pass. star*: $7^h 7^m 30^s$ A.M.

12. *Stars near meridian*: R.A.M. $23^h 25^m 56^s$; *East*, a Andromedæ, Algenib, Achernar. *West*, a Gruis, Fomalhaut, Markab. Approx. mer. alt. 69° 54'.

13. *Star's meridian altitude*: True alt. 69° 52' 45"; dec. 30° 9' 7" S.; lat. 50° 16' 22" S.

14. *Star chron.*: Rate, 8^s·8 gaining; M.T.G. $7^d 17^h 46^m 22^s$; R.A. $19^h 45^m 55^s$; dec. 8° 36' 24" N.; R.A.M.S. $11^h 8^m 18^s$; true alt. 52° 56' 52"; E.H.A. $1^h 52^m 24^s$; long. 165° 17' 15" W.

15. *Star azimuth*: R.A. $22^h 59^m 48^s$; dec. 14° 40' N.; R.A.M.S. $13^h 6^m 31^s$; W.H.A. $3^h 57^m 43^s$; cor. for azi. − 24 + 22 + 2 = 0; true azi. S. 122° 55' W.; error, 3° 39' W.; dev. 17° 59' W.

16. *Ex-meridian*: Time from noon, $32^m 4^s$; true alt. 64° 7' 0"; dec. 9° 48' N.; lat. 14° 50' S.

17. *Sumner.*—*First obs.*: true alt. 15° 50' 49"; dec. 19° 18' 43" S.; eq. time, −14^m 40^s; E.H.A.'s, $2^h 27^m 27^s$·5 and $2^h 23^m 20^s$; long. (a), 125° 56' 45" W.; (b), 124° 54' 45" W.
Second obs.: True alt. 21° 11' 33"; dec. 19° 20' 54" S.; eq. time, −14^m 38^s; W.H.A.'s, $1^h 20^m 54^s$ and $1^h 12^m 32^s$; long. (c), 123° 33' 45" W.; (d), 125° 39' 15" W.; second line of position, S. 70° E. or N. 70° W.; second true azi. S. 20° W.; lat. 47° 24' N.; long. 125° 13½' W.

18. *Polaris*: True alt. 44° 27' 42"; R.A. mer. $5^h 47^m 15^s$; first cor. −29' 19"; second cor. +0' 37"; third cor. +0' 54"; lat. 43° 58' 54" N.

19. *Meridian altitude Moon*: M.T.G. $27^d 4^h 51^m$·5; S.D. 15' 16"; H.P. 55' 20"; dec. 5° 45' 1" S.; appar. alt. 34° 41' 46"; cor. +44' 7"; true alt. 35° 25' 53"; lat. 48° 49' 6" N.

20. *Curve*: Cor. mag. N. 2½° W.; devs. 2½° W., 9½° W., 7½° E., 10½° E., 18½° E., 14½° E., 20½° W., 18½° W.; comp. courses, N. 8° E., S. 71° E., S. 79° W., N. 57° W.; mag. courses, N. 85° E., N. 72° W., S. 65° W., S. 20° E.; bearings, S. 20° W., S. 50° E.

21. *Current*: Dev. 27° E.; comp. course, S. 56½° W.; dist. made good, 22¼ miles.

22. *Sounding*: Table B, −7 ft. 8 in.; 1 ft. 4 in. to subtract.

23. *Lunar.*—*Moon's elements*: S.D. 16' 35"·8; H.P. 60' 13"; R.A. $23^h 7^m 59^s$; dec. 0° 2' 18" S.; R.A.M.S. $8^h 14^m 16^s$; A.A. 27° 31' 49"; T.A. 28° 23' 23"; E.H.A. $3^h 25^m 49^s$.
Star's elements: R.A. $16^h 23^m 17^s$; dec. 26° 12' 38" S.; A.A. 46° 44' 5"; true alt. 46° 43' 11"; W.H.A. $3^h 18^m 53^s$; ap. dist. 100° 47' 13"; angle B, 11° 46' 56"; true dist. 99° 48"; second cor. −3^s; M.T.G. $25^d 14^h 22^m 16^s$·5; error, slow 2^s·5; long. 43° 35' 36" W.

24. *Ivory*: H.E.T. $2^h 2^m 37^s$·5; mid. dec. 0° 1' 16" S.; cor. for run, 0; first T.A. 59° 38' 32"; second T.A. 8° 59' 4"; arc 1, 30° 39' 22"; arc 2, 43° 52' 10";

2 B 2

374   APPENDIX.

arc 3, 34° 45' 20"; arc 4, 90° 1' 28"; arc 5, 55° 16' 8"; cor. for lat. −2' 59"; lat. 24° 12' 3" N.

25. *Simul. altitudes.—Planet's elements:* R.A. $14^h\ 2^m\ 50^s\cdot 6$; dec. 11° 0' 24" S.; R.A.M.S. $3^h\ 50^m\ 50^s\cdot 6$; true alt. 30° 1' 30'; E.H.A.'s, $0^h\ 48^m\ 21^s$ and $0^h\ 27^m\ 10^s$; long. (a), 140° 59' W.; long. (b), 135° 41' 15" W.; bearing, S. $10\frac{1}{2}°$ E.
Star's elements: R.A. $7^h\ 39^m\ 10^s$; dec. 28° 16' 10" N.; true alt. 23° 33' 0"; W.H.A.'s, $5^h\ 41^m\ 28^s$ and $5^h\ 42^m\ 41^s$; long. (c), 139° 27' W.; long. (d), 139° 8' 45" W.; bearing, N. $73\frac{1}{2}°$ W.; lat. $48°\ 12'$ N.; long. 139° 22' W.
26. R.A.M. $21^h\ 14^m\ 30^s\cdot 2$; true alt. 46° 52'; W.H.A. $1^h\ 28^m\ 34^s\cdot 5$; fast M.T.G. 6^s; slow M.T.S. $10^h\ 25^m\ 32^s$.
27. *Great circle:* First course, S. 63° 36' E.; final, N. 65° $10\frac{1}{2}$ E.; long. arrival, 60° 8' E.; long. dep. 79° 2' E.; dist. 4705·6 miles.
28. *Chart:* Lat. in, 68° $10\frac{1}{4}$' N.; long. in, 0° 34' W.
29. *Trig. problems:* Course, N. 35° $6\frac{1}{4}$' E.; dist. 274·5 miles.
30. 86·6 ft.
31. R = 100° 32' 7"; QR = 94° 12' 4"; PR = 3068° 6' 45".
32. AB = 83° 50' 5"; A = 84° 4' 12"; B = 44° 4' 9".

Answers to Paper IX.

1. Log 5·360782; product, 229,499·4.
2. Log 3·630431; quotient, 4270·029.
3. S. 77° E., 48'; S. 33° E., 50'; S. 70° E., 52'; S. 85° E., 50'; N. 58° E., 48'; N. 25° E., 46'; S. 52° E., 44'; N. 54° E., 36'; D. lat. 13'·7 S.; dep. 296·6 E.; course, S. $87\frac{1}{2}°$ E.; dist. 297 miles; lat. in, 51° 39' 18" N.; long. in, 47° 22' 48" W.
4. 35° 39' 30" parallel of latitude.
5. D. lat. 772'; M.D. lat. 1093; D. long. 8537; distance, 6079; true course, N. 82° 42' 15" E.; compass course, S. 79° 32' 45" E.
6. Dublin bar, noon A.M., $0^h\ 13^m$ P.M.; Padstow, $10^h\ 15^m$ A.M., $10^h\ 56^m$ P.M.
7. A.T.G. March $19^d\ 23^h\ 6^m\ 0^s$; dec. 0° 8' 26"·6 S.; true alt. 55° 25' 17"; lat. 34° 43' 9" S.
8. A.T.G. $21^d\ 10^h\ 19^m$; dec. 23° 27' 8"·5 N.; true amp. W. 61° 58' 9" N.; dev. 17° 34' 21" W.
9. Daily rate, 5·5' losing; M.T.G. May $20^d\ 18^h\ 3^m\ 11^s$; dec. 20° 9' 0"·7 N.; eq. time, $-3^m\ 37^s\cdot 14$; true alt. 14° 33' 39"; log 9·718110; H.A. = $6^h\ 10^m\ 20^s$; lat. at sights, 53° 25' N.; long. at sights and noon, 4° 17' W.; azi. log 19·794910; true azi. S. 104° 18' 40" E.; dev. 5° 11' 20" E.
10. A.T.S. $4^h\ 55^m\ 11^s$; dec. 23° 14' 8" N.; eq. time, $+0^m\ 20^s\cdot 54$; cor. −14'+14'+8'; true azi. N. 63° 14' W.; dev. 9° 34' 45" E.
11. A.T.G. June $29^d\ 17^h\ 56^m\ 37^s$; time from noon, $39^m\ 33^s$; dec. 23° 11' 56"·1 N.; true alt. 50° 59' 54"; No. 13,214; zenith dist. 37° 46' 50" S.; lat. at obs. 14° 34' 54" S.; at noon, 14° 32' 6" S.
12. First dec.0° 14' 57"·8 S.; eq. time, $7^m\ 47^s\cdot 04$; second dec.$0°\ 18'\ 51"\cdot 6$ S.; eq. time, $-7^m\ 50^s\cdot 52$; first alt. 32° 59' 45"; second alt. 33° 3' 23"; log (a), 8·830342; H.A. = $2^h\ 0^m\ 37^s$; log (b), 8·798726 = H.A. $1^h\ 56^m\ 12^s$; log (c), 8·830472 = H.A. $2^h\ 0^m\ 38^s$; log (d), 8·798905 = H.A. $1^h\ 56^m\ 14^s$; long. (a), 179° 54' 15" W.; long. (b), 178° 48' W.; long. (c), 179° 35' 30" W.; long. (d), 179° 18' 30" E.; first line, $\begin{Bmatrix}\text{N. }54°\text{ E.}\\\text{S. }54°\text{ W.}\end{Bmatrix}$; azi. S. 36° E.; second line, $\begin{Bmatrix}\text{N. }54°\text{ W.}\\\text{S. }54°\text{ E.}\end{Bmatrix}$; azi. S. 36° W.; position, lat. 51° 19' 30" S., long. 179° 45' 0" W.
13. $6^h\ 5^m\ 34^s\cdot 52$; by inspection, $6^h\ 7^m$ A.T., $6^h\ 4^m$ M.T.
14. R.A.M. $12^h\ 48^m\ 36^s\cdot 19$; δ Centauri, S. $0^h\ 45^m\ 29^s$ W.; γ Corvi, N. $0^h\ 37^m\ 59^s$ W.; α Virginis, N. $0^h\ 44^m\ 53^s$ E.
15. Alt. 62° 22' 45" S.
16. True alt. 15° 15' 25" S.; dec. 3° 41' 41" N.; lat. 71° 2' 54" S.

APPENDIX. 375

17. Daily rate, 10^s·08 gaining; R.A. 16^h 23^m 16^s·84; dec. 26° 12' 37" S.; M.T.G. 19^d 21^h 56^m 10^s; R.A.M.S. 3^h 51^m 21^s·59; true alt. 42° 35' 16"; log 9·089426; H.A. = 2^h 44^m 9^s W.; R.A.M. 19^h 7^m 26^s; long. 100° 1' 30" W.

18. R.A.M. 23^h 41^m 28^s·47; dec. 8° 36' 3" N.; H.A. 3^h 55^m 34^s·79 W.; cor. − 35' + 31' + 15'; true azi. S. 114° 42' W.; dev. 7° 48' W.

19. M.T.G. May 20^d 14^h 28^m 52^s; R.A.M. 20^h 24^m 3^s·99; true alt. 31° 8' 13"; first cor. −19' 31"·5; second cor. +25"·5; third cor. 46"; lat. 30° 49' 53" N.

20. M.T.G. May 3^d 17^h 23^m·83; semidiameter, 16' 21"·96; H. px. 59' 2"·35; dec. 6° 7' 22"·4 S.; true alt. 57° 42' 35"; lat. 26° 10' 13" N.

21. (a) True course, N. 73½° W.; mag. course, N. 55° W.; compass course, N. 40° W.; var. 18½° W.; dev. 14° W.; dist. 16'.
 (b) Lat. 53° 92' N.; long. 4° 0' W.
 (c) Lat. 53° 25' N.; long. 4° 12' W.; dist. 3¼.
 (d) True course, N. 81½° W.; mag. course, N. 63° W.; compass course, N. 47° W.; var. 18½°; dist. 9 miles.

22. Interval, 5^h 56^m; Table B, 21 ft.; correction, 2 in. to add.

23. *Napier's curve*: Co. M.B. S. 8° E.; devs. 10° W., 15° W., 14° W., 8° E., 20° E., 18° E., 3° W., 5° W.
 Courses steered: S. 62° E. N. 87° W. N. 9° E. S. 57½° W.
 Cor. mag. courses: N. 85° W. S. 66° W. S. 22½° E. N. 58° E.
 Dev.: 8° W. S. 64' W. N. 28° W.

24. Moon's semidiameter, 16' 11"·97; H. px. 58' 30"·58; R.A. 23^h 26^m 10^s·19; dec. 2° 2' 2"·6 N. Saturn's R.A. 17^h 21^m 43^s·47; dec. 21° 39' 46"·3 S.; R.A.M.S. 4^h 40^m 28^s·45. Moon's appar. alt. 49° 51' 2"; true alt. 50° 27' 57". Saturn's appar. alt. 12° 12' 57"; true alt. 12° 8' 39"; log 9·707337 = H.A. 6^h 4^m 27^s W.; R.A.M. 23^h 26^m 10^s. Moon's H.A. 0^h 0^m 0^s; log 19·331329 = B. 27° 35' 12'; log 9·856144 = half dist. 45° 53' 32"; interval, 2^h 54^m 51^s; chron. slow 8^s; M.T.S. June, 1^d 18^h 45^m 42^s; long. 147° 43' E.

25. A.T.G. April 16^d 4^h 43^m 10^s; dec. 10° 13' 1"·9 N.; first alt. 51° 35' 43"; second alt. 51° 35' 44"; H.E.T. 1^h 45^m 20^s; angle, 180°; cor. +38' 36" for second alt.; arc 1, 25° 53' 4"; arc 3, 29° 25' 18"; arc 4, 78° 37' 44"; arc 5, 49° 12' 26"; lat. 40° 47' 34" N.

26. R.A.M. 12^h 11^m 35^s·49 and M.T.P.; M.T.G. July 15^d 11^h 13^m 29^s.49; chron. slow on M.T. place, 58^s·6"·49; slow on M.T.G. 0^s·49.

27. Star Antares, R.A. = 16^h 23^m 16^s·8; dec. 26° 12' 37" S.; R.A.M.S. 3^h 54^m 5^s·11; true alt. 13° 4' 10"; log (a), 8·843358 = H.A. 2^h 2^m 29^s E.; log (b), 8·804073 = H.A. 1^h 56^m 57^s; long. (a), 61° 14' 15" W.; long. (b), 59° 51' 15" W.; line, {N. 62° E. / S. 62° W.}; azi. S. 28° E. Star Vega, R.A. 18^h 33^m 33^s·9; dec. 38° 41' 14" N.; true alt. 44° 27' 57"; log (c), 9·426955 = H.A. 4^h 9^m 3^s E. log (d), 9·429249 = H.A. 4^h 9^m 46^s; long. (c), 60° 18' 30" W.; long (d) 60° 29' 15" W.; second line {N. 13° W. / S. 13° E.}; azi. N. 77° E.; position, lat. 45° 48' N., long. 60° 25' W.

28. Great circle sailing: First course, N. 82° 14' W.; last course, S. 64° 8' W.; vertex, lat. 51° 51' N., long. 19° 23' W.; dist. 1712'; points, lats. (1) 51° 26', (2) 51° 44½', (3) 51° 51', (4) 51° 44', (5) 51° 25½', (6) 50° 52½', (7) 50° 6', (8) 49° 5', (9) 47° 47', (10) 46° 39'.

29. (a) Course, N. 49° 42' E.; dist. 348·8 miles.
 (b) 38° 56' 32".
 (c) A = 116° 33' 54"; a = 129° 13' 54"; b = 37° 45' 41".
 (d) Gain = 345 miles.

APPENDIX.

Answers to Paper X.

1. *Logs*: 100 log = 2·000000; 10 log = 1·000000.
2. *Days' work*: N. 75° E., 16'; N. 64° E., 28'; S. 19° W., 42'; S. 5° E., 20'; S. 31° E., 31'; N. 30° E., 30', S. 2° E., 24'; S. 20'; d. lat. 87'·8; dep. 60'·5; course, S. 35° E.; dist. 107 miles; lat. in, 35° 50' N.; long. in, 17° 8' W.
3. *Meridian altitude*: dec. 0° 16' 56"; true alt. 57° 54' 46"; lat. 32° 22' 10" N.
4. *Mercator*: True course, N. 84° 14½' E. or W.; compass course, N. 54° 4½' E. or S. 69° 35½' W.; dist. 7466 miles.
5. *Parallel*: 77·78 miles.
6. *Tides*: No A.M., 0ʰ 12ᵐ P.M.; and 1ʰ 32ᵐ A.M., 2ʰ 2ᵐ P.M.
7. *Amplitude*: Dec. 22° 52' 37" N.; true amp. E. 51° 1½' N.; error, 67° 6' E.; dev. 52° 16' E.
8. *Chron. azimuth*: Rate, 2ˢ·03 gaining; M.T.G. 4ᵈ 20ʰ 2ᵐ 24ˢ; true alt. 19° 44' 48"; dec. 22° 48' 46" N.; H.A. 2ʰ 12ᵐ 59ˢ; eq. time, +4ᵐ 16ˢ; long. 93° 42' 45" E.; true azi. N. 32° 28½' W.; error, 12° 47¼' W.; dev. 5° 42¾' E.
9. *Time azimuth*: A.T.S. 7ʰ 8ᵐ 21ˢ A.M.; dec. 23° 19' 43" N.; cor. for azi. 0 − 19 + 8 = −11'; true azi. N. 78° 40' E.; error, 2° 54' W.; dev. 17° 54' W.
10. *Chart work*: Compass course, N. 36° E.; dev. 28° E.; dist. 41½ miles; var. 18½° W.; lat. 53° 36¾' N.; long. 3° 25½' W.; lat. 53° 47' N.; long. 3° 8' W.; dist. 14½ miles.
11. *Meridian passage star*: 4ʰ 31ᵐ 12ˢ A.M.
12. *Stars near meridian*.—R.A.M. 9ʰ 5ᵐ 53ˢ *East*: Regulus, α Hydræ, β Argus. *West*: Pollux, Procyon. Appar. alt. 39° 31'.
13. *Star's meridian altitude*: True alt. 39° 31' 0"; dec. 5° 28' 52"; lat. 45° 0' 8" S.
14. *Star chronometer*: Rate, 4ˢ·06 gaining; M.T.G. 30ᵈ 18ʰ 0ᵐ 49ˢ; R.A. 18ʰ 33ᵐ 34ˢ; dec. 38° 41' 42" N.; R.A.M.S. 10ʰ 36ᵐ 48ˢ; true alt. 21° 19' 19"; E.H.A. 1ʰ 56ᵐ 41ˢ; long. 179° 49' E.
15. *Star azimuth*: R.A. 7ʰ 34ᵐ 4ˢ; dec. 5° 28' 52" N.; R.A.M.S. 22ʰ 25ᵐ 51ˢ; W.H.A. 1ʰ 21ᵐ 47ˢ; cor. for azi. −24 + 11 + 0 = −13; true azi. S. 154° 34' W.; error, 2° 56' W.; dev. 30° 4' E.
16. *Ex-meridian*: Time from noon, 10ᵐ 50ˢ; true alt. 61° 5' 3"; dec. 15° 11' 29" N.; arc 1, 15° 12½' N.; arc 2, 28° 48¼' S.; lat. 13° 36' S.
17. *Sumner*.—*First obs.*: true alt. 9° 37' 43'; dec. 18° 54' 43"; eq. time, 3ᵐ 49ˢ; E.H.A.'s, 5ʰ 26ᵐ 10ˢ and 5ʰ 26ᵐ 39ˢ; long. (a), 125° 39' 15" W.; (b) 125° 44' 15" W.
Second obs.: True alt. 56° 30' 4'; dec. 18° 57' 16" N.; eq. time, −3ᵐ 49ˢ; E.H.A.'s, 1ʰ 13ᵐ 37ˢ and 1ʰ 8ᵐ 37ˢ; long. (c), 127° 34' 30" W.; (d), 126° 19' 45" W.; first line position, N. 6° W. or S. 6° E.; second azi. S. 33° E.; lat. 49° 28' N.; long. 126° 0' W.
18. *Polaris*: True alt. 36° 9' 16"; R.A.M. 10ʰ 4ᵐ 55ˢ; first cor. +47' 37"; second cor. +20"; third cor. +49"; lat. 36° 57' 2" N.
19. *Meridian altitude Moon*: M.T.G. 12ᵈ 4ʰ 41ᵐ; S.D. 16' 22"; H.P. 59' 14"; dec. 0° 9' 23" N.; Appar. alt. 36° 13' 14"; cor. +46' 29"; T.A. 36° 59' 43"; lat. 52° 50' 54" S.
20. *Curve*: Cor. mag. S. 12° W.; deviations, 20° W., 16° W., 5° E.; 12° E., 12° E., 9° E., 2° E., 4° W.; compass courses, N. 83° W., N. 12° E., S. 26° E., S. 24° W.; mag. courses, N. 79° E., N. 12° W., S. 36° W., S. 72° E.; bearings, N. 59° E., S. 19½° E.
21. *Current*: Dev. 31° E.; compass course, N. 54° E.; dist. made good, 14½ miles.
22. *Sounding*: Table B, −5 ft. 10 in.; cor. 41 ft. 4 in., to be subtracted.
23. *Lunar*.—*Moon's elements*: S.D. 15' 18"; H.P. 55' 14"; R.A. 11ʰ 10ᵐ 48ˢ; dec. 0° 6' 5" S.; R.A.M.S. 2ʰ 0ᵐ 0ˢ·7; A.A. 62° 16' 31"; T.A. 62° 41' 43"; E.H.A. 44ᵐ 50ˢ.

APPENDIX. 377

Star's elements; R.A. 7^h 39^m 10^s; dec. 28° 16' 11" N.; A.A. 23° 13' 24";
T.A. 23° 11' 11"; W.H.A. 2^h 46^m 49^s; app. dist. 57° 51' 38"; angle B.
33° 16' 36"; true dist. 57° 58' 30"·5; second cor. +4'; M.T.G. 21^d 16^h 46^m 30^s;
error, slow 30'; long. 125° 9' 30" W.

24. *Ivory.*—First obs.: R.A. 21^h 42^m 45·1'; dec. 9° 9' 9" S.; S.D. 16' 46";
H.P. 60' 47"; run, +17' 12"; T.A. 30° 16' 11".

Second obs.: R.A. 21^h 53^m 20'; dec. 8° 5' 50" S.; S.D. 16' 44"; H.P.
60' 50"; T.A. 20° 9' 44"; H.E.T. 2^h 10^m $9\frac{1}{4}^s$; arc 1, 32° 7' 33"; arc 2, 8° 37'
11"; arc 3, 59° 33' 21"; arc 4, 100° 11' 59"; arc 5, 40° 38' 38"; cor. for lat.
−13' 20"; lat. 48° 23' 8" N.

25. *Simul. altitudes.*—*Rigel's elements*: R.A. 5^h 9^m 45'; dec. 8° 18' 52" S.;
R.A.M.S. 14^h 3^m 36'; T.A. 26° 12' 21"; E.H.A.'s, 1^h 58^m 53^s and 1^h 51^m 22';
long. (a), 123° 37' E.; long. (b), 125° $29\frac{1}{2}$' E.; bearing S. 32° E.

Markab's elements: R.A. 22^h 59^m 48'; dec. 14° 40' 15" N.; T.A. 27° 9' 5";
W.H.A.'s, 4^h 20^m 11^s and 4^h 19^m 22^s; long. (c), 125° 53' 30" E.; long. (d),
125° 41' 15" E.; bearing, S. $78\frac{1}{2}$° W.; lat. 50° $49\frac{1}{2}$' N.; long. 125° 41' E.

26. *Equal altitudes*: Dec. 26^d $12\frac{1}{2}^h$ S.; R.A. 16^h 23^m 17^s; eq. time, 3^m 44'·5;
slow M.T.S. 7^h 56^m 35^s·68; fast M.T.G. 2^m 32^s·32; slow A.T.G. 1^m 12^s·18.

27. *Great Circle*: First co. S. 38° 9' E.; dist. 3257 miles; lat. vertex,
52° 6' S.; long. vertex, 174° 56' E.; first pt. lat. equator, long. 84° 56' E.;
second pt., lat. 12° $34\frac{1}{2}$' S., long. 94° 56' E.; third pt., lat. 23° 43' S., long.
104° 56' E.; fourth pt., lat. 32° 43' S., long. 114° 56' E.

28. *Chart*: Lat. 53° 38' N.; long. 3° $35\frac{1}{2}$' W.; set, N. 42° W.; drift,
4 miles.

29. *Trig. problems*: Dist. 899·4 ft.
30. Height of staff, 55·9 ft.
31. Time, 11^h 56^m.
32. lat. 74° 26'.
33. ZY = 48° 15' 4"; XY = 73° 16' 7"; X = 51° 10' 25".

ANSWERS TO PAPER XI.

1. Log 9·947983; product, ·887120.
2. Log 1·931488; quotient, 85·405882.
3. S. 12° W., 30'; S. 71° W., 48'; S. 38° W., 52'; S. 27° W., 56'; S. 5° W.,
52'; S. 56° W., 60'; West, 60'; N. 73° E., 10'; D. lat. 217' S.; dep.
213'·6 W.; course, S. 44° 22' 30" W.; dist. 305·4 miles; lat. in, 56°
10' 42" N.; long. in, 50° 37' W.
4. 384·5 miles.
5. D. lat. 1324; M.D. lat. 1352; D. long. 5750; true course, N. 76° 46' E.;
course to steer, E.; distance, 5784.
6. H.W. Cardigan, 11^h 58^m A.M.; no P.M.
7. A.T.G. March 20^d 13^h 58^m 12^s; dec. 0° 6' 15·3" N.; true alt.
56° 35' 24" N.; lat. 33° 18' 21" S.
8. A.T.G. August 1^d 12^h 14^m 57^s; dec. 17° 53' 36·9" N.; true amp. W.
37° 15' 21" N.; dev. 12° 18' 24" W.
9. Daily rate, 10^s·05 losing; M.T.G. June 14^d 5^h 47^m 42'; dec.
23° 17' 15"·6 N.; eq. time, +2'·31; true alt. 46° 38' 31"; log 9·017943
H.A. 2^h 30^m 40^s; long. 49° 15' W.; azi. log 19·896482; azi. S. 125° 9' 32" W.;
deviation, 6° 10' 2" E.
10. M.T.G. July 2^d 12^h 3^m; A.T.S. 8^h 7^m 39'; A.M. Dec. 23° 0' 43"·8 N.;
cor. +2'; true azi. S. 128° 10' E.; dev. 4° 51' 15" W.
11. M.T.G. June 22^d 20^h 30^m 5'; R.A.M. 23^h 34^m 15"·37; dec. 28° 32' 5" N.;
H.A. = 0^h 28^m 57^s·37 E.; log No. 4907; true alt. 72° 5' 47"; zen. dist.
16° 57' 55" N.; lat. 45° 30' 0" N.
12. First dec. 23° 26' 27"·8 N.; eq. time, +1^m 54'·29; true alt. 59° 11' 0";

APPENDIX.

log (a), 8·480546 = H.A. 1ʰ 20ᵐ 7ˢ = long. 0° 24' 15" E.; log (b), 8·380376 ; H.A. 1ʰ 11ᵐ 19ˢ; long. 2° 36' 15" E.; line $\{^{N.\ 54\frac{1}{2}\ E.}_{S.\ 54\frac{1}{2}\ W.}\}$; azi. S. 35½ E. Second dec. 23° 26' 19"·6 N.; eq. time, + 1ᵐ 56ˢ·58; true alt. 46° 26' 24"; log (c), 9·158254 ; H.A. 2ʰ 58ᵐ 23ˢ; long. 0° 6' 0" E.; log (c), 9·146651; H.A. 2ʰ 55ᵐ 54ˢ; long. 0° 31' 15" W.; second line, $\{^{N.\ 22\frac{1}{2}\ W.}_{S.\ 22\frac{1}{2}\ E.}\}$; azi. S. 67½° W. Position, lat. 50° 10' 30" N., long. 0.

13. Mer. pass Antares, 11ʰ 23ᵐ 30ˢ·36 P.M.

14. R.A.M. 20ʰ 36ᵐ 19ˢ·87, α Aquilæ, S. 0ʰ 50ᵐ 29ˢ W.; γ Cygni, S. 0° 17ᵐ 44ˢ W.; α Cygni, S. 0ʰ 1ᵐ 39ˢ E.; α Cephei, N. 0ʰ 39ᵐ 50ˢ E.

15. Obs. alt. 53° 47' 2" S.

16. Dec. 16° 18' 26" N.; true alt. 65° 22' 27"; lat. 8° 19' 7" S.

17. Daily rate, 2ˢ·05 losing; M.T.G., 22ᵈ 5ʰ 46ᵐ 18ˢ; R.A. 19ʰ 45ᵐ 54ˢ·88; dec. 8° 36' 11" N.; R.A.M.S. 6ʰ 2ᵐ 45ˢ·19; R.A.M. 23ʰ 49ᵐ 3ˢ; log 9·408077 H.A. 4ʰ 3ᵐ 8ˢ W.; M.T.S. 17ʰ 46ᵐ 18ˢ; long. 180° 0' 0".

18. R.A.M. 15ʰ 5ᵐ 35ˢ·99; H.A. 3ʰ 21ᵐ 6ˢ·5 W.; dec. 15° 8' 0" N.; cor. − 31' − 8' + 12'; true azi. S. 126° 0' W.; dev. 42° 15' E.

19. A.T.G. June 29ᵈ 4ʰ 48ᵐ 24ˢ; R.A.M. 17ʰ 7ᵐ 24ˢ·01; true alt. 15° 14' 8"; first cor. +40' 27"; second cor. +8"; third cor. +1' 36"; lat. 15° 56' 19" N.

20. M.T.G. June 22ᵈ 23ʰ 53ᵐ·77; semidiameter, 16' 33"·42; H. P. 60' 15"·28; dec. 23° 22' 14"·9 S.; true alt. 15° 53' 39"; lat. 50° 44' 6" N.

21. (a) True course, S. 15° W.; mag. course, S. 33½° W.; compass course, S. 1½° E.; dist. 45 miles; dev. 35° E.; var. 18½° W.

(b) Lat. 53° 52' N.; long. 3° 15' W.

(c) Lat. 53° 27' N.; long. 3° 27' W.; dist. 4¾ miles.

(d) True course, S. 1° W.; mag. course, S. 19½° W.; compass course, S. 12° E.; D.M.G. 34 miles.

22. Interval 6ʰ 0ᵐ; Table B, 24 ft. 10 in.; cor. 2 ft. 6 in. to add.

23. *Napier's curve:* C.M.B. N.; dev. 6° E., 7° W., 17° W., 16° W., 8° W., 6° E., 18° E., 20° E.
Courses to steer: N. 17° W. S. 57½° W. S. 9½° E. S. 84° E.
Cor. mag. courses: N. 83° W. N. 1° W. N. 61° W. N. 25° W.
Dev.: 15° E. Bearings, N. 51½° E. S. 24½° E.

24. Moon's semidiameter, 16' 31"·59 ; H.P. 59' 58"·86 ; R.A. 23ʰ 30ᵐ 32ˢ·64; dec. 2° 27' 21"·1 N. Sun's R.A. 8ʰ 22ᵐ 13ˢ·61; dec. 19° 26' 22"·43 N.; R.A.M.S. 8ʰ 15ᵐ 56ˢ·27. Moon's true alt. 32° 0' 0"; appar. alt. 31° 10' 16"; log 9·370803; H.A. 3ʰ 51ᵐ 54ˢ W.; R.A.M. 3ʰ 22ᵐ 27ˢ. Sun's H.A. = 4ʰ 59ᵐ 47ˢ; appar. alt. 14° 17' 37"; true alt. 14° 14' 4"; log 18·508215, B = 10° 20' 30"; log 19·910424; ½ dist. 64° 25' 26"; interval 30ᵐ 31ˢ; cor. −3ˢ; chron. fast 2ˢ; M.T.S. July 25ᵈ 19ʰ 6ᵐ 31ˢ; long. 80° 59' 15" W.

25. A.T.G. March 20ᵈ 7ʰ 38ᵐ; H.E.T. 2ʰ 48ᵐ; dec. 0° 0' 0"·4; angle, 6°; dist. 44'·8; cor. −44' 33" to first alt.; first alt. 45° 58' 2"; second alt. 49° 26' 41"; arc 1, 42° 0' 0"; arc 2, 1° 44' 56"; arc 3, 5° 30' 45"; arc 4, 90° 0' 0"; arc 5, 84° 29' 15"; approx. lat. 5° 30' 36" N.; second cor. +7"·601; lat. 5° 30' 44" N.

26. M.T.G. September 22ᵈ 18ʰ 29ᵐ 38ˢ·32; dec. 0° 0' 0"; eq. time, −7ᵐ 33ˢ·68; interval, 9ʰ 30ᵐ 5ˢ; two-day change, 2806·2; first log, 0·7375; second log, 0; part 1, +5"·46; part 2, 0; error on A.T.S. slow 7ᵐ 31ˢ·04; on M.T.S. fast 2ᵐ·64; part 3 fast 5ʰ 32ᵐ 50ˢ·64.

27. Dec. 0° 2' 8" N.; eq. −7ᵐ 31ˢ·8; T.A. 41° 7' 5"; fast of M.T.G. 2ᵐ 16ˢ.

28. α Cygni, R.A. 20ʰ 38ᵐ 2ˢ·42; dec. 44° 55' 11" N., R.A.M.S. 6ʰ 6ᵐ 14ˢ·97; alt. 81° 56' 33"; log (a), 7·899294; H.A. 0ʰ 40ᵐ 52ˢ·5 long. 177° 38' 15" W.; log (b), 7·816119 H.A. 0ʰ 36ᵐ 8ˢ long.; 178° 34' 30" W. first line, $\{^{N.\ 32°\ W.}_{S.\ 32°\ E.}\}$; azi. S. 58° W. Altair, R.A. 19ʰ 45ᵐ 54ˢ·88; dec. 8° 36' 11" N.; alt. 44° 47' 1"; log (c), 8·642332; H.A. 1ʰ 36ᵐ 44ˢ long.

APPENDIX. 379

176° 42' 15" W. ; log (d), 8·553912 ; H.A. 1ʰ 27ᵐ 15ˢ ; long. 179° 4' 30" W. ;
second line, $\begin{Bmatrix} \text{N. } 57° \ 30' \ \text{W.} \\ \text{S. } 57° \ 30' \ \text{E.} \end{Bmatrix}$; azi. S. 32° 30' W. ; position, lat. 49° 39' 30" N.,
long. 178° 14' 0" W.
  29. First course, S. 51° 31' E. ; final, N. 43° 10' E. ; first dist. 2029·5 ;
second dist. 2897 ; third dist. 986·5 ; long. of first position, 163° 13' W. ;
long. where parallel is left, 134° 33' W.
  30. (a) Dist. 74·7 miles ; lat. in, 50° 0'·4 S.
     (b) P = 45° 4' 43" ; R = 69° 15' 17" ; PR = 1019 ft.
     (c) B = 54° 44' 9" ; A = 125° 15' 51" ; b = 45°.
     (d) S. 34° 30" E. or W.

Answers to Paper XII.

  1. *Logs*: ·2202 log = 9·342749 ; ·01591 log = 8·201676.
  2. *Days' Work*: S. 27° W., 12' ; S. 50° W., 29' ; S. 29° E., 28' ; S. 34° W., 30' ;
N. 39° E., 20' ; S. 79° W., 35' ; N. 47° E., 27' ; S. 18° E., 42' ; diff. lat. 91'·4 ;
dep. 19'·9 ; course, S. 12° W. ; dist. 94' ; lat. in, 49° 13' N. ; long. in, 0° 18' W.
  3. *Meridian altitude*: Dec. 14° 46' 52" N. ; true alt. 35° 35' 36" ; lat.
39° 37' 32" S.
  4. *Mercator*: True course, S. 86° 39' 46" W. ; comp. course, N.
89° 50' 14" W. ; dist. 8538 miles.
  5. *Parallel*: Lat. 48° 11½'.
  6. *Tides*: 11ʰ 42ᵐ A.M., no P.M. ; and 9ʰ 43ᵐ A.M., 10ʰ 9ᵐ P.M.
  7. *Amplitude*: Dec. 0° ; true amp. W. ; error, 30° 56' E. ; dev. 17° 11' E.
  8. *Chron. azimuth*: Rate, ·7ˢ losing ; M.T.G. 15ᵈ 9ʰ 10ᵐ 15ˢ ; true alt.
16° 18' 29" ; dec. 18° 36' 47" S. ; H.A. 5ʰ 30ᵐ 40ˢ ; eq. time, – 15ᵐ 13ˢ·5 ; long.
58° 42' 15" W. ; true azi. N. 101° 40' W. ; error, 5° 13' E. ; dev. 6° 47' W.
  9. *Time azimuth*: A.T.S. 6ʰ 28ᵐ 47ˢ P.M. ; dec. 20° 13' 55" N. ; cor. for azi.
– 12½ – 7½ + 3 = – 17' ; true azi. N. 73° 34' W. ; error, 2° 22' E. ; dev.
21° 37' E.
  10. *Chart work*: Compass course, S. 77½° E. ; dev. 2½° E. ; var. 19° W. ;
dist. 93 miles ; lat. 53° 28' N., long. 4° 51' W. ; lat. 53° 29¼' N., long.
4° 12½' W. ; dist. 5 miles.
  11. *Meridian passage Star*: 11ʰ 14ᵐ 38ˢ A.M.
  12. *Stars near Meridian*.—*East*: α Orionis ; Rigel ; Capella ; Aldebaran.
West: α Persei ; α Arietis. Approx. mer. alt. 44° 6'.
  13. *Meridian altitude star*: True alt. 44° 7' 15" ; dec. 45° 53' 39" N. ; lat.
0° 0' 54" N.
  14. *Star chron.*: Rate, 2ˢ·4 gaining ; M.G.T. 15ᵈ 10ʰ 2ᵐ 24ˢ ; R.A. 4ʰ 30ᵐ 10ˢ ;
dec. 16° 18' 27" N. ; R.A.M.S. 19ʰ 40ᵐ 32ˢ ; true alt. 57° 28' 33" ; E.H.A.
1ʰ 41ᵐ 14ˢ ; long. 43° 30' W.
  15. *Star azimuth*: R.A. 5ʰ 9ᵐ 43ˢ ; dec. 8° 18' 52" S. ; R.A.M.S. 14ʰ 17ᵐ 20ˢ ;
E.H.A. 4ʰ 46ᵐ 8ˢ ; cor. for azi. + ·5 – 20·5 + 0 = – 19 ; true azi. S. 81° 14' E. ;
error, 3° 9' E. ; dev. 23° 24' E.
  16. *Ex-meridian*: R.A.M.S. 20ʰ 52ᵐ 59ˢ ; R.A. 10ʰ 3ᵐ 2ˢ ; dec.
12° 27' 25" N. ; W.H.A. 32ᵐ 28ˢ ; true alt. 56° 5' 24" ; lat. 20° 30½' S.
  17. *Summer*.—*First obs.*: True alt. 36° 40' 45" ; dec. 2° 34' 57" S. ; eq.
time, – 9ᵐ 48ˢ ; W.H.A.'s, 1ʰ 17ᵐ 16ˢ and 1ʰ 10ᵐ 16ˢ ; long. (a), 124° 41' W. ;
long. (b), 126° 26' W.
  Second obs.: True alt. 18° 44' 50" ; dec. 2° 37' 28" ; eq. time, – 9ᵐ 50ˢ ;
W.H.A.'s, 3ʰ 51ᵐ 31ˢ and 3ʰ 49ᵐ 59ˢ ; long. (c), 125° 7' 15" W. ; long. (d), 125° 30'
15" W. ; first line position, N. 65° W. or S. 65° E. ; second azi., S. 63° W. ; lat.
48° 36' N. ; long. 125° 34' W.
  18. *Polaris*: True alt. 51° 14' 41" ; R.A. mer. 19ʰ 13ᵐ 56ˢ ; first cor.
+ 2' 35" ; second cor. + 0' 58" ; third cor. + 0' 57" ; lat. 51° 18' 11" N.
  19. *Meridian altitude Moon*: M.T.G. 5ᵈ 9ʰ 42ᵐ·4 ; S.D. 16' 16" ; H.P.

58' 39"; dec. 3° 23' 16" N.; appar. alt. 70° 7' 13"; cor. +19' 37"; T.A. 70° 26' 50"; lat. 16° 9' 54" S.

20. *Curve:* Cor. mag. E. 5° S.; devs. 5° E., 5° E., 2° W., 13° W., 15° W., 3° W., 13° E., 8° E.; compass courses, S. 30° E., S. 51° W., N. 5½° W., N. 65° E.; mag. courses, N. 62° W., N. 57° E., S. 12° E., N. 10° E. bearings, S. 87° E., S. 15° E.

21. *Current:* Dev. 2½° E.; compass course, S. 77½° E.; dist. made good, 52 miles.

22. *Sounding:* Table B. +2 ft. 6 in.; cor. 29 ft., to be subtracted.

23. *Lunar.—Moon's elements:* S.D. 16' 59"; H.P. 61' 16"; R.A. $20^h 47^m 52^s$; dec. 14° 17' 56" S.; R.A.M.S. $9^h 52^m 41^s$; A.A. 57° 57' 16"; T.A. 58° 29' 12"; W.H.A. $1^h 19^m 47^s$.

Star's elements: R.A. $22^h 52^m 9^s$; dec. 30° 8' 58" S.; A.A. 48° 14' 17'; T.A. 48° 13' 26"; E.H.A. $44^m 29^s$; appar. dist. 32° 43' 11"; angle B, 31° 45' 51·5"; true dist. 32° 40' 15"; second cor. $-1^m 15^s$; M.T.G. $19^h 13^h 24^m 47^s$; error, 8^s slow; long. 17° 27' W.

24. *Ivory:* H.E.T. $1^h 18^m 0^s$; mid. dec. 2° 36' 11" S.; cor. for run, −18' 41"; first T.A. 36° 22' 4"; second T.A. 18° 44' 50"; arc 1, 19° 28' 45"; arc 2, 24° 1' 45"; arc 3, 57° 55' 50"; arc 4, 92° 45' 41"; arc 5, 34° 49' 51"; cor. for lat. +2' 20"; lat. 48° 36' 13" N.

25. *Simul. altitudes.—Aldebaran:* R.A. $4^h 30^m 9^s$; dec. 16° 18' 25" N.; R.A.M.S. $23^h 49^m 5^s$; T.A. 34° 0' 29"; W.H.A.'s, $3^h 44^m 20^s$ and $3^h 43^m 25^s$; long. (a), 40° 22' 45" W.; long. (b), 40° 36' 30" W.; bearing S. 76° W.

Regulus: R.A. $10^h 3^m 3^s$; dec. 12° 27' 24" N.; T.A. 46° 17' 18"; E.H.A.'s, $1^h 49^m 20^s$ and $1^h 45^m 34^s$; long. (c), 40° 34' 15" W.; long. (d), 39° 37' 45" W.; bearing, S. 39° E.; lat. 50° 4½' N.; long. 40° 26' W.

26. *Equal altitudes:* Eq. time, $10^m 37^s$; slow M.T.S. $1^h 9^m 57^s \cdot 2$; fast M.T.G. $2^m 16^s \cdot 8$; slow A.T.G. $8^m 20^s \cdot 2$.

27. *Great circle:* First course, N. 49° 23½' E.; dist. 4957 miles; lat. vertex, 50° 18¼' N.; long. 172° 27' W.; lat. points from Towson.

28. *Chart:* Lat. 68° 44' N.; long. 8° 20' E.; course, South; dist. 53 miles.

29. *Trig. problems:* First boats from cliff, 491·8 ft. and 656·5 ft.; dist. between boats, 820·2 ft.; bearing second boat, N. 53° 9¼' E. First course, S. 60° 11' E.; second course, N. 18° 35' W. BC = 53° 53' 38"; AB = 100° 17' 28"; A = 55° 11' 58". Diff. long. 76° 51'.

www.ingramcontent.com/pod-product-compliance
Lightning Source LLC
Chambersburg PA
CBHW030405230426
43664CB00007BB/761